Extreme Chromatography

Faster, Hotter, Smaller

Editors
William Craig Byrdwell and Michal Holčapek

AOCS PRESS

AOCS Mission Statement

To be a global forum to promote the exchange of ideas, information, and experience, to enhance personal excellence, and to provide high standards of quality among those with a professional interest in the science and technology of fats, oils, surfactants, and related materials.

Library of Congress Cataloging-in-Publication Data

Extreme chromatography : faster, hotter, smaller / editors, William Craig Byrdwell, Michal Holcapek.
p. cm.
Includes bibliographical references and index.
1. High performance liquid chromatography. 2. Chromatographic analysis.
I. Byrdwell, William Craig. II. Holcapek, Michal.

QP519.9.H53E97 2011
543'.8--dc22
2010052414

Printed in the United States of America
15 14 13 12 11 5 4 3 2 1

The paper used in this book is acid-free, and falls within the guidelines established to ensure permanence and durability.

Contents

Preface

In the past decade, many new technologies have been developed to provide new tools for analytical chemists to use, while other relatively new technologies are coming to maturity. Since both gas chromatography and conventional high performance liquid chromatography were already mature fields, the new developments have often focused on expanding the field of chromatography by expanding the range of conditions (i.e., pressure, temperature, particle size, column dimensions, etc.) that can routinely be used. Many of the conditions used nowadays are considered quite extreme compared to traditional chromatographic techniques. Some of these extreme conditions are now possible only because of improvements in fabrication technologies (e.g., for column packing materials) and commercial instrument designs (e.g., allowing the use of extremely high pressures in ultra high performance liquid chromatography). Thus, techniques that were demonstrated only in principle as long as decades ago have now come to commercial availability because of technological advancements in instrument design, software development, and manufacturing and production processes. Many techniques that were previously demonstrated only as proofs of concept are now within reach of the typical research or commercial laboratory.

While these developments offer exciting possibilities to solve long-intractable problems and to address extremely complex separations as never before possible, they also present difficulties in determining which techniques offer concrete benefits and which ones are merely 'flashes in the pan'. A dizzying array of specialty applications has appeared in the peer-reviewed literature, which can create a degree of information overload, as today's chemist grapples with decisions as to which techniques are worthy of adoption (at the cost of substantial capital investment) and which are merely interesting novelties.

There are, however, clear trends in the dramatically increasing number of literature reports for some techniques, and these trends run in parallel with the commercial availability of instruments, columns, and stationary phases that manufacturers have determined have reached a critical threshold of implementation justifying their mass production. Because it has become apparent that some techniques are here to stay, this moment in the history of chromatography represents an excellent time to assess the state of the art and to provide a resource for analytical chemists in the decision making process, and for students and professionals interested in learning more about these techniques. While reviews of the techniques included here are certainly extant in the literature, uncovering the information that describes both the fundamental principles and practical applications of the wide variety of up-and-coming techniques involves locating the respective 'needles' for each technique in the literature 'haystack'. Therefore, we hope that a single resource that draws together detailed descriptions of many of the most pertinent analytical options, presented by the authors who helped to lay the foundations in their respective areas will prove a valuable tool and a concise source for today's analyst. The chapters herein describe the fundamental principles behind many of the techniques, their benefits and difficulties, and then practical applications in a wide range of areas. This information will help determine which techniques are most appropriate for which areas.

The common element that runs through these chapters is that they employ conditions, instruments, or components that would be considered extreme by conventional standards. Whether it is the extreme pressures allowed by UHPLC, the extreme temperatures used in high-temperature separations, the extreme columns used in nanochromatography, monolithic column chromatography, silver-ion chromatography, chiral separations, and hydrophobic interaction liquid chromatography, the extreme instrumentation used in multi-dimensional separations, or the extreme mass spectrometric detection techniques used in combination with chromatography, these chapters provide insight into the steps necessary to implement in a practical manner the techniques that have risen above others as the most useful and practical new and innovative techniques. By including extreme techniques that have a proven track record, with chapters written by recognized experts in the field, we hope to provide a compendium that will have enduring usefulness to act as an ongoing resource, while other techniques come and go.

William Craig Byrdwell

1

The Theory and Practice of UHPLC and UHPLC-MS

Davy Guillarme and Jean-Luc Veuthey
School of Pharmaceutical Sciences, University of Geneva, University of Lausanne, Boulevard d'Yvoy 20, 1211 Geneva 4, Switzerland

Introduction

Reversed-phase liquid chromatography (RPLC) is nowadays one of the most widely used separation techniques. Indeed, despite a lower chromatographic efficiency in comparison with capillary GC, the interaction of the analytes with both the stationary phase and the mobile phase provides an important selectivity, and thus RPLC can be applied to solve numerous analytical problems. During the last few years, some substantial improvements, such as innovative supports and instrumentation, helped to achieve high throughput analyses and highly efficient separations (Guillarme et al., 2007b; Novakova et al., 2006). Such advances were mainly driven by the need to handle either a growing number of analyses or more complex samples.

Regarding high throughput separations, there is a growing demand in numerous fields, including toxicology, doping, forensic, clinical chemistry and environmental analyses, where the delivery time response must be reduced as much as possible. The pharmaceutical field, with its need for enhanced productivity and reduced costs, is the main driving force for faster separations (Wren & Tchelitcheff, 2006). Due to the high number of analyses required for common pharmaceutical applications, such as purity assays, pharmacokinetic studies, and quality control, rapid analytical procedures (less than 5 minutes including equilibration time) are often mandatory (Al-Sayah et al., 2008).

Highly efficient separations are also necessary for many applications, including genomics, proteomics and metabolomics, which all deal with very complex samples, such as biological samples, tryptic digests, or natural plant extracts (Grata et al., 2008; Petricoin et al., 2004). With such difficult samples, conventional HPLC systems present some obvious limitations, thus demanding analytical procedures to yield high resolution within an acceptable analysis time, even when a large number of compounds need to be separated.

Among the different strategies used to achieve fast and high resolution separations, ultra high pressure liquid chromatography (UHPLC) has

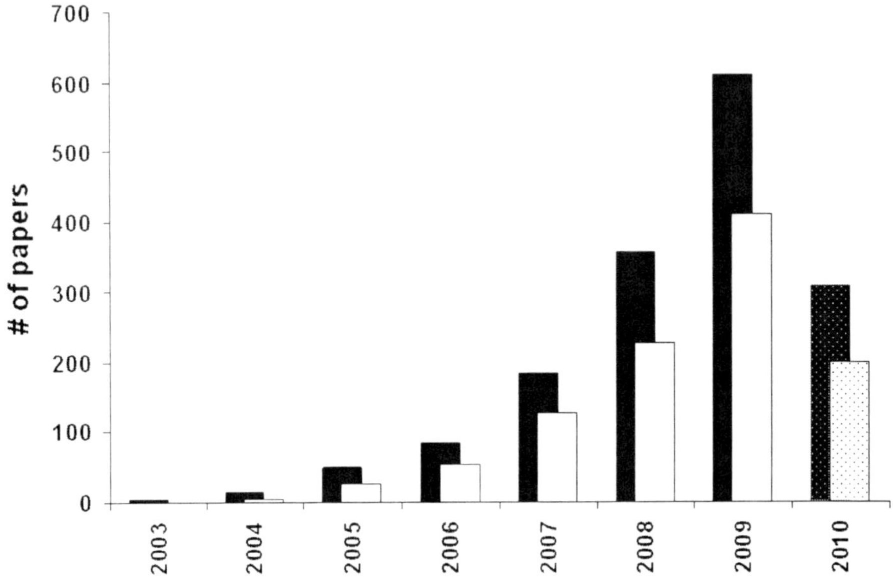

Fig. 1.1. Number of papers published each year in the field of UHPLC and UHPLC-MS, since 2003. Before this date, few papers were published by Jorgenson's and Lee's groups. Black bars were obtained with keywords "UPLC" and "UHPLC", while white bars were obtained with an additional filter (keyword "MS"). Source: Scifinder scholar 2007 search of the Chemical Abstracts database from 2003 to 2010. Date of information gathering: June 2010.

been rapidly recognized as a powerful and robust analytical tool. As shown in Fig. 1.1, the number of articles published in the fields of UHPLC and UHPLC-MS have risen rapidly since 2004. Thus, the present chapter will demonstrate the possibility to speed up and/or attain highly efficient separations using the UHPLC strategy in combination with UV as well as MS detectors. In addition, this chapter will discuss the advantages and drawbacks of this approach, in comparison with other existing techniques.

Brief Description of UHPLC and Historical Background

Interest in Small Particles in Liquid Chromatography

In liquid chromatography, a reduction of particle size (d_p) provides an important gain in chromatographic performance (Knox, 1977; Knox & Saleeem, 1969; Poppe, 1997). Indeed, the reduction of d_p allows faster separations as well as higher plate counts. Since the beginning of LC, there has been a continuing progression in the reduction of particle size that has culminated in the recent commercialization of columns packed with sub-2 µm particles (Table 1.A). According to equations 1 and 2, these small particles lead to significant improvements in terms of i) efficiency; since N is inversely proportional to d_p and ii) time reduction; because the mobile phase linear velocity is inversely proportional to particle diameter.

$$N = \frac{L}{h \cdot d_p} \qquad (1)$$

Where, L is the column length, h the reduced plate height (generally between 2 and 3) and d_p the particle size.

$$u = \frac{\nu \cdot D_m}{d_p} \qquad (2)$$

Where, ν is the reduced linear velocity, u the linear velocity, and D_m the diffusion coefficient of the solute into the mobile phase.

At the end of the 60s, Horváth and co-workers introduced columns packed with rigid pellicular particles (40–50 µm) compatible with high pressures (Horvath et al., 1967). The thin porous coating allowed a rapid solute mass transfer into and out of the packing, producing a significant improvement in terms of column efficiency compared to the large porous particles commonly employed at that time. Nevertheless, this pellicular packing had a too limited surface area and therefore low sample capacity.

The transition from large porous and pellicular particles to smaller particles (in the range of 10 µm) occurred during the 70s (Majors, 2003). However, particles of silica smaller than 40 µm have demonstrated some difficulties with packing reproducibility, since irregular shapes of micro-porous particles were used (Kirkland, 1972; Majors, 1972; Snyder 1969; Asshauer & Halasz, 1974), until spherical materials were developed and improved (Endele et al., 1974; Vivilecchia et al., 1974).

Table 1.A. Evolution of Particles Size in Liquid Chromatography. Adapted from (Majors, 2003).

Years	Particle Size	Plates/15cm
1950s	100 µm	200
1967	50 µm	1000
1972	10 µm	6000
1985	5 µm	12000
1992	3–3.5 µm	22000
1996*	1.5 µm	30000
2000	2.5 µm	25000
2004	1.7 µm	30000

* Non-porous silica or resins

In 1977, Knox (1969) stated that ultra-fast LC would require a new generation of particles and instrumentation. Particles of 1 or 2 μm should be used to obtain $t_0 \approx 10$ s with reasonable pressures, in combination with column lengths of 20 to 40 mm. However, due to the dramatic reduction of the retention volume, and the use of elevated mobile phase flow rate, the instrumentation would become critical, and contribute strongly to additional peak broadening. Specifically, the injector (e.g., precise injection of small amounts) and detector devices (e.g., reduction of cell volume and improvement of electronics, such as acquisition rate and time constant) were not initially adapted. For this reason, 20 to 30 years have been spent to develop instrumentation compatible with sub-2 μm particles and short columns.

In the 80s, columns packed with 5 μm particles became the standard, and in the early 90s, 3–3.5 μm particle diameters were also commercially available. The latter demonstrated 30–50% faster analysis times and higher efficiencies compared to 5 μm particles. An additional advantage is that methods developed on columns packed with 5 μm can be easily transferred to a similar 3 or 3.5 μm stationary phase (Phillips et al., 1997). To further improve chromatographic performance, small non-porous supports of 1.5 μm were introduced in 1996 (Majors, 2003). These supports minimize the pore diffusion and mass transfer resistance effects. Therefore, non-porous silica columns can work in a much wider flow rate range without any loss of chromatographic performance and provide high efficiency (equivalent to 200,000 plates/m). Non-porous silica supports are also more durable and remain stable at high temperatures and pressures (Kirkland, 2000). However, due to the reduction in surface area, non-porous supports exhibit lower retention times and too limited loading capacity in comparison with porous columns. Nowadays, such materials are only proposed for separating very large molecules that slowly diffuse in the mobile phase (Lommen & Snyder, 1993).

In 2004, the first available porous silica with sub-2μm particle size was commercialized (1.7 μm), allowing a better resolution compared to the current 5 or 3.5 μm. As shown in Fig. 1.2, the kinetic performance can be drastically improved when decreasing the particle size from the conventional columns packed with 5μm particles to the recently introduced sub-2μm particles. With the latter, it is possible to attain an H value of less than 5 μm, meaning that a column of only 50 mm packed with sub-2μm particles can provide an efficiency of 10,000 plates (i.e., equivalent to a 150 mm column length packed with 5 μm particles). In addition, the mobile phase linear velocity should be increased with smaller particles (see equation 2), and thus it is possible to work 3 times faster when using sub-2μm *vs.* 5μm particles. Finally, the mass transfer resistance (right side of the van Deemter curve) is drastically reduced, allowing columns to work at a linear velocity higher than the optimal one, with only a limited impact on efficiency. Because of these kinetic characteristics, the analysis times can be reduced by a factor 9, for a similar efficiency

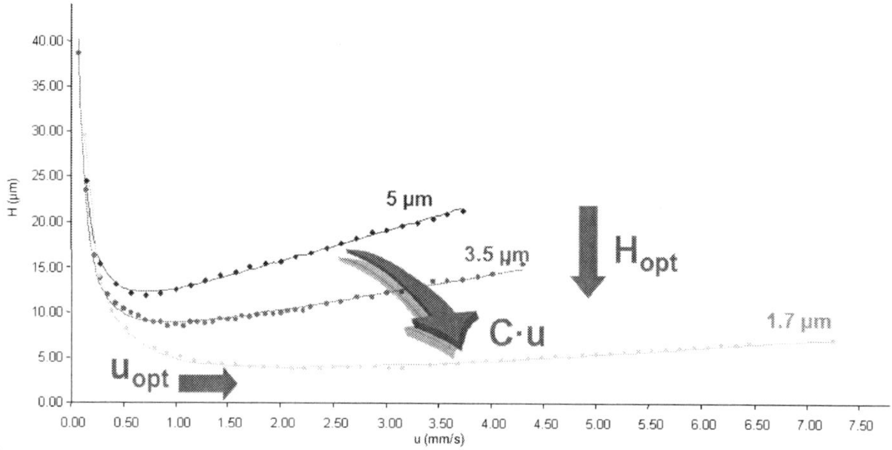

Fig. 1.2. Impact of particle size reduction on the van Deemter curves. These experimental curves were obtained with a mobile phase: H2O/MeCN (60:40 v/v), with butylparaben 25 ppm in H2O, UV detection: 254nm, Columns: XTerra RP18 4.6 × 150mm, 5µm; XTerra RP 18 4.6 × 50mm, 3.5µm; AcquityShield BEH C18 2.1 × 50mm, 1.7µm. Adapted from (Nguyen et al., 2006), with permission.

between HPLC (150 mm, 5 µm) and UHPLC (50 mm, 1.7 µm) at the optimal linear velocity. On the other hand, if the critical factor is resolution, maintaining the equivalent column length in UHPLC as in HPLC allows a higher resolution with an equivalent analysis time.

In gradient mode, several authors have demonstrated studies where decreasing particle size was used for improving peak capacity and thus chromatographic resolution (Neue & Mazzeo, 2001; Gilar et al., 2004). Indeed, particle size reduction has more impact than reducing column length, gradient time or flow rate to improve peak capacity in gradient mode.

However, these columns packed with small particles generate an elevated backpressure (> 400 bar) often incompatible with conventional instrumentation. Darcy's law (equation 3), shows the dependence of column inlet pressure on the particle diameter d_p.

$$\Delta P = \frac{u \cdot L \cdot \eta \cdot \phi}{d_p^2} \tag{3}$$

Where, L is the column length, η the mobile phase viscosity, and ϕ the flow resistance.

At the optimal linear velocity (u_{opt}), the pressure drop is inversely proportional to the cube of particle diameter (according to equations 2 and 3). Thus, under optimal flow rate conditions, 1.7 µm particles should generate a pressure 27 times higher than 5 µm particles for an identical column length. Then, the backpressure limitation of conventional HPLC at 400 bar can turn out to be an issue, and there is a need to employ dedicated LC systems that withstand ultra high pressures (up to 1300 bar nowadays).

Interest in Very High Pressures in Liquid Chromatography

As described by equation 3, reducing the particle size rapidly increases the backpressure. Early in the sixties, Knox showed that the speed limitation of chromatographic systems would be related to the maximum operating pressure. In addition, when there is a need to improve efficiency with long columns, an increase in pressure drop is mandatory to work within acceptable flow rate conditions (Knox & Saleem, 1969; Giddings, 1965; Knox, 1961).

Recently, Desmet et al. described a useful approach illustrating the influence of maximal back-pressure (ΔP_{max}) on achievable N and t_0, based on the van Deemter data (u, H), column permeability (K_{v0}), mobile phase viscosity (η) and equations 4 and 5 (Desmet et al., 2006; Desmet et al., 2005a; Billen, 2007; Desmet et al., 2005b; Desmet & Cabooter 2009; Desmet, 2008).

$$N = \frac{\Delta P_{max}}{ç}\left(\frac{K_{v,0}}{u_0 H}\right) \quad (4)$$

$$t_0 = \frac{\Delta P_{max}}{ç}\left(\frac{K_{v,0}}{u_0^2}\right) \quad (5)$$

The kinetic plot representation (which corresponds to the representation of the kinetic performance of a chromatographic system, considering i) mobile phase flow rate and column length, as variables and ii) mobile phase nature and composition as well as maximal pressure supported by the system, as constants) is advantageous because the performance of different LC supports can be directly compared whatever their lengths, particle sizes, and pressure drops. Fig. 1.3 shows kinetic plots representing the achievable efficiency on the X-axis as a function of the time required to attain a given efficiency on the Y-axis. The curves were constructed for a column packed with sub-2μm particles and using different maximal operating pressures, namely 400, 600, 1000, and 1200 bar. This representation shows that, when dealing with high throughput separations requiring short columns (N in the range 5000–15000 plates), the corresponding analysis times are almost identical for systems with a maximal pressure drop of 600, 1000 or 1200 bar. However, the performance was reduced with a conventional HPLC system compatible with only 400 bar (logarithmic scale). These observations are related to the achievable mobile phase flow rate and the fact that optimal linear velocity (u_{opt}) cannot always be reached, depending on the maximal pressure drop.

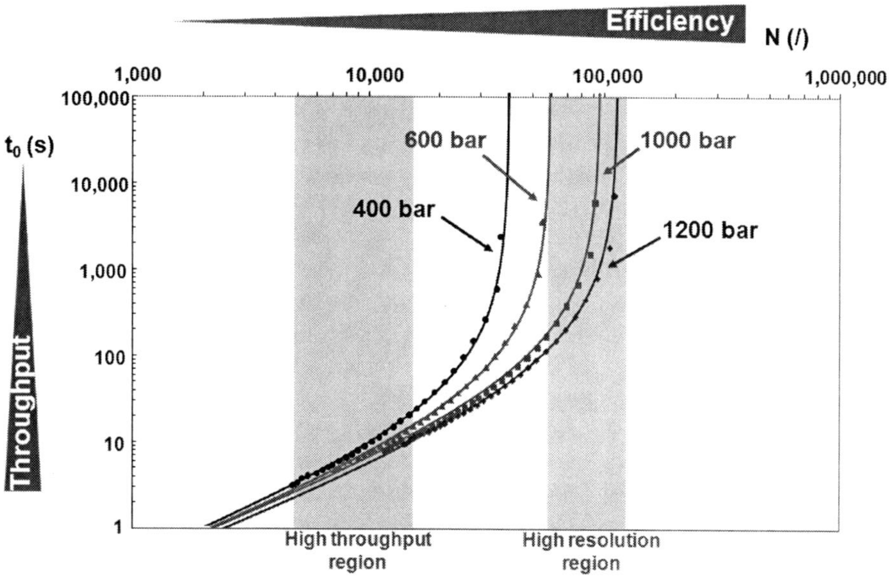

Fig. 1.3. Representation illustrating the importance to increase the maximum pressure of the system in UHPLC. Kinetic plots were constructed for a Waters Acquity BEH C18, 1.7μm column, considering a mobile phase viscosity of 0.85cp (mobile phase ACN:H2O 40:60 at 30°C).

On the other hand, when a very high efficiency is required (right side of the kinetic curves, N ranging between 50,000–100,000 plates), the maximal operating pressure of the system becomes a critical parameter. Indeed, for a system that withstands 400 bar, the maximal achievable efficiency would only be equal to 40,000 plates, because a compromise should be found between column length and mobile phase flow rate, to maintain acceptable backpressure. Then, the N value increases proportionally with the maximal pressure drop, up to 120,000 plates for a system compatible with a maximal pressure of 1200 bar. To attain such efficiency, a column of 45 to 60 cm packed with sub-2μm particles could be employed, as reported in the literature (Cabooter et al., 2008). Thus, there is a need to work with a dedicated system compatible with ultra high pressure when using columns packed with sub-2μm particles. This statement is particularly valid when dealing with high resolution separations.

Preliminary Works of J.W. Jorgenson and M.L. Lee in UHPLC

The first report on UHPLC was published as early as 1972 by Bidlingmeyer et al. (Bildingmeyer & Rogers, 1972), who employed long, thick columns packed with sub-2μm particles. However, irreproducible results were obtained due to the inherent difficulty in packing such columns at that time.

The first remarkable separation performed at a pressure up to 4100 bar in a fused-silica capillary column (52 cm length, 30 µm I.D., 1.5 µm non-porous particles) was described in 1997 by Jorgenson et al. For an analysis time of less than 10 min, a number of plates in the range of 140,000 to 190,000 was obtained with small molecules (MacNair et al., 1997). In 2003, the maximal pressure drop of their system was extended and the particle size decreased to obtain outstanding chromatographic performance. As shown in Fig. 1.4.A, a separation of four model compounds under isocratic conditions was obtained at a pressure of 7200 bar using a column of 43 cm length, 30 µm I.D., packed with 1.0 µm non-porous particles. For this separation, analysis time was reduced to less than 4 min with plate number between 196,000 and 310,000 (Jerkovich et al., 2003).

Later, Lee et al. also investigated UHPLC with capillary columns, but with the aim of reducing analysis times as much as possible. For this purpose, they developed a system able to work up to a pressure of 3600 bar, and a separation of benzodiazepines was performed in less than 60 s using a column of 125 mm length, 29 µm I.D., and packed with 1.5 µm non-porous particles (Fig. 1.4.B) (Lippert et al., 1999). Their UHPLC setup was also successfully coupled to ESI-TOF via a sheath liquid interface to carry out high speed and high resolution analyses of pharmaceutical compounds and herbicides. Columns of different geometries (29–100 µm I.D.) were used to obtain separations in less than 100 s with efficiencies ranging from 20,000 to 30,000 plates

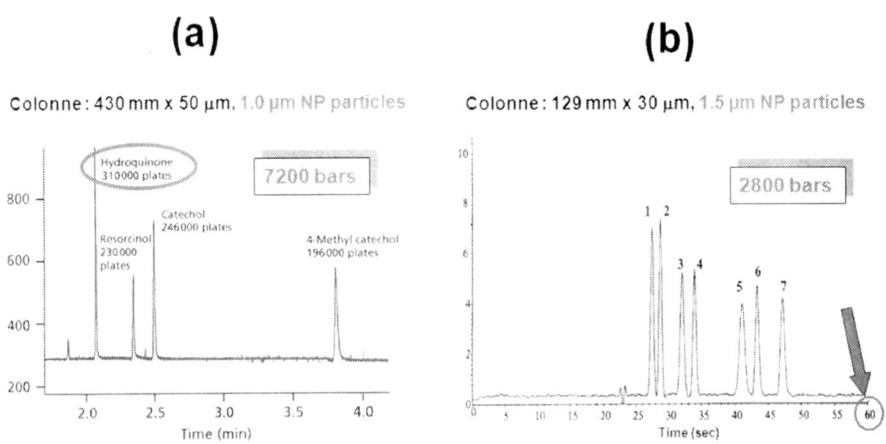

Fig. 1.4. Illustration of the first UHPLC experiments made by the groups of Jorgenson et al. and Milton Lee et al. (a) Ultra efficient separation of small organic test compounds obtained; Conditions: 7200 bar inlet pressure, capillary column (430 mm × 50 µm I.D., packed with 1.0 µm non-porous particles) (Jerkovich et al., 2003). (b) Ultra-fast separation of a triazine herbicides mixture. Conditions: 2800 bar inlet pressure; capillary column (129 mm × 30 µm I.D. packed with 1.5 µm Kovasil MS-H nonporous particles.) Adapted from (Lippert et al., 1999), with permission.

(Wu et al., 2001). Furthermore, elevated mobile phase temperature can be employed to reduce the pressure drop and make the use of longer columns or smaller particles possible. Lee et al. (Xiang et al., 2003a; Xiang et al., 2003b) investigated this approach on a capillary column (145 mm length, 50 μm I.D., 1.0 μm non-porous particles) packed with zirconia material, due to its chemical stability at elevated temperature. A separation of various benzodiazepines was carried out in less than 1.2 min at 100°C using this column, at a pressure of 1480 bar. Five herbicides were also resolved with excellent efficiency in only 60 s at 90°C at 1800 bar.

As described, most of Jorgenson's publications in UHPLC were dedicated to complex mixtures requiring a high number of plates ($N > 100{,}000$) while Lee et al. mainly used UHPLC for performing high throughput analysis of small molecules with a limited efficiency ($N < 30{,}000$).

Kinetic Comparison of UHPLC with Other Existing Technologies for Fast and High Resolution Liquid Chromatography

Brief Presentation of the Alternative Approaches to UHPLC

Table 1.B summarizes the main characteristics (advantages and limitations) of alternative approaches to obtain fast separations and high resolution.

Monolithic supports consist of a single rod of porous material that presents unique features in terms of permeability and efficiency. These materials were mainly developed by Hjerten et al. (1989), Svec and Frechet (1992) and Tanaka, Nakanishi and co-workers (Minakuchi et al., 1996). Silica-based monolithic supports are the most interesting ones for conventional RPLC and have been available from Merck and Phenomenex since 2000 and 2006, respectively. Their bimodal structure is characterized by 2-μm macropores and 13-nm mesopores, leading to efficiency equivalent to that of porous 3.5 μm silica particles (Cabrera, 2004). Tallarek et al. demonstrated that this material exhibits elevated permeability, equivalent to that of a column packed with 11 μm particles (Leinweber & Tallarek, 2003). The low generated backpressure and good mass transfer allow the application of elevated flow rates (3 to 10 times higher), thus enabling ultra fast separations, down to only a few seconds for the separation of several substrates and metabolites (van Nederkassel et al., 2003). Alternatively, it is also possible to use very long monoliths at a reasonable flow rate to reach an elevated resolution in a practical analysis time. For example, Tanaka et al. (Miyamoto et al., 2008) constructed a 11.4 m column by coupling numerous monolith columns which provide 1,000,000 theoretical plates for an analysis time around 16 h. Despite these exceptional properties, monoliths are not widely used and less than 1% of chromatographers routinely use silica-based monolithic columns (Majors, 2008). Several explanations for their limited use include patent exclusivity (limited number of suppliers), column chemistry and geometry (columns are now available in

Table 1.B. Comparison of Selected Fast Chromatographic Approaches. Adapted from (Guillarme et al., 2010).

Approaches	Advantages	Limitations
Monoliths	Very low backpressure because of the elevated permeability	Lack of chemistries (C18, C18 endcapped, C8) and providers
	Approach compatible with a conventional HPLC system	Direct method transfer impossible between conventional HPLC and monolithic supports
	Different geometries (e.g. 2.1 mm I.D.) are available	Limited resistance in terms of backpressure (< 200 bar) and pH (2 < pH < 8)
High Temperature Liquid Chromatography (HTLC)	Green chemistry: decrease of the organic modifier amount at elevated temperature	Stability of the solutes and silica-based stationary phases can be critical at T > 100°C
	Improvement of peak shape for basic drugs and large molecules (e.g. peptides and proteins)	Need to use dedicated instrumentation (preheating and cooling devices + backpressure regulator)
	Possible to use this strategy in conjunction with UHPLC to further improve performance	Method transfer difficult because of changes in selectivity with temperature
Ultra high pressure liquid chromatography (UHPLC)	Easy method transfer between HPLC and UHPLC	Need to use dedicated instrumentation (low σ^2_{ext}, elevated acquisition rate, fast injection) and column compatible with ultra high pressures
	Important decrease in analysis time	Cost of instrumentation and consumables higher than conventional HPLC
	Large variety of columns packed with sub-2μm particles (more than 10 providers)	Solvent compressibility and frictional heating are issues for ΔP close to 1000 bar
Fused-core	Interesting approach to limit diffusion of large molecules in pores	Lack of chemistries (C18, C8, HILIC) and providers
	The quality of the packing is excellent (h ≈ 1.5) compared to other materials (h ≈ 2–2.5)	Retention and loading capacity slightly lower than conventional HPLC (particularly for 1.7μm fused core)
	Approach potentially compatible with a conventional HPLC system	Lower resistance in terms of backpressure (< 600 bar) and pH (2 < pH < 9) compared to UHPLC

2, 3 and 4.6 mm I.D., but only with a maximal length of 100 mm), and the limited resistance of the support in terms of pH and, more importantly, backpressure (ΔP_{max} = 200 bar) (Brice et al., 2009).

High temperature liquid chromatography (HTLC) is a simple and elegant strategy to improve chromatographic performance. An elevated mobile phase temperature (beyond 60°C and up to 200°C) could be valuable for improving chromatographic performance. Indeed, a temperature increase results in a significant reduction of mobile phase viscosity, leading to higher diffusion coefficients and improved mass transfer, thereby increasing the optimal linear velocity towards higher values (u_{opt} proportional to the ratio T/η) (Heinisch & Rocca, 2009; Li et al., 1997; Guillarme et al., 2007c). Temperature, which is directly proportional to the solvent viscosity, also causes a significant reduction in column backpressure at a constant flow rate. Due to these features, it is possible to maintain resolution and accelerate separations by a factor of 3 to 5 (at 90°C) and up to a factor of 20 (at 200°C), with hydro-organic mixtures containing MeOH and water (Guillarme et al., 2004). Alternatively, longer columns with acceptable backpressures can be employed at elevated temperatures, although it becomes difficult to work under optimal flow rate conditions. In addition to its kinetic performance, HTLC has some additional benefits, summarized in Table 1.B. First, both the dielectric constant and surface tension of water decrease at elevated temperature, and thus water can replace a significant proportion of the organic solvent in the mobile phase (5–10% less organic solvent for each 30°C change) (Guillarme & Heinisch, 2005; Hartonen & Riekkola, 2008). Second, an improvement in peak shape has been reported for basic compounds, due to a decrease of pKa and thus a reduction of secondary interactions with residual silanol groups (Heinisch et al., 2006; Albert et al., 2005). For large molecules, such as peptides, the peak shapes can also be improved since diffusion coefficients strongly increase with temperature. Finally, as temperature is a thermodynamic parameter, it can be used to tune selectivity during method development (McNeff et al., 2007). Although HTLC has been deeply investigated in various academic laboratories, it remains seldom used in an industrial environment. The major limitations of this approach are related to the relatively limited number of stable stationary phases compatible with elevated temperature (Teutenberg et al., 2007), the required modification of LC equipment to adequately control the mobile phase temperature (Thompson et al., 2001), and most importantly, the putative thermal degradation of compounds (Thompson et al., 2002).

The most recently introduced strategy to improve chromatographic performance is fused-core technology (also called core-shell or superficially porous particles), which became commercially available in 2007 under the trademarks Halo from Advanced Materials Technology (Wilmington, DE, USA), Ascentis from Supelco (Bellefonte, PA, USA), Kinetex from Phenomenex (Torrance, CA, USA) and more recently Poroshell from Agilent (Waldbronn, Germany). This technology was originally developed by J.J. Kirkland in the 1990s to limit diffusion of macromolecules into the pores (Kirkland, 1992), but smaller particle sizes (1.7, 2.6 or 2.7 μm) have now been incorporated. Nowadays, this approach consists of using sub-3μm superficially porous particles composed of a solid inner core and a thin porous outer core. In comparison with totally porous particles of similar diameters, the diffusion path is much shorter since the inner core is solid fused silica, which is impenetrable by analytes (thus decreasing the C-term of the van Deemter curve, mass transfer resistance) (Salisbury et al., 2008; Cavazzini et al., 2007). This minimizes peak broadening, especially at elevated linear velocities (Destefano et al., 2008). This characteristic is particularly important for the separation of large molecules (i.e. peptides or intact proteins) because slow mass transfer induces a loss of efficiency with rapid separations on porous particles (Gritti et al., 2007). Additionally, this material presents an extremely narrow particle size distribution and high packing density compared with porous particles, leading to a smaller A term in the van Deemter curve (i.e. Eddy diffusion) (Gritti et al., 2007). Indeed, various authors have determined h values down to 1.5 (Gritti et al., 2007) or even 1 (Gritti et al., 2010) for such columns, in contrast to values of 2–3 for columns packed with porous particles. Thus, for an identical column length, the semi-porous particles maintain around 80% of the efficiency of sub-2 μm particles, but with a 2-fold lower backpressure (Cunliffe et al., 2007; Fekete et al., 2009). This promising approach has grown quickly in recent years, but today the number of suppliers for fused-core particle columns is still restricted because of patent reasons. In addition, a limited number of support chemistries are available, but this tends to change rapidly as these products become more widely accepted by the scientific community (Majors, 2008; Novakova et al., 2009). Finally, even if the generated backpressure is two times less than that of columns packed with sub-2 μm particles, the resistance of the support to pressure claimed by providers is also almost twice lower (600 *vs.* 1000 bar).

Best LC Approach in Isocratic Mode—Theory and Applications

In isocratic mode, the throughput and resolving power of the different strategies were evaluated using a 2D-map reported in Fig. 1.5. In this representation, two important parameters were selected. For the separation speed, the column dead time ($t_{10,000}$) required to attain an efficiency of 10,000 plates was estimated. This plate number is generally sufficient for high throughput separations of conventional samples that contain a limited number of compounds. For the resolving power, the maximal achievable efficiency N_{30min} was calculated for a column dead time of 30 min (equivalent to an analysis time of 3 hours for k = 5). This analysis time is quite long, but not prohibitive when dealing with very complex samples (e.g. proteomics, metabolomics...). Basically, the values of $t_{10,000}$ and N_{30min} were calculated using the van Deemter data (H, u) and permeability values K_{v0} experimentally determined for each analytical strategy. Then, the data were computed considering the maximal pressure drop of the system (ΔP_{max} of 200 bar for monoliths; 400 bar for HPLC, sub-2μm and HTLC; 600 bar for fused-core and 1000 bar for UHPLC and HT-UHPLC). The corresponding column lengths and mobile phase flow rates required were calculated.

Then, the column dead times and plate numbers presented in Fig. 1.5 always correspond to a pressure drop equal to ΔP_{max} with variable column lengths and mobile phase flow rates. For a better understanding of the employed procedure to construct this 2D-map, readers can refer to a few didactic papers on kinetic plot methodology (Desmet et al., 2006; Desmet et al., 2005; Billen, 2007; Desmet et al., 2005b; Desmet & Cabooter 2009; Desmet, 2008).

According to Fig. 1.5, in terms of throughput (Y-axis), conventional HPLC ($t_{10,000}$ = 55 s for N = 10,000 plates) is clearly not competitive with the other strategies. As expected from theory, the analysis time can be significantly reduced with monoliths ($t_{10,000}$ = 15 s) because of the very low generated backpressure, but this material suffers from higher H values (H = ~ 8 µm, corresponding to 120,000 plates/m) compared to fused-core columns or those packed with sub-2µm particles (H = ~ 4–5 µm, corresponding to more than 200,000 plates/m). HTLC appeared beneficial for reducing the analysis time ($t_{10,000}$ = 11 s), because of the improvement of diffusion coefficients related to the mobile phase viscosity decrease with temperature. For example, Yang et al. (2000) showed that ultra-fast separations could be achieved at very high temperatures. A separation of five alkylphenones was carried out with a

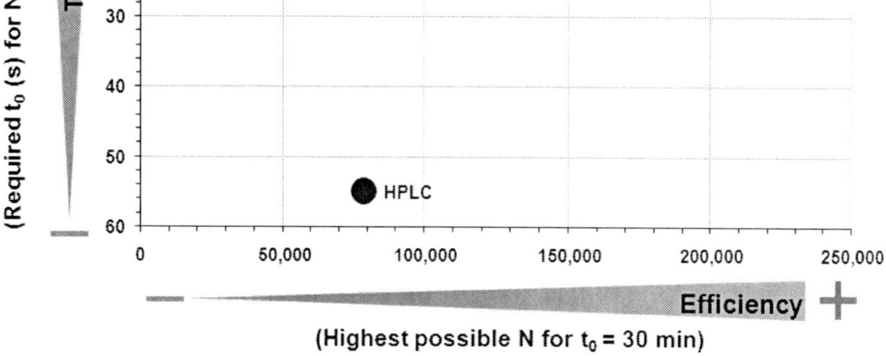

Fig. 1.5. Performance comparison of LC strategies in isocratic mode in terms of throughput and maximal resolution for a model compound, butylparaben with MW of 200 g.mol-1. Isocratic comparison in terms of throughput (t0 for N = 10000 plates) and maximal efficiency (Nmax for t0 = 30 min). The data were gathered using the kinetic plot methodology, considering a maximal pressure of 200 bar for monoliths, 400 bar for HPLC, HTLC and sub-2µm, 600 bar for fused-core and 1000 bar for UHPLC and HT-UHPLC. Adapted from (Guillarme et al., 2010), with permission.

conventional 50 × 4.6 mm column packed with 2.5-μm particles in only 20 s (ΔP of 360 bar), instead of 20 min at ambient temperature. The use of columns packed with sub-2μm particles remains the most efficient strategy to improve throughput, particularly when small particles are combined with high pressure and elevated temperature, where a 20-fold increase of throughput compared to that of conventional HPLC ($t_{10,000}$ = 3 s for N = 10,000 plates) is possible. As shown in Fig. 1.6a, a mixture of four preservatives was separated in less than 10 s by HT-UHPLC using a 50 × 2.1 mm column packed with 1.7 μm particles at 1.8 mL/min and 90°C (Nguyen et al., 2007).

The maximal efficiency that can be reached for a t_0 = 30 min (X-axis, N_{30min}) is comprised between 31,000 and 208,000 plates. Unfortunately, long columns operating in the B term-dominated region of the van Deemter curve are required (low mobile phase flow rate zone) (Guillarme et al., 2009; Desmet et al., 2006b). To reach such elevated efficiencies, the required column was between 0.6 and 3.4 m length, while the mobile phase flow rates ranged between 50 and 270 μL/min for a 2.1-mm I.D. column. Therefore, the strategies involving the use of small particles (i.e., sub-2μm, UHPLC and HT-UHPLC) are less potent (31,000 < N_{30min} < 73.000 plates) because of the elevated generated backpressure, thereby limiting the column length that can be employed. This observation has been experimentally confirmed by Sandra et al. (De Villiers et al., 2006b) who demonstrated that an N_{max} of 74,000 plates was possible for a test mixture under isocratic conditions at 40°C, using a 450 mm column length packed with 1.7 μm particles at 1000 bar. In contrast, the maximal efficiency of columns packed with 5-μm particles reported in Fig. 1.5 was around 80,000 plates and can be increased by 2.6-fold at elevated temperatures. Because of their elevated permeability, monolithic supports also represents a valuable strategy for increasing the plate count compared to conventional HPLC (N_{30min} is 2-fold higher) (Tanaka et al., 2002). However, the column length needs to be around 1.7 m, so numerous commercially available 100 mm columns length would need to be coupled in series. Although some authors have coupled up to 10 columns, the cost becomes rapidly prohibitive (Bones et al., 2008). Despite the elevated maximal efficiency observed with monoliths, this approach is not competitive with HTLC, as the lower maximal pressure drop capability (ΔP_{max} = 200 bar) limits the length of monolith that can be employed. In HTLC, more than 200,000 plates for a t_0 = 30 min can be attained, but with a 3 m column, which can be very expensive (twelve 250-mm columns). Experimentally, Sandra et al. coupled in series eight 250-mm columns packed with 5-μm particles, corresponding to a total length of 2 m. With this configuration, efficiencies as high as 180,000 plates were attained at 80°C for a test mixture, with t_0 values of 20 min and analysis times around 100 min (Lestremau et al., 2006; Lestremeau et al., 2007). Finally, fused-core technology provides N_{30min} values similar to monoliths, but with more acceptable column lengths and mobile phase flow rates, such as 1 m and 90 μL/min for a 2.1-mm I.D. column, respectively. These results can be attributed to the elevated plates/m values of the fused-core column in conjunction with a backpressure around 2-fold lower than that of columns packed with sub-2μm particles.

In conclusion, high-throughput separations require the use of columns packed with small particles (UHPLC) and should be carried out preferentially at increased temperatures (HT-UHPLC). It was also demonstrated that temperature and maximal system pressure drop should be increased as much as possible, since both parameters are beneficial for increasing the plate count as well as the throughput.

Best LC Approach in Gradient Mode—Theory and Applications

In the case of complex analyses, the separation is generally carried out in gradient mode to handle compounds possessing very different physicochemical properties and/or to improve the resolving power. In gradient mode, peak capacity must be calculated to estimate the performance (Neue, 2008). Peak capacity describes the number of peaks that can be separated with a resolution of 1 during a definite gradient duration and depends mainly on the isocratic efficiency, column dead time and gradient time. Two parameters were selected to evaluate the throughput and resolving power in gradient mode, namely the gradient time required to attain a peak capacity of 100 (t_{100}) and the maximal peak capacity for a gradient time of 3 hours (P_{3h}).

In recent papers (Guillarme et al., 2009; Guillarme et al., 2010a; Guillarme et al., 2010b), we applied the strategy developed by Schoenmakers et al. (Wang et al., 2006) to construct kinetic plots in gradient mode. In the graphical representation proposed in Fig. 1.7, a similar approach was employed, using data previously obtained in isocratic mode (H, u and K_{v0}) for each analytical strategy. Both the lowest gradient times, t_{100}, and highest peak capacities, P_{3h}, correspond to a pressure drop equal to ΔP_{max} and consequently to different column lengths and mobile phase flow rates.

According to Fig. 1.7, in terms of throughput, the ranking was similar for isocratic and gradient modes, and columns packed with small particles were clearly advantageous (i.e., sub-2μm, UHPLC and HT-UHPLC). Indeed, it is theoretically possible to attain a P = 100 in only 30 s with HT-UHPLC, whereas 7 min are required in conventional HPLC at 400 bar. Columns packed with superficially porous sub-3μm particles (fused-core technology) performed almost equivalently to columns packed with porous sub-2μm particles (t_{grad} of 1.4 min for P = 100), but monoliths and HTLC were less powerful (t_{grad} around 2.5 min for P = 100). These calculated values were in agreement with examples from the literature. For the separation of various pharmaceutical compounds, 50 mm columns packed with porous sub-2μm or superficially porous sub-3μm particles produce experimental P values higher than 70 in less than 2 min at ambient temperature (Zhang et al., 2009). Fig. 1.6.B shows

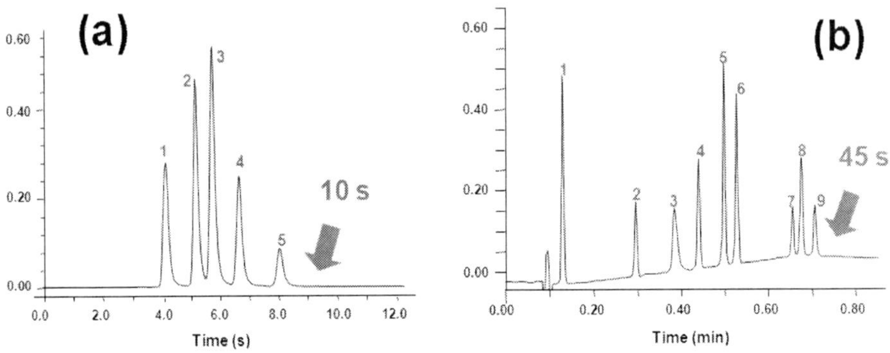

Fig. 1.6. Performance comparison of LC strategies in gradient mode in terms of throughput and maximal resolution for a model compound, butylparaben with MW of 200 g.mol-1. Gradient comparison in terms of throughput (tgrad for P = 100) and maximal peak capacity (Pmax for tgrad = 3 hours). The data were gathered using the kinetic plot methodology applied to gradient elution, and considering the same maximal pressure as for isocratic mode. Adapted from (Guillarme et al., 2010), with permission.

the potential of HT-UHPLC, with a gradient separation of various steroids in less than 45 s, at a temperature of 90°C. Heinisch et al. (Heinisch & Rocca, 2009; Barrioulet et al., 2007) also reported an impressive chromatogram of nine small aromatic compounds separated in less than 15 s by HT-UHPLC. This separation was performed with a 7.8 s gradient, using a 50 × 2.1 mm, 1.7 μm column at 2 mL/min and 90°C.

In terms of the resolving power, monoliths, fused-core, UHPLC, HT-UHPLC and HTLC offer almost the same peak capacities, ranging between 415 and 480 for a gradient time of 3 hours. Compared to the isocratic mode, UHPLC and HT-UHPLC generate higher resolution separations in gradient mode. This is because peak capacity is not strictly related to the chromatographic efficiency but also to the column dead time in gradient mode. Since the latter is strongly reduced in UHPLC and HT-UHPLC compared to the traditional approaches, the maximal peak capacity for a 3 h gradient is enhanced. Conventional HPLC at 400 bar offers around 30% less peak capacity compared to the other approaches, demonstrating that 5 μm particles do not present any practical benefit in gradient mode. Additionally, the sub-2μm strategy, with a maximal pressure of only 400 bar, was also significantly less practical, due to the low permeability of the packing and small backpressure limitation that reduces the column length and mobile phase flow rate that can be employed.

In conclusion, the fused-core and UHPLC technologies are very attractive for maximizing both the throughput and resolution in gradient mode. Whatever the selected strategy, the use of elevated mobile phase temperature is an additional parameter to improve gradient performance.

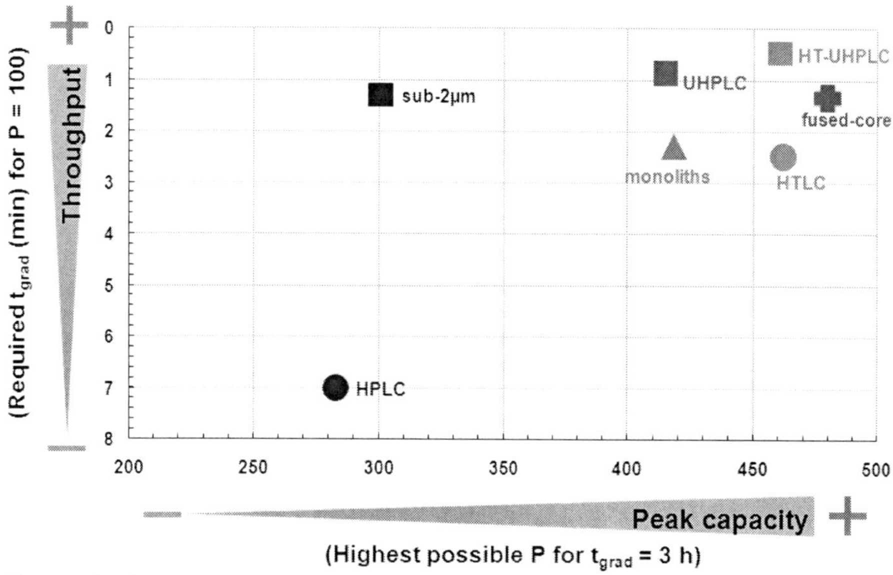

Fig. 1.7. Ultra fast separations carried out in HT-UHPLC. (a) Isocratic separation of various preservatives and uracil. Column: Acquity BEH C18 (50 × 2.1 mm I.D., 1.7 µm); mobile phase: water–acetonitrile (50:50%, v/v); flow-rate: 1800 µL/min; temperature: 90°C; (b) Gradient separation of several doping agents. Column: Acquity BEH Shield RP18 (50 × 2.1 mm I.D., 1.7 µm); mobile phase: 0.1% formic acid in water–0.1% formic acid in acetonitrile; flow-rate: 1800 µL/min; temperature: 90°C. Adapted from (Nguyen et al., 2007), with permission.

Problems Related to UHPLC

According to Table 1.B, the main drawbacks of UHPLC are i) the need to acquire a dedicated system, optimized in terms of high backpressure pumps and injector, acquisition rate of the detector, injection cycle time, dwell volume and system dead volume (Fountain et al., 2009), ii) the need to work with a column that withstands ultra high pressures, iii) the elevated backpressure that can produce some important changes in solvent properties (frictional heating and solvent compressibility effects). All these limitations are discussed in detail below.

The Need to Work with a Dedicated Instrumentation

As shown in Table 1.C, there is now a wide choice of instruments that withstand pressures above 400 bar, from various suppliers. Regarding the selection of a UHPLC system, the cost is certainly a decisive consideration, but it is also important to analyse specifications of all available instruments on the market, as none of them are equivalent. The most important feature is certainly the

Table 1.C. Summary of UHPLC Systems, with the Launch Year, Pressure Tolerance, Maximal Flow Rate, Oven Temperature and UV Acquisition Rate. The information presented was gathered from advertising and from the provider's Web Sites in June 2010. Adapted from (Eugster et al., 2010).

Launch year	Provider	Name of the system	Pumping system	Pressure tolerance	Maximal flow rate (corresponding pressure)	Maximal oven temperature	UV acquisition rate
2004	Waters[1]	Acquity UPLC	High pressure	1000 bar	2 mL/min (at 600 bar)	90°C	80 Hz
2006	Agilent[2]	Series 1200	High pressure	600 bar	5 mL/min (at 600 bar)	100°C	80 Hz
2006	Jasco[3]	Xtreme-LC	High pressure	1000 bar	3.5 mL/min (at 1000 bar)	65°C	100 Hz
2006	Rheos[4]	Allegro UHPLC	Low pressure	1000 bar	1 mL/min (at 1000 bar)	95°C	20 Hz
2006	Thermo[5]	Accela 1000 system	Low pressure	1000 bar	2 mL/min (at 1000 bar)	95°C	80 Hz
2007	Shimadzu[6]	Prominence UFLCxr	High pressure	660 bar	5 mL/min (at 440 bar)	85°C	50 Hz
2007	VWR[7]	Lachrom Ultra	High pressure	600 bar	5 mL/min (at 600 bar)	85°C	100 Hz
2008	Dionex[8]	Ultimate 3000	High / Low pressure	800 bar	5 mL/min (at 800 bar)	110°C	100 Hz
2008	Knauer[9]	PLATINblue	High / Low pressure	1000 bar	5 mL/min (at 800 bar)	140°C	200 Hz
2008	Perkin-Elmer[10]	Flexar FX-10 (eq. Series 275Hres)	High pressure	690 bar	3 mL/min (at 600 bar)	90°C	100 Hz
2009	Agilent[2]	Series 1290 Infinity	High pressure	1200 bar	5 mL/min (at 800 bar)	100°C	160 Hz
2009	Perkin-Elmer[10]	Flexar FX-15	High pressure	1240 bar	5 mL/min (at 1240 bar)	90°C	100 Hz
2009	Thermo[5]	Accela 600 system	Low pressure	600 bar	5 mL/min (at 600 bar)	95°C	80 Hz
2010	Waters[1]	Acquity UPLC H-Class	Low pressure	1000 bar	2 mL/min (at 600 bar)	90°C	80 Hz
2010	Shimadzu[6]	Nexera	High pressure	1300 bar	5 mL/min (at 800 bar)	150°C	100Hz

[1]Waters Corporation, Milford, MA. [2]Agilent Technologies, Waldbronn, Germany. [3]Jasco Corporation, Tokyo, Japan. [4]Flux Instruments, Reinach BL, Switzerland. [5]Thermo Fisher Scientific Inc., Waltham, MA. [6]Shimadzu Corporation, Kyoto, Japan. [7]VWR International, West Chester, PA. [8]Dionex Corporation, Sunnyvale, CA. [9]Knauer, Berlin, Germany. [10]Perkin-Elmer, Inc., Waltham, MA.

maximal available pressure and corresponding flow rate, which mostly defines the price of a UHPLC system. For commercial apparatuses, the ΔP_{max} varies between 600 and 1300 bar. It has been demonstrated that for fast or ultra-fast separations of simple mixtures, the use of small particles was obvious, but there was no need to work with very elevated pressures (Eugster et al., 2010; Grata et al., 2008). For high throughput experiments, UHPLC instruments with pressure limits around 600 bar could provide a suitable solution at a reasonable price. On the other hand, for high resolution separation, a system with an elevated maximal pressure is mandatory to work at acceptable flow rates (Eugster et al., 2010; Grata et al., 2008). Other than the pressure capability of the apparatus, it is also important that the instrument was adapted to operate in fast and ultra-fast modes with reduced column volumes. Indeed, small diameter columns (1 and 2.1 mm I.D.) commonly employed in UHPLC require limited extra-column volumes of detection, tubing, and injection volume. The following criteria have to be fulfilled to perform suitable separations.

The connecting tubes should be drastically reduced. For this reason, the tubing length should be as short as possible and its diameter selected as a compromise between an acceptable pressure drop and a low volume. For this reason, a system plumbed with 0.005″ I.D. stainless steel tubing and zero-dead volume fittings should be preferred for UHPLC experiments.

The injection volume should be selected in agreement with column geometry. A rule of thumb is to maintain the injected volume between 1 and 5% of the column dead volume in isocratic mode. As most of the experiments carried out in UHPLC are performed with a 50 × 2.1 mm column (V_0 = 120 µL), the injected volume should be between 1 and 5 µL, to limit band broadening. In addition, a fast injection cycle time is mandatory for analysis times lower than one or two minutes.

Last but not least, the detector cell volume, time constant and acquisition rate should be carefully selected. The detector used in UHPLC should ideally possess a low cell volume (2 µL or lower), while the sensitivity shouldn't be lower compared to that of a conventional HPLC instrument (equivalent path length of 10 mm). The detector time constant has to be fast enough ($\tau \leq 100$ ms), because peak widths are very small in UHPLC (only a few seconds). Finally, the UV detector sampling rate must be sufficiently high to acquire a suitable amount of data points for each peak (> 20 Hz).

Finally, to reduce the detrimental of extra-column volumes and avoid unacceptable loss in efficiency, chromatographic conditions leading to high retention factors (k or ke at least equal to 3) are often recommended in UHPLC.

The last critical parameter to perform ultra-fast separations in gradient mode is the system dwell volume, which corresponds to the time necessary for the gradient to reach the column inlet. To work with ultra-fast separations, a small gradient delay volume is required. With a large dwell volume, fast separations are compromised because an isocratic hold is generated at the beginning of the gradient, inducing potential changes in selectivity and longer analysis times.

To check the compatibility of a UHPLC instrument with a given column geometry, it is recommended to characterize the chromatographic system by determining the extra-column and dwell volumes. For optimal compatibility with ultra-fast separations, the former should be lower than 20 µL while the latter should be reduced to a few hundred µL. Finally, a comparative study made by a pharmaceutical company on various UHPLC systems can be found elsewhere, for additional information (Cunliffe et al., 2007).

The Need for Specific Columns Compatible with Ultra High Pressures

An important aspect when selecting a UHPLC set-up is the selection of a stationary phase that should provide sufficient selectivity as well as acceptable performance and lifetime. Before the commercialization of UHPLC technology, only non-porous particles were used. Jorgenson et al. were the first to compare 1.5 µm porous hybrid particles packed into a 30 µm I.D. fused silica capillary column supplied by Waters (Milford, USA) to 1.0 µm non-porous silica material (Mellors & Jorgenson, 2004). Chromatographic performance was evaluated at pressures up to 4500 bar, and hybrid particles were similar to 1.0 µm non-porous silica particles, in terms of pressure resistance.

In 2004, Waters launched a new generation of hybrid columns packed with 1.7 µm particles that were stable up to 1000 bar. A bridged ethylsiloxane/silica hybrid (BEH) particle has mechanical and chemical resistance in extreme conditions of pH (1 to 12), pressure, and temperature (up to 180°C) (Gika et al., 2008b). Numerous other providers have also made available stationary phases packed with sub-2µm particles. A list of such stationary phases is given in Table 1.D. As shown, the number of columns is quite extensive, with around 80 supports and numerous chemistries accessible from more than 10 different providers, demonstrating the opportunity to transfer almost all existing methods from HPLC to UHPLC. Indeed, the variety of phase chemistries can resolve almost all analytical issues: C8 and C18 for compounds of average polarity; C4 and cyano for the most apolar anlaytes; diol, amino, silica and HILIC for the most polar molecules and biphenyl, perfluorophenyl (PFP) or zirconia for alternative selectivity. All these stationary phases are not equivalent in terms of pressure tolerance (from 600 to 1200 bar) and particle size (from 1.5 to 2 µm), but also pH and temperature ranges. Some performance comparisons between the different phases can be found in the literature, and data for column lifetime have also been published.

One of the main criticisms made by early UHPLC users has been the reduced lifetime of columns packed with sub-2µm particles, compared to conventional columns. UHPLC columns are systematically exposed to very high pressures, but the packing pressure has been increased in the meantime. In our laboratory, we have observed that lifetimes of UHPLC and regular HPLC

Table 1.D. Summary of Providers for Columns Packed with Sub-2μm, with the Number of Available Chemistries, the Main Types of Chemistry, Pressure Tolerance and Particle Size. The information presented here was gathered from advertising and from the providers' Web Sites in June 2010. Adapted from (Eugster et al., 2010).

Provider	Name	Number of chemistries	Main types of chemistry	Pressure tolerance	Particle size
			500 bar < ΔP_{max} < 800 bar		
Agilent[1]	RRHT	8	C18, C8, CN	600 bar	1.8 μm
Grace-Davison[2]	Vision-HT	6	C18, HILIC, Silica	800 bar	1.5 μm
Sepax Technologies[3]	Sepax UHPLC	11	C18, C8, C4, Phenyl, Amino, CN, SCX, SAX, HILIC	600 bar	1.8 μm
VWR4	LaChromUltra	1	C18	600 bar	2 μm
YMC5	UltraHT	2	C18	500 bar	2 μm
Zirchrom Technologies[6]	Zirchrom-PBD & -Phase	2	Zirconia-based material	700 bar	2 μm
			$\Delta P_{max} \geq$ 1000 bar		
Agilent[1]	RRHD	2	C18	1200 bar	1.8 μm
ES Industries[7]	Epic	7	C18, PFP, HILIC, Diol, Silica + 3 phases SFC	1000 bar	1.8 μm
Interchim[8]	Strategy	2	C18, HILIC	1000 bar	1.7μm
Knauer[9]	BlueOrchid	7	C18, C8, PFP, Phenyl, CN, Silica	1000 bar	1.8 μm
Macherey-Nagel[10]	Nucleodur	5	C18, C8	1000 bar	1.8μm
Restek[11]	PinnacleDB and Ultra II	12	C18, C8, PFP, Biphenyl, CN, Silica	1000 bar	1.9 μm
Thermo[12]	Hypersil GOLD	11	C18, C8, C4, PFP, CN, Phenyl, Amino, AX, SAX, Silica	1000 bar	1.9 μm
Waters[13]	Acquity BEH	6	C18, C8, C4, Phenyl, HILIC, Amide	1000 bar	1.7 μm
Waters[13]	Acquity HSS	3	C18	1000 bar	1.8 μm

[1]Agilent Technologies, Waldbronn, Germany. [2]Grace Davison Discovery Sciences, Deerfield, IL. [3]Sepax Technologies, Inc., Newark, DE. [4]VWR International, West Chester, PA. [5]YMC, Kyoto, Japan. [6]ZirChrom Separations, Inc., Anoka, MN. [7]ES Industries, West Berlin, NJ. [8]Interchim, Montluçon, France. [9]Knauer, Berlin, Germany. [10]Macherey-Nagel GmbH & Co. KG, Düren, Germany. [11]Restek, Bellefonte, PA. [12]Thermo Fisher Scientific Inc., Waltham, MA. [13]Waters Corporation, Milford, MA.

columns were comparable. However, column lifetime could be expressed as the number of injections, number of column volumes or period of time used. With the latest generation of columns packed with sub-2µm particles commercialized by different suppliers, it is possible to perform between 500 and 2000 injections or even more on a single column (King et al., 2005; Grumbach et al., 2005). Such values correspond to about 5,000–20,000 column volumes, which are fully comparable to those obtained on standard HPLC columns. However, when considering the corresponding period of time, it is significantly reduced compared to conventional HPLC because of the higher throughput in UHPLC. For example, in a routine laboratory that performs a UHPLC analysis in 1 to 5 min, one thousand injections can be performed in a very short period of time. This time is significantly reduced compared to that using conventional instrumentation (10-fold longer), although a similar amount of work has been carried out.

The Changes in Solvent Properties with Pressure

Martin and Guiochon (Martin & Guiochon, 2005) recently demonstrated that beyond 100 bar, experimental parameters which are considered as constant by chromatographers (e.g., column porosity, mobile phase density and viscosity, diffusion coefficient, retention factors, and efficiency) could depend on pressure to some extent. Consequently, irreproducible retention factors or loss in efficiency can be observed.

Frictional heating of the mobile phase is certainly the most critical effect that can be observed in UHPLC conditions (Halasz et al., 1975) and has been largely investigated in recent years by Colon et al., Guiochon et al., Sandra et al. and Desmet et al. Columns packed with small particles present low permeability, and thus generate a considerable amount of frictional heating under high pressure drops and elevated flow rates. The heat generation, or power dissipation, is the product of pressure drop (ΔP) and flow rate (F) as reported in equation 6 (MacNair et al., 1997).

$$Power = F \cdot \Delta P \tag{6}$$

Frictional heating and poor heat dissipation cause significant radial and axial temperature gradients across the column. An irregular radial temperature profile produces solute retention alteration, unexpected changes in selectivity and additional band broadening (Jerkovich et al., 2003; Martin & Guiochon, 1983). Reducing the column diameter to capillary dimensions largely eliminates this effect (Jerkovich et al., 2003). Therefore, only capillary columns (e.g., 30–150 µm I.D.) were used before commercialization of UHPLC systems, but Jorgenson et al. (MacNair et al., 1997) recognized that columns of larger diameter could be used with a reasonable value of generated power (i.e., <1 W).

In 2004, Colón et al. investigated in detail the power dissipation of column diameter in the mm range, and confirmed that the latter, packed with small particles, would generate acceptable frictional heating. A column of 150 × 1 mm I.D. packed with 1.5 μm particles size with a flow rate of 85 μL/min generated about 1380 bar. Under these conditions, the power dissipation was 195 mW, which corresponded to values encountered in conventional HPLC. Similar results were obtained with a 2.1 mm I.D. column packed with 1.7 μm particles at a linear velocity of 2.5 mm/s (Colon et al., 2004). Thus, the frictional heating generated with the 1 to 2.1 mm I.D. UHPLC columns employed nowadays was not detrimental to the separation (Nguyen et al., 2006).

Sandra et al. more recently investigated the effects of the temperature gradient across the column length due to frictional heating. Experimental data from a 2.1 mm I.D. column packed with 1.7 μm particles, at 1000 bar with a conventional still-air column heater showed no significant efficiency loss (De Villiers et al., 2006).

On the other hand, Guiochon et al. demonstrated the importance of the column external environment on its performance and particularly that frictional heating effects could be detrimental with a 2.1 mm I.D. column. This is illustrated in Fig. 1.8 where an experimental setup allows the measure of the temperature at the surface of the column and the temperature of the liquid exiting the column outlet. With this strategy, complete temperature profiles along and across various UHPLC columns under still air conditions at room temperature were constructed. For this purpose, four different column lengths were considered, namely 150, 100, 50 and 30 mm lengths, in 1 and 2.1 mm I.D. Only the results for the 2.1 mm I.D. columns of 50, 100 and 150 mm are presented in Fig. 1.8. The authors concluded that, with columns of 2.1 mm I.D., the difference between the inlet and outlet temperature (longitudinal gradient) was comprised between 10 and 20 K while it varied between 3 and 16 K with 1 mm I.D. column, depending on the considered column length. To reduce the effect of thermal gradient, Desmet et al. recently split up a column with given length L into n segments with length L/n, while providing an active cooling to the capillaries connecting the segments (Gritti & Guiochon, 2008). In this way, the viscous heat was removed at a location where the radial heat removal does not lead to an efficiency loss (i.e., in the thin connection capillaries), while the column segments can be operated under near-adiabatic conditions without suffering from an unacceptable rise of the mobile phase temperature (Broeckhoven et al., 2010).

Besides frictional heating effects, mobile phase viscosity is also dependent on pressure, and the organic solvents such as methanol or acetonitrile are more compressible than water (Foley et al., 1989). Jorgenson et al. investigated the solvent compressibility effects in gradient mode and noticed a reduction of reproducibility in peptide retention times from run to run in UHPLC conditions (for pressures up to 5000 bar), but also indicated that this problem could

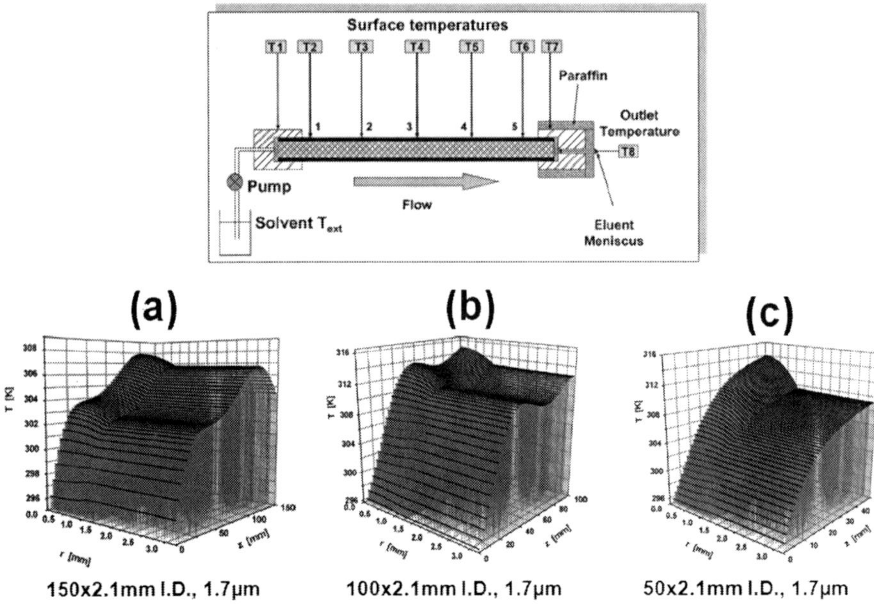

Fig. 1.8. 3-D representation of the temperature profiles along and across various chromatographic columns (A) 150 mm × 2.1 mm BEH-C18 column, flow rate = 0.95 mL/min, Pinlet = 979 bar. (B) 100 mm × 2.1 mm BEH-C18 column, flow rate = 1.45 mL/min, Pinlet = 1005 bar. (C) 50 mm × 2.1 mm BEH-C18 column, flow rate = 2 mL/min, Pinlet = 775 bar. The experimental setup used to measure the temperature of the surface of the column and the temperature of the liquid exiting the column outlet is also shown. Adapted from (Gritti & Guiochon, 2008), with permission.

be resolved with a second-generation system (MacNair et al., 1999). In another study performed in isocratic mode, the surge in linear velocity caused by mobile phase compression at the start of isocratic UHPLC runs (with pressures up to 6300 bar) was found to induce a 50% increase in the measured Knox C-term (Jerkovich et al., 2005). However, because the maximal pressure of commercial UHPLC instrument is limited to 1300 bar, the negative impacts of solvent compressibility (i.e., lack of reproducibility and increase of C-term) are less critical.

Method Transfer from HPLC to UHPLC

In various fields of application (e.g., pharmaceutical, environment, food...), it is essential to be able to transfer existing methods (performed in conventional HPLC conditions) to faster separations involving the use of columns packed with sub-2μm particles. As most providers now offer equivalent stationary phases packed with 5, 3 and sub-2μm particles (see Table 1.D), a geometrical transfer can be performed if the stationary phase chemistry remains identical between the original and final sets of conditions. For this purpose, some rules have to be strictly applied in both isocratic and gradient modes.

The Rules for Isocratic Mode – Theory and Applications

For an isocratic transfer between conventional HPLC and UHPLC, two important parameters have to be adjusted, namely the injection volume and the mobile phase flow-rate (Guillarme et al., 2007).

To avoid a detrimental extra-column band broadening and maintain equivalent sensitivity, it is necessary to adapt the injection volume in agreement with the change of column dimensions. In liquid chromatography, the injected volume should represent only 1 to 5% of the column volume. The latter should be adjusted proportionally to the column internal diameter (d_c) and length (L). Therefore, the injection volume is independent on the particle size and only proportional to the column volume. The new injected volume (V_{inj_2}) can be determined simply by maintaining the ratio of column dead volume and injected volume constant between regular HPLC and UHPLC.

$$V_{inj_2} = V_{inj_1} \cdot \frac{d_{c_2}^2}{d_{c_1}^2} \cdot \frac{L_2}{L_1} \tag{7}$$

In this equation, subscripts 1 and 2 are related to HPLC and UHPLC column dimensions, respectively. For example, from a conventional 150 × 4.6 mm, 5 μm column to a UHPLC 50 × 2.1 mm, 1.7 μm column, the injected volume should be decreased by 14-fold. To maximize sensitivity, it is possible to inject larger volume, but in this case the sample should be dissolved in a solvent of weaker eluent strength than the initial mobile phase composition (sample focusing).

Regarding mobile phase flow rate, this parameter should be adapted to remain close to maintaining a constant reduced mobile phase linear velocity (ν). In HPLC, it is well-known that the mobile phase linear velocity (u) is directly proportional to the square of column diameter, and also depends on the particle size of the support. It is, however, completely independent on the column length. For a successful method transfer, it is mandatory to maintain the product $u*d_p$ constant, to take into account simultaneous changes in column diameter and particle size of the support. Therefore, for a geometrical transfer, the UHPLC flow rate (F_2) can be calculated with the following equation:

$$F_2 = F_1 \cdot \frac{d_{c_2}^2}{d_{c_1}^2} \cdot \frac{d_{p_1}}{d_{p_2}} \tag{8}$$

As an example, from a regular 150 × 4.6 mm, 5 μm column to a UHPLC 50 × 2.1 mm, 1.7 μm column, the mobile phase flow rate should be decreased by 1.6-fold.

The expected analysis time of the transferred method (t_{ana_2}) is directly proportional to the change in column dead time and can be estimated according to:

$$t_{ana_2} = t_{ana_1} \cdot \frac{d_{p_2}}{d_{p_1}} \cdot \frac{L_2}{L_1} \tag{9}$$

The expected backpressure (ΔP_2) can be calculated from Darcy's law, which shows that ΔP is inversely proportional to d_p^3 (at the optimal linear velocity) and is strictly related to the column length:

$$\Delta P_2 = \Delta P_1 \cdot \frac{L_2}{L_1} \cdot \frac{d_{p_1}^3}{d_{p_2}^3} \tag{10}$$

Finally, the expected solvent consumption of the transferred method (V_2) can be calculated by taking into account the change in internal diameter and column length.

$$V_2 = V_1 \cdot \frac{d_{c_2}^2}{d_{c_1}^2} \cdot \frac{L_2}{L_1} \tag{11}$$

Therefore, from a regular 150 × 4.6 mm, 5 µm column to a UHPLC 50 × 2.1 mm, 1.7 µm column, the analysis time is reduced by 9-fold. For the abovementioned transfer, the efficiency would be identical, while the backpressure should be 9-fold higher and the solvent consumption reduced by 14-fold. This shows the obvious benefits of the UHPLC strategy.

It is possible to find in the literature a large quantity of applications showing the possibility to transfer isocratic HPLC methods to columns packed with sub-2µm particles, providing that the chemistry of the two analytical supports remains identical. One example is reported in Fig. 1.9 and presents a method transfer from a conventional 150 × 4.6 mm, 5 µm column to a UHPLC 50 × 2.1 mm, 1.9 µm column (Russo et al., 2008). Both columns provide an equivalent efficiency of around 10,000 plates (similar *L/dp* ratio). As shown in the chromatograms, efficiency, selectivity and resolution remain equivalent for the separation of seven common anxiolitic agents. After adjustment of mobile phase flow rate, the analysis time could be decreased by a factor of 7 (22 *vs.* 3.2 min), as expected from theory for a transfer from 5 to 1.9 µm particles.

The Rules for Gradient Mode—Theory and Applications

The rules for gradient method transfer between conventional HPLC and UHPLC are more complex than isocratic ones, but are also based on the basic principles of chromatography. First, the injection volume and mobile phase

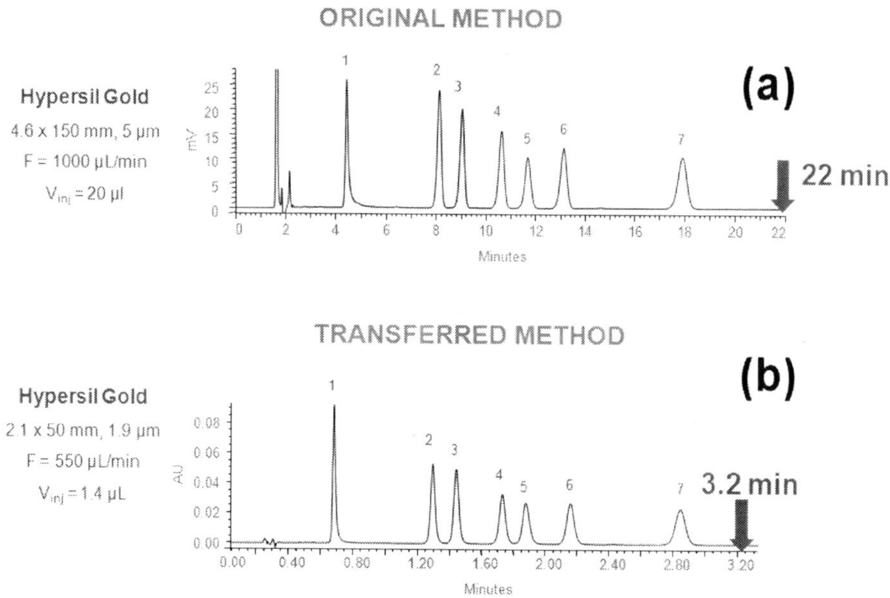

Fig. 1.9. Isocratic method transfer from regular HPLC to UHPLC. Separation of a benzodiazepines mixture in isocratic mode with a mobile phase containing ACN-water (31:69, v/v) with 0.1% formic acid, T = 30°C and λ = 254 nm. (a) Hypersil GOLD 150 × 4.6 mm, 5 µm, F = 1000 µL/min, Vinj = 20 µL. (b) Hypersil GOLD 50 × 2.1 mm, 1.9 µm, F=550 µL/min, Vinj=1.4 µL. Adapted from (Russo et al., 2008).

flow rate should be adapted in a similar way as the isocratic mode (see equations 7 and 8) (Guillarme et al., 2008).

In linear or multi-linear gradient elution, the gradient profile can be decomposed as the combination of various isocratic and gradient segments. The rules for efficient gradient transfer originally established by Snyder and Dolan (Dolan & Snyder, 1998) should be strictly followed. For both parts, it is important to scale the gradient volume in proportion to the number of column volumes to yield identical elution patterns, while the initial and final compositions should be constant. In fact, the number of column volumes percolated during the gradient in the regular HPLC system should be equivalent to that of the UHPLC set-up.

For any isocratic step within the gradient (i.e. initial isocratic step, isocratic step during a multi-linear gradient and also re-equilibrating time), the ratio between the isocratic step time (t_{iso}) and the column dead time should be maintained equivalent between conventional HPLC and UHPLC conditions. Therefore, the UHPLC isocratic step (t_{iso_2}) can be determined using:

$$t_{iso_2} = t_{iso_1} \cdot \frac{d_{p_2}}{d_{p_1}} \cdot \frac{L_2}{L_1} \tag{12}$$

As an example, from a regular 150 × 4.6 mm, 5 μm column to a UHPLC 50 × 2.1 mm, 1.7 μm column, the isocratic steps which occurred during the gradient process should be reduced by 9-fold.

For slope segments, it is mandatory to keep the initial and final gradient composition (*%B*) constant. The new gradient time (t_{grad_2}) can be expressed as:

$$t_{grad_2} = \frac{(\%B_{final_1} - \%B_{initial_1})}{slope_2} \tag{13}$$

The gradient slope ($slope_2$) should be calculated to maintain the product of gradient slope and column dead time constant. The new gradient slope ($slope_2$) can be expressed as:

$$slope_2 = slope_1 \cdot \frac{d_{c_1}^2}{d_{c_2}^2} \cdot \frac{L_1}{L_2} \cdot \frac{F_2}{F_1} \tag{14}$$

As an example, from a regular 150 × 4.6 mm, 5 μm column to a UHPLC 50 × 2.1 mm, 1.9 μm column, the gradient slope during the gradient process should be increased by 9-fold.

When transferring a gradient method from regular HPLC to UHPLC, some changes in selectivity could occur during the gradient run because of differences in dwell volume between the original and the UHPLC configuration. The system dwell volume (V_d) refers to the volume between the mixing point of solvents and the head of the analytical column (Dolan, 2006). After starting the gradient, it will take time until the selected proportion of solvent reaches the column. Since the gradient dwell volume may differ from one system to another, this extra isocratic step would be different and could result in retention time variations affecting resolution for early eluting peaks when transferring a method. To overcome this problem, the ratio of system dwell time (t_d) and column dead time (t_0) must be held constant while changing column dimensions, particle size or mobile phase flow rate.

As the column dead time is reduced by around 9-fold between a regular 150 × 4.6 mm, 5 μm column and a 50 × 2.1 mm, 1.7 μm column, the system dwell time should be reduced by the same factor. As mentioned previously, it is then mandatory to work in UHPLC with a system possessing a low dwell volume (no more than a few hundred μL) to limit its influence.

Again, many applications of the above discussed approach can be found in the literature. One example has been selected, and is presented in Fig. 1.10. The original separation of 12 pharmaceutical compounds was achieved using a 150 × 4.6 mm, 5 μm, C_{18} column and subsequently transferred to UHPLC with a 50 × 2.1 mm, 1.7 μm column possessing strictly similar chemistry (Guillarme et al., 2008). The original separation was performed in approximately 27 min, and was efficiently transferred to UHPLC, giving a separation that was complete in less than 3 min (reduction by a factor of 9, as expected from theory). In addition, both separations were equivalent in terms of sensitivity, peak capacities and resolution, mainly because of an adequate reduction of system dwell volume (from 1 mL to 100 μL for HPLC and UHPLC, respectively). Finally, the mobile phase flow rate can be increased up to the maximal operating pressure of instrumentation to further reduce analysis time. As illustrated in Fig. 1.10.C, at 1000 μL/min the analysis time was cut to only 1.6 min, but with a slight loss (about 10%) of chromatographic performance.

Fields of Application for UHPLC-MS and Related Issues

The coupling of UHPLC with MS appears as the ultimate approach in terms of sensitivity, selectivity, and peak assignment for the determination of analytes at low concentrations in complex matrices. Two main types of MS analyzers were coupled with UHPLC, namely quadrupole-based and time-of-flight (TOF-MS) instruments. The former, operating in the selected ion monitoring (SIM) or selected reaction monitoring (SRM) modes, was preferentially selected for targeted analysis (e.g., bioanalysis, multi-residue screening), while the latter was particularly useful for non-targeted analysis (e.g., drug metabolism studies, metabolomics).

Regarding quadrupole-based analyzers, the sampling rate can be an issue and modern instruments possessing improved acquisition rates should be selected for hyphenation with UHPLC. With the new generation of analyzers, dwell times have been reduced, down to 1 ms for some providers, in SIM as well as in SRM modes (Schappler et al., 2009; Guillarme et al., 2010b). Time-of-flight instruments are also well adapted to record and store data over a broad mass range without compromising sensitivity. With the latest generation of TOF-MS, high mass resolution (e.g., higher than 10,000 FWMH) can be attained at speeds of up to 40 full spectra per second.

Aside from the acquisition rate, it has been demonstrated that MS instruments could represent a non-negligible source of extra-column band broadening in UHPLC compared to a UV detector (Grata et al., 2009). In gradient mode, the peak capacity was reduced by 15 to 30% with a TOF-MS compared to UV because of the important ionization chamber volume, transfer capillary volume and critical electronic signal treatment (Grata et al., 2009). In addition, even if fast polarity switching (i.e., 20 ms) and/or fast ESI/APCI mode switching (i.e., 20 ms) are available from several providers to increase productivity, it always compromised sensitivity, peak width and sampling rate in UHPLC, and should then be used with caution (Schappler et al., 2009; Guillarme et al., 2010b; Grata et al., 2009).

Fig. 1.10 shows that MS(/MS) is the detector of choice for more than 60% of separations carried out by UHPLC. The current state of the art of UHPLC-MS(/MS) is reported in the present chapter, with emphasis on the throughput

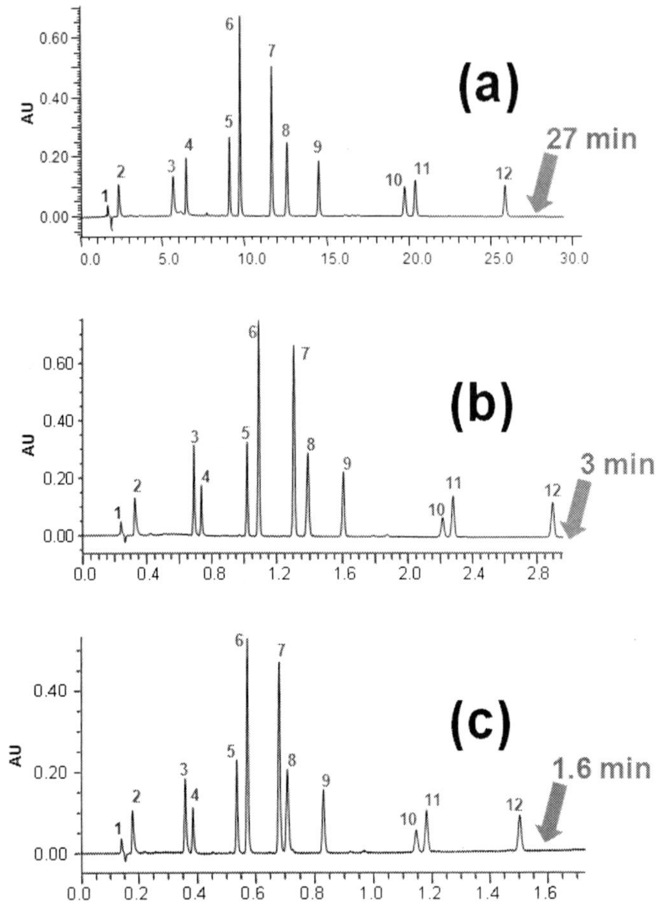

Fig. 1.10. Gradient method transfer from regular HPLC to UHPLC. Separation of a pharmaceutical mixture containing the main product (6) and 11 impurities in gradient mode with HPLC and UHPLC system: (a) original HPLC method: column: XBridge C18 150 × 4.6 mm, 5 μm; flow rate: 1000 μL/min; injected volume: 20 μL; total gradient time: 45 min. (b) Transferred UHPLC method: column: ACQUITY BEH C18 50 × 2.1 mm, 1.7 μm; flow rate: 610 μL/min; injected volume: 1.4 μL; total gradient time: 5.1 min. (c) Transferred and optimized UHPLC method: column: Acquity BEH C18 50 × 2.1 mm, 1.7 μm; flow rate: 1000 μL/min; injected volume: 1.4 μL; total gradient time: 3.1 min. Adapted from (Guillarme et al., 2008).

increase and/or resolution improvement afforded by UHPLC technology, and its impact on MS detection capabilities. For this purpose, three promising application fields of the technique were considered: (i) bioanalysis and drug metabolism with tandem MS operating in the SRM mode on QqTOF analyzers; (ii) rapid multi-residue screening, using either quadrupole-based instruments in SRM mode or TOF analyzers; and (iii) metabolomics, taking advantage of the very high chromatographic resolution of UHPLC in combination with the exact mass measurement of TOF/MS (Guillarme et al., 2010b).

UHPLC-MS/MS for High Throughput in Bioanalysis

Challenges in bioanalytical laboratories include the development of fast LC-MS methods able to separate closely related compounds, such as analytes and metabolites, from endogenous components.

Since UHPLC greatly enhances the separation throughput and resolution, base peaks as narrow as (or less than) 1 s can be obtained. This induces practical issues for bioanalytical applications using MS, as sufficient data points (e.g., >15 points per peak) are essential to ensure reliable quantitation. Several critical applications can be found in the literature. For example, Petsalo et al. (2008) recently published a UHPLC-MS/MS procedure for analyzing nine drugs and their respective metabolites in urine, using a 4 min gradient. An ESI source operating sequentially in positive and negative polarity modes was employed, and dwell times (DT) of 20–30 ms were applied for each SRM transition. Because peak widths of 4 s were experimentally obtained, only 6 points were acquired to define peaks, which could limit performance, particularly at the LLOQ. To accommodate the small UHPLC peak widths, DT can be reduced when many SRM transitions have to be monitored (Schappler et al., 2009), but this can result in sensitivity loss. An alternative approach consists of using various time windows during the acquisition, as recently proposed by Berg et al. (2009). Finally, to circumvent DT reduction, Li et al. suggested a useful "peak parking" strategy, which consisted of reducing the flow rate during peak elution, and thus extending the MS acquisition window for quantitative bioanalytical assays (Li et al., 2008). However, the latter strategy is only suitable when a limited number of targeted analytes are analyzed.

In addition, various studies have demonstrated that a significant reduction of matrix effects was brought about by UHPLC technology, compared to regular HPLC. As an example, Chambers et al. (2007) proved that polymeric mixed-mode SPE, combined with UHPLC technology and appropriate mobile phase pH, provided significant benefits for reducing matrix effects from plasma matrix components, and improving ruggedness and sensitivity of bioanalytical methods.

Considering the analysis time reduction offered by UHPLC technology, the sample preparation procedure becomes the limiting step in terms of total analysis time. Numerous UHPLC bioanalytical applications still involve traditional sample preparation procedures, which drastically increase the total analysis time. A few authors have suggested solutions to this issue, while maintaining sufficient sample preparation selectivity. To date, only a few studies involving the 96-well plate format were reported prior to UHPLC-MS/MS bioanalysis, allowing for selective, sensitive, and, above all, high-throughput analyses (Licea-Perez et al., 2007; Yadav et al., 2008).

High Resolution Drug Metabolism by UHPLC-MS using QqTOF Analyzers
In drug metabolism experiments, there are also many challenges due to the complex nature of biological matrices and the large diversity of produced metabolites. To identify unexpected metabolites, high chromatographic resolution and mass accuracy for fragmentation patterns are key requirements. UHPLC with high-resolution analyzers such as QqTOF are particularly useful to fulfill both tasks in a high-throughput environment.

Castro-Perez et al. (2005) were the first to report the use of UHPLC technology with a QqTOF analyzer in the early drug discovery process. This study emphasized improved resolution, in terms of chromatographic and mass spectral quality, and the associated gain in sensitivity afforded by the UHPLC-QqTOF/MS system. These features were explained by the combination of reduced peak width and low ion suppression that may result from the coelution of metabolites and endogenous compounds.

Walles et al. and others have investigated the benefits and drawbacks of three UHPLC-QqTOF/MS methods for fast metabolite identification using alternative MS/MS experiments (MSE) (Walles et al., 2007; Plumb et al., 2006; Crockford et al., 2008). The high efficiency attributable to UHPLC was the key to the successful identification of isobaric metabolites. In fact, they could not be distinguished with accurate mass of QqTOF, as they had an identical elemental composition and often similar MS/MS fragmentation patterns. It can be noted that the time spent for structure elucidation created additional bottlenecks and becomes the limiting step when UHPLC-QqTOF experiments are performed.

UHPLC-MS for Multi-residue Screening
Multi-residue screening techniques are generally developed to quickly assess the presence of contaminants in a complex sample. Thus, the developed method should be able to detect as many components as possible in a single analytical run. In this context, UHPLC coupled with tandem MS or TOF/MS remains the gold standard. Multi-residue screening methods with UHPLC technology have been applied to a wide variety of analytes and matrices, including: (i) doping agents (Thorngren et al., 2008; Badoud et al., 2009; Badoud et al., 2010) and veterinary drugs (Kaufmann et al., 2007) in biological matrices; (ii) drugs (Kasprzyk-Hordern et al., 2008), pesticides (Gervais et al., 2008), and herbicides (Pastor Montoro et al., 2007) in environmental matrices; (iii) veterinary drugs (Stolker et al., 2008), drugs (Cai et al., 2008), and pesticides (Romero-Gonzalez et al., 2008; Taylor et al., 2008; Garrido Frenich et al., 2008) in food samples.

Due to the high number of investigated compounds, conventional HPLC runs can be relatively long, particularly to avoid peak coelution. It is indeed important to attain sufficient chromatographic resolution, to minimize coelution of compounds with close m/z ratios and similar fragmentation pathways.

As shown in Fig. 1.11, for the separation of 103 doping agents in urine sample, gradient conditions were selected to elute analytes presenting a wide polarity range, and formic acid is generally the preferred additive. To screen compounds by UHPLC, 50 to 100 mm column lengths were used with gradient times ranging between 5 and 15 min, followed by a re-equilibration time of 2 to 4 min. The column length should be selected in agreement with the gradient time, since the longest column did not always provide the highest peak capacity in UHPLC. Indeed, a 50 mm column was found to be optimal for gradient times shorter than 7 min, while a 100 mm column was only beneficial for longer gradients. Finally, the 2.1 mm I.D. column was often preferred, to limit extra-column band broadening contributions. Only Kasprzyk-Hordern et al. (2008a; 2008b) reported the successful screening of about 50 pharmaceuticals in wastewater using a 1 mm I.D. column. Even if consumption of mobile phase and analyte were drastically reduced, peaks were notably broader and distorted with a 1 mm I.D. column, as expected from the influence of external volume contributions.

From the above papers, an approximate increase in throughput by 3 to 5 fold was observed in UHPLC compared to conventional HPLC methods. In addition to the analysis time decrease, an equivalent or higher chromatographic resolution was reported (Pastor Montoro et al., 2007; Petrovic et al., 2006; Farre et al., 2008). Such improvements could be attributed to a probable

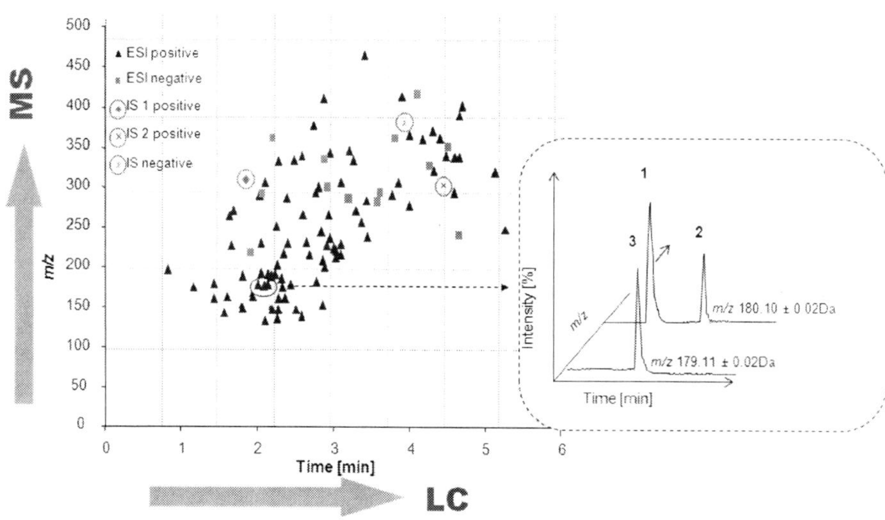

Fig. 1.11. Separation of 103 doping agents in urine sample according to their m/z and tR. Data from ESI positive and negative mode are plotted together. The three I.S are circled with a continuous line, A zone (dashed line) is magnified to show the selectivity of coupling UHPLC to QTOF mass spectrometer. In the magnified zone, the compounds (1) methylephedrine, (2) MDA and (3) nikethamide are separated in function of time, intensity and m/z. Adapted from (Badoud et al., 2009), with permission.

increase in peak capacity, but also to column selectivity changes. Indeed, comparisons were often not supported by the use of strictly similar column chemistries.

UHPLC-MS in Metabolomics

Due to the inherent complexity of metabolomic samples and because metabolites can be found at low concentrations, there is a need for analytical systems providing high resolution and increased sensitivity. For this reason, the value of UHPLC-TOF/MS and -QqTOF/MS platforms has been demonstrated in a number of studies. The strong reduction of analysis time provided by UHPLC vs. HPLC opens up the possibility of high-throughput screening for metabolomic fingerprinting. On the other hand, a longer UHPLC run can be employed to increase the amount of information, essentially for metabolomic profiling. The UHPLC platform has been applied for the global metabolic profiling of (i) human and animal biological fluids, including rat urine (Gika et al., 2008b; Plumb et al., 2005; Gika et al., 2008c), human urine (Gika et al. 2008; Guy et al., 2008; Wong et al., 2008), and human serum (Dunn et al., 2008), as well as (ii) plant extracts, such as *Arabidopsis thaliana* (Grata et al., 2008; Glauser et al., 2008) and *Panax* herbs (Xie et al., 2008a; Xie et al., 2008b; Dan et al., 2008).

Wilson and co-workers have used UHPLC for the profiling of rat and mouse urine since 2005 (Gika et al., 2008a; Plumb et al., 2005; Gika et al., 2008b; Gika et al., 2008c). Initially, biological fluids were analyzed on a 50 mm column packed with 1.7 μm particles in combination with TOF/MS. In terms of chromatography, the average peak widths were around 1 s, thus generating a peak capacity of 60 for UHPLC runs of only 1 min. With the additional TOF/MS information, a total of 1000 features (i.e., signals observed with specific *m/z* and retention time that can be considered as a variable for data treatment) were determined in rat urine. This number was equivalent, or even better than, that achieved on a conventional HPLC instrumentation, but with a 10-fold reduction in analysis time. The study performed by Nordstrom et al. (2006) on the quantitative analysis of endogenous and exogenous metabolites in human serum confirmed these results. Indeed, UHPLC provided 20% more detected components in comparison with HPLC. Finally, the study demonstrated that UHPLC displayed some additional advantages over HPLC, such as better retention time reproducibility and signal-to-noise ratios. Finally, two interesting approaches, namely, the application of elevated temperature in UHPLC (up to 180°C) and the use of HILIC (i.e., hydrophilic interaction liquid chromatography) columns packed with sub-2μm particles, were proposed by Wilson and colleagues to further extend the applicability of UHPLC in metabolomics (Gika et al., 2008b; Gika et al., 2008c).

Because of the complexity and chemical diversity of metabolites present in natural plant extracts, metabolomics is also gaining interest in the field of phytochemistry. Two research groups have recently reported the use of

UHPLC-TOF/MS as well as QqTOF/MS for untargeted metabolic profiling. Jia and coworkers reported the profiling of several medicinal *Panax* herbs (Xie et al., 2008a; Xie et al., 2008b; Dan et al., 2008), while Wolfender and coworkers evaluated a UHPLC-TOF/MS platform for the analysis of a model plant, *Arabidopsis thaliana* (Grata et al., 2008; Glauser et al., 2008). As shown in Fig. 1.12, they proposed a useful and sensitive multi-step strategy for the detection, isolation, and identification of stress-induced metabolites in *Arabidopsis thaliana* after leaf wounding, which mimicked herbivore attack (Grata et al., 2008; Glauser et al., 2008). In the first step, a rapid screening gradient was carried out by UHPLC-TOF/MS using a short column of 50 mm. This metabolite fingerprinting was performed on numerous plant specimens to evaluate the intra-sample variability and achieve adequate pool formation (Grata et al., 2008). The second step consisted of high-resolution metabolite profiling of selected pool samples using a UHPLC column of 150 mm. Gradient conditions used in the metabolomic fingerprinting were adequately transferred to the new column geometry, and analysis times were increased up to 100 min. This profiling allowed confirmation of the presence of different stress-related compounds. The high peak capacity afforded by long columns packed with sub-2μm was indeed essential to obtain a complete deconvolution of the

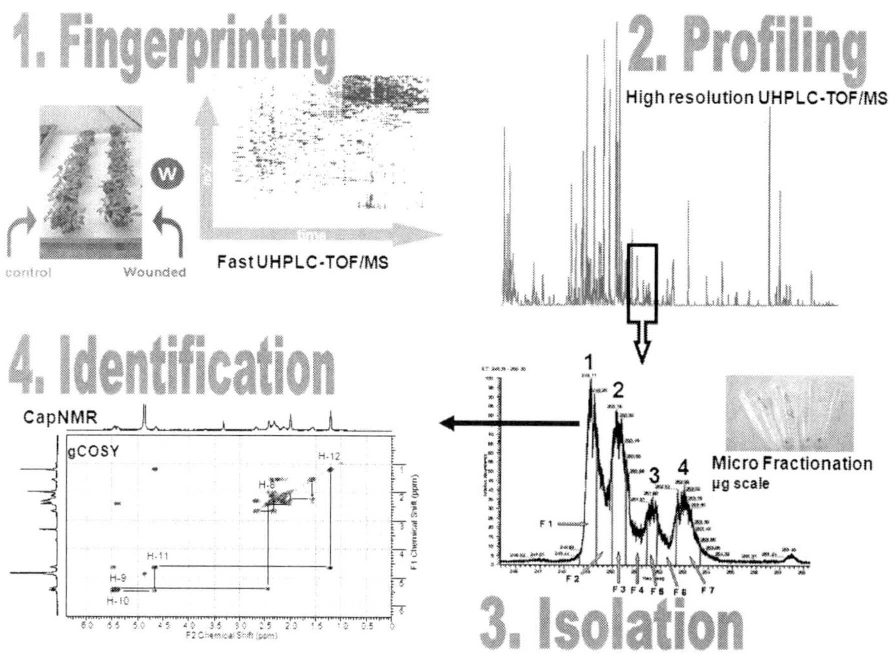

Fig. 1.12. Plant metabolomics based on a four-step strategy: fingerprinting, profiling, isolation, and identification of stress biomarkers. Adapted from (Guillarme et al., 2010), with permission.

biomarkers, and resolution of numerous closely related isomers (Grata et al., 2008). The last step of the process was the complete structural determination of minor biomarkers in plants using LC-MS-triggered preparative isolation. For this purpose, the UHPLC separation obtained during the metabolic profiling was transferred to semi-preparative conditions, using a 19 mm I.D. column packed with 5 µm particles of the same chemistry. Based on the use of a capillary-NMR probe, 1D and 2D-spectra of good quality were obtained at the µg level, allowing unambiguous structural elucidation of the isolated wound-biomarkers (including known signaling molecules, as well as original oxylipins and jasmonates) (Glauser et al., 2008). This generic analytical platform can be used to screen various other plant extracts.

Conclusion/Perspectives

As shown in this chapter, UHPLC is a powerful technology that is easy to implement. This strategy makes it possible to increase drastically the throughput while maintaining equivalent performance and/or to increase the resolution within an acceptable analysis time. However, to take full advantage of this platform, it is important to keep in mind that the performance of UHPLC is based not only on the column packing, but also on the quality of the chromatographic system. For this reason, UHPLC experiments should be performed on a system compatible with ultra-high pressure, and which possesses reduced system and dwell volumes.

The coupling of UHPLC with MS appears to be the ultimate approach, in terms of sensitivity, selectivity, and peak assignment for the determination of analytes at low concentrations in complex matrices. This strategy has become very popular, and has now been applied in numerous fields of application such as bioanalysis, drug metabolism, multi-residue screening and metabolomics. However, it is recommended to work with an MS device of the latest generation (quadrupole-based and time-of-flight instruments are the most appropriate) that possesses a sufficient data acquisition rate.

In the future, if 2.1 mm I.D. becomes the standard dimension for UHPLC columns, there is certainly not much interest in reducing the particle size of the support even more. Indeed, the backpressure generated by such packing would be detrimental for the chromatographic separation, because of the axial and longitudinal temperature gradients within the column (frictional heating effects). To limit these negative effects, the solution would be to work with reduced columns I.D. (500 µm – 1 mm). But in this case, the current chromatographic system needs to be strongly improved in terms of extra-column and dwell volumes. From our point of view, the use of small particles, in conjunction with ultra high pressure and elevated temperature, up to 90°C (HT-UHPLC) is certainly more promising both for throughput and resolution, but also for selectivity and peak shape.

References

Albert, M.; Cretier, G.; Guillarme, D.; Heinisch, S.; Rocca, J.L. Some advantages of high temperature for the separation of pharmaceutical compounds with mass spectrometry detection. J. Sep. Sci. 2005, 28, 1803-1811.

Al-Sayah, M.A.; Rizos, P.; Antonucci, V.; Wu, N. High throughput screening of active pharmaceutical ingredients by UPLC. J. Sep. Sci. 2008, 31, 2167-2172.

Asshauer, J.; Halasz, I. Reproducibility and efficiency of columns packed with 10μm silica in liquid chromatography. J. Chromatogr. Sci. 1974, 12, 139-147.

Badoud, F.; Grata, E.; Perrenoud, L.; Avois, L.; Saugy, M.; Rudaz, S.; Veuthey, J.L. Fast analysis of doping agents in urine by ultra-high-pressure liquid chromatography–quadrupole time-of-flight mass spectrometry: I. Screening analysis. J. Chromatogr. A 2009, 1216, 4423-4433.

Badoud, F.; Grata, E.; Perrenoud, L.; Saugy, M.; Rudaz, S.; Veuthey, J.L. Fast analysis of doping agents in urine by ultra-high-pressure liquid chromatography–quadrupole time-of-flight mass spectrometry. II: Confirmatory analysis. J. Chromatogr. A 2010, 1217, 4109-4119.

Barrioulet, M.P.; Heinisch, S.; Rocca, J.L. L'intérêt de la haute température en chromatographie en phase liquide. Spectra Analyse 2007, 36, 38-44.

Berg, T.; Lundanes, E.; Christophersen, A.S.; Strand, D.H. Determination of opiates and cocaine in urine by high pH mobile phase reversed phase UPLC–MS/MS. J. Chromatogr. B 2009, 877, 421-432.

Bidlingmeyer, B.A.; Rogers, L.B. Investigation of pressure-induced changes in chromatographic selectivity of methyl and ethyl orange on silica-gel. Sep. Sci. 1972, 7, 131-158.

Billen, J.; Guillarme, D.; Rudaz, S.; Veuthey, J.L.; Ritchie, H.; Grady, B.; Desmet, G. Relation between the particle size distribution and the kinetic performance of packed columns: Application to a commercial sub-2μm particle material. J. Chromatogr. A 2007, 1161, 224-233.

Bones, J.; Duffy, C.; Macka, M.; Paull, B. Using coupled monolithic rods for ultra-high peak capacity LC and LC-MS under normal LC operating pressures. Analyst 2008, 133, 180-183.

Brice, R.W.; Zhang, X.; Colon, L.A. Fused-core, sub-2μm packings, and monolithic HPLC columns: a comparative evaluation. J. Sep. Sci. 2009, 32, 2723-2731.

Broeckhoven, K.; Billen, J.; Verstraeten, M.; Choikhet, K.; Dittman, M.; Rozing, G.; Desmet, G. Towards a solution for viscous heating in ultra-high pressure liquid chromatography using intermediate cooling. J. Chromatogr. A 2010, 1217, 2022-2031.

Cabooter, D.; Lestremau, F.; Lynen, F.; Sandra, P.; Desmet, G. Kinetic plot method as a tool to design coupled column systems producing 100,000 theoretical plates in the shortest possible time, J. Chromatogr. A 2008, 1212, 23-34.

Cabrera, K. Applications of silica-based monolithic HPLC columns, J. Sep. Sci. 2004, 27, 843-852.

Cai, Z.; Zhang, Y.; Pan, H.; Tie, X.; Ren, Y. Simultaneous determination of 24 sulfonamide residues in meat by ultra-performance liquid chromatography tandem mass spectrometry. J. Chromatogr. A 2008, 1200, 144-155.

Castro-Perez, J.; Plumb, R.S.; Granger, J.H.; Beattie, I.; Joncour, K.; Wright, A. Increasing throughput and information content for in vitro drug metabolism experiments using ultra-performance liquid chromatography coupled to a quadrupole time-of-flight mass spectrometer. Rapid Comm. Mass Spectrom. 2005, 19, 843-848.

Cavazzini, A.; Gritti, F.; Kaczmarski, K.; Marchetti, N.; Guiochon, G. Mass-transfer kinetics in a shell packing material for chromatography. Anal. Chem. 2007, 79, 5972-5979.

Chambers, E.; Wagrowski-Diehl, D.M.; Lu, Z.; Mazzeo, J.R. Systematic and comprehensive strategy for reducing matrix effects in LC-MS/MS analyses. J. Chromatogr. B 2007, 852, 22-29.

Colon, L.A.; Cintron, J.M.; Anspach, J.A.; Fermier, A.M.; Swinney, K.A. Very high pressure HPLC with 1 mm id columns. Analyst 2004, 129, 503-504.

Crockford, D.J.; Maher, A.D.; Ahmadi, K.R.; Barrett, A.; Plumb, R.S.; Wilson, I.D.; Nicholson, J.K. 1H NMR and UPLC-MSE statistical heterospectroscopy: characterization of drug metabolites (xenometabolome) in epidemiological studies. Anal. Chem. 2008, 80, 6835-6844.

Cunliffe, J.M.; Maloney, T.D. Fused-core particle technology as an alternative to sub-2-μm particles to achieve high separation efficiency with low backpressure. J. Sep. Sci. 2007, 30, 3104-3109.

Cunliffe, J.M.; Adams-Hall, S.B.; Maloney, T.D. Evaluation and comparison of very high pressure liquid chromatography systems for the separation and validation of pharmaceutical compounds. J. Sep. Sci. 2007, 30, 1214-1223.

Dan, M.; Su, M.; Gao, X.; Zhao, T.; Zhao, A.; Xie, G.X.; Qiu, Y.; Zhou, M.; Liu, Z.; Jia, W. Metabolite profiling of Panax notoginseng using UPLC-ESI-MS, Phytochem. 2008, 69, 2237-2244.

De Villiers, A.; Lauer, H.; Szucs, R.; Goodall, S.; Sandra, P. Influence of frictional heating on temperature gradients in ultra-high-pressure liquid chromatography on 2.1 mm I.D. columns. J. Chromatogr. A 2006a, 1113, 84-91.

De Villiers, A.; Lestremau, F.; Szucs, R.; Gélébart, S.; David, F.; Sandra, P. Evaluation of ultra performance liquid chromatography: Part I. Possibilities and limitations. J. Chromatogr. A 2006b, 1127, 60-69.

Desmet, G.; Cabooter, D. Are short columns always the best option? LC-GC Eur. 2009, 22, 70-77.

Desmet, G.; Clicq, D.; Gzil, P. Geometry-independent plate height representation methods for the direct comparison of the kinetic performance of LC supports with a different size or morphology. Anal. Chem. 2005a, 77, 4058-4070.

Desmet, G.; Gzil, P.; Clicq, D. Kinetic plots to directly compare the performance of LC supports. LC-GC Eur. 2005b, 18, 403-409.

Desmet, G.; Cabooter, D.; Gzil, P.; Verelst, H.; Mangelings, D.; van der Heyden, Y.; Clicq, D. Future of high pressure liquid chromatography: Do we need porosity or do we need pressure? J. Chromatogr. A 2006a, 1130, 158-166.

Desmet, G.; Clicq, D.; Nguyen, D.T.T.; Guillarme, D.; Rudaz, S.; Veuthey, J.L.; Vervoort, N.; Torok, G.; Cabooter, D.; Gzil, P. Including practical constraints in the kinetic plot representation of chromatographic performance data: theory and application to experimental data. Anal. Chem. 2006b, 78, 2150-2162.

Desmet, G. Comparison techniques for HPLC column performance. LC-GC Eur. 2008, 21, 310-320.

Destefano, J.J.; Langlois, T.J.; Kirkland, J.J. Characteristics of superficially-porous silica particles for fast HPLC: Some performance comparisons with sub-2-mu m particles. J. Chromatogr. Sci. 2008, 46, 254-260.

Dolan, J.W.; Snyder, L.R. Maintaining fixed band spacing when changing column dimensions in gradient elution. J. Chromatogr. A 1998, 799, 21-34.

Dolan, J.W. Dwell volume revisited. LC GC N. A. 2006, 24, 458-466.

Dunn, W.B.; Broadhurst, D.; Brown, M.; Baker, P.N.; Redman, C.W.G.; Kenny, L.C.; Kell, D.B. Metabolic profiling of serum using Ultra Performance Liquid Chromatography and the LTQ-Orbitrap mass spectrometry system. J. Chromatogr. B 2008, 871, 288-298.

Endele, R.; Halasz, I.; Unger, K. Influence of the particle size (5–35 μm) of spherical silica on column efficiencies in high-pressure liquid chromatography. J. Chromatogr. 1974, 99, 377-393.

Eugster, P.J.; Guillarme, D.; Rudaz, S.; Veuthey, J.L.; Carrupt, P.A.; Wolfender, J.L. Ultra high pressure liquid chromatography for crude plant extracts profiling. J. AOAC int. 2010, In press.

Farre, M.; Gros, M.; Hernandez, B.; Petrovic, M.; Hancock, P.; Barcelo, D. Analysis of biologically active compounds in water by ultra-performance liquid chromatography quadrupole time-of-flight mass spectrometry. Rapid Comm. Mass Spectrom. 2008, 22, 41-51.

Fekete, S.; Fekete, J.; Ganzler, K. Facts and myths about columns packed with sub-3 mu m and sub-2 mu m particles. J. Pharm. Biomed. Anal. 2009, 49, 56-64.

Foley, J.P.; Crow, J.A.; Thomas, B.A.; Zamora, M. Unavoidable flow-rate errors in high-performance liquid chromatography. J. Chromatogr. 1989, 478, 287-309.

Fountain, K.J.; Neue, U.D.; Grumbach, E.S.; Diehl, D.M. Effects of extra-column band spreading, liquid chromatography system operating pressure, and column temperature on the performance of sub-2-μm porous particles. J. Chromatogr. A 2009, 1216, 5979-5988.

Garrido Frenich, A.; Martinez Vidal, J.; Pastor-Montoro, E.; Romero-Gonzalez, R. High-throughput determination of pesticide residues in food commodities by use of ultra-performance liquid chromatography-tandem mass spectrometry. Anal. Bioanal. Chem. 2008, 390, 947-959.

Gervais, C.; Brosillon, S.; Laplanche, A.; Helen, C. Ultra-pressure liquid chromatography–electrospray tandem mass spectrometry for multiresidue determination of pesticides in water. J. Chromatogr. A 2008, 1202, 163-172.

Giddings, J.C. Comparison of Theoretical Limit of Separating Speed in Gas and Liquid Chromatography. Anal. Chem. 1965, 37, 60-63.

Gika, H.G.; Macpherson, E.; Theodoridis, G.A.; Wilson, I.D. Evaluation of the repeatability of ultra-performance liquid chromatography–TOF-MS for global metabolic profiling of human urine samples. J. Chromatogr. B 2008a, 871, 299-305.

Gika, H.G.; Theodoridis, G.; Extance, J.; Edge, A.M.; Wilson, I.D. High temperature-ultra performance liquid chromatography–mass spectrometry for the metabonomic analysis of Zucker rat urine. J. Chromatogr. B 2008b, 871, 279-286.

Gika, H.G.; Theodoridis, G.A.; Wilson, I.D. Hydrophilic interaction and reversed-phase ultra-performance liquid chromatography TOF-MS for metabonomic analysis of Zucker rat urine. J. Sep. Sci. 2008c, 31, 1598-1608.

Gilar, M.; Daly, A.E.; Kele, M.; Neue, U.D.; Gebler, J.C. Implications of column peak capacity on the separation of complex peptide mixtures in single- and two-dimensional high-performance liquid chromatography. J. Chromatogr. A 2004, 1061, 183-192.

Glauser, G.; Guillarme, D.; Grata, E.; Boccard, J.; Thiocone, A.; Carrupt, P.A.; Veuthey, J.L.; Rudaz, S.; Wolfender, J.L. Optimized liquid chromatography-mass spectrometry approach for the isolation of minor stress biomarkers in plant extracts and their identification by capillary nuclear magnetic resonance at the microgram scale. J. Chromatogr. A 2008, 1180, 90-98.

Grata, E.; Boccard, J.; Guillarme, D.; Glauser, G.; Carrupt, P.A.; Farmer, E.E.; Wolfender, J.L.; Rudaz, S. UPLC-TOF-MS for plant metabolomics: a sequential approach for wound markers analysis in Arabidopsis Thaliana. J. Chromatogr. B 2008, 871, 261-270.

Grata, E.; Guillarme, D.; Glauser, G.; Boccard, J.; Carrupt, P.A.; Veuthey, J.L.; Rudaz, S.; Wolfender, J.L. Metabolite profiling of plant extracts by high temperature UHPLC-TOF/MS. J. Chromatogr. A 2009, 1216, 5660-5668.

Gritti, F.; Guiochon, G. Complete Temperature profiles in ultra-high-pressure liquid chromatography columns. Anal. Chem. 2008, 80, 5009-5020.

Gritti, F.; Cavazzini, A.; Marchetti, N.; Guiochon, G. Comparison between the efficiencies of columns packed with fully and partially porous C18-bonded silica materials. J. Chromatogr. A 2007, 1157, 289-303.

Gritti, F.; Leonardis, I.; Shock, D.; Stevenson, P.; Shalliker, A.; Guiochon, G. Performance of columns packed with the new shell particles, Kinetex-C18. J. Chromatogr. A 2010, 1217, 1589-1603.

Grumbach, E.S.; Wheat, T.E.; Kele, M.; Mazzeo, J.R. Developing columns for UPLC: Design considerations and recent developments. LC-GC Eur. 2005, 12, 40-44.

Guillarme, D.; Heinisch, S.; Rocca, J.L. Effect of temperature in reversed phase liquid chromatography. J. Chromatogr. A 2004, 1052, 39-51.

Guillarme, D.; Heinisch, S. Detection modes with high temperature liquid chromatography, a review. Sep. Purif. Rev. 2005, 34, 181-216.

Guillarme, D.; Nguyen, D.T.T.; Rudaz, S.; Veuthey, J.L. Method transfer for fast liquid chromatography in pharmaceutical analysis: Application to short columns packed with small particles – Part I, isocratic separation. Eur. J. Pharm. Biopharm. 2007a, 66, 475-482.

Guillarme, D.; Nguyen, D.T.T.; Rudaz, S.; Veuthey, J.L. Recent developments in liquid chromatography – impact on qualitative and quantitative performance. J. Chromatogr. A 2007b, 1149, 20-29.

Guillarme, D.; Russo, R.; Rudaz, S.; Bicchi, C.; Veuthey, J.L. Chromatographic performance of silica-based stationary phases in high temperature liquid chromatography: pharmaceutical applications. Curr. Pharm. Anal. 2007c, 4, 221-229.

Guillarme, D.; Nguyen, D.T.T.; Rudaz, S.; Veuthey, J.L. Method transfer for fast liquid chromatography in pharmaceutical analysis: Application to short columns packed with small particles – Part II, gradient separation. Eur. J. Pharm. Biopharm. 2008, 68, 430-440.

Guillarme, D.; Grata, E.; Glauser, G.; Wolfender, J.L.; Veuthey, J.L.; Rudaz, S. Some solutions to obtain very efficient separations in isocratic and gradient modes using small particles size and ultra high pressure. J. Chromatogr. A 2009, 1216, 3232-3243.

Guillarme, D.; Ruta, J.; Rudaz, S.; Veuthey, J.-L. New trends in fast and high-resolution liquid chromatography: A critical comparison of existing approaches. Anal. Bioanal. Chem. 2010a, 397, 1069-1082.

Guillarme, D.; Schappler, J.; Rudaz, S.; Veuthey, J.L. Coupling ultra high pressure liquid chromatography with mass spectrometry. Trends Anal. Chem. 2010b, 29, 15-27.

Guy, P.A.; Tavazzi, I.; Bruce, S.J.; Ramadan, Z.; Kochhar, S. Global metabolic profiling analysis on human urine by UPLC–TOFMS: Issues and method validation in nutritional metabolomics. J. Chromatogr. B 2008, 871, 253-260.

Halasz, I.; Endele, R.; Asshauer, J. Ultimate limits in high-pressure liquid chromatography. J. Chromatogr. 1975, 112, 37-60.

Hartonen, K.; Riekkola, M.L. Liquid chromatography at elevated temperatures with pure water as the mobile phase. Trends Anal. Chem. 2008, 27, 1-14.

Heinisch, S.; Rocca, J.L. Sense and nonsense of high-temperature liquid chromatography. J. Chromatogr. A 2009, 1216, 642-658.

Heinisch, S.; Puy, G.; Barrioulet, M.P.; Rocca, J.L. Effect of temperature on the retention of ionizable compounds in reversed-phase liquid chromatography: Application to method development. J. Chromatogr. A 2006, 1118, 234-243.

Hjerten, S.; Liao, J.L.; Zhang, R. High-performance liquid chromatography on continuous polymer beds. J. Chromatogr. 1989, 473, 273-275.

Horvath, C.G.; Preiss, B.A.; Lipsky, S.R. Fast liquid chromatography. Investigation of operating parameters and the separation of nucleotides on pellicular ion exchangers. Anal. Chem. 1967, 39, 1422-1428.

Jerkovich, A.D.; Mellors, J.S.; Jorgenson, J.W. The use of micron-sized particles in ultrahigh-pressure liquid chromatography. LC-GC Eur. 2003, 16, 20-23.

Jerkovich, A.D.; Mellors, J.S.; Thompson, J.W.; Jorgenson, J.W. Linear velocity surge caused by mobile-phase compression as a source of band broadening in isocratic ultrahigh-pressure liquid chromatography. Anal. Chem. 2005, 77, 6292-6299.

Kasprzyk-Hordern, B.; Dinsdale, R.M.; Guwy, A.J. Multiresidue methods for the analysis of pharmaceuticals, personal care products and illicit drugs in surface water and wastewater by solid-phase extraction and ultra performance liquid chromatography-electrospray tandem mass spectrometry. Anal. Bioanal. Chem. 2008a, 391, 1293-1308.

Kasprzyk-Hordern, B.; Dinsdale, R.M.; Guwy, A.J. The effect of signal suppression and mobile phase composition on the simultaneous analysis of multiple classes of acidic/neutral pharmaceuticals and personal care products in surface water by solid-phase extraction and ultra performance liquid chromatography-negative electrospray tandem mass spectrometry. Talanta 2008b, 74, 1299-1312.

Kaufmann, A.; Butcher, P.; Maden, K.; Widmer, M. Ultra-performance liquid chromatography coupled to time of flight mass spectrometry (UPLC-TOF): A novel tool for multiresidue screening of veterinary drugs in urine. Anal. Chim. Acta 2007, 586, 13-21.

King, S.; Stoffolano, P.J.; Robinson, E.; Eichhold, T.E.; Hoke, S.H.; Baker, T.R.; Richarson, E.C.; Wehmeyer, K.R. The evaluation and application of UPLC for the rapid analysis of dose formulations. LC-GC Eur. 2005, 12, 36-39.

Kirkland, J.J. High performance liquid chromatography with porous silica microspheres. J. Chromatogr. Sci. 1972, 10, 593-599.

Kirkland, J.J. Superficially porous silica microspheres for the fast high-performance liquid chromatography of macromolecules. Anal. Chem. 1992, 64, 1239-1245.

Kirkland, J.J. Ultrafast reversed-phase high-performance liquid chromatographic separations: An overview. J. Chromatogr. Sci. 2000, 38, 535-544.

Knox, J.H.; Saleem, M. Kinetic conditions for optimum speed and resolution in column chromatography. J. Chromatogr. Sci. 1969, 7, 614-622.

Knox, J.H. Speed of analysis by gas chromatography. J. Chem. Soc. 1961, 1, 433-441.

Knox, J.H. Practical aspects of LC theory. J. Chromatogr. Sci. 1977, 15, 352-364.

Leinweber, F.C.; Tallarek, U. Chromatographic performance of monolithic and particulate stationary phases: Hydrodynamics and adsorption capacity. J. Chromatogr. A 2003, 1006, 207-228.

Lestremau, F.; De Villiers, A.; Lynen, F.; Cooper, A.; Szucs, R.; Sandra, P. High efficiency liquid chromatography on conventional columns and instrumentation by using temperature as a variable: Kinetic plots and experimental verification. J. Chromatogr. A 2007, 1138, 120-131.

Lestremau, F.; Cooper, A.; Szucs, R.; David, F.; Sandra, P. High-efficiency liquid chromatography on conventional columns and instrumentation by using temperature as a variable: I. Experiments with 25 cm × 4.6 mm I.D., 5 µm ODS columns. J. Chromatogr. A 2006, 1109, 191-196.

Li, J.; Hu, Y.; Carr, P.W. Fast separations at elevated temperatures on polybutadiene-coated zirconia reversed-phase material. Anal. Chem. 1997, 69, 3884-3888.

Li, F.; Maguigad, J.; Peizer, M.; Jiang, X.; Ji, Q.C. A novel "peak parking' strategy for ultra-performance liquid chromatography/tandem mass spectrometric detection for enhanced performance of bioanalytical assays. Rapid Comm. Mass Spectrom. 2008, 22, 486-494.

Licea-Perez, H.; Wang, S.; Bowen, C.L.; Yang, E. A semi-automated 96-well plate method for the simultaneous determination of oral contraceptives concentrations in human plasma using ultra performance liquid chromatography coupled with tandem mass spectrometry. J. Chromatogr. B 2007, 852, 69-76.

Lippert, J. A.; Xin, B.; Wu, N.; Lee, M.L. Fast ultrahigh-pressure liquid chromatography: On-column UV and time-of-flight mass spectrometric detection. J. Microcol. Sep. 1999, 11, 631-643.

Lommen, D.C.; Snyder L.R. Fast HPLC separations with small non porous particles. LC-GC Eur. 1993, 11, 222-232.

MacNair, J.E.; Lewis, K.C.; Jorgenson, J.W. Ultrahigh-pressure reversed-phase liquid chromatography in packed capillary columns. Anal. Chem. 1997, 69, 983-989.

MacNair, J.E.; Patel, K.D.; Jorgenson, J.W. Ultrahigh-pressure reversed-phase capillary liquid chromatography - Isocratic and gradient elution using columns packed with 1.0-µm particles. Anal. Chem. 1999, 71, 700-708.

Majors, R.E. High-performance liquid chromatography on small particle silica gel. Anal. Chem. 1972, 44, 1722-1726.

Majors, R.E. HPLC column packing design. LC-GC Eur. 2003, 16, 8-13.

Majors, R.E. Highlights of HPLC 2008: Part I. LC-GC N. A. 2008, 26, 676-691.

Martin, M.; Guiochon, G. Pressure dependence of diffusion coefficient and effect on plate height in liquid chromatography. Anal. Chem. 1983, 55, 2302-2309.

Martin, M.; Guiochon, G. Effects of high pressure in liquid chromatography. J. Chromatogr. A 2005, 1090, 16-38.

McNeff, C.V.; Yan, B.; Stoll, D.R.; Henry, R.A. Practice and theory of high temperature liquid chromatography. J. Sep. Sci. 2007, 30, 1672-1685.

Mellors, J.S.; Jorgenson, J.W. Use of 1.5-µm porous ethyl-bridged hybrid particles as a stationary-phase support for reversed-phase ultrahigh-pressure liquid chromatography. Anal. Chem. 2004, 76, 5441-5450.

Minakuchi, H.; Nakanishi, K.; Soga, N.; Ishizuka, N.; Tanaka, N. Octadecylsilylated porous silica rods as separation media for reversed-phase liquid chromatography. Anal. Chem. 1996, 68, 3498-3501.

Miyamoto, K.; Hara, T.; Kobayashi, H.; Morisaka, H.; Tokuda, D.; Horie, K.; Koduki, K.; Makino, S.; Nunez, O.; Yang, C.; et al. High-efficiency liquid chromatographic separation utilizing long monolithic silica capillary columns. Anal. Chem. 2008, 80, 8741-8750.

Neue, U.D.; Mazzeo, J.R. A theoretical study of the optimization of gradients at elevated temperature. J. Sep. Sci. 2001, 24, 921-929.

Neue, U.D. Peak capacity in unidimensional chromatography. J. Chromatogr. A 2008, 1184 107-130.

Nguyen, D.T.T.; Guillarme, D.; Rudaz, S.; Veuthey, J.L. Fast analysis in liquid chromatography using small particles size and ultra high pressure. J. Sep. Sci. 2006, 29, 1836-1848.

Nguyen, D.T.T.; Guillarme, D.; Rudaz, S.; Veuthey, J.L. Chromatographic behaviour and comparison of column packed with sub-2µm stationary phases in liquid chromatography. J. Chromatogr. A 2006, 1128, 105-113.

Nguyen, D.T.T.; Guillarme, D.; Heinisch, S.; Barrioulet, M.P.; Rocca, J.L.; Rudaz, S.; Veuthey J.L. High throughput liquid chromatography with sub-2µm particles at very high pressure and high temperature. J. Chromatogr. A 2007, 1167, 76-84.

Nordstrom, A.; O'Maille, G.; Qin, C.; Siuzdak, G. Nonlinear data alignment for UPLC-MS and HPLC-MS based metabolomics: quantitative analysis of endogenous and exogenous metabolites in human serum. Anal. Chem. 2006, 78, 3289-3295.

Novakova, L.; Vickova, H. A review of current trends and advances in modern bioanalytical methods: Chromatography and sample preparation. Anal. Chim. Acta 2009, 656, 8-35.

Novakova, L.; Solichova, D.; Solich, P. Advantages of ultra performance liquid chromatography over high-performance liquid chromatography: Comparison of different analytical 2 approaches during analysis of diclofenac gel. J. Sep. Sci. 2006, 29, 2433-2443.

Pastor Montoro, E.; Romero Gonzalez, R.; Garrido Frenich, A.; Hernandez Torres, M.E.; Martinez Vidal, J.L. Fast determination of herbicides in waters by ultra-performance liquid chromatography/tandem mass spectrometry. Rapid Comm. Mass Spectrom. 2007, 21, 3585-3592.

Petricoin, E.F.; Liotta, L.A. Proteomic approaches in cancer risk and response assessment. Trends Mol. Med. 2004, 10, 59-64.

Petrovic, M.; Gros, M.; Barcelo, D. Multi-residue analysis of pharmaceuticals in wastewater by ultra-performance liquid chromatography–quadrupole–time-of-flight mass spectrometry. J. Chromatogr. A 2006, 1124, 68-81.

Petsalo, A.; Turpeinen, M.; Pelkonen, O.; Tolonen, A. Analysis of nine drugs and their cytochrome P450-specific probe metabolites from urine by liquid chromatography–tandem mass spectrometry utilizing sub 2 μm particle size column. J. Chromatogr. A 2008, 1215, 107-115.

Plumb, R.S.; Granger, J.H.; Stumpf, C.L.; Johnson, K.A.; Smith, B.W.; Gaulitz, S.; Wilson, I.D.; Castro-Perez, J. A rapid screening approach to metabonomics using UPLC and oa-TOF mass spectrometry: application to age, gender and diurnal variation in normal/Zucker obese rats and black, white and nude mice. Analyst 2005, 130, 844-849.

Plumb, R.S.; Johnson, K.A.; Rainville, P.; Smith, B.W.; Wilson, I.D.; Castro-Perez, J.M.; Nicholson, J.K. UPLC/MSE: a new approach for generating molecular fragment information for biomarker structure elucidation. Rapid. Comm. Mass Spec. 2006, 20, 1989-1994.

Poppe, H. Some reflections on speed and efficiency of modern chromatographic methods. J. Chromatogr. A 1997, 778, 3-21.

Romero-Gonzalez, R.; Garrido Frenich, A.; Martinez Vidal, J.L. Multiresidue method for fast determination of pesticides in fruit juices by ultra performance liquid chromatography coupled to tandem mass spectrometry. Talanta 2008, 76, 211-225.

Russo, R.; Guillarme, D.; Nguyen, D.T.T.; Bicchi, C.; Rudaz, S.; Veuthey, J.L. Pharmaceutical applications on columns packed with sub-2μm particles. J. Chromatogr. Sci. 2008, 46, 199-208.

Ruta, J.; Guillarme, D.; Rudaz, S.; Veuthey, J.L. Comparison of columns packed with porous sub-2μm and superficially porous sub-3μm particles for peptides analysis at ambient and high temperature. J. Sep. Sci. 2010, In press.

Salisbury, J.J. Fused-core particles: A practical alternative to sub-2 micron particles. J. Chromatogr. Sci. 2008, 46, 883-886.

Schappler, J.; Nicoli, R.; Nguyen, D.T.T.; Rudaz, S.; Veuthey, J.L.; Guillarme, D. Coupling ultra high pressure liquid chromatography with single quadrupole mass spectrometry for the analysis of a complex drug mixture. Talanta 2009, 78, 377-387.

Snyder, L.R. Column efficiencies in liquid adsorption chromatography: past, present and future. J. Chromatogr. Sci. 1969, 7, 352-360.

Stolker, A.A.M.; Rutgers, P.; Oosterink, E.; Lasaroms, J.J.P.; Peters, R.J.B.; van Rhijn, J.A.; Nielen, N.W.F. Comprehensive screening and quantification of veterinary drugs in milk using UPLC-ToF-MS. Anal. Bioanal. Chem. 2008, 391, 2309-2322.

Svec, F.; Frechet, J.M.J. Continuous rods of macroporous polymer as high-performance liquid chromatography separation media. Anal. Chem. 1992, 64, 820-822.

Tanaka, N.; Kobayashi, H.; Ishizuka, N.; Minakuchi, H.; Nakanishi, K.; Hosoya, K.; Ikegami, T. Monolithic silica columns for high-efficiency chromatographic separations. J. Chromatogr. A 2002, 965, 35-49.

Taylor, M.J.; Keenan, G.A.; Reid, K.B.; Uria Fernandez, D. The utility of ultra-performance liquid chromatography/electrospray ionisation time-of-flight mass spectrometry for multi-residue determination of pesticides in strawberry. Rapid Comm. Mass Spectrom. 2008, 22, 2731-2746.

Teutenberg, T.; Tuerk, J.; Holzhauser, M.; Giegold, S. Temperature stability of reversed phase and normal phase stationary phases under aqueous conditions. J. Sep. Sci. 2007, 30, 1101-1114.

Thompson, J.D.; Brown, J.S.; Carr, P.W. Dependence of thermal mismatch broadening on column diameter in high-speed liquid chromatography at elevated temperatures. Anal. Chem. 2001, 73, 3340-3347.

Thompson, J.D.; Carr, P.W. A study of the critical criteria for analyte stability in high-temperature liquid chromatography. Anal. Chem. 2002, 74, 1017-1023.

Thorngren, J.O.; Ostervall, F.; Garle, M. A high-throughput multicomponent screening method for diuretics, masking agents, central nervous system (CNS) stimulants and opiates in human urine by UPLC-MS/MS. J. Mass Spectrom. 2008, 43, 980-992.

van Nederkassel, A.M.; Aerts, A.; Dierick, A.; Massart, D.L.; van der Heyden, Y. Fast separations on monolithic silica columns: method transfer, robustness and column ageing for some case studies. J. Pharm. Biomed. Anal. 2003, 32, 233-249.

Vivilecchia, R.V.; Cotter, R.L.; Limpert, R.J.; Thimot, N.Z.; Little, J.N. Considerations of small particles in different modes of liquid chromatography. J. Chromatogr. 1974, 99, 407-424.

Walles, M.; Gauvin, C.; Morin, P.E.; Panetta, R.; Ducharme J. Comparison of sub-2-μm particle columns for fast metabolite ID. J. Sep. Sci. 2007, 30, 1191-1199.

Wang, X.; Stoll, D.R.; Carr, P.W.; Schoenmakers, P.J. A graphical method for understanding the kinetics of peak capacity production in gradient elution liquid chromatography. J. Chromatogr. A 2006, 1125, 177-181.

Wong, M.C.Y; Lee, W.T.K.; Wong, J.S.Y.; Frost, G.; Lodge, J. An approach towards method development for untargeted urinary metabolite profiling in metabonomic research using UPLC/QToF MS. J. Chromatogr. B 2008, 871, 341-348.

Wren, S.A.C; Tchelitcheff, P. Use of ultra-performance liquid chromatography in pharmaceutical development. J. Chromatogr. A 2006, 1119, 140-146.

Wu, N.; Lippert, J.A.; Lee, M.L. Practical aspects of ultrahigh pressure capillary liquid chromatography. J. Chromatogr. A 2001, 911, 1-12.

Xiang, Y.; Yan, B.; McNeff, C.V.; Carr, P.W.; Lee, M.L. Synthesis of micron diameter polybutadiene-encapsulated non-porous zirconia particles for ultrahigh pressure liquid chromatography. J. Chromatogr. A 2003a, 1002, 71-78.

Xiang, Y.; Yan, B.; Yue, B.; McNeff, C.V.; Carr, P.W.; Lee, M.L. Elevated-temperature ultrahigh-pressure liquid chromatography using very small polybutadiene-coated nonporous zirconia particles. J. Chromatogr. A 2003b, 983, 83-89.

Xie, G.X.; Ni, Y.; Su, M.M.; Zhang, Y.Y.; Zhao, A.H.; Gao X.F.; Liu, Z.; Xiao, P.G.; Jia, W. Application of ultra-performance LC-TOF MS metabolite profiling techniques to the analysis of medicinal Panax herbs. Metabolomics 2008a, 4, 248-260.

Xie, G.X.; Plumb, R.S.; Su, M.; Xu, Z.; Zhao, A.; Qiu, M.; Long, X.; Liu, Z.; Jia, W. Ultra-performance LC/TOF MS analysis of medicinal Panax herbs for metabolomic research. J. Sep. Sci. 2008b, 31, 1015-1026.

Yadav, M.; Contractor, P.; Upadhyay, V.; Gupta, A.; Guttikar, S.; Singhal, P.; Goswami, S.; Shrivastav, P.S. Automated liquid–liquid extraction based on 96-well plate format in conjunction with ultra-performance liquid chromatography tandem mass spectrometry (UPLC–MS/MS) for the quantitation of methoxsalen in human plasma. J. Chromatogr. B 2008, 872, 167-171.

Yang, B.; Zhao, J.; Brown, J.S.; Blackwell, J.; Carr, P.W. High-temperature ultrafast liquid chromatography. Anal. Chem. 2000, 72, 1253-1262.

Zhang, Y.; Wang, X.; Mukherjee, P.; Petersson, P. Critical comparison of performances of superficially porous particles and sub-2μm particles under optimized ultra-high pressure conditions. J. Chromatogr. A 2009, 1216, 4597-4605.

2

Hydrophilic Interaction Chromatography—An Excellent Method for Separation of Polar Samples

Pavel Jandera
Department of Analytical Chemistry, Faculty of Chemical Technology, University of Pardubice, Studentská 573, CZ-53210 Pardubice, Czech Republic

Introduction

In reversed-phase (RP) systems, most frequently used in contemporary HPLC practice, the stationary phase is non-polar, usually an alkyl-silica type bonded phase, while the mobile phase is a polar mixture of one or more organic solvents and water or a buffer. As a rule, the retention increases with more lipophilic stationary phases and with decreasing concentration of the organic solvent(s) in the mobile phase; polar solutes are less strongly retained than the non-polar ones. On the other hand, the stationary phase in normal-phase (NP) chromatography is more polar than the mobile phase, and opposite to RP HPLC, the retention increases with increasing polarity of the sample and of the stationary phase and in less polar mobile phases. In non-aqueous mobile phases traditionally used in NP chromatography, the retention mechanism is based on the competition between the sample and the mobile phase for localized polar adsorption centres on the adsorbent surface (Snyder et al., 2009; Snyder, 1968; Syder, 1974; Soczewinski, 1969; Nawrocki, 1997).

Weakly or moderately polar compounds can be separated either in RPLC or in NPLC, with significant differences in the elution order and separation selectivity. Very hydrophilic samples such as carbohydrates or small strongly polar compounds are weakly retained in reversed-phase LC systems and often elute close to the column hold-up volume, so that their separation from one another and from polar matrix interferences may be difficult to accomplish, even in highly aqueous mobile phases (Pereira et al., 2009). Strongly polar compounds are often excessively retained in non-aqueous normal-phase systems or are poorly soluble in non-polar or in weakly polar organic solvents. However they can be separated on polar stationary phases with water added to the mobile phase (Huber et al., 1984). The aqueous normal-phase liquid chromatography (ANPLC) had been occasionally used long time before Alpert introduced the name "Hydrophilic Interaction Liquid Chromatography" (HILIC) for this separation mode (Alpert, 1983; Alpert, 1990), in analogy to "Hydrophobic Interaction Liquid Chromatography" (HIC) of biopolymers on mildly hydrophobic stationary phases in aqueous mobile phases containing salts (Tiselius, 1948). The term "hydrophilic" refers to the affinity to water. Essentially, HILIC employs an NP stationary phase in combination with an RP mobile phase, usually containing 50% or more organic solvent in water and provides appropriate retention and resolution of many polar compounds, often with better separation efficiency in comparison to the RP chromatography (Gritti et al., 2010). The diffusion coefficients of ionized basic compounds in less viscous organic-rich mobile phases under HILIC conditions are approximately twice those under RP conditions, leading to a lower mass transfer contribution to the height equivalent of a theoretical plate (HETP) (McCalley, 2007; McCalley, 2008). Another reason for increasing popularity of HILIC is its excellent suitability for coupling to mass spectrometry (LC/MS).

First, HILIC was applied mainly in the carbohydrate and peptide analysis in proteomics and glycomics (Zhu et al., 1991; Feste & Kahn, 1992; Churms, 1996; Lin & Lee, 1998; Strege, 1998). Later, HILIC has gradually found useful applications for separations of drugs, toxins, plant extracts and other small polar compounds in clinical, food and pharmaceutical analysis (Tolstikov & Fiehn, 2002; Garbis et al., 2001).

Growing interest in HILIC separations resulted in several recently published review articles. Excellent overview of the recent progress in the development of polar stationary phases until 2006 was written by Hemström and Irgum (2006). Other reviews focused on the effects of the operating conditions on HILIC separations (Hao et al., 2008), separation efficiency (Ikagami et al., 2008), HILIC method development (Dejaegher et al., 2008), coupling HILIC systems with MS and MS/MS (Nguyen & Schug, 2008; Hsieh, 2008), implementation of HILIC systems in two-dimensional separation modes (Jandera, 2008) and on HILIC applications in biological (Yoshida, 2004; Jian et al., 2010), pharmaceutical (Dejaegher & Vander Heyden, 2010) and metabolite (Iwasakiet al., 2007; Spagou et al., 2010) analysis. Two special thematic issues of *Journal of Separation Science* were devoted to various aspects of HILIC separations in the last two years: volume 31, No. 9 (2008) and volume 33, No. 6–7 (2010).

Principles of HILIC Separations

Phase Systems in HILIC

Like non-aqueous normal-phase (adsorption) LC, the retention in the HILIC mode increases proportionally to the polarity of the solute and is higher on polar stationary phases and in less polar mobile phases. Hydrogen bonding, hydrophobic interactions and electrostatic (ion-exchange or ion repulsion) interactions may be more or less involved in the HILIC separation mechanism, depending on the type of the analyte and of the stationary phase and on the composition of the mobile phase (Hao et al., 2008; Wu et al., 2008). Hydrogen bonding is most important at very low concentrations of water (Berthod et al., 1998). Electrostatic interactions affect mainly the retention of ionizable polar analytes on HILIC stationary phases containing anionic, cationic or zwitterionic groups.

Together with the stationary phase, the type and concentration of the organic solvent in the mobile phase control the separation in the HILIC mode. Increasing concentration of the organic solvent (acetonitrile) in the mobile phase enhances the retention of polar compounds. Changes in separation selectivity and even in the order of elution may occur at changing the % of water (or aqueous buffer) in the mobile phase.

HILIC Retention Mechanism—Partition or Adsorption?

In non-aqueous normal-phase liquid chromatography, the retention is based mainly on localized adsorption on polar centres on the adsorbent surface. The adsorption model yields, with some simplifications, Eq. (1) to describe the relationship between the concentration (volume fraction, φ) of the stronger (more polar) solvent in the two-component organic mobile phase on the retention factor of the analyte, $k = t_R/t_0 - 1$ (Snyder, 1974; Jandera & Churáček, 1974a):

$$\log k = \log k_0 - m \log \varphi \qquad (1)$$

Here, t_R is the sample elution time, t_0 is the column hold-up time, i.e., the elution time of an un-retained inert compound, φ is the volume fraction of the more polar (stronger) organic solvent in the mobile phase, k_0 is the retention factor extrapolated to the pure polar solvent ($\varphi = 1$) and m is the solvent elution strength factor, characterizing the ability of the polar solvent to displace the adsorbed analyte.

Eq. (1) describes satisfactorily the retention in many non-aqueous normal-phase systems (Snyder & Poppe, 1980). If Eq. (1) is applied to HILIC systems, φ is used for the volume fraction of water (or of an aqueous buffer) and k_0 is the (extrapolated) retention factor in pure water or aqueous buffer, as the more polar solvent in aqueous-organic mobile phases.

Because of its extreme polarity, water is preferentially adsorbed on the surface of polar adsorbents from mixed aqueous-organic mobile phases used in the HILIC systems. In aqueous-organic NP systems containing more than 0.5–1% water, a water-rich adsorbed diffuse liquid multi-layer forms, which is thick enough to induce liquid-liquid partitioning between the more organic bulk mobile phase and the adsorbed aqueous liquid layer (Fig. 2.1).

In LC systems where the sample distribution is controlled by partition (or non-localized adsorption) such as in the reversed-phase systems, the effects of the concentration (volume fraction), φ, of the organic solvent in a binary

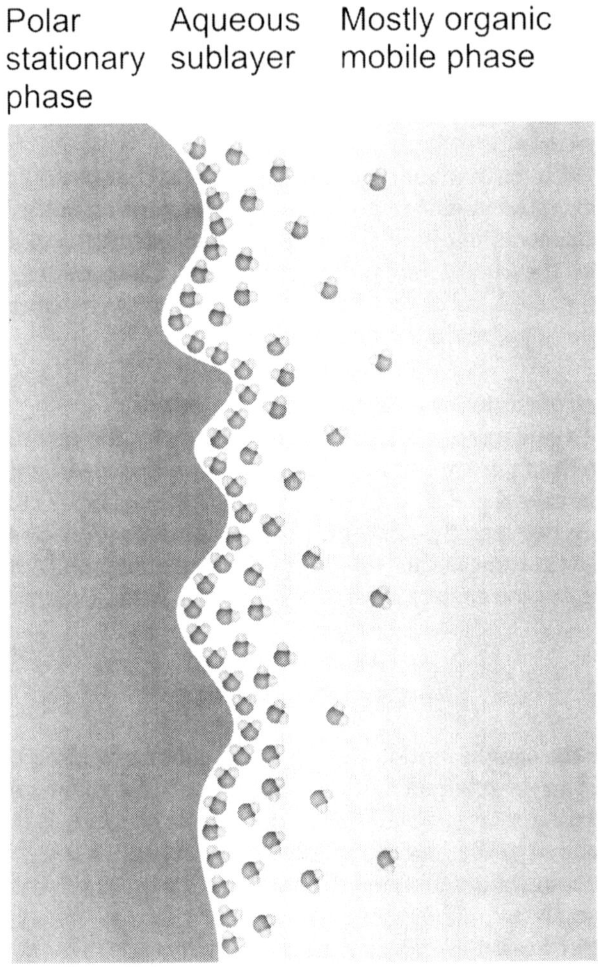

Fig. 2.1. Diffuse water layer at the surface of a polar stationary phase in highly organic environment.

mobile phase on the retention factor, k, is often satisfactorily described by Eq. (2), at least over a limited concentration range (Snyder et al., 1979; Jandera et al., 1979):

$$\log k = \log k_0 - m_1 \cdot \varphi = a - m_1 \cdot \varphi \tag{2}$$

The constant a in Eq. (2) theoretically should be equal to the logarithm of the solute retention factor in pure water, k_0; the parameter m_1 increases with decreasing polarity of the organic solvent and with increasing size of the sample molecule.

In non-aqueous normal-phase chromatography, another equation was derived to describe the retention of moderately polar compounds over a broad range of mobile phase composition (Jandera et al., 1978):

$$k = (a + b\,\varphi)^{-m} \tag{3}$$

a, b and m are experimental constants depending on the solute and on the chromatographic system. $a = 1 / (k_a)^{(1/m)}$; k_a is the retention factor in pure non-polar solvent. In non-aqueous normal-phase chromatography Eq. (3) often provides better fit to the experimental data than Eq. (1) for various low-molecular weight (Jandera & Kučerová, 1997) and polymer (Schoenmakers et al., 2002) samples.

Some studies attempted to attribute the mechanism controlling the sample retention in a particular HILIC separation system on the basis of the goodness of the fit of the experimental retention data to Eq. (1) or to Eq. (2). As Eq. (1) describes in a simplified way the model of localized adsorption and Eq. (2) that of non-localized distribution between the stationary and the mobile phase (Jandera &. Churáček, 1974), linearity of the experimental log k *versus* log φ plots observed with some polar stationary phase in a more or less limited high organic HILIC mobile phases was considered as the proof of the adsorption or ion-exchange nature of the HILIC retention mechanism (Alpert, 1990; Hemström & Irgum, 2006; Iwasaki et al., 2007).

On the other hand, good linear fit of the log k *versus* φ plots was believed to prove the partitioning retention mechanism, and the lack of fit of either Eq. (1) or Eq. (2) to the retention data observed on some other polar stationary phases in HILIC systems was attributed to mixed-mode HILIC mechanism (Nguyen et al., 2010; Chauve et al., 2010).

With some polar samples, the log k *versus* log φ plots are close to straight lines and the data fit to Eq. (1) often improves with increasing polarity of stationary phases. In mobile phases containing 5–40% v/v water, solutes with a larger number of polar functional groups show higher slope parameters m of Eq. (1), in agreement with the surface adsorption model (Wu et al., 2008). The slope parameter m often decreases at decreasing difference in polarities between water and the organic solvent, such as with methanol (a hydrogen donor solvent) with respect to acetonitrile (showing predominantly dipole-dipole interactions) (Škeříková & Jandera, 2010). Eq. (1) usually fails at low concentrations of water (< 2%) under HILIC conditions (Jandera & Hájek, 2009).

The goodness of the fit of Eq. (1) or Eq. (2) to the experimental log k *versus* φ or log k *versus* log φ plots alone may not be convincing proof of the validity of the adsorption, partition, or any mixed mode HILIC retention mechanism. Because of the simplifying assumptions adopted in the derivation, Eq. (1) or Eq. (2) are valid only over a more or less limited mobile phase composition range, and the quality of the fit may depend on the selection of

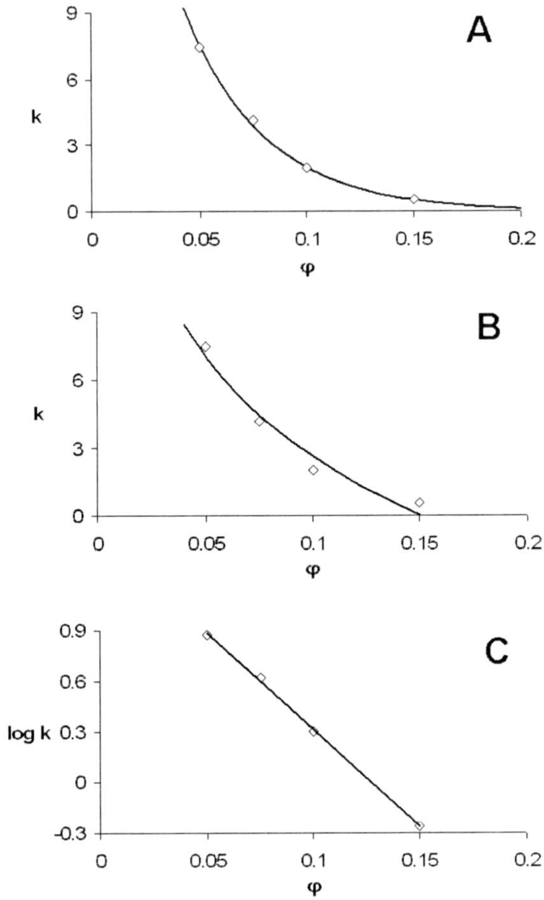

Fig. 2.2. Comparison of the equations describing the retention factors, *k*, of vanillic acid on an UDC cholesterol column in the HILIC range of buffered aqueous-organic mobile phases with various volume fractions of acetonitrile, φ. Points - experimental data, lines - best-fit plots determined using and Eq. (3) – 2.2.A, Eq. (1) – 2.2.B .and Eq. (2) – 2.2.C (J. Soukup, P. Jandera unpublished results).

the range of experimental retention data. Like in non-aqueous NP, the retention data in HILIC systems may fit better Eq. (3) than either Eq. (1) or Eq. (2), depending on the column, solute and organic solvent. Fig. 2.2 shows an illustrative example of mobile phase effects on the retention of vanillic acid on an UDC cholesterol column in buffered aqueous acetonitrile. Both Eq. (3) (Fig. 2.2.C) and Eq. (2) (Fig. 2.2.A) describe satisfactorily the experimental retention factors, *k*, whereas Eq. (1) does not fit the data very well (Fig. 2.2.B).

Regardless of the retention mechanism controlling the retention, Eq. (1), Eq. (2) or Eq. (3) can be considered as empirical equations useful for prediction of retention, method development and optimization in HILIC systems.

Ion-exchange Contribution to the Retention in HILIC Systems

The retention in the HILIC systems with "bare" silica, bonded amine, ion exchange and zwitterionic stationary phases in organic-rich mobile phases is often governed by a combination of mechanisms involving polar and ion-exchange interactions (Strege et al., 2000; Naidong et al., 2001). Organic solvents often promote ion-exchange interactions between the sample and the charged surface of the stationary phase, resulting in significant improvement of separation of complex mixtures such as peptides with large differences in hydrophobicity.

In ion-exchange (IEX) chromatographic systems, which must contain a salt (or a buffer) in the mobile phase, the retention can also be formally described by Eq. (1), with φ characterizing the molar concentration of the electrolyte in the mobile phase (Jandera & Churáček, 1974a; Baba, 1989). k_0 is the retention factor in the mobile phase with 1 mol/L electrolyte, m is the stoichiometric coefficient of the ion exchange of the sample ions and counter-ions in the mobile phase; (m is close to 1 for the exchange of uni-valent ions). This applies for both water-rich and highly-organic mobile phases. Hence, the ion-exchange contribution in mixed HILIC/IEX systems can be expected to not significantly affect the goodness of the fit of Eq. (1) or of Eq. (3) to the experimental retention data; however in such cases the meaning of the equation parameters cannot be clearly defined.

Mixed-mode HILIC/IEX chromatography provides important advantages for the separation of peptides with low differences in hydrophilicity/hydrophobicity (by a single carbon atom) or of peptides differing only in the sequences of amino acids. The greater the charge on the peptide molecules, the better the separation achieved by HILIC/IEX. In addition, HILIC/IEX separation of peptides on columns with cation-exchange functionalities is often superior to that of the more commonly applied RP mode (Mant & Hodges, 2008).

So-called "Electrostatic-Repulsion Hydrophilic-Interaction Chromatography" (ERLIC) on weak anion exchangers in highly organic mobile phases offers possibilities for independent adjusting of separation selectivity by promoting either the HILIC or the ionic interactions. Samples can be retained via hydrophilic interactions even if they have the same (positive) charge polarity as the stationary phase. Electrostatic repulsion decreases the retention of basic peptides in the low pH range, but at increased concentration of acetonitrile, adequate retention and resolution of basic and acidic peptides can be achieved in a single run (Alpert, 2008).

Temperature Effects

The equilibrium distribution constant, $K = e^{-(\Delta G/RT)}$, of a sample depends on the change in the Gibbs energy connected with the solute retention, ΔG^0, and is related to the corresponding changes in standard partial molar entropy, ΔS^0, and in the standard partial molar enthalpy, ΔH^0, of the transfer of the solute from the mobile to the stationary phase:

$$\Delta G^0 = -RT \ln K = \Delta H^0 - T\Delta S^0 \tag{4}$$

Here, T is the thermodynamic temperature (in Kelvins) and R is the gas constant. If a single retention mechanism controls the retention, the effect of temperature on the retention factor, k, is described by the van't Hoff equation, Eq. (5) (Melander & Horvath, 1980):

$$\ln k = \ln K + \ln \frac{V_S}{V_M} = -\frac{\Delta G^0}{RT} + \ln \frac{V_S}{V_M} = \frac{\Delta S^0}{R} + \ln \frac{V_S}{V_M} - \frac{\Delta H^0}{RT} = A_i + \left(\frac{B_i}{T}\right) \tag{5}$$

The retention in liquid chromatography usually decreases at elevated temperatures, presumably by reducing the difference in cohesive energy, hydrogen bonding and other polar interactions between the mobile and the stationary phases. A change in temperature may to some extent influence the retention and separation selectivity of compounds that differ in size, shape, or hydrogen-bonding basicity, especially partially ionized solutes (Zhu et al., 1996).

Evaluating the experimental plots of $\ln k$ versus $1/T$ over a sufficiently broad temperature range, the enthalpic contributions to the chromatographic retention and selectivity, $-\Delta H^0$, can be calculated from the slope B_i and the entropic contributions, ΔS^0, from the intercept A_i of Eq. (5). Besides the entropic term, the experimental value of the intercept includes also the ratio of the volumes of the stationary phase, V_S, and of the mobile phase, V_M, in the column. Deviations from the linear van't Hoff plots may indicate that the retention mechanism may change over the studied temperature range (Melander et al., 1978).

In HILIC systems, the column temperature usually affects the retention less than the composition of the mobile phase (Guo et al., 2007), however the separation selectivity may be significantly affected by temperature. The example in Fig. 2.3 shows the HILIC retention data of four phenolic acids on a silica hydride column and on a hydrosilated silica column modified with cholesterol. The data fit reasonably well the van't Hoff plots [Eq. (5)] and are steeper for more retained syringic (1) and ferulic (2) acids with one or two methoxy groups than for less retained *p*-coumaric (3) and salicylic (4) acids. Consequently, these four acids are better separated at ambient temperatures than elevated temperatures (J. Soukup, P. Jandera, unpublished data). Usually, a negative value is found for the retention enthalpy in HILIC systems, which means that the retention process is exothermic. For example, on 2-mercaptoethanol and 1-thioglycerol stationary phases, linear van't Hoff plots were found in the range 15–45°C, with retention enthalpy values $-\Delta H$

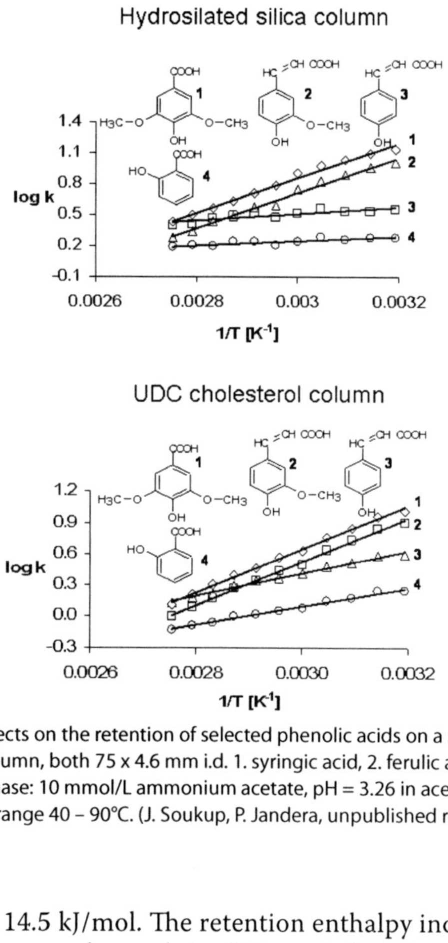

Fig. 2.3. Temperature effects on the retention of selected phenolic acids on a silica hydride column and a UDC cholesterol column, both 75 x 4.6 mm i.d. 1. syringic acid, 2. ferulic acid, 3. *p*-coumaric acid, 4. salicylic acid. Mobile phase: 10 mmol/L ammonium acetate, pH = 3.26 in acetonitrile 5 : 95 % (*v/v*), 0.5 ml/min; Temperature range 40 – 90°C. (J. Soukup, P. Jandera, unpublished results).

in between 4.5 and 14.5 kJ/mol. The retention enthalpy increases in more polar phases with more polar analytes (Wu et al., 2008).

However, in some other HILIC systems the retention increases at higher temperatures, or more often, is almost independent of temperature. Some phenolic acids (salicylic, acetylsalicylic) show negative slopes, B_i, of van't Hoff plots and positive retention enthalpy, $-\Delta H$, on bonded aminopropyl silica columns (Guo & Gaiki, 2005), whereas the retention on other types of HILIC columns such as on a poly(methacrylate) monolithic sulfobetaine column, the retention is more or less independent of temperature and the slopes of the van't Hoff plots are close to zero (Škeříková & Jandera, 2010). This behavior suggests very low enthalpic contribution to the retention and major role of entropic effects on the zwitterionic column in the HILIC mode, possibly due to different hydrogen-bonding sample solvation in the occluded stationary liquid phase and the bulk mobile phase.

Increased column temperature often positively affects the separation efficiency due to lower viscosity and higher diffusivity. The bandwidths of the peaks of monosaccharides and uronic acids improved, and their retention on an Amide-80 column slightly decreased as the temperature changed from 40 to 60°C (Karlsson et al., 2005). Also, the efficiency of both the isocratic and the gradient separation of phenolic acids on a poly(methacrylate) monolithic sulfobetaine column significantly improved at 60–80°C in comparison to the separation at 30°C (Škeříková & Jandera, 2010).

Silica Gel-based Stationary Phases

Normal-phase separations on bare silica and amino-silica columns in aqueous-organic mobile phases were reported as early as the mid seventies (Linden & Lawhead, 1975); since then a large variety of silica-based columns suitable for HILIC separations have been developed, including hydrosilated silica (silica hydride), hybrid silica-organic phases. Many of them are commercially available, however new polar stationary phases, potentially useful in the HILIC mode, are reported in the literature by various research groups.

Silica-based polar bonded phases are generally prepared by chemical modification of the silica gel surface by reactions with trialkoxysilanes containing polar and alkyl groups. Some moderately polar stationary phases with chemically bonded cyano-, Diol-, amino-, cyclodextrin-, polyethylene glycol (PEG), pentafluorophenylpropyl, alkyls with embedded amide or carbamate groups and other functionalities were originally intended for reversed-phase applications in water-rich mobile phases, but many show typical normal-phase HILIC behavior in organic-rich mobile phases, where the retention of polar compounds increases proportionally to the concentration of acetonitrile. Some silica-based stationary phases were developed specifically for use under HILIC conditions, e.g., poly(succinimide)-, or sulfoalkylbetaine- silica.

The bonded moieties can often form a more or less thick polymeric network able to incorporate a water-enriched liquid layer, providing partitioning equilibria. Generally, the retention of polar analytes increases in the order: cyanopropyl < diol < aminopropyl < silica stationary phases, but selective interactions may strongly affect the selectivity of separation and the elution order of polar compounds. Fig. 2.4 shows the structures of some silica-based stationary phases used in HILIC applications.

Bare Silica

Unlike chemically bonded stationary phases, non-modified bare silica gel does not gradually lose its performance by "bleeding" the bonded stationary phase. Commercially available silica gel materials may differ in purity, depending on the technology of preparation. The "type A" silica gel is the original all-purpose chromatographic material produced by precipitation from alkali silicate solutions. It is usually highly contaminated with metals and is not

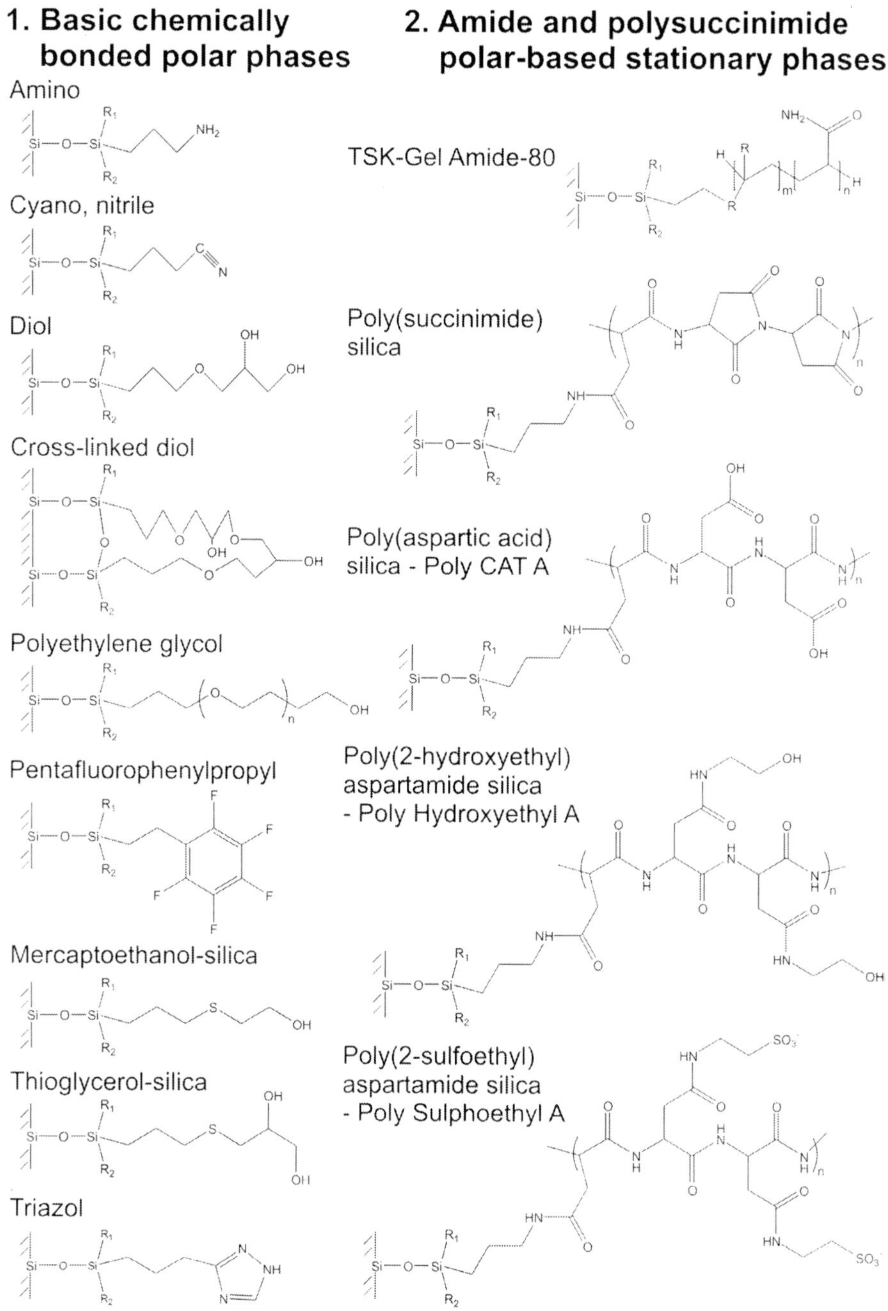

Fig. 2.4. Structures of some simple polar and poly(succinimide) HILIC stationary phases chemically bonded on silica gel surface.

suitable for HILIC applications. Nowadays, silica "type B" materials are used in most conventional HPLC separations, including HILIC applications. The spherical Silica type B particles are formed by aggregation of silica sols in the air, contain very low amounts of metal impurities and are relatively stable at intermediate and higher pH to at least pH 9. Some silica materials, such as Atlantis HILIC silica, are intended specifically for applications with aqueous-organic mobile phases (Grumbach et al., 2004).

On bare silica columns, the retention in aqueous-organic mobile phases may be contributed to by partitioning, adsorption and ion exchange, depending on the sample type and on the composition of the mobile phase (Chauve et al., 2010). Some analytes such as carbohydrates may show tailing peaks or even irreversible adsorption on bare silica under HILIC conditions (Li & Huang, 2004). At increasing pH, the ionization of surface silanol groups is enhanced and cation exchange interactions may play a significant role in retention, especially basic compounds may be strongly retained on silica gel by hydrogen bonding and ion-exchange interactions (Naidong, 2003). Peak symmetry of basic compounds on "Type B" silica gel columns in HILIC mobile phases (acetonitrile-ammonium formate buffer) often improves in comparison to reversed-phase HPLC (McCalley, 2007). Addition of trifluoroacetic acid to HILIC mobile phases generally suppresses ion-exchange interactions on the silica surface, but may promote an ion-pairing mechanism.

Separations of polar drugs can often be accomplished on bare silica gel columns in aqueous-organic mobile phases with acetonitrile concentrations ranging from 75 to 95%, usually containing 10–100 mmol/L ammonium acetate or formate buffers. HILIC applications of silica gel columns in pharmaceutical analysis were recently reviewed in detail (Dejaeghar & Vander Heyden, 2010).

Hydrosilated Silica and Hybrid Organic-silica Materials

The silica hydride "type C" has up to 95% of the original high population of silanol (Si-OH) groups replaced by non-polar Si-H groups by hydrosilation (Sandoval & Pesek, 1989). Fig. 2.5 shows the structure of the hydrosilated silica surface. The hydrosilation process changes fundamental properties of the silica surface, so that silica hydride materials are less polar, attract less water and show improved reproducibility of retention.

The un-modified silica hydride shows very low hydrophobic selectivity and retention under RP conditions (Molíková & Jandera, 2010). Modification of the silica hydride surface with hydrocarbon or lipophilic ligands increases the non-polar interactions on the adsorbent surface and induces new selectivity properties for HILIC separations of less polar compounds so that such materials can be used either for separations in the aqueous normal-phase (HILIC) mode in buffered mobile phases containing more than 50–70%

1. Silica hydride

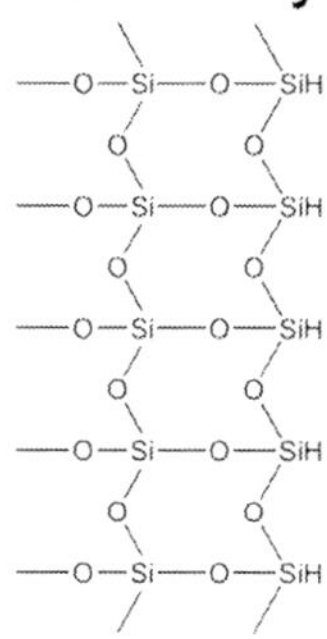

2. Cholesterol hydride

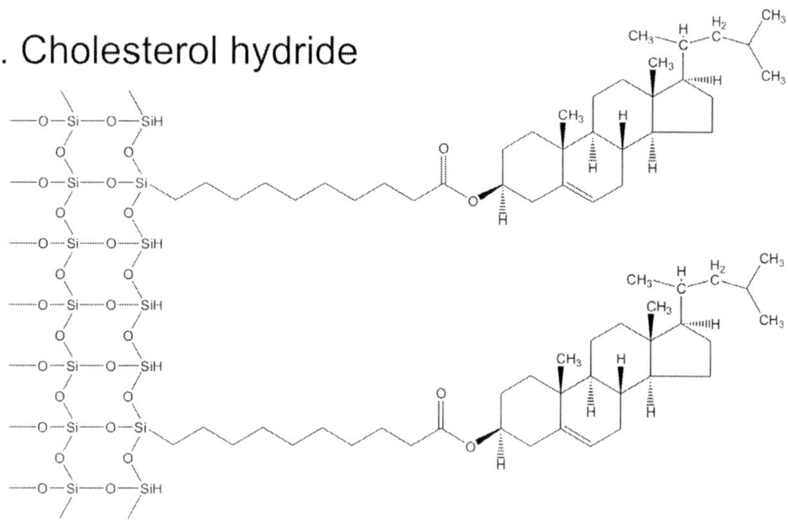

3. Bidentate C18 hydride

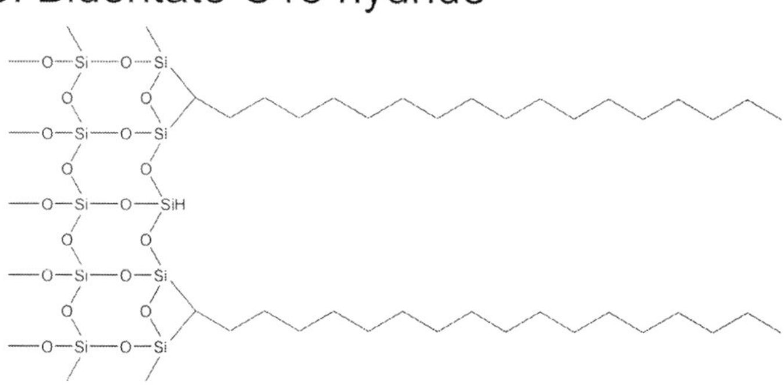

Fig. 2.5. Structures of un-modified hydrosilated silica and hydrosilated silica modified with non-polar ligands.

acetonitrile, or in the RP mode with highly aqueous mobile phases (Pesek & Matyska, 2007). "Diamond hydride" stationary phases containing ca 2.5% carbon were introduced on the market; chemically modified bidentate C18 and cholesterol columns are also commercially available (Fig. 2.5). Silica hydride modified with undecanoic acid (UDA silica) was used for separation of mono-, di- and tri-phosphate nucleotides under HILIC conditions using gradient elution with increasing concentration of ammonium formate buffer in acetonitrile. Improved separation selectivity with respect to unmodified material was attributed to enhanced ion-exchange properties (Matyska et al., 2010).

Fig. 2.6 shows the HILIC separation of eight phenolic acids on unmodified silica hydride, cholesterol and bidentate C_{18} columns in 95% acetonitrile containing 10 mmol/L ammonium acetate buffer. The elution order in the HILIC mode is similar, but the retention increases and the separation selectivity improves at increasing polarity of the adsorbent surface (C_{18} < cholesterol < bare silica hydride). The silica hydride column provides best separation of vanillic and ferulic acids (J. Soukup, P. Jandera, unpublished results).

Bridged ethylene hybrid silica (BEH) particles show improved chemical resistance with respect to the pure silica materials (Grumbach et al., 2008). An interesting example of using BEH HILIC silica is the determination of melamine illegally added to milk and infant formulas to boost the apparent nitrogen content. The method combines solid-phase extraction with rapid HILIC gradient separation (from 95% to 78% acetonitrile with 10 mM ammonium acetate) and MS/MS identification.

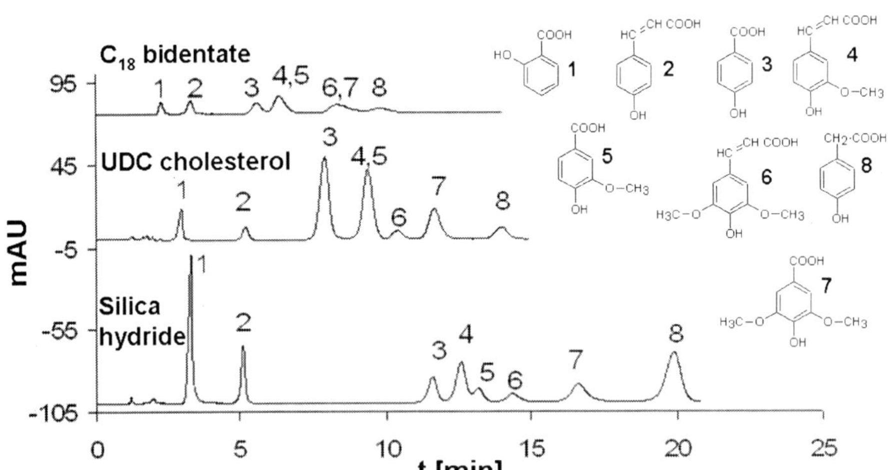

Fig. 2.6. Separation of phenolic acids on an un-modified hydrosilated silica column and on hydrosilated silica columns modified with non-polar C_{18} bidentate and cholesterol ligands (J. Soukup, P. Jandera, unpublished results).

Chemically Bonded Amine, Amide, and Poly(succinimide) Stationary Phases

Aminopropyl silica has been traditionally used for normal-phase separations of sugars, amino acids, peptides, carboxylic acids, nucleosides and some pharmaceuticals in aqueous-organic mobile phases. However, various brands of amino-silica columns may show significant differences in retention and selectivities (Olsen, 2001).

Bonded stationary amino phases show increased affinities for acidic compounds, due to weak ion-exchange interactions. Due to the contribution of the ion-exchange mechanism, the retention of ionic samples decreases at increasing ionic strength of the mobile phase (Oyler et al., 1996). Relatively low efficiency, long equilibration times and irreversible adsorption due to mixed-mode retention mechanism may be observed with some bonded amino phases in mobile phases containing some types of buffers. The primary amino group is relatively reactive, which may cause bleeding of the bonded phase by hydrolysis, or by formation of Schiff bases with aldehydes, which may cause problems in HILIC separations of carbohydrates (Ikegami et al., 2008). Aminopropyl-silica columns provide an increased rate of anomer mutarotation so that monosaccharide diastereomers co-elute and do not form doublet peaks which may complicate the interpretation of the chromatograms of sugars on other columns. Non-ionic (Jandera et al., 1998; Jandera et al., 2000) and ionic (Jandera & Prokeš, 1996) ethylene oxide co-oligomer surfactants can be separated according to the distribution of oxyethylene units on aminopropyl-silica columns in the HILIC mode.

Stationary phases with secondary or tertiary amine groups, such as YMC Pack Polyamine II, do not form Schiff bases with carbonyl compounds, which improves the column stability and lifetime with respect to aminopropyl silica. Aromatic amine stationary phases bonded on silica (Epic HILIC-PI) were reported to provide symmetrical peak shapes and short separation times in HILIC chromatography of amines or organic acids.

The amide stationary phases contain a carbamoyl or an amide group bonded to the silica gel surface *via* a short alkyl spacer (Fig. 2.4) and, in contrast to amino-silica or poly(2-sulphoethyl aspartamide) columns, do not possess basic properties so that the retention of ionizable analytes is unaffected by ion-exchange interactions and ionic mobile phase additives usually are not necessary. Hence, the column effluent does not need to be de-salted before the introduction into a mass spectrometer. Further, irreversible sample adsorption on the amide columns is less probable than on the aminopropyl silica, resulting in better stability and prolonged column lifetime (Hemstrom & Irgum, 2006).

Some amide stationary phases have been specifically designed for HILIC applications, such as carbamoyl-silica HILIC TSK-gel Amide-80 (Tosoh, Tokyo, Japan, Fig. 2.4), which shows higher retention for many polar compounds than silica gel (Guo & Gaiki, 2005). The amide-silica columns are generally recommended for efficient and fast separations of highly polar samples, first of all peptides and amino acids. Hydrophilic peptides are more strongly retained on the amide stationary phases than the hydrophobic peptides. The amide silica/bonded columns are suitable also for HILIC separations of other hydrophilic or amphiphilic high-molecular weight samples, such as mono- and oligosaccharides, glycoproteins, or glycosides (Karlsson et al., 2005; Yoshida, 1997).

1.7 μm ethylene bridged hybrid (BEH) amide columns strongly retain polar basic pteridines, for which they provided more improved separation under HILIC UHPLC conditions than un-derivatized BEH silica, especially at a high pH (Nováková et al., 2010).

Alpert (Alpert, 1983; Alpert, 1990) introduced a series of HILIC stationary phases, prepared by the reaction of aminopropyl-silica with poly(succinimide), followed by a second-step reaction: Alkalic hydrolysis yields poly(aspartic acid)-silica; reaction with 2-aminoethanol provides poly(2-hydroxyethyl)aspartamide-silica and modification with 2-aminoethylsulfonic acid leads to poly(2-sulfoethyl)aspartamide-silica (Fig. 2.4). The poly(succinimide)-silica based materials are produced and supplied by PolyLC (Columbia, MD, USA) under trademarks PolyCAT A, PolyHydroxyethyl A, PolyGlycoplex and PolySulfoethyl A. (Poly)succinimide type silica bonded phases offer suitable retention for a wide range of polar analytes such as peptides, proteins, nucleic acid constituents, oligosaccharides, carbohydrates, etc. Successful separations with symmetrical peaks may require mobile phases containing at least 7–10 mM electrolyte (such as triethylammonium buffer), especially for ionized solutes.

PolySulfoethyl A has strong cation-exchange properties and shows a mixed-mode HILIC/cation-exchange mechanism. This column can be used for separations of hydrophilic peptides (Hartmann et al., 2003). Aqueous-organic mobile phases containing 50% or more acetonitrile provide complementary selectivity to reversed-phase separations (Alpert & Andrews, 1988). Often, elution with increasing salt gradients is used. Separation efficiency may be poor in mobile phases containing less than 90% acetonitrile. Hence the PolySulfoethyl A column is recommended for not too strongly retained samples.

PolyHydroxyethyl A can be applied for separations of phosphorylated and non-phosphorylated amino acids, peptides and glycopeptides, proteins, oligonucleotides, carbohydrates and glycosides and small polar solutes (Oyler et al., 1996). Plant metabolites elute in the order: polar lipids, flavonoids, glucosinolated, saccharides and amino acids (Tolstikov & Fiehn, 2002).

PolyGlycoplex columns are suitable for separations of monosaccharides, oligosaccharides and sialyl sugars or their *p*-nitrobenzyloxy derivatives (Alpert et al., 1994).

Other Types of Silica-based Chemically Bonded Stationary Phases

Chemically bonded Diol phases, containing neutral hydrophilic 2,3-dihydroxypropyl ligands, are prepared by bonding glycidoxypropyltrimethoxy silane to the silica gel surface. (Fig. 2.4). They show high polarity and hydrogen bonding properties and do not contain ionizable groups other than non-reacted residual silanols which can be partially blocked by silylating reagents to avoid irreversible adsorption of polar compounds. Diol-silica columns have been used for HILIC separations of proteins (Regnier & Noel, 1976) but more recently were employed also for separations of low-molecular weight phenolic compounds, both in low organic solvent RP LC mode and high organic solvent HILIC mode (Jandera & Hájek, 2009). The Diol columns may slowly release the bonded phase under acidic conditions. Unlike amino-bonded phases, anomers and different cyclic forms of monosaccharides can be resolved on Diol columns. This allows monitoring the rate of the transition between the individual forms, i.e., mutarotation (Pazourek, 2010).

Cross-linked Diol stationary phase Luna HILIC 200 (Phenomenex, Torrance, CA, USA, Fig. 2.4) shows increased stability against hydrolysis, stronger hydrophobic interactions and better peak shape and resolution compared to the non-cross-linked Diol-silica phases (Jandera et al., 2010a). The Luna HILIC 200 column contains both oxyethylene and hydroxy bonded groups and combines to a certain extent the properties of the polyethylene glycol (PEG) and the Diol columns (Jandera & Hájek, 2009). For example, nicotine and its principal metabolites cotinine, trans-3-hydroxycotinine, nicotine-*N*-oxide and cotinine-*N*-oxide in urine were separated in 13 min, after liquid–liquid extraction, on a Luna HILIC 200 column using gradient elution from 98% to 35% acetonitrile, with mobile phase buffered at pH 3.0 with 10mmol/L ammonium formate buffer. The sample compounds were identified using tandem mass spectrometry (Marclay & Saugy, 2010). Six basic hydrophilic sympathomimetic drugs were separated by nano-LC at HILIC conditions on the aminopropyl-silica and Luna HILIC 200 columns (Aturki et al., 2011).

Less polar polyethylene glycol (PEG) bonded stationary phases (Fig. 2.4) show stronger hydrophobic interactions than the Diol columns (Blahová et al., 2006). The separations on Diol and (PEG) stationary phases differ significantly in selectivities from bonded alkyl-silica columns, both in reversed-phase mode (Jandera et al., 2008) and at HILIC conditions (Wang et al., 2005; Jandera & Hájek, 2009).

Cyanopropyl-silica bonded phases (Fig. 2.4) lack hydrogen bonding capabilities and often retain too weakly polar compounds such as peptides in mobile phases with high concentrations of organic solvents in water (Yoshida & Okada, 1999). Hence, their use in HILIC separations is limited to specific applications, such as the determination of common denaturants (denatonium benzoate, crystal violet and methylene blue) in denaturated alcohol samples.

The cyanoethyl phase provided stronger retention of organic cations than aminopropyl or silica columns in mobile phases containing 60% (v/v) acetonitrile and 10 mmol/L $HClO_4$ as the anionic ion-pairing reagent under HILIC conditions (Daunoravicius et al., 2006).

Cyclodextrin stationary phases were bonded on silica *via* amine, urethane or ether linkages, or "click chemistry" was used for immobilization of alkyne-modified β-cyclodextrin on the surface of azide-modified silica particles (Fig. 2.7). Cyclodextrins are cyclic oligosaccharides that possess six (α-CD), seven (β-CD), or eight (γ-CD) D-glucose units in toroid ring structures with relatively hydrophobic cavities. Bonded CD stationary phases designed for HILIC separations show chiral recognition properties useful for chromatographic separations of enantiomers in highly organic mobile phases (Liu et al., 2008). β-cyclodextrin stationary phases have been used for separation of sugar alcohols, monosaccharides and oligosaccharides with 1-8 monosaccharide units in 70% acetonitrile as the mobile phase (Armstrong & Jin, 1989). The retention increases with increasing number of monosaccharide units which is attributed to the interactions with the hydrophilic aqueous layer on the exterior of the CD molecules rather than to the penetration of sample molecules inside the cavity (Berthod et al., 1998). Cyclodextrin-bonded columns retain more strongly amino acids and some other polar samples than the TSKgel Amide 80 phase, are more stable and provide more reproducible separations than aminopropyl silica stationary phases (Risley & Strege, 2000).

Some flavones and isoflavones co-elute under RP-HPLC conditions, however were well retained and could be separated on the Atlantis HILIC silica and cyclodextrin columns under HILIC conditions (Zhang et al., 2008). Silica gel with bonded Teicoplanin, a macrocyclic antibiotic chiral selector can be also employed for enantiomer separations under HILIC conditions.

Several other polar stationary phases bonded on silica gel were used for separations at HILIC conditions (Fig. 2.4). Cosmosil HILIC column with 1,2,4-triazol groups bonded on silica (Naclai Tesque, Kyoto, Japan), has basic character and was reported to provide highly efficient separations of water soluble vitamins, carboxylic acids, amino acids, peptides and polar pharmaceuticals (Ikegami et al., 2008). Pentafluorophenyl silica can be applied for HILIC separations of aromatic amines, halogenated and other polar aromatic compounds.

2-mercaptoethanol and 1-thioglycerol groups were attached onto vinylised silica, followed by oxidation with excess hydrogen peroxide in aqueous medium. The stationary phases prepared in this way can be used both in the HILIC and in the RP mode (Fig. 2.4) and differ significantly from Diol bonded phases in separation selectivity and even in the elution order of various samples such as vitamins, nucleosides and bases (Wu et al., 2008; Bicker et al., 2008).

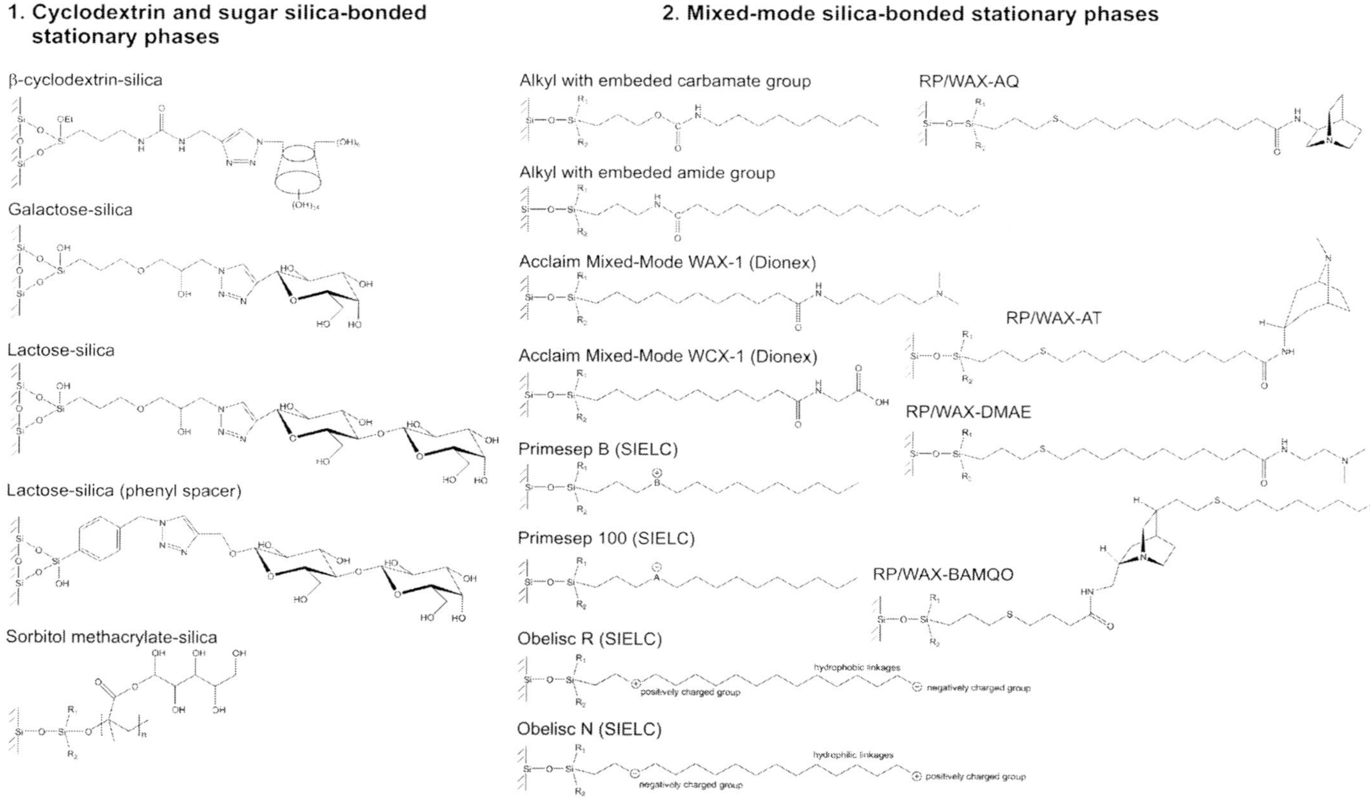

Fig. 2.7. Structures of some cyclodextrin, sugar and mixed-mode HILIC/ion-exchange stationary phases chemically bonded on silica gel surface.

A variety of sugar stationary phases covalently bonded on silica gel supports were prepared using the "click" chemistry approach, in which sugar alkynes are covalently coupled to the azido-activated silica gel surface in the presence of a copper catalyst (Fig. 2.7) (Guo et al., 2007; Yu et al., 2009; Santoyo-Gonzalez & Hernandez-Mateo, 2009). The bonded carbohydrates retain their configuration so that these highly polar stationary phases can be applied to stereoselective separations of monosaccharides and for HILIC separations of highly polar amino acids, glycopeptides, oligonucleotides and natural products such as flavonoids (Moni et al., 2010). Separation selectivity different from that of other polar bonded columns and of bare silica was found also for a hydrophilic stationary phase prepared by graft polymerization of sorbitol methacrylate on the surface of silica particles (Fig. 2.7) (Persson et al., 2008). Columns packed with carbohydrate-modified silica particles may show fair efficiencies for separation of oligosaccharides and for some chiral separations in the HILIC mode.

Mixed-mode HILIC/ion-exchange columns chemically bonded on the silica gel support have been introduced to enable separations of a wide range of strongly and weakly polar, or even non-polar, compounds in organic-rich mobile phases, such as the Acclaim mixed-mode HILIC column (Dionex) and some other commercial columns, possessing a bonded ligand with a long alkyl chain and a hydrophilic polar terminal group (Fig. 2.7) (Wu et al., 2008). A weak anion exchanger PolyWAX LP, prepared by modifying silica with a cross-linked coating of linear poly(ethyleneimine), can be also used for mixed mechanism HILIC/ion-exchange separations in highly organic mobile phases.

Mixed anion-exchange/cation-exchange/HILIC mechanism on silica-based weak ion exchangers was found useful for the analysis and purification of di-peptides and other compounds in biological samples and in the synthetic combinatorial chemistry products, using combined gradients of pH and increasing concentration of water (Strege et al., 2000). A trimodal stationary phase prepared by coating porous spherical silica particles with charged organic polymer nano-beads (0.1 μm) has the inner-pore area covalently modified with an organic layer, showing both reversed-phase and weak anion-exchange properties. The outer surface is modified with strong cation-exchange groups, hence the spatial separation of the anion-exchange and cation-exchange regions provide possibilities for simultaneous separations of acids, bases and neutral compounds. At high acetonitrile concentrations, the material exhibits mixed HILIC/ion-exchange retention mechanism for charged analytes, which can be used for simultaneous separation of ionized hydrophilic drugs and their counter-ions such as penicillin G and its potassium salt (Liu & Pohl, 2010).

Sub-2 μm, Porous Shell and Monolithic Silica Materials

HILIC columns packed with totally porous sub-2 μm silica gel particles or with porous shell particles of 2.7 μm I.D are very promising for HILIC applications, as they allow more rapid and efficient HILIC separations than columns packed with fully porous 5 μm particles. Because of low viscosity of acetonitrile-rich mobile phases, fast and efficient separations of strongly and moderately polar compounds on bare silica columns packed with sub-2 μm particles are often possible at pressures below 400 bar using flow rates usual in ultra-pressure liquid chromatography (UPLC) on reversed-phase columns in more aqueous rich media (Hsieh et al., 2009). UPLC on HILIC columns packed with sub-2 μm particles enabled significant increasing the number of detected metabolites in biological samples, such as in urine samples obtained from male Zucker rats (Gika et al., 2008). Very fast separations in the HILIC mode can be accomplished on short columns in highly organic mobile phases, such as baseline separation of uracil and cytosine in less than 4s (Appelblad et al., 2008).

Fused-core superficially porous shell particles (Halo HILIC or Kinetex HILIC) have become available for fast HILIC separations. These particles can be operated at backpressures two times lower than the sub-2-μm particles in columns of the same dimensions, however at the cost of approximately 30% lower efficiency (Chauve et al., 2010).

In contrast to the columns packed with particles, monolithic continuous flow-through separation media are prepared as single-piece blocks in a confining container (a tube or a capillary). Silica-based monoliths form a skeleton with bi-modal porous structure, combining high permeability of the flow-through pores (typically 2 μm in average diameter), with the surface area of up to 300 m^2/g of skeletal meso-pores (2–50 nm), which provide the desired retention, selectivity and efficiency. These materials are well suited for fast separations of small molecules and peptides under moderate operating pressures (Siouffi, 2003).

Silica-based monolithic columns generally show lower separation efficiencies in HILIC applications than alkyl silica monoliths in reversed-phase HPLC. A Chromolith Performance Si monolithic column (Merck, Darmstadt, Germany) was used in the HILIC mode in 60–80% acetonitrile as the mobile phase for separation of small inorganic ions and drugs such as naproxen or warfarin (Pack & Risley, 2005). Monolithic bare silica capillary columns were used for fast and efficient capillary electrochromatography (CEC) separation of xanthine alkaloids in 95% acetonitrile containing 1 mmol/L Tris buffer (Puy et al., 2006).

Polymer-coated monolithic silica columns for HILIC applications were prepared by direct on-column polymerization of vinyl monomers with acrylamide (Ikegami et al., 2006), acrylic acid (Horie et al., 2007), 3-diethylamino-2-hydroxypropyl methacrylate, or *p*-styrenesulfonic acid on the silica surface modified with 3-(methacryloxypropyl) trialkoxysilane or 3-(methacrylamidopropyl) trialkoxysilane (Ikegami et al., 2007b), 2-acrylamido-2-methyl-propanesulfonic acid or other functionalities (Fig. 2.8). Monolithic silica gel columns coated with a polyacrylamide layer show improved chemical stability

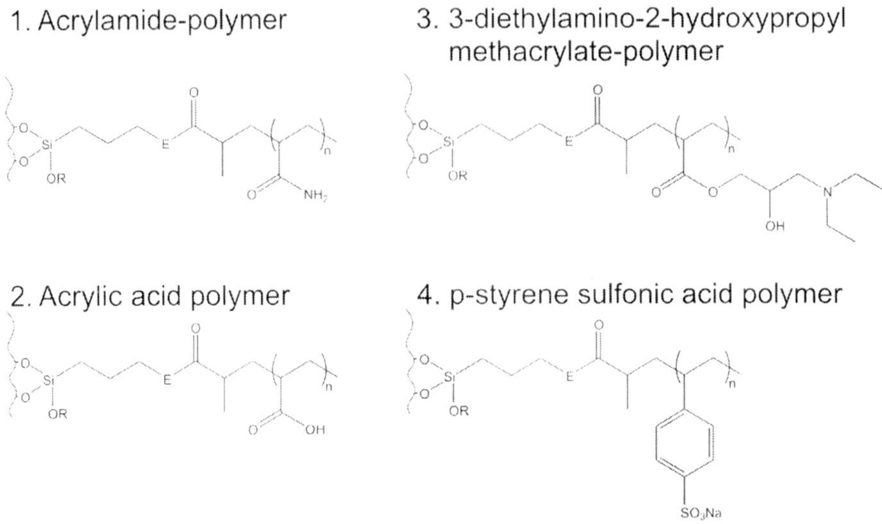

Fig. 2.8. Structures of polymer stationary phases coated on monolithic silica columns.

and higher retention in comparison to non-coated monolithic silica and were successfully applied for HILIC separation of nucleic bases in 90% acetonitrile. A weak cation/exchange stationary phase prepared by *in-stu* polymerization of acrylic acid in a monolithic silica gel capillary column was employed in both HILIC and cation-exchange mode for separations of oligosaccharides, peptides, proteins and nucleosides (Ikegami et al., 2007a).

Agglomerated monolithic columns, prepared by electrostatically attaching polyionic latex particles onto a silica surface, retain the high efficiency and permeability of the native silica monolith and allow fast HILIC separations of benzoates, nucleotides and amino acids at high %ACN in less than 15 s. The latex-coated monoliths show chromatographic retention behavior controlled by mixed electrostatic repulsion–hydrophilic liquid interactions (ERLIC) (Ibrahim et al., 2010).

Organic Polymer Stationary Phases

Particulate Organic Polymer HILIC Materials

Organic polymer particles applied in HILIC usually contain ion-exchange or zwitterionic charged surface groups. Separations of saccharides and other neutral compounds on ion exchangers in organic-rich mobile phases have been practiced for a very long time on both cation-exchange and anion-exchange styrene-divinylbenzene resins. The retention of non-ionic polar compounds increases with increasing concentration of ethanol in aqueous-organic mobile phases, showing typical HILIC behavior (Samuelson et al.,

1963). These early low-efficient, time-consuming, but surprisingly highly selective separations were later reviewed (Jandera & Churáček, 1974b).

Monosaccharides, glycols and glycerol were separated in acetonitrile-rich aqueous-organic mobile phases on efficient ion-exchange porous styrene-divinylbenzene cation-exchange resins with only slightly sulphonated surface (Chambers & Fritz, 1998). Polymer stationary phases with amino group functionalities were employed, e.g., for separations of oligosaccharides (Makino et al., 1998), taurine and methionine in carbohydrate-rich matrices (de Person et al., 2005), or for the separation and determination of 5-fluorouracil in biological samples (Pisano et al., 2005). Some amine polymer particle columns show stronger anion-exchange properties, improved retention time reproducibility, more stable baseline and a longer lifetime than aminopropyl silica columns. However, columns packed with polymer particles still provide significantly lower separation efficiency under HILIC conditions than silica-based columns, so that they can be recommended mainly for separations requiring highly acidic or basic conditions, where silica gel-based materials are not stable enough.

Organic Polymer HILIC Monolithic Columns

Organic polymer monolithic columns possess flow-through pores in an inter-adhered polymer globule structure, however they contain only a very low proportion of meso-pores containing stagnant mobile phase in the polymer skeleton. In contrast to the separation on silica-based monoliths, these materials generally show rather poor efficiency for the separation of small molecules which can only slowly diffuse into and out of the nano-scale pores, due to restricted fluid transport. The nano-pores are not accessible to large molecules, so their mass transfer to the stationary phase is attributed to the convection rather than to the diffusion phenomena. Hence organic-polymer monolithic phases are excellently suited for fast gradient separations of proteins and other biopolymers (Nischang & Bruggemann, 2010).

Monolithic polymer-based columns for HILIC separation mode can be prepared by incorporating polar (hydrophilic) moieties in the polymer matrix. Polar capillary monolithic columns with large pores were prepared by *in-situ* reaction of tris(2,3-epoxypropyl) isocyanurate (TEPIC) with 4-[(4-aminocyclohexyl)methyl] cyclohexylamine (BACM) in the presence of poly(ethylene glycol) as a component of the porogen solvent (Courtois et al., 2006). Nucleic bases and nucleosides could be separated on a TEPIC-BACM capillary column under HILIC conditions in 90% aqueous acetonitrile (Hosoya et al., 2006).

Co-polymerization of glycidyl methacrylate and ethylene dimethacrylate with subsequent hydrolysis of the oxirane group yielded a polymeric monolithic DIOL phase suitable for HILIC separations (Xu et al., 2003). A monolithic (poly)hydroxymethacrylate capillary column was employed for HILIC of oligonucleotides (Holdšvendová et al., 2007).

Polymer monolithic columns employed in capillary electrochromatography (CEC) under HILIC conditions at high concentrations of acetonitrile in the working buffer may provide high efficiencies due to the flat flow profile generated by electroosmotic flow (EOF) in the capillaries. Using 2-cyanoethyl acrylate or 3-amino-1-propyl vinyl ether as a component of the polymerization mixture, 0.1 mm I.D. capillary columns containing cyano- or amine-functionalities were prepared and employed for efficient HILIC CEC/MS analysis of bile acids, sugar alcohols (Que et al., 2000) and maltooligosaccharides, which elute in the order of increasing number of monosaccharide units (Que & Novotny, 2002).

Another type of monolithic capillary column for CEC under HILIC conditions was prepared by free radical copolymerization of *N*-acryloxysuccinimide and ethylene dimethacrylate, subsequent grafting of hexyldiamine provides zwitterionic character and positive surface charge, which can be optimized by adjusting the pH of the mobile phase to tune the intensity and direction of the electroosmotic flow (EOF) to suit specific applications (Guerrouache et al., 2010).

Zwitterionic Stationary Phases

Stationary phases with zwitterionic functionalities were originally intended for ion-exchange separations. Irgum's group introduced sufoalkylbetaine stationary phases for HILIC separations of inorganic salts, small organic ionic compounds and proteins (Jiang & Irgum, 1999; Viklund & Irgum, 2000; Jiang & Irgum, 2002; Wikberg et al., 2009). Zwitterionic columns are commercially available under the trademarks ZIC-HILIC (silica gel support) and ZIC-*p*HILIC (polymer support) from SeQuant (Umea, Sweden). The active layer grafted on wide-pore silica gel or polymer support contains both strongly acidic sulfonic acid groups and strongly basic quaternary ammonium groups separated by a short alkyl spacer (Fig. 2.9). The two oppositely charged groups are present in the 1:1 molar ratio, so that there is only a very low net negative surface charge of the bonded layer, attributed to larger distance of the sulphonic groups from the silica gel surface (Guo & Gaiki, 2005).

Sulfobetaine-bonded silica is suitable for separation of a wide range of polar analytes. Polar (hydrogen-bonding and dipole-dipole) interactions in the stationary phase are of primary importance, even though weak electrostatic interactions affect the separation of analytes carrying either positive or negative charges. The chromatographic properties of zwitterionic materials significantly differ from other HILIC phases. Simultaneous separations of anionic and cationic compounds are possible. ZIC-HILIC columns are widely used for separations of small polar compounds (Appelblad & Abrahamsson, 2005), metabolites (Idborg et al., 2005a), glucosinolates (Wade et al., 2007), aminoglycosides (Oertel et al., 2004), peptides (Boersema et al., 2007), glycopeptides (Takegawa et al., 2006) and other compounds.

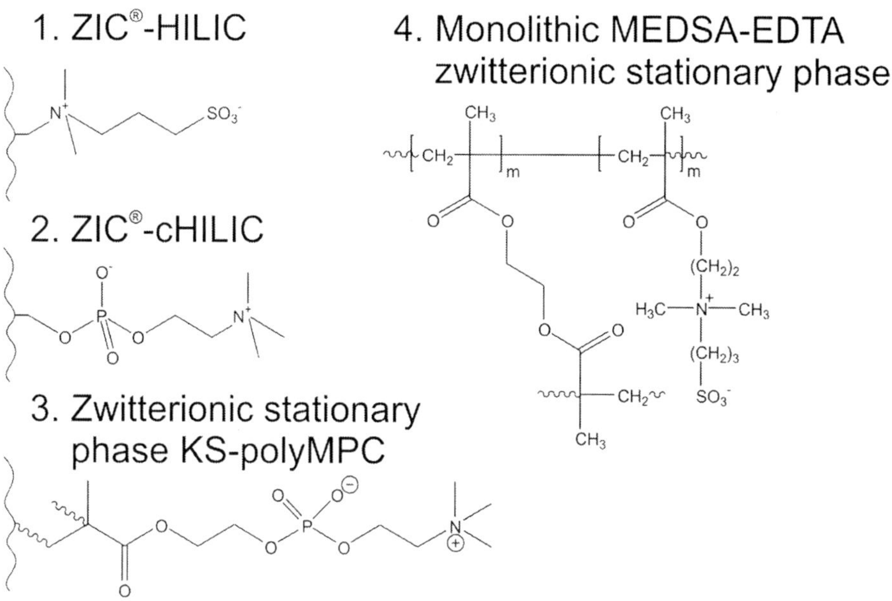

Fig. 2.9. Structures of some zwitterionic HILIC stationary phases.

The HILIC separation of peptides on ZIC-HILIC columns strongly depends on pH. At pH=3, it resembles separations on strong cation exchangers (SCX). At a higher pH (7–8), it shows better chromatographic resolution, especially of prevalent +2 and +3 charged peptides in comparison to SCX separations. The resolution was the best at pH 6.8, but the orthogonality against a C_{18} phase was better at pH 3 (Boersema et al., 2007). HILIC on a capillary zwitterionic ZIC-HILIC column in combination with mass spectrometry using electrospray ionization (ESI-MS) has been applied to the detection and identification of more than 100 *N*-glycopeptides and *O*-glycopeptides in a single run (Takegawa et al., 2008b). Glycopeptides could be separated by their glycan composition on a monolithic silica gel capillary surface-modified with sulfobetaine stationary phase, whereas such separation could not be achieved in RP HPLC (Wohlgemuth et al., 2010).

Purine and pyrimidine bases and nucleosides can be separated using gradient elution with decreasing concentration of acetonitrile in buffered aqueous-organic mobile phase on sulfobetaine ZIC-HILIC or on TSKgel Amide columns. The compounds elute in order of decreasing hydrophobicities, in agreement with generally observed HILIC behavior (Marrubini et al., 2010).

Phosphorylcholine stationary phases for HILIC separations are prepared by graft polymerization of 2-methacryloyloxyethyl phosphorylcholine onto the surface of silica gel support (Fig. 2.9) (Jiang et al., 2006). They are available under the trade names ZIC-cHILIC (SeQuant) and KS-polyMPC (PolyLC) and differ from the sulfobetaine ZIC-HILIC material not only by the nature of the negatively charged group and by the charge arrangement, causing significant differences in the elution order and in the separation selectivity for peptides and other samples, e.g., free amino acid and carboxylic acid ligands and their metal complexes in plant samples (Weber et al., 2008).

Polymer monolithic columns containing sulfoalkylbetaine moiety were prepared by photo-induced (Viklund et al., 2001) or thermal (Jiang et al., 2007) co-polymerization of ethylene dimethacrylate (EDMA) and *N,N*-dimethyl-*N*-methacryloxyethyl-*N*-(3-sulfopropyl)ammonium betaine (MEDSA) inside fused silica capillaries (Fig. 2.9). The zwitterionic capillary monolithic columns originally intended for cation-exchange LC of proteins are suitable for HILIC separations of neutral, basic and acidic polar compounds in aqueous-organic mobile phases with 60% or more acetonitrile. Porosity, permeability, selectivity and retention characteristics of monolithic sulfobetaine columns depend on the concentration of MEDSA and improve when using water-containing porogen solvents in the polymerization process (Urban et al., 2009).

Hybrid particle-monolithic columns with MEDSA-EDMA monolithic phase in the interparticle volume of 3–5 μm C_{18} and aminopropyl-silica packed capillary columns were prepared, showing separation efficiency and selectivity in between purely particle-packed and purely monolithic columns (Jandera et al., 2010b).

Practical Aspects of HILIC Method Development

As usual in the development of other HPLC methods, the experimental factors that can affect sample behavior under HILIC conditions should be adjusted to provide desired separation. Systematic method development was suggested, based on univariate chemometric optimization approach, where the experimental conditions are examined sequentially, one after another (Dejaegher et al., 2008). However, empirical trial-and-error method development is still used more frequently. Optimum mobile phase composition and temperature can be predicted using Eq. (1)–Eq. (5), presented in the HILIC Retention Mechanism: Partition or Adsorption? and the Temperature Effects sections. In the sections, The Column, The Sample Solvent, and The Mobile Phase, several empirical rules and practical hints are presented for selecting the HILIC operation conditions.

The Column

Usually a 100–150 mm long, 4.6 mm I.D. column packed with 3–5 μm particles is selected for standard operation; smaller diameter columns are preferred for HILIC coupled on-line with MS detection. The column length and flow-rate can be changed for increasing resolution or decreasing run time, with pressure as the limiting factor. Working at very high pressures considerably increases the speed of separation, however secondary impacts on HILIC separations can be observed. The retention of sugars in hydrophilic interaction chromatography decreases at high pressures, probably due to the changes in the hydration of strongly polar analytes as they move from the bulk highly organic mobile phase into the water-rich layer at the surface of the polar adsorbent (Neue et al., 2010).

As discussed in the Silica Gel-based Stationary Phases and Organic Polymer Stationary Phases sections, various types of HILIC columns differ in retention, chromatographic selectivity and efficiency. Ikegami et al. (2008) characterized and compared separation properties of the most frequently used types of particle-packed and monolithic polar columns (bare silica, amino-silica, amide-silica, diol-silica, cyano-silica, polysuccinimide-silica, sulfobetaine-silica, triazol-silica and cyclodextrin-silica) for HILIC separations of carboxylic- and amino-acids, peptides, amines, amides and substituted urea derivatives, nucleic bases and nucleosides, poly-alcohols, oligosaccharides and carbohydrates. Some HILIC columns may require lower flow rate for optimum performance. Like in RP HPLC, the efficiency and column back-pressure increase with decreasing particle size; organic polymer columns usually show higher HETP than silica-based columns (Gritti et al., 2010).

Guo and Gaiki (2005) compared the retention of representative carboxylic acids, nucleosides and nucleotides on bare silica and polar silica-bonded amide-, amino-, aspartamide-, and sulfobetaine-stationary phases commonly used in the HILIC mode. Silica gel showed the least retention, but the highest selectivity differences with respect to other polar bonded stationary phases.

In another study, chromatographic behavior of ten model peptides on bare silica, amide, poly-hydroxyethyl aspartamide, diol and zwitterionic HILIC columns was investigated. The Alltima-Alltech silica column showed the best performance for the selected peptide set, followed by the zwitterionic ZIC-HILIC column (Van Dorpe et al., 2010).

Recently, a generic screening protocol was suggested for the selection of column packed with bridged ethylene hybrid (BEH) organic-silica gel 1.7 μm particles, suitable for fast HILIC separations. The selection can be made from among three stationary phases, namely bonded amide-, diol- and cyanophenyl-polar ligands, which differ from one another and from un-derivatized BEH HILIC and is based on the selectivity correlations obtained in the experiments with 28 polar test probes at weakly acidic (pH = 3) and at weakly basic (pH = 9) mobile phases (Fountain et al., 2010).

The Sample Solvent

In contrast to RP HPLC, samples that are dissolved in pure organic solvents usually can be directly injected into a HILIC system because organic solvents are weak eluents under HILIC conditions, so that polar analytes are accumulated in a narrow zone on the top of the column (sample on-column focusing). This simplifies the analysis of basic drugs in plasma samples, as plasma proteins are precipitated in acetonitrile solutions and matrix interferences are removed. The acetonitrile supernatant can be directly analyzed at HILIC conditions, as extract evaporation and reconstitution are usually unnecessary.

On the other hand, direct injections of aqueous samples into highly organic HILIC systems may cause problems because the high elution strength of water may limit the partitioning of solutes into the stationary phase, easily causing column overload and poor separation with broad, distorted or even split peaks (Ali et al., 2008; Guinebault & Broquaire, 1981; Vukmanic & Chiba, 1989; Layne et al., 2001; Loeser & Drumm, 2006). This problem can sometimes be remedied by using an injection loop only partially filled by aqueous sample solution, while the rest of the loop volume contains a diluting organic solvent (Johnson et al., 2010). Generally, it is recommended to use sample solvents with at least 50% organic solvent to achieve good performance in HILIC systems. Unfortunately, extraction of samples containing polar compounds with organic solvents is often impeded by solubility limitations.

Some polar sample matrix components may be strongly retained under HILIC conditions, interfere with the target sample peaks and (or) gradually deteriorate the column performance. The polar matrix problem often can be overcome by adjusting the pH of the mobile phase, or by sample clean-up on an off-line or on-line trapping column before injection (Jian et al., 2010).

The Mobile Phase

HILIC of polar compounds on polar stationary phases employs aqueous-organic mobile phases, usually containing 60–95% organic solvent (acetonitrile) in water or a buffer (often volatile ammonium formate or ammonium acetate for MS detection and on-line identification). It is recommended to perform initial experiments at a relatively high concentration (e.g., 40%) of aqueous buffer and then increase the concentration of acetonitrile until acceptable sample retention and resolution are achieved. Alternatively, a scouting gradient of decreasing acetonitrile concentration can be used. The optimum composition of mobile phase for particular HILIC separations can be predicted using Eq. (1), Eq. (2), or Eq. (3). Many HILIC separations, especially on bare silica columns, employ a gradient of increasing concentration of water, or, more often, a buffer, usually starting from 95% acetonitrile. To remove strongly polar sample matrix impurities, the gradient can be terminated at a low acetonitrile concentration.

The selection of the organic solvent has a strong effect on the retention and overall separation performance in the HILIC mode. Increasing polarity and ability to participate in proton-donor/proton-acceptor interactions enhance the solvent strength, which generally decreases in the order: water >> methanol > ethanol > 2-propanol > tetrahydrofuran > acetonitrile (Quiming et al., 2007b).

Acetonitrile, which does not show proton-donor interactions, is a strongly preferred organic solvent in most HILIC applications, as mobile phases containing methanol (and to less extent ethanol, 2-propanol and other protic solvents) often provide insufficient sample retention and broad or asymmetrical peak shapes. The poor performance of alcohols as organic components of HILIC mobile phases is attributed to their similarity to water in providing strong hydrogen-bonding interactions. The retention is obviously promoted by increasing differences in selective polar interactions between the highly aqueous liquid layer occluded on the surface of polar adsorbents and the bulk organic-rich mobile phase. These differences are higher with acetonitrile-rich mobile phases than with the mobile phases containing protic organic solvents.

The attempts to replace acetonitrile with a less toxic solvent have not been very successful so far. Acetone has similar polarity as acetonitrile, but shows lower retention under HILIC conditions. Moreover, it cannot be used with UV detection and provides lower intensity MS signals (Fountain et al., 2010). At high concentrations of carbon dioxide added to ethanol-water mobile phases, HILIC separations of nucleic bases and other simple polar compounds on bare silica were reported as similar to those in acetonitrile-water mobile phases (Pereira et al., 2010). However, the routine use of such mobile phases will obviously be subject to serious practical problems.

HILIC mobile phases usually contain a buffer whose pH and ionic strength are usually selected to enhance the sample ionization, retention and separation selectivity, i.e., pH > 7 for acids and pH < 7 for bases (Jandera, 2008; Hao et al., 2008). The retention in HILIC systems with uncharged stationary phases usually increases at rising salt (buffer) concentrations, probably because of enhanced hydrogen-bonding interactions between the analyte and the stationary phase facilitated by including larger proportion of water into the solvated ion population in the bulk mobile phase. On the other hand, the retention of ionic samples on the stationary phases with ionic or ionizable functionalities may decrease at increasing ionic strength as the salt counter-ions displace the retained ionized sample molecules by ion-exchange interactions. That is why the retention of acidic compounds on bonded amino phases decreases at higher salt concentrations, in contrast to other simple polar bonded silica-based stationary phases (Nguyen et al., 2010).

At high ionic strengths, the peak tailing in HILIC systems often improves. Some samples may require buffer concentration 100 mmol/L or higher to achieve acceptable peak shapes. Addition of trifluoroacetic acid (TFA) to the mobile phase may also improve peak shapes of basic compounds by formation of ion associates. However, high ionic strength or TFA additives are not recommended in HILIC/MS applications, as they may suppress sample ionization in the mass spectrometer.

Because of the high elution strength of water under HILIC conditions; even a small increase in the water content in the mobile phase may very significantly decrease the retention and deteriorate the sample resolution. To increase the retention and to improve the separation of weakly retained compounds, the content of water in the mobile phase should be reduced to 5% or even less. Very low water concentrations in the mobile phase sometimes improve HILIC separations of moderately polar samples. Polyoxyethylene alkyl ethers (Jandera et al., 1998) or nonylphenyl ether (PONPE) surfactants could be separated according to the number of repeat oxyethylene (EO) units on an aminopropyl-silica column in acetonitrile–dichloromethane mobile phases containing 0.7–1.8% water. After addition of 0.04 mol/L cetyltributyl ammonium bromide, sulfated and non-sulfated PONPEs could be separated according to the EO distribution in a single run (Jandera & Prokeš, 1996). Poly(oxyethylene)-poly(oxypropylene) co-oligomer (EO-PO) surfactants could be separated according to the distribution of EO units on an aminopropyl silica column in mobile phase comprised of 98.5% acetonitrile, 1.48% dichloromethane and only 0.02% water (Jandera et al., 2000).

In some applications, water can be partly or fully replaced by a weaker polar protic solvent. The retention increases in the order of decreasing polarity and elution strength of the protic organic solvent. For example, the retention of methacrylic acid, cytosine, nortriptyline and nicotinic acid on a BEH HILIC column increased considerably and the separation improved when substituting 5% of water with a second organic solvent (methanol, ethanol, or 2-propanol) in the original buffered mobile phase containing 90% acetonitrile (Grumbach et al., 2008).

Water can be even fully replaced by a polar organic solvent in so-called "non-aqueous HILIC chromatography" (NA-HILIC) employing mixed polar organic mobile phases. There is no water in the mobile phase and the liquid diffuse layer adsorbed on the stationary phase contains instead increased concentration of an organic "protic modifier". NA-HILIC may fill the gap between the non-aqueous NP (containing a non-polar and one or more polar organic solvents) and aqueous-organic HILIC systems. It is potentially useful in the analysis of some poorly soluble oligomers or weakly polar compounds, which may precipitate in water-containing HILIC mobile phases. By selecting the type of the protic solvent added to acetonitrile-ethylene diol, methanol or ethanol (with elution strength decreasing in this order), the retention and separation selectivity of nucleic bases, nucleosides and deoxynucleosides could be adjusted in HILIC on bare silica, DIOL (Luna HILIC), bonded thioglycerol and oxidized thioglycerol polar bonded phases. In the non-aqueous HILIC chromatography of some polar compounds with various organic protic solvents, good validity of van't Hoff Eq. (5) was observed, with the slopes, B_i, of the ln k versus $1/T$ plots indicating an enthalpy-controlled exothermic retention mechanism (Bicker et al., 2008).

Coupling HILIC Separations with Spectroscopic Methods

All general-purpose HPLC detectors, such as UV-absorption, fluorimetric, evaporative light-scattering, etc., can be used with HILIC separations. However, one of the main reasons for increasing popularity of the HILIC mode is its excellent suitability for direct coupling with MS or MS/MS detection. Low retention of polar samples on alkyl silica columns may result in poor resolution and unfavorable matrix effects in RP-HPLC/MS. Even more important is that highly aqueous mobile phases can negatively affect the electrospray (ESI) signal intensity and stability. In acetonitrile-rich mobile phases typical for HILIC separations, spraying conditions improve and enhance the efficiency of desolvation and ionization in the ESI ion source, thus increasing signal sensitivity in comparison to the RP conditions (Brown et al., 2002; Oertel et al., 2004; Tostikov & Fiehn, 2002; Garbis et al., 2001; Weng et al., 2004).

HILIC separations coupled on-line with MS sample identification are ideally suited for the analysis of polar pharmaceuticals and for profiling of endogenous human and animal drug metabolites in physiological fluids. A variety of HILIC stationary phases, especially silica gel columns, were applied for this purpose. Aqueous-organic mobile phases should be buffered with volatile buffers (such as ammonium acetate or formate), to avoid problems with solid deposits and (or) signal suppression of ionization in the MS ion source. A few recent examples illustrate the advantages of coupling HILIC with MS.

In comparison to RP ion pair liquid chromatography, HILIC with mass spectrometric detection provided 80 to 210 times greater sensitivity and lower LODs for the determination of streptomycin and dihydrostreptomycin residues in milk (Gremilogianni et al., 2010). Using ternary mobile phase containing 85% methyltributylether, 13.5% acetonitrile and 1.5% water, non-aqueous extracts of muraglitazar could be directly injected and separated on a silica gel column with MS signal intensity four times higher in comparison to RPLC/MS analysis (Xue et al., 2006). Doxazosin was determined in human plasma on an Atlantis HILIC Silica column with ESI MS/MS detection, which gave anLOQ of 0.2 ng/mL (Ji et al., 2008).

Matrix interferences can seriously affect the results of an LC/MS analysis, as interfering compounds may co-elute with the analytes of interest and suppress the ionization in the ion source and consequently decrease the intensity of the MS signals. Some polar compounds may be strongly retained and interfere with target analytes under HILIC conditions. The best way to compensate for matrix effects is by using internal standards labeled with stable isotopes such as deuterium, which are subject to the same ion suppression or enhancement (Jian et al., 2010). The chromatographic deuterium isotope effect is negligibly small under the HILIC conditions, avoiding thus undesirable doublet peaks often found in high-resolution RP HPLC and facilitating precise comparison of MS signal intensities. This advantage was used in comparative quantitative analysis of glycans and glycopeptides, based on the MS signal intensities (Takegawa et al., 2008a).

Atmospheric pressure chemical ionization (APCI) can also be used with HILIC-HPLC/MS for less polar compounds, if they are stable enough in the ion source. Generally, APCI is less subject to matrix effects than ESI ionization.

HILIC can be coupled to NMR using D_2O instead of water and deuterated acetonitrile. A peak sampling unit used for the transfer of the column effluent into the flow cell of an NMR spectrometer enables using longer signal acquisition times for enhanced signal sensitivity (Godejohann, 2007). Fractionation by gel permeation chromatography followed by HILIC separation on a TSKgel Amide column coupled with MS and NMR spectrometry allowed identification of peptides responsible for the taste properties of aqueous bean extracts (Dunkel et al., 2007).

Sample Structure and Retention in HILIC Systems

Higher number of polar groups in sample molecules enhances the sample retention in HILIC separation systems. This behavior was observed, e.g., for increasing numbers of -OH groups in polyphenolic catechins and flavanols, of monomer sugar units in oligosaccharides, of polar amino acid units in peptides, or of epicatechin units in oligomeric proanthocyanidins, etc. (Yanagida et al., 2007). The retention of estrogen metabolites on amide-, Diol- and cyclodextrin-silica polar phases in the HILIC mode increases proportionally to the number of hydroxy- groups in the molecules, while the peak efficiency decreased (Nguyen et al., 2010). This behavior confirms strong hydrogen bonding effects on the sample retention.

The Linear Solvation Energy Relationships (LSER) model represents a QSRR (Quantitative Structure Retention Relationships) approach describing the structural effects on the retention, assuming additive contributions of the individual selective polar and non-polar interactions to the retention in terms of the logarithm of the retention factor, which is proportional to the energy of sample transition from the mobile to the stationary phase (Park et al., 1988; Abraham & Rosés, 1994; Abraham et al., 1997):

$$\log k = (\log k)_0 + m' V_X/100 + s'\pi_2 + a'\alpha_2 + b'\beta_2 \quad (6)$$

The molecular structural descriptors in Eq. (6) include the molar volume of solvated solute, V_X, and constants describing the dipole-dipole, $\pi_2{}^*$, hydrogen-bonding basicity, β_2, and hydrogen-bonding acidity, α_2, interactions. The equation parameters m', s', a' and b', characterize the response of the separation system (the specific combination of the stationary and the mobile phase) to the selective solute structural descriptors. The contributions of the selective interactions to the retention in a particular separation system can be estimated as the products of the system parameters and the corresponding molecular structural descriptors in Eq. (6). The characteristic parameters of the separation system can be obtained using multi-linear regression of the experimental retention data for a set of suitable representative solutes. They can be used to compare the suitability of various separation systems for separation of particular samples and to predict the retention on the basis of the molecular structural descriptors of analytes which can be found in literature.

Even though the LSER and related QSRR models have been used mainly in reversed-phase systems to characterize and classify various columns on the basis of their hydrophobic and specific polar interactions, they can be applied also for HILIC systems. Quiming et al. used a multivariate QSRR approach to compare the retention of adrenoreceptor agonists and antagonists on silica gel and diol columns in the HILIC mode at pH 3.0, 4.0 and 5.0 (Quiming et al., 2007a; Quiming et al., 2007c). Multiple linear regression showed that the number of hydrogen-bond donors and the shape of the molecules govern the retention of ginsenosides on a poly(vinyl alcohol) (PVA) bonded stationary phase in the HILIC chromatographic mode (Quiming et al., 2007b).

Prediction models taking into account the mobile phase composition were suggested, such as the retention model based on five predictors, including the %ACN in the mobile phase and four structural descriptors, namely the logarithm of the partition coefficient (log D), the number of hydrogen bond donors (HBD), the desolvation energy for octanol (FOct), and the total absolute atomic charge (TAAC), which was applied to predict the retention of adrenoreceptor agonists and antagonists on a polyvinyl alcohol-bonded stationary phase under HILIC conditions (Quiming et al., 2008a). A model with the %ACN as a parameter and the local dipole index (LDI) as the only structural descriptor was used to characterize the HILIC retention of seven ginsenosides on a polyamine-bonded stationary phase (Quiming et al., 2008b).

Eq. (6) was applied to characterize the structural effects on the retention of phenolic acids and flavones on five polar columns, both under HILIC and RP conditions (Jandera et al., 2010a). In the HILIC mode, good validity of the LSER model was found for the ZIC-HILIC, PEG and Diol columns, slightly worse for the Luna HILIC 200 column, but the retention data on the Silica hydride column do not fit the model. The basic properties of the column (positive proton acceptor parameters, a') enhance the retention of phenolic acids on the PEG, Diol, Luna HILIC 200 and zwitterionic ZIC HILIC columns. The proton acceptor contributions to the retention of flavones, a', on the PEG column (low surface concentration of the -OH groups) are more positive than on the Luna HILIC column, whereas the proton-donor parameters b' are negative on the PEG column, but positively contribute to the retention on the diol column and on the ZIC HILIC column (probably due to the dual charge character of the zwitterionic stationary phase). The Silica

Hydride column shows negative proton-acceptor contributions (a'), but positive proton-donor contributions (b') to the retention of phenolic acids. These results illustrate significant differences in polar selectivities for various polar columns employed in HILIC systems. Increasing concentration of acetonitrile in the mobile phase affects more significantly the system parameters of Eq. (6) for the PEG, Diol and Luna HILIC 200 columns than for the sulfobetaine ZIC HILIC column.

Combined HILIC-RP Retention Mechanism in Aqueous-organic Mobile Phases

On the stationary phases containing both polar groups and non-polar hydrocarbon moieties, combined reversed-phase and normal-phase (HILIC) behavior may contribute to the retention to different degrees, depending on the sample, stationary and mobile phases. This behavior was observed for crown ethers and polar peptides on alkyl silica stationary phases as early as almost 30 years ago (Nahum & Horvath, 1981). Generally, the RP behavior on dual-mode stationary phases predominates in water-rich mobile phases and the retention decreases at rising concentration of the organic solvent until it drops to a minimum value. Further increasing the concentration of the organic solvent causes the polar interactions to become stronger than the solvophobic ones and HILIC retention is promoted in highly organic mobile phases. Fig. 2.10 shows an example of the mobile phase effects on the elution order of uracil, thiourea, phenol and toluene on a monolithic sulfobetaine zwitterionic polymethacrylate monolithic capillary column, resulting in gradual transition between the predominating HILIC and RP mechanisms. In 40% acetonitrile, the elution order follows typical RP behavior, with polar uracil eluting first and non-polar toluene last. In 80 and 90% acetonitrile, toluene and phenol elute before uracil and thiourea under HILIC conditions, whereas in 50% acetonitrile both HILIC and RP mechanism affect the retention simultaneously (Urban et al., 2009).

Assuming additivity of the RP, k_{RP}, and of the HILIC, k_{HILIC}, contributions to the retention, Eq. (1) can be combined with Eq. (2), to describe the effects of the mobile phase in the presence of a dual HILIC/RP retention mechanism (Urban et al., 2009):

$$\log k = \log k_{RP} + \log k_{HILIC} = a_1 + m_{RP}\,\varphi(H_2O) - m_{HILIC} \log \varphi(H_2O) \qquad (7)$$

$\varphi(H_2O)$ in Eq. (7) is the concentration of water (or of aqueous buffer) in the mobile phase. The parameter m_{RP} characterizes the effect of increasing concentration of water on the RP contribution to the retention, whereas the parameter m_{HILIC} is a measure of the effect of water concentration on the HILIC retention; a_1 is the system constant. The parameters a_1, m_{RP} and m_{HILIC} of Eq. (7) can be determined by non-linear regression of the experimental

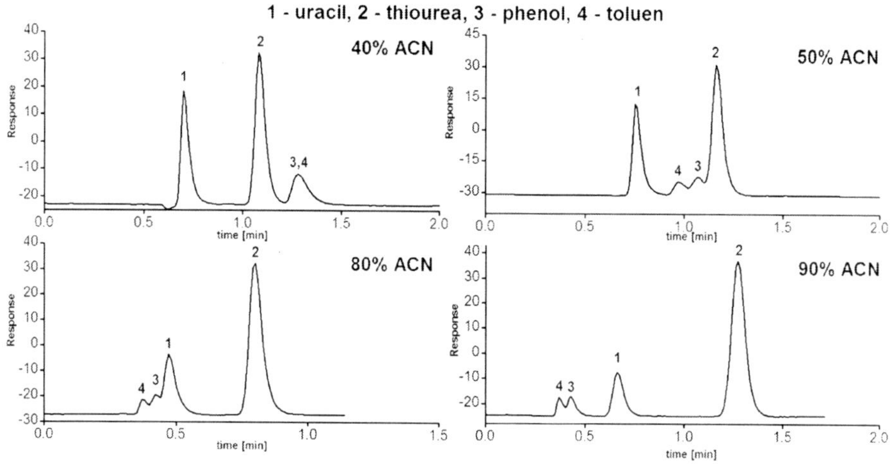

Fig. 2.10. Effects of the sample polarity and % acetonitrile in the mobile phase on the RP and HILIC contributions to the retention on a sulfobetaine polymethacrylate monolithic MEDSA-EDMA capillary column, 191x0.32 mm i.d. Flow rate 30 μL/min, UV detection at 220 nm. (Redrawn on the basis of the data in: Urban et al. (2009).

retention factors measured at varying volume fractions of water in the mobile phase. From Eq. (7), the "U-turn" volume fraction of water (or an aqueous buffer), φ_{min}, corresponding to the minimum retention and to the transition from the predominating RP to the HILIC mechanism can be predicted as (Urban et al., 2009):

$$\varphi_{min} = 0.434\, m_{HILIC} / m_{RP} \tag{8}$$

Using Eq. (3) to describe more accurately the HILIC contribution to the retention in the high organic solvent concentration range and combine it with Eq. (2), we obtain a four-parameter Eq. (9), describing the U-shape retention plots (Jandera &. Hájek, 2009):

$$\log k = a_2 + m_{RP} \cdot \varphi(H_2O) - m_{HILIC} \cdot \log\,[1 + b \cdot \varphi(H_2O)] \tag{9}$$

The parameter b (usually $b > 1$) is the correction term for the HILIC retention in mobile phases with very low concentrations of water.

Many samples show U-shape plots of k *versus* the concentration of the organic solvent over the full composition range of a binary aqueous-organic mobile phase on a variety of polar columns (Pesek & Matyska, 2007). U-shape profiles of the retention *versus* the volume fraction of aqueous buffer, corresponding to the dual HILIC-RP retention mechanism were observed for phenolic acids and flavones on a PEG column with bonded (poly)oxyethylene groups, a Diol column with bonded glycerol groups and a Luna HILIC 200A column, containing a cross-linked diol stationary phase, stabilized by oxyethylene bridges. The transition between the HILIC and the RP behavior is observed at lower volume fractions of aqueous buffer on the less polar PEG column (in 70–90% acetonitrile), in comparison to more polar Diol and Luna HILIC columns, which contain hydroxy groups in the bonded ligands and show similar separation selectivity in the HILIC mode (in 40–70% acetonitrile). Fig. 2.11 shows the U-shape plot for myricetin in aqueous-organic mobile phases containing 2–98% acetonitrile in 0.01 M acetate buffer, pH = 3.0, on the Luna HILIC 200A column. The experimental retention data (points) show markedly better fit to Eq. (9) (full lines) than to Eq. (7) (broken lines) (Jandera & Hájek, 2009).

Monolithic sulfobetaine polymethacrylate capillary columns provide a dual HILIC/RP retention mechanism for phenolic compounds (Urban et al., 2009), like columns on the basis of hydrosilated silica. Hydrophobic modification of the hydrosilated silica column surface enhances the retention in the RP mode, which increases from the bare silica hydride to cholesterol-bonded and to bidentate C_{18}-modified hydrosilated silica columns, as illustrates Fig. 2.12.

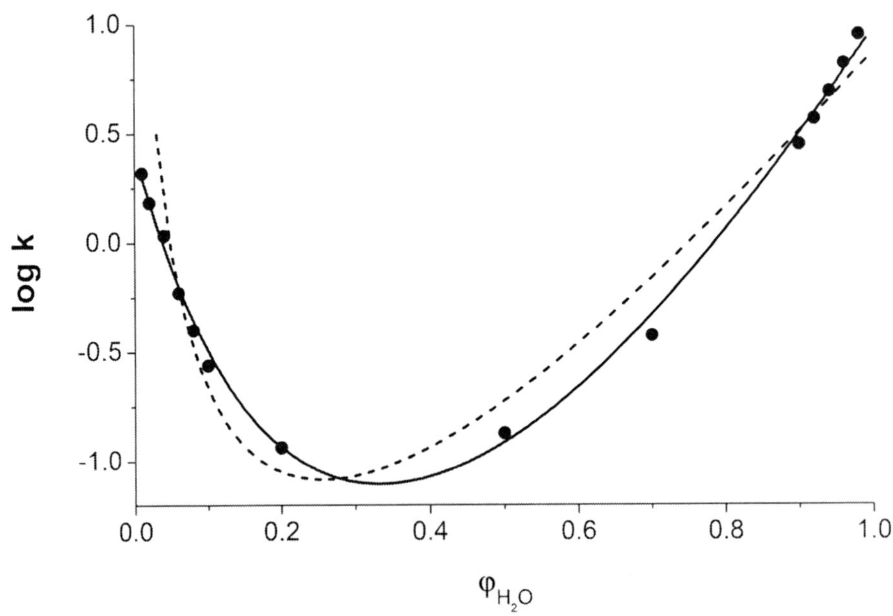

Fig. 2.11. Effects of the volume fraction of aqueous buffer, φ_{H2O}, (10 mmol/L ammonium acetate, pH 3), on the retention factors, *k*, of myricetine on a Luna HILIC 200 column over the full range of aqueous acetonitrile mobile phase. Temperature 40°C, Points – experimental data, Full line – best fit plot of Eq. (9), dotted line – best fit plot of Eq. (7). (Redrawn on the basis of the data in: Jandera & Hájek (2009).

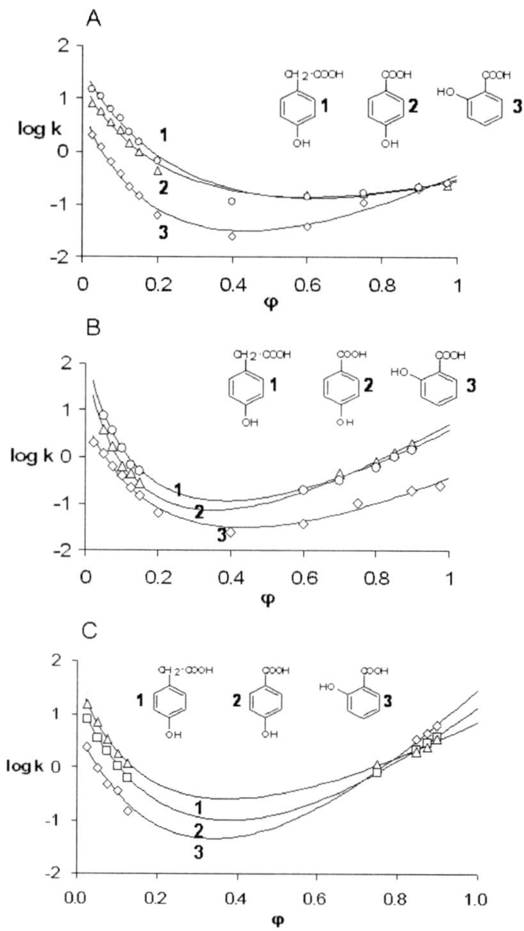

Fig. 2.12. Effects of the type of hydrosilated column on HILIC and RP contributions to the retention factors, *k*, of phenolic acids over the full range of aqueous acetonitrile mobile phases. φ, - volume fraction of aqueous buffer (10 mmol/L ammonium acetate, pH = 3.26). 1. 4-hydroxyphenylacetic acid, 2. 4-hydroxybenzoic acid, 3. salicylic acid. Columns: A – Silica hydride, 75 × 4.6 mm i.d.; B – UDC cholesterol, 75 × 4.6 mm i.d., C - C_{18} bidentate, 50 × 2.1 mm i.d. (J. Soukup, P. Jandera, unpublished results).

(The structures of the stationary phases are shown in Fig. 2.5.) The figure also shows good fit of the Eq. (9) lines to the experimental data points, whereas the three-parameter Eq. (7) did not fit well the experimental data (J. Soukup, P. Jandera, unpublished results).

A Diol column in aqueous acetonitrile with 10 mmol/L ammonium chloride as the mobile phase was used for the quantitative analysis of polar active pharmaceutical ingredients in drug formulations in the HILIC separation mode, providing results complementary to gradient reversed-phase chromatography on a C18 column (Wang et al., 2005).

The stationary phase with undecyl-1,2-diol ligands bonded on silica gel shows a dual RP/HILIC retention mechanism because of the presence of a long non-polar alkyl and a polar diol group and can be operated both in the low organic solvent RP LC mode and in the high organic solvent HILIC mode. It was applied, e.g., for the determination of dual structural distribution in ethoxylated alcohol surfactants, which are separated according to the alkyl lengths in the RP mode and according to the number of oxyethylene groups at HILIC conditions in acetonitrile-rich mobile phases. The transition between the HILIC and the RP modes was observed in mobile phases with 75% acetonitrile (Liu & Pohl, 2008).

Mixed-mode reversed phase/weak anion-exchangers (RP/WAX) prepared by chemical modification of thiol-modified silica, or the commercial Acclaim Mixed-Mode WAX-1 material (Fig. 2.7), have lower hydrophobic selectivity in the RP mode than Synergi Fusion RP, Obelisc R A, Primesep B2 A, or Ultisphere MM3 polar embedded RP columns. In the HILIC mode, the RP/WAX phases differ from TSKgel Amide-80, ZIC-HILIC, or polysulfoethyl A columns, to which they provide a certain degree of complementary application possibilities (Lammerhofer et al., 2008).

As the separation selectivities in the HILIC and in the RP mode are more or less complementary, a sample can be injected in subsequent experiments at alternating HILIC and RP conditions. Subsequent gradients of increasing (RP) and decreasing (HILIC) concentration of acetonitrile in an acetate buffer were employed for separation of natural antioxidants in tea and other beverages on a Luna HILIC column in subsequent alternating HILIC and RP gradient runs (Jandera et al., 2010a).

HILIC in Two-dimensional Separations

To increase the number of compounds resolved in a complex sample, HILIC may be combined with reversed-phase or other LC modes in two-dimensional (2D) LC-LC separation systems, either in on-line or off-line setups. In the "heart-cut" 2D, only a few fractions from the first-dimension column are selected for separation on a second-dimension column, whereas in comprehensive 2D LCxLC chromatographic techniques, the whole effluent from the first-dimension column is stepwise transferred to the second-dimension column, so that all sample compounds are subject to the separation in the two dimensions, without loss in resolution (Schoenmakers et al., 2003). When a 2D system is designed to combine complementary structural effects on the retention (so-called orthogonal systems), the number of compounds separated in a single run considerably increase. Two-dimensional combinations of HILIC and RP systems may represent a very efficient way to this aim (Jandera, 2008).

However, when combining RP and HILIC separations in on-line comprehensive 2D systems, serious mobile phase compatibility problems may arise due to the differences in the elution strength in the two modes. Mobile phases with high concentrations of acetonitrile employed under HILIC conditions usually are too strong eluents in RP systems, whereas mobile phases more rich in water usual in RP HPLC are too strong HILIC eluents. If the mobile phase from the first dimension is used for fraction transfer to the second dimension, either in HILICxRP or in RPxHILIC systems, the retention of early eluting compounds may significantly decrease, un-symmetrical or even split peaks may appear, with detrimental effects on the separation. Hence, off-line HILICxRP or RPxHILIC 2D combinations are still more frequent than comprehensive on-line methods. Off-line heart-cutting approach was successfully applied in plasma metabolite profiling. Two-step plasma solvent extraction yielded a highly polar fraction, which was analyzed on a BEH HILIC column, whereas the other, less polar fraction on a C_{18} BEH column under RP conditions (Cai et al., 2009).

Several solutions were suggested to solve the elution strength compatibility problem:

1) The first-dimension mobile phase contained in small transferred fraction volumes (e.g., 2 μL) often can be dispersed in the second dimension without affecting significantly the separation, especially when using a larger-diameter column. Low fraction volumes, however, may negatively affect the sensitivity of two-dimensional separations.
2) Strong dilution of the first-dimension fractions with a weak solvent decreases the elution strength before introduction into the second dimension, but requires more complex instrumentation with an additional pump delivering the diluting solvent.
3) Fractions from the first dimension can be trapped on a suitable high-retention small column, while the solvent is directed to waste. In the second step, the trapped analytes are released and back-flushed onto the second column in a small volume of a stronger mobile phase used in the second dimension. This approach increases the second-dimension operation time and it is not always easy to find a suitable trapping column.

Recently, solvent evaporation combined with low-volume injection was applied for two-dimensional separation of phenolic compounds in green tea. In the first dimension, a Diol column was operated at HILIC conditions, using a ternary gradient of water and methanol in acetonitrile containing acetic acid. 50 μL fractions were collected each in 1-min and then evaporated under nitrogen to 2 μL, before injection onto a second dimension C_{18} column (Kallili & de Villiers, 2010).

A combination of the fraction dilution and trapping approach was reported for simultaneous separations of hydrophilic and hydrophobic compounds on a silica gel HILIC column used in the first dimension, connected to a C_{18} column in the second-dimension *via* two switching valves and another short C_{18} fraction trapping column. First, acetonitrile-rich mobile phase flushed non-polar sample solutes from the HILIC column to a small mixer chamber, where a highly aqueous transfer solvent was added to the HILIC effluent. Then the valve was switched and the content of the mixer was flushed onto a trapping column. At the same time, polar compounds were separated on the first-dimension HILIC column. When the second valve was switched, the trapped less-polar compounds were back-flushed from the trapping column into the second-dimension C_{18} column and separated using gradient of methanol in water, while the separation on the HILIC column continued. This method was applied to the analysis of a traditional Chinese medicine (TCM) extract (Wang et al., 2008c).

The fraction dilution and trapping approach was also employed for online HILIC-SPE-RP separation of tryptic digest peptides. An RP solid-phase enrichment column was connected to the first dimension HILIC column, to allow direct injection of large volumes of aqueous samples with high salt concentrations. Fractions from the first-dimension ZIC-HILIC column were diluted with water and trapped subsequently each on a separate RP column cartridge. The trapped analytes were subsequently transferred to a polymeric PLRP column for second-dimension separation with UV/MS detection (Wilson et al., 2007).

Strong cation-exchange (SCX) chromatography combined with reversed-phase LC is widely employed for peptide analysis. Because of high correlations between the peptide charge and the HILIC retention on bare silica gel, which is due to ionic interactions of peptides with charged silanol groups, the HILIC mode resembles to some extent the SCX mode. Hence, HILIC can substitute SCX in the first dimension for 2D peptide separations (Boersema et al., 2007). The data obtained in the HILIC and in the RP modes coupled to MS or to MS/MS can be used for monitoring hydrophilicity/hydrophobicity variations induced by substitution of single amino-acids in peptides (Hodges et al., 2004).

High degree of orthogonality was found between the retention of almost 200 peptides carrying 1–5 positive charges on a silica gel column in aqueous-organic mobile phase at pH = 4.5 and on a C_{18} column operated at pH = 2.6 (Gilar et al., 2005). Combination of the data acquired on a sulfobetaine zwitterionic column with reversed-phase chromatographic data enabled identification of more than 1000 proteins after the analysis of complex peptide mixtures in cellular nuclear lysates (Boersema et al., 2007).

HILIC provides valuable complementary tool to RP HPLC also for metabolite profiling using LC/MS/MS techniques. An Atlantis HILIC silica column (Wang et al., 2008a), or a Diamond hydride column (Callahan et al., 2009) in the HILIC mode can be coupled to the RP separation mode on a C_{18} column. On-line 2D LC/MS (ESI) with a ZIC-HILIC column in the first dimension and a C_{18} column in the second dimension allowed metabolic fingerprinting of rat urine samples for the diagnosis of particulate diseases, based on the presence of selected biomarkers (Idborg et al., 2005a; Idborg et al., 2005b). Early diagnosis of kidney cancer was possible by detecting the biomarkers in urine using a combination of an array of hyphenated analytical techniques: GC/MS (TOF), reversed-phase LC/MS and HILIC LC/MS on a bonded amine column, with electrospray ionization and a linear ion trap analyzer (Kind et al., 2007).

HILIC systems were incorporated in two-dimensional and three-dimensional separation methods used in glycoproteomic research for elucidation of the structures of oligosaccharides. A HILIC amide column can be used for the separation of derivatized (pyridylaminated) oligosaccharides according to the number of monosaccharide units, whereas a C_{18} column provides reversed-phase separation according to the "functionality type distribution" of oligosaccharides with the same elution times on the HILIC column. The structure of an unknown oligosaccharide can be determined from the coordinates in the ODS–HILIC two-dimensional retention plane, calibrated by oligosaccharide standards (Takahashi et al., 1993). Later, a third dimension was added to characterize sialyl *N*-linked oligosaccharides enzymatically released from glycopeptides (Takahashi, 1996). Neutral, mono-, di-, tri and tetra-sialoyl oligosaccharides were separated on a weak anion-exchange diethylaminoethyl (DEAE) column in the first dimension; the individual first-dimension fractions were subjected to the separation of a C_{18} column in the second dimension and, after evaporating the mobile phase, each second-dimension fraction was injected onto a HILIC column for separation in the third dimension. 2D maps could be constructed for each sialylated fraction and included into a three-dimensional oligosaccharide database.

On-line combinations of HILIC and ion-exchange systems or of two different HILIC systems usually are free of compatibility problems characteristic for direct coupling of HILIC and RP systems. An on-line two-dimensional system was developed for combined anion-exchange (DEAE column) and HILIC mode (Amide-80, or ZIC-HILIC sulfobetaine column) serial separations of 2-pyridylamino derivatives of neutral and mono- to tetrasialylated human serum protein *N*-glycans (PA-*N*-glycans), using combined fluorimetric and mass spectrometric detection. Using a ternary gradient of simultaneously decreasing concentration of acetonitrile and increasing concentration of a salt (ammonium acetate) on the two columns connected in series, the neutral PA-*N*-glycans and mono-, di-, tri-, and tetrasialylated PA-*N*-glycans elute in turn according to the number of sialic acids (Deguchi et al., 2008).

A HILIC-HILIC 2D system combining a TSKgel Amide-80 column in the first dimension and a short Poly(hydroxyethyl) A column in the second dimension, connected to a high-speed Q-TOF mass spectrometer, was employed for comprehensive two-dimensional separation of saponins in plant extracts. Gradients of decreasing concentration of acetonitrile in mobile phases buffered with formic acid and ammonium acetate - acetic acid were used in both dimensions. The two HILIC columns show relatively high differences in selectivity, so that some pairs of saponin isomers, differing only in the disaccharide type at the terminal position, were well separated in the 2D HILIC-HILIC system. The identification was based on the MS/MS spectra and the retention time coordinates in the 2D contour chromatogram. (Wang et al., 2008b).

Two-dimensional separations combining HILIC and RP systems often offer a high degree of orthogonality for separations of samples of synthetic products containing structural constituents differing in polarities. (Poly)oxyethylene alkyl or alkylaryl ethers, used as industrial or household surfactants, show dual distribution of the alkyl lengths and of the number of oxyethylene (EO) units. The dual monomer distribution could be characterized using a comprehensive HILICxRPLC system. A silica gel column and a gradient of increasing concentration of water in acetonitrile was used in the first dimension for the determination of the EO distribution and a short 3 µm C_{18} column with aqueous methanol for rapid isocratic determination of the alkyl length distribution in the second dimension (Murphy et al., 1998).

The distribution of EO and PO monomer units in oxyethylene–oxypropylene (EO-PO) (co)oligomers affects the properties of these surfactants. Using a highly orthogonal comprehensive 2D RPxHILIC system, the sample was separated according to the distribution of less polar PO units in the first, RP, dimension with a C_{18} column and a gradient of acetonitrile in water. In the second, (almost) non-aqueous HILIC, dimension with an aminopropyl silica column and mobile phase comprised of acetonitrile, dichloromethane and 0.023% water, the first-dimension fractions were separated according to the number of more polar EO units (Fig. 2.13) (Jandera et al., 2006).

These few examples illustrate so far largely unexplored potentials of two-dimensional LC-LC separations with implemented HILIC systems.

Concluding Remarks

The last decade has witnessed steeply growing interest in the HILIC techniques, because of increasing demands on the separation of samples containing strongly polar compounds, which are difficult to achieve in other HPLC modes. HILIC provides valuable information complementary to those obtained using established RP methods, simplifies sample treatment by directly removing some matrix interferences and offers high sensitivity and selectivity when directly coupled with mass spectrometry.

Bare silica has probably been the most frequently used stationary phase for HILIC separations, however the list of useful HILIC stationary phases for separation of various applications is rapidly expanding, including new silica-

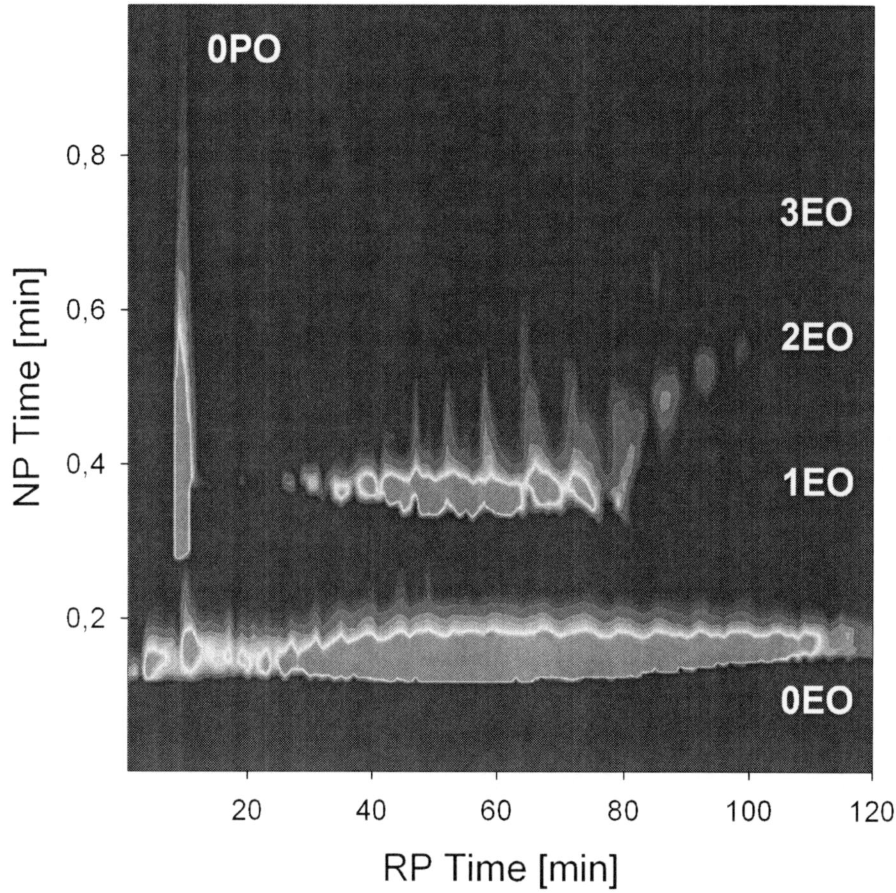

Fig. 2.13. Comprehensive 2D RPxHILIC separation of an EO-PO (co)oligomer sample (Slovanik). Contour plot: x-axis - RP retention times, y-axis - NP (HILIC) retention times. Conditions: First dimension: RP column: Zorbax SB-C18, 5 μm, 150x1 mm I.D, gradient: 50-100% acetonitrile in water in 120 min, 10 μL/min. Second dimension: APS column: Polaris NH_2 column, 3 μm, 50x1 mm I.D., mobile phase: dichloromethane - ethanol - water 98.5 : 1.48 : 0.02, 0.5 mL/min Sample: 1 μL 3% Slovanik in acetonitrile, switching valve cycle frequency 1 min, ELSD detection. (Reprinted from: Jandera et al., 2006, with permission by Elsevier).

bonded or organic-polymer polar stationary phases. So far amine-, amide-, polysuccinimide-, diol-, zwitterionic-, and mixed-mode WAX-HILIC materials have proven their utility for various HILIC applications. HILIC columns with hydrosilated, monolithic or hybrid supports, sub-2 μm fully-porous or fused-core superficially porous particles have become available for fast, efficient and selective HILIC separations.

Mixed HILIC/ion-exchange mode systems are useful for separations of ionic samples. On most HILIC columns, reversed-phase interactions may play also a more or less important role, increasing as the concentration of the organic solvent decreases. Many polar columns can be operated either in the HILIC, or in the RP mode, extending thus the information on sample that can be obtained on a single column. Combinations of HILIC and (not only) reversed-phase systems show great potentials for the analysis of complex samples containing peptides, drugs, metabolites, saccharides, and other types of natural or synthetic compounds. The development of on-line comprehensive RPxHILIC systems is still hampered by the mobile phase incompatibility. This problem has not been satisfactorily solved and requires further research.

Acetonitrile is the commonly used organic component of the mobile phase offering the best selectivity in the HILIC mode, probably because of selective polar interactions largely differing from water, the other component of the mobile phase. Non-aqueous HILIC separation systems with mobile phases comprised of acetonitrile and an alcohol as a polar protic solvent may extend HILIC applications to less polar samples, which are too weakly retained at even low concentrations of water. Ternary aqueous-organic mobile phases show so far largely unexplored possibilities for further development of new selective HILIC methods.

In conclusion, HILIC separations have found a stable place among the family of HPLC techniques. In the future, we can expect further development of new HILIC materials and applications for various types of polar samples and for simultaneous separations of ionic, strongly and weakly polar and even non-polar sample components in a single run.

References

Abraham, M.H.; Rosés, M. Hydrogen bonding. 38. Effect of solute structure and mobile phase composition on reversed-phase high-performance liquid chromatographic capacity factors. J. Phys. Org. Chem. 1994, 7, 672-684.

Abraham, M.H.; Rosés, M.; Poole, C.F.; Poole, S.K. Hydrogen bonding. 42. Characterization of reversed-phase high-performance liquid chromatographic C18 stationary phases. J. Phys. Org. Chem. 1997, 10, 358-368.

Ali, M.S.; Rafiuddin, S.; Al-Jawi, D.A.; Al-Hetari, Y; Ghori, U.M.H.; Khatri A.J. Stability-indicating assay of sodium cromoglicate in ophthalmic solution using mixed-mode hydrophilic interaction chromatography. J. Sep. Sci. 2008, 31, 1645-1650.

Alpert, A.J. Cation-exchange high-performance liquid chromatography of proteins on poly(aspartic acid)-silica. J. Chromatogr. A 1983, 266, 23-37.

Alpert, A.J.; Andrews, P.C. Cation-exchange chromatography of peptides on poly(2-sulfoethyl aspartamide)-silica. J. Chromatogr. 1988, 443, 85-96.

Alpert, A.J. Hydrophilic-interaction chromatography for the separation of peptides, nucleic acids and other polar compounds. J. Chromatogr. 1990, 449, 177-196.

Alpert, A.J. Electrostatic repulsion hydrophilic interaction chromatography for isocratic separation of charged solutes and selective isolation of phosphopeptides. Anal. Chem. 2008, 80, 62-76.

Alpert, A.J.; Shukla, M.; Shukla, A.K.; Zieske, L.R.; Yuen, S.W.; Ferguson M.A.J.; Mehlert, A.; Pauly, M.; Orlando, R. Hydrophilic-interaction chromatography of complex carbohydrates. J. Chromatogr. A 1994, 676, 191-202.

Appelblad, P.; Abrahamsson, P. MS detection of homocysteine, methyl-malonic acid, and succinic acid using HILIC separation on a covalently bonded zwitterionic stationary phase. LCGC N. Am. Suppl. S. 2005, 24-25.

Appelblad, P.; Jonsson, T.; Jiang, W.; Irgum, K. Fast hydrophilic interaction liquid chromatographic separations on bonded zwitterionic stationary phase. J. Sep. Sci. 2008, 31, 1529-1536.

Armstrong, D.W.; Jin, H.L. Evaluation of the liquid chromatographic separation of monosaccharides, disaccharides, trisaccharides, tetrasaccharides, deoxysaccharides and sugar alcohols with stable cyclodextrin bonded phase columns. J. Chromatogr. 1989, 462, 219-232.

Aturki, Z.; D'Orazio, G.; Rocco, A.; Si-Ahmed, K.; Fanali S. Investigation of polar stationary phases for the separation of sympathomimetic drugs with nano-liquid chromatography in hydrophilic interaction liquid chromatography mode. Anal. Chim. Acta. 2011, 685, 103-110.

Baba, Y. Computer-assisted retention prediction for high-performance liquid chromatography in the ion-exchange mode. J. Chromatogr. 1989, 485, 143-168.

Berthod, A.; Chang, S.S.C.; Kullman, J.P.S.; Armstrong, D.W. Practice and mechanism of HPLC oligosaccharide separation with a cyclodextrin bonded phase. Talanta 1998, 47, 1001-1012.

Bicker, W.; Wu, J.Y.; Lammerhofer, M.; Lindner, W. Hydrophilic interaction chromatography in nonaqueous elution mode for separation of hydrophilic analytes on silica-based packings with noncharged polar bondings. J. Sep. Sci. 2008, 31, 2971-2987.

Blahová, E.; Jandera, P.; Cacciola, F.; Mondello, L. Two-dimensional and serial column reversed-phase separation of phenolic antioxidants on octadecyl-, polyethyleneglycol-, and pentafluorophenylpropyl-silica columns. J. Sep. Sci. 2006, 29, 555-566.

Boersema, P.J.; Divecha, N.; Heck, A.J.R., Mohammed, S. Evaluation and optimization of ZIC-HILIC-RP as an alternative MudPIT strategy. J. Proteome Res. 2007, 6, 937-946.

Brown, S.D.; White, C.A.; Bartlett, M.G. Hydrophilic interaction liquid chromatography/electrospray mass spectrometry determination of acyclovir in pregnant rat plasma and tissues. Rapid Commun. Mass Spectrom. 2002, 16, 1871-1876.

Cai, X.; Zou, L.; Dong, J.; Zhao, L.; Wang, Y.; Xu, Q.; Xue, X.; Zhang, X.; Liang, X. Analysis of highly polar metabolites in human plasma by ultra-performance hydrophilic interaction liquid chromatography coupled with quadrupole-time of flight mass spectrometry. Anal. Chim. Acta 2009, 650, 10-15.

Callahan, D.L.;. Souza, D.; Bacic, A.; Rossner, U. Profiling of polar metabolites in biological extracts using diamond hydride-based aqueous normal phase chromatography. J. Sep. Sci. 2009, 32, 2273-2280.

Chambers, T.K.; Fritz, J.S. Effect of polystyrene-divinylbenzene resin sulfonation on solute retention in high-performance liquid chromatography. J. Chromatogr. A 1998, 797, 139-147.

Chauve, B.; Guillarme, D.; Cléon, P.; Veuthey, J.-L. Evaluation of various HILIC materials for the fast separation of polar compounds. J. Sep. Sci. 2010, 33, 752-764.

Churms, S.C. Recent progress in carbohydrate separation by high-performance liquid chromatography based on hydrophilic interaction. J. Chromatogr. A 1996, 720, 75-91.

Courtois, J.; Byström, E.; Irgum, K. Novel monolithic materials using poly(ethylene glycol) as porogen for protein separation. Polymer 2006, 47, 2603-2611.

Daunoravicius, Z.; Juknaite, I.; Naujalis, E., Padarauskas, A. Simple and rapid determination of denaturants in alcohol formulations by hydrophilic interaction chromatography. Chromatographia 2006, 63, 373-377.

Deguchi, K.; Keira, T.; Yamada, K.; Ito, H.; Takegawa, Y.; Nakagawa, H.; Nishimura, S.I. Two-dimensional hydrophilic interaction chromatography coupling anion-exchange and hydrophilic interaction columns for separation of 2-pyridylamino derivatives of neutral and sialylated N-glycans. J. Chromatogr. A 2008, 1189, 169-174.

Dejaegher, B.; Mangelins, D.; Vander Heyden, Y. Method development for HILIC assays. J. Sep. Sci. 2008, 31, 1438-1448.

Dejaegher, B.; Vander Heyden, Y. HILIC methods in pharmaceutical analysis. J. Sep. Sci. 2010, 33, 698-715.

de Person, M.; Hazotte, A.; Elfakir, C.; Lafosse, M. Development and validation of a hydrophilic interaction chromatography-mass spectrometry assay for taurine and methionine in matrices rich in carbohydrates. J. Chromatogr. A 2005, 1081, 174-181.

Dos Santos Pereira, A.; David, F.; Vanhoenacker, G.; Sandra, P. The acetonitrile shortage: Is reversed HILIC with water an alternative for the analysis of highly polar ionizable solutes? J. Sep. Sci. 2009, 32, 2001-2007.

Dunkel, A. Köster, J.; Hofmann, T. Molecular and sensory characterization of γ-glutamyl peptides as key contributors to the kokumi taste of edible beans (Phaseolus vulgaris L.). J. Agric. Food Chem. 2007, 55, 6712-6719.

Feste, A.S.; Khan, I. Separation of glucooligosaccharides and polysaccharide hydrolysates by gradient elution hydrophilic interaction chromatography with pulsed amperometric detection. J. Chromatogr. 1993, 630, 129-139.

Fountain, K.J.; Xu, J.; Diehl, D.M.; Morrison, D. Influence of stationary phase chemistry and mobile-phase composition on retention, selectivity, and MS response in hydrophilic interaction chromatography. J. Sep. Sci. 2010, 33, 740-751.

Garbis, S.D.; Melse-Boonstra, A.; West, C.E.; van Breemen, R.B. Determination of folates in human plasma using hydrophilic interaction chromatography - Tandem mass spectrometry. Anal. Chem. 2001, 73, 5358-5364.

Gika, H.G.; Theodoridis, G.A.; Wilson, I.D. Hydrophilic interaction and reversed-phase ultra-performance liquid chromatography TOF-MS for metabonomic analysis of Zucker rat urine. J. Sep. Sci. 2008, 31, 1598-1608.

Gilar, M.; Olivova, P.; Daly, A.E.; Gebler, J.C. Orthogonality of separation in two-dimensional liquid chromatography. Anal. Chem. 2005, 77, 6426-6434.

Godejohann, M. Hydrophilic interaction chromatography coupled to nuclear magnetic resonance spectroscopy and mass spectroscopy-A new approach for the separation and identification of extremely polar analytes in bodyfluids. J. Chromatogr. A 2007, 1156, 87-93.

Gremilogianni, A.M.; Megoulas, N.C.; Koupparis, M.A. Hydrophilic interaction vs ion pair liquid chromatography for the determination of streptomycin and dihydrostreptomycin residues in milk based on mass spectrometric detection. J. Chromatogr. A 2010, 1217, 6646-6651.

Gritti, F.; dos Santos Pereira, A.; Sandra, P.; Guiochon, G. Efficiency of the same neat silica column in hydrophilic interaction chromatography and per aqueous liquid chromatography. J. Chromatogr. A 2010, 1217, 683-688.

Grumbach, E.S.; Diehl, D.M.; Neue, U.D. The application of novel 1.7 μm ethylene bridged hybrid particles for hydrophilic interaction chromatography. J. Sep. Sci. 2008, 31, 1511-1518.

Grumbach, E.S.; Wagrowski-Diehl, D.M.; Mazzeo, J.R.; Alden, B.; Iraneta, P.C. Hydrophilic interaction chromatography using silica columns for the retention of polar analytes and enhanced ESI-MS sensitivity. LCGC N. Am. 2004, 22, 1010-1023.

Guinebault, P.; Broquaire, M. Large-volume injection of samples dissolved in a noneluting solvent. Application to the determination of antipyrine using normal-phase high-performance liquid chromatography. J. Chromatogr. 1981, 217, 509-522.

Guerrouache, M.; Pantazaki, A.; Millot, M.-C.; Carbonnier, B. Zwitterionic polymeric monoliths for HILIC/ RP mixed mode for CEC separation applications. J. Sep. Sci. 2010, 33, 787-792.

Guo, Y.; Gaiki, S. Retention behavior of small polar compounds on polar stationary phases in hydrophilic interaction chromatography. J. Chromatogr. A 2005, 1074, 71-90.

Guo, Y.; Srinivasan, S.; Gaiki, S. Investigating the effect of chromatographic conditions on retention of organic acids in hydrophilic interaction chromatography using a design of experiment. Chromatographia 2007, 66, 223-229.

Guo, Z.; Lei, A.; Zhang, Y.; Xu, Q.; Xue, X; Zhang, F.; Liang, X. „Click saccharides": Novel separation materials for hydrophilic interaction liquid chromatography. Chem. Commun. 2007, 2491-2493.

Hartmann, E.; Chen, Y.; Mant, C.T.; Jungbauer, A.; Hodges, R.S. Comparison of reversed-phase liquid chromatography and hydrophilic interaction/cation-exchange chromatography for the separation of amphipathic α-helical peptides with L- and D-amino acid substitutions in the hydrophilic face. J. Chromatogr. A 2003, 1009, 61-71.

Hao, Z.; Xiao, B.; Weng, N. Impact of column temperature and mobile phase components on selectivity of hydrophilic interaction chromatography (HILIC). J. Sep. Sci. 2008, 31, 1449-1464.

Hemström, P.; Irgum, K. Hydrophilic interaction chromatography. J. Sep. Sci. 2006, 29, 1784-1821.

Holdšvendová, P.; Suchánková, J.; Bunček, M.; Bačkovská, V. Hydroxymethyl methacrylate-based monolithic columns designed for separation of oligonucleotides in hydrophilic-interaction capillary liquid chromatography. J. Biochem. Biophys. Methods 2007, 70, 23-29.

Hodges, R.S.; Chen, Y.; Kopecky, E.; Mant, C.T. Monitoring the hydrophilicity/hydrophobicity of amino acid side-chains in the non-polar and polar faces of amphipathic α-helices by reversed-phase and hydrophilic interaction/cation-exchange chromatography. J. Chromatogr. A 2004, 1053, 161-172.

Horie, K.; Ikegami, T.; Hosoya, K.; Saad, N.; Fiehn, O.; Tanaka, N. Highly efficient monolithic silica capillary columns modified with poly(acrylic acid) for hydrophilic interaction chromatography. J. Chromatogr. 2007, 1164, 198-205.

Hosoya, K.; Hira, N.; Yamamoto, K.; Nishimura, M.; Tanaka, N. High-performance polymer-based monolithic capillary column. Anal. Chem. 2006, 78, 5729-5735.

Hsieh, Y. Potential of HILIC-MS in quantitative bioanalysis of drugs and drug metabolites. J. Sep. Sci. 2008, 31, 1481-1491.

Hsieh, Y.; Galviz, G.; Long, B. Ultra-performance hydrophilic interaction liquid chromatography/tandem mass spectrometry for the determination of everolimus in mouse plasma. Rapid Com. Mass Spectrom. 2009, 23, 1461-1466.

Huber, J.F.K; Pawlowska, M.; Markl, P. Solvent-generated liquid-liquid chromatography with aqueous ternary systems. Chromatographia 1984, 19, 19-28.

Ibrahim, M.E.A.; Zhou, T.; Lucy, C.A. Agglomerated silica monolithic column for hydrophilic interaction LC. J. Sep. Sci. 33, 2010, 773-778.

Idborg, H.; Zamani, L.; Edlund, P.-O.; Schuppe-Koistinen, I.; Jacobsson, S.P. Metabolic fingerprinting of rat urine by LC/MS: Part 1. Analysis by hydrophilic interaction liquid chromatography-electrospray ionization mass spectrometry. J. Chromatogr. B 2005a, 828, 9-13.

Idborg, H.; Zamani, L.; Edlund, P.-O.; Schuppe-Koistinen, I.; Jacobsson, S.P. Metabolic fingerprinting of rat urine by LC/MS: Part 2. Data pretreatment methods for handling of complex data. J. Chromatogr. B 2005b, 828, 14-20.

Ikegami, T.; Fujita, H.; Horie, K.; Hosoya, K.; Tanaka, N. HILIC mode separation of polar compounds by monolithic silica capillary columns coated with polyacrylamide. Anal. Bioanal. Chem. 2006, 386, 578-585.

Ikegami, T.; Horie, K.; Jaafar, J.; Hosoya, K.; Tanaka, N. Preparation of highly efficient monolithic silica capillary columns for the separations in weak cation-exchange and HILIC modes. J. Biochem. Biophys. Methods 2007a, 70, 31-37.

Ikegami, T.; Ichimaru, J.; Kajiwara, W.; Nagasawa, N.; Hosoya, K.; Tanaka, N. Anion- and cation-exchange microHPLC utilizing poly(methacrylates)-coated monolithic silica capillary columns. Anal. Sci. 2007b, 23, 109-113.

Ikegami, T.; Tomomatsu, K.; Takubo, H.; Horie, K.; Tanaka, N. Separation efficiencies in hydrophilic interaction chromatography. J. Chronatogr. A 2008, 1184, 474-503.

Iwasaki, Y.; Ishii, Y.; Ito, R.; Saito, K.; Nakazawa, H. New approaches for analysis of metabolism compounds in hydrophilic interaction chromatography. J. Liq. Chromatogr. & Rel. Technol. 2007, 30, 2117-2126.

Jandera, P. Stationary phases for hydrophilic interaction chromatography, their characterization and implementation into multidimensional chromatography concepts. J. Sep. Sci. 2008, 31, 1421-1437.

Jandera, P.; Churáček, J. Gradient elution in liquid chromatography. I. The influence of the composition of the mobile phase on the capacity ratio (retention volume, band width, and resolution) in isocratic elution - theoretical considerations. J. Chromatogr. 1974a, 91, 207-221.

Jandera, P.; Churáček, J. Ion exchange chromatography of aldehydes, ketones, ethers, alcohols, polyols and saccharides. J. Chromatogr. 1974b, 98, 55-104.

Jandera, P.; Churáček, J.; Svoboda, L. Gradient elution in liquid chromatography. X. Retention characteristics in reversed-phase gradient elution chromatography. J. Chromatogr. 1979, 174, 35-50.

Jandera, P.; Fischer, J.; Lahovská, H.; Novotná, K.; Česla, P.; Kolářová, L. Two-dimensional liquid chromatography normal-phase and reversed-phase separation of (co)Oligomers. J. Chromatogr. A 2006, 1119, 3-10.

Jandera, P.; Hájek, T. Utilization of dual retention mechanism on columns with bonded PEG and diol stationary phases for adjusting the separation selectivity of phenolic and flavone natural antioxidants. J. Sep. Sci. 2009, 32, 3603-3619.

Jandera, P.; Hájek, T.; Škeříková, V.; Soukup, J. Dual hydrophilic interaction-RP retention mechanism on polar columns: Structural correlations and implementation for 2-D separations on a single column. J. Sep. Sci. 2010a, 33, 841-852.

Jandera, P.; Holčapek, M.; Theodoridis, G. Investigation of chromatographic behaviour of ethoxylated alcohol surfactants in normal-phase and reversed-phase systems using high-performance liquid chromatography-mass spectrometry. J. Chromatogr. A 1998, 813, 299-311.

Jandera, P.; Holčapek, M.;. Kolářová, L. Retention mechanism, isocratic and gradient-elution separation and characterization of (co)polymers in normal-phase and reversed-phase high-performance liquid chromatography. J. Chromatogr. A 2000, 869 65-84.

Jandera, P.; Holčapek, M.; Kolářová, L. Retention Behavior of Oligomers and Cooligomers in Reversed-phase and in Normal-phase Interactive Liquid Chromatographic Systems. Int. J. Polym. Anal. Charact. 2001, 6, 261-294.

Jandera, P.; Janderová, M.; Churáček, J. Gradient elution in liquid chromatography. VIII. Selection of the optimal composition of the mobile phase in liquid chromatography under isocratic conditions. J. Chromatogr. 1978, 148, 79-97.

Jandera, P.; Kučerová, M. Prediction of retention in gradient-elution normal-phase high performance liquid chromatography with binary solvent gradients. J. Chromatogr. A 1997, 759, 13-25.

Jandera, P.; Prokeš, B. Chromatographic behaviour of un-sulphated and sulphated oligoethyleneglycol nonylphenyl ether surfactants in normal-phase liquid chromatographic systems containing water. Chromatographia 1996, 42, 539-546.

Jandera, P.; Urban, J.; Škeříková, V.; Langmaier, P.; Kubíčková, R.; Planeta, J. Polymethacrylate monolithic and hybrid particle-monolithic columns for reversed-phase and hydrophilic interaction capillary liquid chromatography. J. Chromatogr. A 2010b, 1217, 22-33.

Jandera, P.; Vyňuchalová, K.; Hájek, T.; Česla, P.; Vohralík, G. Characterization of HPLC columns for two-dimensional LC × LC separations of phenolic acids and flavonoids. J. Chemometrics 2008, 22, 203-217.

Ji, H.Y.; Park, E.J.; Lee, K.C.; Lee, H.S. Quantification of doxazosin in human plasma using hydrophilic interaction liquid chromatography with tandem mass spectrometry. J. Sep. Sci. 2008, 31, 1628-1633.

Jian, W.; Edom, R.W.; Xu, Y.; Weng, N. Recent advances in application of hydrophilic interaction chromatography for quantitative bioanalysis. J. Sep. Sci 2010, 33, 681-697.

Jiang, W.; Fischer, G.; Girmay, Y.; Irgum, K. Zwitterionic stationary phase with covalently bonded phosphorylcholine type polymer grafts and its applicability to separation of peptides in the hydrophilic interaction liquid chromatography mode. J. Chromatogr. A 2006, 1127, 82-91.

Jiang, W.; Irgum, K. Covalently bonded polymeric zwitterionic stationary phase for simultaneous separation of inorganic cations and anions. Anal. Chem. 1999, 71, 333-344.

Jiang, W.; Irgum, K. Tentacle-type zwitterionic stationary phase prepared by surface-initiated graft polymerization of 3-[N,N-dimethyl-N-(methacryloyloxyethyl)-ammonium] propanesulfonate through peroxide groups tethered on porous silica. Anal. Chem. 2002, 74, 4682-4687.

Jiang, Z.; Smith, N.W.; Ferguson, P.D.; Taylor, M.R. Hydrophilic interaction chromatography using methacrylate-based monolithic capillary column for the separation of polar analytes. Anal. Chem. 2007, 79, 1243-1250.

Johnson, J.R.; Karlsson, D.; Dalene, M.; Skarping, G. Determination of aromatic amines in aqueous extracts of polyurethane foam using hydrophilic interaction liquid chromatography and mass spectrometry. Anal. Chim. Acta 2010, 678, 117-123.

Kallili, K.M.; de Villiers, A. Off-line comprehensive two-dimensional hydrophilic interaction × reversed phase liquid chromatographic analysis of green tea phenolics. J. Sep. Sci. 2010, 33, 853-863.

Karlsson, G.; Winge, S.; Sandberg, H. Separation of monosaccharides by hydrophilic interaction chromatography with evaporative light scattering detection. J. Chromatogr. A 2005, 1092, 246-249.

Kind, T.; Tolstikov, V.; Fiehn, O.; Weiss, R.H. A comprehensive urinary metabolomic approach for identifying kidney cancer. Anal. Biochem. 2007, 363, 185-195.

Lammerhofer, M.; Richter, M.; Wu, J.; Nogueira, R.; Bicker, W.; Lindner, W. Mixed-mode ion-exchangers and their comparative chromatographic characterization in reversed-phase and hydrophilic interaction chromatography elution modes. J. Sep. Sci. 2008, 31, 2572-2588.

Layne, J.; Farcas, Y.; Rustamov, I.; Ahmed, F. Volume-load capacity in fast-gradient liquid chromatography: Effect of sample solvent composition and injection volume on chromatographic performance. J. Chromatogr. A 2001, 913, 233-242.

Li, R.; Huang, J.X. Chromatographic behavior of epirubicin and its analogues on high-purity silica in hydrophilic interaction chromatography. J. Chromatogr. A 2004, 1041, 163-169.

Lin, S.C.; Lee, W.C. Separation of a fructo-oligosaccharide mixture by hydrophilic interaction chromatography using silica-based micropellicular sorbents. J. Chromatogr. A 1998, 803, 302-306.

Linden, J.C.; Lawhead, C.L. Liquid chromatography of saccharides. J. Chromatogr. 1975, 105, 125-133.

Liu, X.D.; Pohl, C. New hydrophilic interaction/reversed-phase mixed-mode stationary phase and its application for analysis of nonionic ethoxylated surfactants. J. Chromatogr. A 2008, 1191, 83-89.

Liu, X.; Pohl, C.A. HILIC behavior of a reversed-phase/cationexchange/ anion-exchange trimode column. J. Sep. Sci. 2010, 33, 779-786.

Liu, Y.; Xue, X.; Guo, Z.; Xu, Q.; Zhang, F.; Liang, X. Novel two-dimensional reversed-phase liquid chromatography/hydrophilic interaction chromatography, an excellent orthogonal system for practical analysis. J. Chromatogr. A 2008, 1208, 133-140.

Loeser, E.; Drumm, P. Using strong injection solvents with 100% aqueous mobile phase in RP-LC. J. Sep. Sci. 2006, 29, 2847-2852.

Makino, Y.; Omichi, K.; Hase, S. Analysis of oligosaccharide structures from the reducing end terminal by combining partial acid hydrolysis and a two-dimensional sugar map. Anal. Biochem. 1998, 264, 172-179.

Mant, C.T.; Hodges, R.S. Mixed-mode hydrophilic interaction/cation-exchange chromatography: Separation of complex mixtures of peptides of varying charge and hydrophobicity. J. Sep. Sci. 2008, 31, 1573-1584.

Marclay, F.; Saugy, M. Determination of nicotine and nicotine metabolites in urine by hydrophilic interaction chromatography-tandem mass spectrometry: Potential use of smokeless tobacco products by ice hockey players. J. Chromatogr. A 2010, 1217, 7528-7538.

Marrubini, G.; Castillo-Mendoza, B.E.; Massolini, G. Separation of purine and pyrimidine bases and nucleosides by hydrophilic interaction chromatography. J. Sep. Sci. 2010, 33, 803-816.

Matyska, M.T.; Pesek, J.J.; Duley, J.; Zamzami, M.; Fischer, S.M. Aqueous normal phase retention of nucleotides on silica hydride-based columns: Method development strategies for analytes revelant in clinical analysis. J. Sep. Sci. 2010, 33, 930-938.

McCalley, D.V. Is hydrophilic interaction chromatography with silica columns a viable alternative to reversed-phase liquid chromatography for the analysis of ionisable compounds? J. Chromatogr. A, 2007, 1171, 46-55.

McCalley, D.V. Evaluation of the properties of a superficially porous silica stationary phase in hydrophilic interaction chromatography. J. Chromatogr. A 2008, 1193, 85-91.

Melander, W.; Campbell, D.E.; Horváth, C. Enthalpy—entropy compensation in reversed-phase chromatography. J. Chromatogr. 1978, 158, 215-225.

Melander, W.; Horvath, C. Reversed-Phase Chromatography. in: High-Performance Liquid Chromatography, 2, C. Horvath, Ed.; Academic Press: New York, 1980, 114-320.

Molíková, M.; Jandera, P. Characterization of stationary phases for reversed-phase chromatography. J. Sep. Sci. 2010, 33, 453-463.

Moni, L.; Ciogli, A.; Acquarica, I.D.; Dondoni, A.; Gasparrini, F.; Marra, A. Synthesis of sugar-based silica gels by copper-catalysed azide-alkyne cycloaddition via a single-step azido-activated silica intermediate and the use of the gels in hydrophilic interaction chromatography. Chem. Eur. J. 2010, 16, 5712-5722.

Murphy, R.E.; Schure, M.R.; Foley, J.P. One- and Two-Dimensional Chromatographic Analysis of Alcohol Ethoxylates. Anal. Chem. 1998, 70, 4353-4360.

Nahum, A.;. Horvath, C. Surface silanols in silica-bonded hydrocarbonaceous stationary phases. I. Dual retention mechanism in reversed-phase chromatography. J. Chromatogr. A 1981, 203, 53-63.

Naidong, W. Bioanalytical liquid chromatography tandem mass spectrometry methods on underivatized silica columns with aqueous/organic mobile phases. J. Chromatogr. B 2003, 796, 209-224.

Naidong, W.; Shou W.; Chen, Y.L.; Jiang, X.Y. Novel liquid chromatographic-tandem mass spectrometric methods using silica columns and aqueous-organic mobile phases for quantitative analysis of polar ionic analytes in biological fluids. J. Chromatogr. B 2001, 754, 387-399.

Naidong, W.; Zhou, W.; Song, Q.; Zhou, S. Direct injection of 96-well organic extracts onto a hydrophilic interaction chromatography/tandem mass spectrometry system using a silica stationary phase and an aqueous/organic mobile phase. Rapid Commun. Mass Spectrom. 2004, 18, 2963-2968.

Nawrocki, J. The silanol group and its role in liquid chromatography. J. Chromatogr. A 1997, 779, 29-71.

Neue, U.D.; Hudalla, C.J.; Iraneta, P.C. Influence of pressure on the retention of sugars in hydrophilic interaction chromatography. J. Sep. Sci. 2010, 33, 838-840.

Nguyen, H.P.; Schug, K.A. The advantages of ESI-MS detection in conjunction with HILIC mode separations: Fundamentals and applications. J. Sep. Sci. 2008, 31, 1465-1480.

Nguyen, H.P.; Yang, S.H.; Wigginton, J.G.; Simpkins, J.W.; Schug, K.A. Retention behavior of estrogen metabolites on hydrophilic interaction chromatography stationary phases. J. Sep. Sci. 2010, 33, 793-802.

Nischang, I.; Bruggemann, O. On the separation of small molecules by means of nano-liquid chromatography with methacrylate-based macroporous polymer monoliths. J. Chromatogr. A 2010, 1217, 5389-5397.

Nováková, L.; Kaufmannová, I.; Jánská, R., Evaluation of hybrid hydrophilic interaction chromatography stationary phases for ultra- HPLC in analysis of polar pteridines. J. Sep. Sci. 2010, 33, 765-772.

Oertel, R.; Neumeister, V.; Kirch, W. Hydrophilic interaction chromatography combined with tandem-mass spectrometry to determine six aminoglycosides in serum. J. Chromatogr. 2004, 1058, 197-201.

Olsen, B.A. Hydrophilic interaction chromatography using amino and silica columns for the determination of polar pharmaceuticals and impurities. J. Chromatogr. A 2001, 913, 113-122.

Oyler, A.R.; Armstrong, B.L.; Cha, J.Y.; Zhou, M.X.; Yang, Q.; Robinson, R.I.; Dunphy, R.; Burinsky, D.J. Hydrophilic interaction chromatography on amino-silica phases complements reversed-phase high-performance liquid chromatography and capillary electrophoresis for peptide analysis. J. Chromatogr. A 1996, 724, 378-383.

Pack, B.W.; Risley, D.S. Evaluation of a monolithic silica column operated in the hydrophilic interaction chromatography mode with evaporative light scattering detection for the separation and detection of counter-ions. J. Chromatogr. A 2005, 1073, 269-275.

Park, J.H.; Carr, P.W.; Abraham, M.H.; Taft, R.W.; Doherty, R.M.; Kamlet, M.J. Some observations regarding different retention properties of HPLC stationary phases. Chromatographia 1988, 25, 373-381.

Pazourek, J. Monitoring of mutarotation of monosaccharides by hydrophilic interaction chromatography. J. Sep. Sci. 2010, 33, 974-981.

Pereira, A.S.; Girón, A.J.; Admasu, E.; Sandra, P. Green hydrophilic interaction chromatography using ethanol-water-carbon dioxide mixtures. J. Sep. Sci. 2010, 33, 834-837.

Persson, J.; Hemström, P.; Irgum, K. Preparation of a sorbitol methacrylate grafted silica as stationary phase for hydrophilic interaction chromatography. J. Sep. Sci. 2008, 31, 1504-1510.

Pesek, J.J.; Matyska, M.T.; Larrabee, S. HPLC retention behavior on hydride-based stationary phases. J. Sep. Sci. 2007, 30, 637-647.

Pisano, R.; Breda, M.; Grassi, S.; James, C.A. Hydrophilic interaction liquid chromatography-APCI-mass spectrometry determination of 5-fluorouracil in plasma and tissues. J. Pharm. Biomed. Anal. 2005, 38, 738-745.

Puy, G.; Demesmay, C.; Rocca, J.L.; Iapichella, J.; Galarneau, A.; Brunel, D. Electrochromatographic behavior of silica monolithic capillaries of different skeleton sizes synthesized with a simplified and shortened sol-gel procedure. Electrophoresis 2006, 27, 3971-3980.

Que, A.H.; Konse, T.; Baker, A.G.; Novotny, M.V. Analysis of bile acids and their conjugates by capillary electrochromatography/electrospray ion trap mass spectrometry. Anal. Chem. 2000, 72, 2703-2710.

Que, A.H.; Novotny, M.V. Separation of neutral saccharide mixtures with capillary electrochromatography using hydrophilic monolithic columns. Anal. Chem. 2002, 74, 5184-5191.

Quiming, N.S.; Denola, N.L.; Saito, Y.; Jinno, K. Retention prediction of adrenoreceptor agonists and antagonists on unmodified silica phase in hydrophilic interaction chromatography. Anal. Bioanal. Chem., 2007a, 388, 1693-1706.

Quiming, N.S.; Denola, N.,L.; Soliev, A.B.; Saito, Y.; Jinno, K. Retention behavior of ginsenosides on a poly(vinyl alcohol)-bonded stationary phase in hydrophilic interaction chromatography. Anal. Bioanal. Chem., 2007b, 389, 1477-1488.

Quiming, N.S.; Denola, N.L.; Ueta, I.; Saito, Y.; Tatematsu, S.; Jinno, K. Retention prediction of adrenoreceptor agonists and antagonists on a diol column in hydrophilic interaction chromatography. Anal. Chim. Acta 2007c, 598, 41-50.

Quiming, N.S.; Denola, N.L.; Bin Samsuri, S.R.; Saito, Y.; Jinno, K. Development of retention prediction models for adrenoreceptor agonists and antagonists on a polyvinyl alcohol-bonded stationary phase in hydrophilic interaction chromatography. J. Sep. Sci. 2008a, 31, 1537-1549.

Quiming, N.S.; Denola, N.L.; Saito, Y.; Jinno, K. Multiple linear regression and artificial neural network retention prediction models for ginsenosides on a polyamine-bonded stationary phase in hydrophilic interaction chromatography. J. Sep. Sci. 2008b, 31, 1550-1563.

Regnier, F.E.; Noel, R. Glycerolpropylsilane Bonded Phases in the Steric Exclusion Chromatography of Biological Macromolecules. J. Chromatogr. Sci. 1976, 14, 316-320.

Risley, D.S.; Strege, M.A. Chiral separations of polar compounds by hydrophilic interaction chromatography with evaporative light scattering detection. Anal. Chem. 2000, 72, 1736-1739.

Samuelson, O.; Swenson, B. Partition chromatography on ion-exchange resins separation of sugars. Anal. Chim. Acta 1963, 28, 426-432.

Sandoval, J.E.; Pesek, J.J. Synthesis and characterization of a hydride-modified porous silica material as an intermediate in the preparation of chemically bonded chromatographic stationary phases. Anal. Chem. 1989, 61, 2067-2075.

Santoyo-Gonzalez, F.; Hernandez-Mateo, F. Silica-based clicked hybrid glyco materials. Chem. Soc. Rev. 2009, 38, 3449-3462.

Schoenmakers, P.; Fitzpatrick, F.; Grothey, R. Predicting the behaviour of polydisperse polymers in liquid chromatography under isocratic and gradient conditions. J. Chromatogr. A 2002, 965, 93-107.

Schoenmakers, P.J.; Mariott, P.; Beens, J. Nomenclature and conventions in comprehensive multidimensional chromatography. LCGC Eur. 2003, 16, 335-339.

Škeříková, V.; Jandera, P. Effects of the operation parameters on Hydrophilic Interaction Liquid Chromatography separation of phenolic acids on zwitterionic monolithic capillary columns. J. Chromatogr. A 2010, 1217, 7981-7989.

Siouffi, A.M. Silica gel-based monoliths prepared by the sol-gel method: Facts and figures. J. Chromatogr. A 2003, 1000, 801-818.

Snyder, L.R. Principles of Adsorption Chromatography, Marcel Dekker New York, 1968.

Snyder, L.R. Role of the solvent in liquid-solid chromatography. Review. Anal. Chem. 1974, 46, 1384-1393.

Snyder, L.R.; Dolan, J.W.; Gant, J.R. Gradient elution in high-performance liquid chromatography : I. Theoretical basis for reversed-phase systems. J. Chromatogr. 1979, 165, 3-30.

Snyder, L.R.; Kirkland, J.R.; Dolan, J.W. Introduction to Modern Liquid Chromatography, 3rd ed., Wiley: Hoboken, NJ, 2009.

Snyder, L.R.; Poppe, H. Mechanism of solute retention in liquid—solid chromatography and the role of the mobile phase in affecting separation : Competition versus "sorption". J. Chromatogr. 1980, 184, 363-413.

Soczewiński, E. Solvent composition effects in thin-layer chromatography systems of the type silica gel-electron donor solvent. Anal. Chem. 1969, 41, 179-182.

Spagou, K.; Tsoukali, H.; Raikos, N.; Gika, H.; Wilson, I.D.; Theodoridis, G. Hydrophilic interaction chromatography coupled to MS for metabonomic/metabolomic studies. J. Sep. Sci. 2010, 33, 716-727.

Strege, M.A. Hydrophilic Interaction Chromatography-Electrospray Mass Spectrometry Analysis of Polar Compounds for Natural Product Drug Discovery. Anal. Chem. 1998, 70, 2439-2445.

Strege, M.A.; Stevenson, S.; Lawrence, S.M. Mixed-mode anion-cation exchange/hydrophilic interaction liquid chromatography-electrospray mass spectrometry as an alternative to reversed phase for small molecule drug discovery. Anal. Chem. 2000, 72, 4629-4633.

Takahashi, N. Three-dimensional mapping of N-linked oligosaccharides using anion-exchange, hydrophobic and hydrophilic interaction modes of high-performance liquid chromatography. J. Chromatogr. A 1996, 720, 217-225.

Takahashi, N.; Wada, Y.; Awaya, J.; Kurono, M.; Tomiya, N. Two-dimensional elution map of GalNAc-containing N-linked oligosaccharides. Anal. Biochem. 1993, 208, 96-109.

Takegawa, Y.; Deguchi, K.; Keira, T.; Ito, H.; Nakagawa, H.; Nishimura, S. Separation of isomeric 2-aminopyridine derivatized N-glycans and N-glycopeptides of human serum immunoglobulin G by using a zwitterionic type of hydrophilic-interaction chromatography. J. Chromatogr. A 2006, 1113, 177-181.

Takegawa, Y.; Hato, M.; Deguchi, K.; Nakagawa, H.; Nishimura, S.-I. Chromatographic deuterium isotope effects of derivatized N-glycans and N-glycopeptides in a zwitterionic type of hydrophilic interaction chromatography. J. Sep. Sci. 2008a, 31, 1594-1597.

Takegawa, Y.; Ito, H.; Keira, T.; Deguchi, K.; Nakagawa, H.; Nishimura, S.-I. Profiling of N- and O-glycopeptides of erythropoietin by capillary zwitterionic type of hydrophilic interaction chromatography/electrospray ionization mass spectrometry. J. Sep. Sci., 2008b, 31, 1585-1593.

Tiselius, A.; Ark. Kem. Min. Geol. 1948, 26B.

Tolstikov, V.V.; Fiehn, O. Analysis of highly polar compounds of plant origin: Combination of hydrophilic interaction chromatography and electrospray ion trap mass spectrometry. Anal. Biochem. 2002, 301, 298-307.

Urban, J.; Škeříková, V.; Jandera, P.; Kubíčková, R.; Pospíšilová, M. Preparation and characterization of polymethacrylate monolithic capillary columns with dual hydrophilic interaction reversed-phase retention mechanism for polar compounds. J. Sep. Sci. 2009, 32, 2530-2543.

Van Dorpe, S.; Vergote, V.; Pezeshki, A.; Burvenich, C.; Peremans, K.;. De Spiegeleer, B. Hydrophilic interaction LC of peptides: Columns comparison and clustering. J. Sep. Sci., 2010, 33, 728-739.

Viklund, C.; Irgum, K. Synthesis of porous zwitterionic sulfobetaine monoliths and characterization of their interaction with proteins. Macromolecules 2000, 33, 2539-2544.

Viklund, C.; Sjorgen, A.; Irgum, K.;. Nes, I. Chromatographic interactions between proteins and sulfoalkylbetaine-based zwitterionic copolymers in fully aqueous low-salt buffers. Anal. Chem. 2001, 73, 444-452.

Vukmanic, D.; Chiba, M. Effect of organic solvents in sample solutions and injection volumes on chromatographic peak profiles of analytes in reversed-phase high-performance liquid chromatography. J. Chromatogr. 1989, 483, 189-196.

Wade, K.L.; Garard, I.J.; Fahey, J.W. Improved hydrophilic interaction chromatography method for the identification and quantification of glucosinolates. J. Chromatogr. A 2007, 1154, 469-472.

Wang, X.; Li, W.; Rasmussen, H.T. Orthogonal method development using hydrophilic interaction chromatography and reversed-phase high-performance liquid chromatography for the determination of pharmaceuticals and impurities. J. Chromatogr. A 2005, 1083, 58-62.

Wang, Y.; Lehmann, R.; Lu, X.; Zhao, X.; Xu, G. Novel, fully automatic hydrophilic interaction/reversed-phase column-switching high-performance liquid chromatographic system for the complementary analysis of polar and apolar compounds in complex samples. J. Chromatogr. A 2008a, 1204, 28-34.

Wang, Y.; Lu, X.; Xu, G. Development of a comprehensive two-dimensional hydrophilic interaction chromatography/quadrupole time-of-flight mass spectrometry system and its application in separation and identification of saponins from Quillaja saponaria. J. Chromatogr. A 2008b, 1181, 51-59.

Wang, Y.; Lu, X,; Xu, G. Simultaneous separation of hydrophilic and hydrophobic compounds by using an online HILIC-RPLC system with two detectors. J. Sep. Sci. 2008c, 31, 1564-1572.

Weber, G.; von Wiren, N.; Hayen, H.; Hydrophilic interaction chromatography of small metal species in plants using sulfobetaine- and phosphorylcholine-type zwitterionic stationary phases. J. Sep. Sci. 2008, 31, 1615-1622.

Wilson, S.R.; Jankowski, M.; Pepaj, M.; Mihailova, A.; Boix, F.; Truyols, G.V.; Lundanes, E.; Greibrokk, T. 2D LC separation and determination of bradykinin in rat muscle tissue dialysate with on-line SPE-HILIC-SPE-RP-MS. Chromatographia 2007, 66, 469-474.

Wikberg, E.; Verhage, J.J.; Viklund, C.; Irgum, K. Grafting of silica with sulfobetaine polymers via aqueous reversible addition fragmentation chain transfer polymerization and its use as a stationary phase in HILIC. J. Sep. Sci. 2009, 32, 2008-2016.

Wohlgemuth, J.; Karas, M.; Jiang, W.; Hendriks, R.; Andrecht, S. Enhanced glycoprofiling by specific glycopeptide enrichment and complementary monolithic nano-LC (ZIC-HILIC/RP18e)/ESI-MS analysis. J. Sep. Sci. 2010, 33, 880-890.

Wu, J.Y.; Bicker, W. Lindner, W. Separation properties of novel and commercial polar stationary phases in hydrophilic interaction and reversed-phase liquid chromatography mode. J. Sep. Sci. 2008, 31, 1492-1503.

Xu, M.C.; Peterson, D.S.; Rohr, T.; Švec, F.; Frechet, J.M.J. Polar polymeric stationary phases for normal-phase HPLC based on monodisperse macroporous poly(2,3-dihydroxypropyl methacrylate-co-ethylene dimethacrylate) beads. Anal. Chem. 75, 2003, 1011-1021.

Xue, Y.; Liu, J.; Unger, S. A 96-well single-pot liquid-liquid extraction, hydrophilic interaction liquid chromatography-mass spectrometry method for the determination of muraglitazar in human plasma. J. Pharm. Biomed. Anal. 2006, 41, 979-988.

Yanagida, A.; Murao, H.; Ohnishi-Kameyama, M.; Yamakawa, Y.; Shoji, A.; Tagashira, M.; Kanda, T.; Shindo, H.; Shibusawa, Y. Retention behavior of oligomeric proanthocyanidins in hydrophilic interaction chromatography. J. Chromatogr. A 2007, 1143, 153-161.

Yoshida, T. Peptide Separation in Normal Phase Liquid Chromatography. Anal. Chem. 1997, 69, 3038-3043.

Yoshida, T. Peptide separation by Hydrophilic-Interaction Chromatography: A review. J. Biochem. Biophys. Met. 2004, 60, 265-280.

Yoshida, T.; Okada, T. Peptide separation in normal-phase liquid chromatography. Study of selectivity and mobile phase effects on various columns. J. Chromatogr. A 1999, 840, 1-9.

Yu, L.; Li, X.; Guo, Z.; Zhang, X.; Liang, X. Hydrophilic interaction chromatography based enrichment of glycopeptides by using click maltose: a matrix with high selectivity and glycosylation heterogeneity coverage. Chem. Eur. J., 2009, 15, 12618-12626.

Zhang, H.; Guo, Z.; Zhang, F.; Xu, Q.; Liang, X. HILIC for separation of co-eluted flavonoids under RP-HPLC mode. J. Sep. Sci. 2008, 31, 1623-1627.

Zhu, B.Y.; Colin, C.T.; Hodges, R.S. Hydrophilic-interaction chromatography of peptides on hydrophilic and strong cation-exchange columns. J. Chromatogr. 548, 1991, 13-24.

Zhu, P.L.; Dolan, J.W.; Snyder, L.R.; Hill, D.V.; Van Heukelem, L.; Waeghe, T.J. Combined use of temperature and solvent strength in reversed-phase gradient elution. III. Selectivity for ionizable samples as a function of sample type and pH. J. Chromatogr. 1996, 756, 51-62.

3

Chiral Separations

Chiral Dynamic Chromatography Pushed to Its Extreme Low Temperature or High Speed Limits (Chiral DUHPLC): A Very Effective Tool in the Study of Stereolabile Compounds

Francesco Gasparrini, Ilaria D'Acquarica, Marco Pierini, and Claudio Villani
Dipartimento di Chimica e Tecnologie del Farmaco, Sapienza Università di Roma, Roma (Italy)

Introduction

The wide variety of chromatographic techniques developed and improved over the past three decades (Ward & Ward, 2010) allows analysts nowadays to address and resolve most problems related to the separation of molecular or ionic species, under both preparative and analytical conditions. Unusual and interesting examples of preparative separations have been performed for synthetic products, bioactive molecules extracted from biological sources and for racemates of chiral drugs. Typical analytical separations have been applied for the quantification of compounds in complex mixtures, the determination of enantiomeric excess of scalemic mixtures, and the identification of organic compounds by hyphenated chromatographic techniques (Wilkins, 1983; Holt et al., 1997; Ellis & Roberts, 1997; Guetens et al., 2002a; Guetens et al., 2002b). A peculiar and elegant application of chromatography has also been devoted to supramolecular chemistry where it allows calculating the thermodynamic stability of supramolecular adducts and the association constants of host-guest complexes. Moreover, chromatography has been recently used for the study of internal molecular dynamics of a range of stereochemically labile organic compounds and for the determination of kinetic parameters of the pertinent equilibrium (i.e., the reversible isomerization of one enantiomer into the other) (D'Acquarica et al., 2006).

In this context, the implementation of highly efficient chromatographic techniques pushed to their extreme limits may represent a very effective tool to achieve thermodynamic and kinetic data (Krupcik et al., 2003; Katsanos et al., 1998; Wolf, 2005a; Wolf, 2008), when experimental conditions make the use of alternative methods difficult or impossible, such as dynamic nuclear magnetic resonance (DNMR) spectroscopy and batch-wise classical approaches.

This chapter provides a basic overview of the concepts behind the determination of reaction rate constants and activation barriers by dynamic chromatography (DC). Several examples of the application of such technique in the study of model stereochemically labile compounds have been provided as well.

Dynamic Chromatography (DC): General Principles

Typically, efficient chromatographic separation of complex mixtures results in a series of baseline well-resolved peaks. However, if some of the species involved are subjected to a secondary dynamic equilibrium concomitant with the chromatographic distribution equilibria, band spreading and peak distortion may be observed, and the peaks may be joined by an elevated baseline, commonly referred to as a plateau. Such a type of dynamic profile (also called *dynamic chromatogram*) is depicted in Fig. 3.1, for the simple instance of a couple of isomers I_A and I_B, which may interconvert one to each other according to the chemical equilibrium (1).

$$I_A \underset{Kv_{-1}}{\overset{Kv_1}{\rightleftharpoons}} I_B \qquad K_{eq} = \frac{kv_1}{kv_{-1}} = \frac{I_A}{I_B} \tag{1}$$

As a function of the time spent by the two isomers inside the chromatographic column and of the operative temperature set during separation, three line-shapes (*a*, *b*, and *c*) may be in principle registered in the final chromatograms, shown in Fig. 3.1.

Line-shape (*a*) refers to a situation where no appreciable interconversion occurred between I_A and I_B during the chromatographic run, and two baseline resolved peaks are observed. Line-shape (*b*) refers to an active interconversion between I_A and I_B species, which happened faster than the separation of the single species during their elution through the chromatographic column. In other words, all molecules of I_A and I_B have undergone at least one interconversion cycle during elution. As a result, the peaks relevant to I_A and I_B coalesced to a single one, having an intermediate retention time (tr) between those of the single species (i.e., tr_{I_A} and tr_{I_B}). Such a dynamic profile is not suited to obtain kinetic information about the isomerization process that occurred. Finally, line-shape (*c*) refers to an intermediate situation between

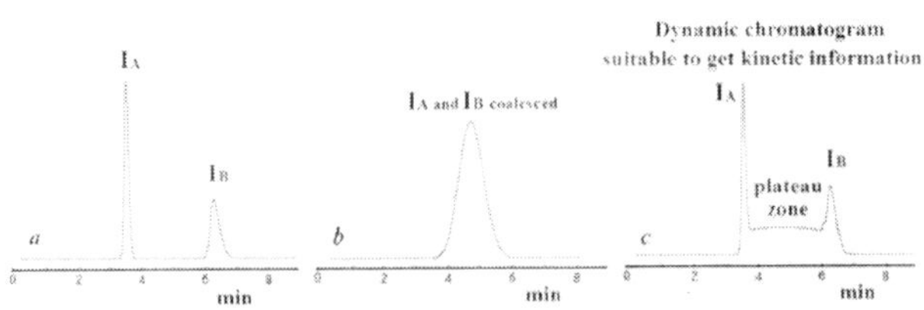

Fig. 3.1. Line-shape of chromatograms showed by isomeric I_A and I_B species that, in principle, may undergo interconversion during elution through the chromatographic column. (*a*) the $I_A \rightleftharpoons I_B$ process occurs much slower than the chromatographic separation. (*b*) the $I_A \rightleftharpoons I_B$ process occurs much faster than the chromatographic separation. (*c*) the $I_A \rightleftharpoons I_B$ process occurs in the time-scale of the chromatographic separation.

(*a*) and (*b*), as pointed out by an elevated baseline between the two well resolved I_A and I_B peaks (plateau zone). This is diagnostic for an active, but only partial, interconversion that occurred between I_A and I_B in the time scale of the chromatographic separation. In other words, in the scenario depicted in line-shape (*c*), only a fraction of I_A and I_B molecules underwent isomerization during the separation process, giving rise to the interference region lying between the couple of classical peaks which, in turn, are generated by the other fraction of molecules never involved in the dynamic process. Both shape and height of such a plateau inside a dynamic chromatogram contain all the kinetic information useful to describe the secondary equilibrium (an isomerization process, in the proposed example) that occurred during separation. Thus, rate constants of chemical processes can be investigated by suitable line-shape analysis of the dynamic profiles, that is to say, by simulating/analyzing dynamic chromatograms according to one of the several mathematical models available in the literature (see the Models Available to Simulate/Analyze Dynamic Chromatograms). In the most frequently employed approach, the iterative comparison of simulated and experimental chromatograms until reaching a good similarity, affords the desired kinetic data. The implementation of such an experimental approach is commonly known as *Dynamic Chromatography* (DC). A suitable choice of the operative conditions to set in the DC analysis may yield appropriate dynamic chromatograms. Effective modulation of the plateau height can be tuned by both temperature and residence time of the interconverting species inside the column, which can be optimized through a timely selection of mobile phase (MP) and flow-rate. In such a way, errors associated with the determination of rate constants can be minimized. However, a likely drawback of the method is that the stationary phase (SP) may have a perturbing effect on the kinetics of the dynamic process. In fact, the forward and backward rate constants experimentally determined by the interconverting I_A and I_B species during a DC experiment are expressed by apparent rate constants (kv_1^{app} and kv_{-1}^{app}) that result from a double weighed contribution coming from the same process occurring in both mobile (kv_1^{m} and kv_{-1}^{m}) and stationary (kv_1^{s} and kv_{-1}^{s}) phases, according to the thermodynamic cycle and equations reported in Fig. 3.2.

In a typical DC experiment, the forward rate constant k_{v1} matches with that in the MP, kv_1^{m}, and is approximated by the apparent rate constant kv_1^{app}, the backward rate constant kv_{-1} matches with kv_{-1}^{m} and is approximated by the kv_{-1}^{app}. The factors employed to modulate the kinetic contributions from the mobile (coefficients XI_A^{m} and XI_B^{m}) and stationary (coefficients XI_A^{s} and XI_B^{s}) phases are represented by the molar fractions that I_A and I_B assume in the two chromatographic phases (Fig. 3.2). In turn, these latter can be directly calculated from the capacity factors ($k'I_{A/B}$) of compounds I_A and I_B (see the relevant equations in Fig. 3.2).

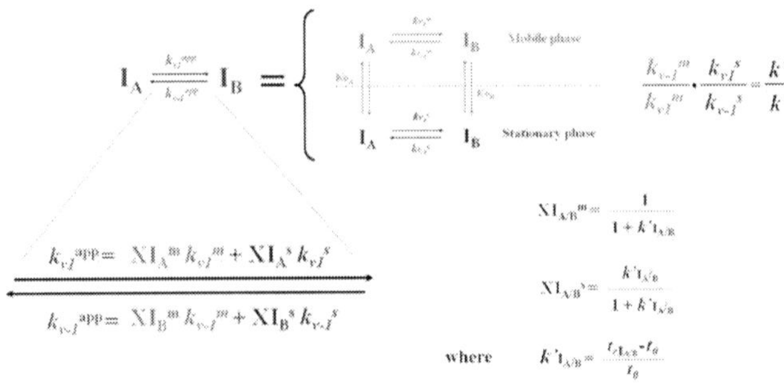

Fig. 3.2. Thermodynamic cycle involved during DC experiments (for the meaning of symbols, see text).

Models Available to Simulate/Analyze Dynamic Chromatograms

Several mathematical models have been developed in the last three decades to extract kinetic data from dynamic chromatograms. The *Continuous Flow Model* (CFM) is the oldest model, developed by Horváth and co-workers in 1984 (Melander et al., 1984; Jacobson et al., 1984). Because of its limited flexibility for the development of the required iterative procedures, it is currently out-of-date. In fact, it involves rather complex mathematics, *i.e.*, the coupled chromatographic and secondary chemical equilibria are described by suitable differential equations that have to be resolved by difficult numerical solutions.

Much more important, from an operative point of view, is the approach known as the *Theoretical Plates Model* (TPM) (Bürkle et al., 1984). It portrays the chromatographic column as a discontinuous entity constituted by a defined number, N_{th}, of elementary chemical reactors (called *theoretical plates*), and approximating the concurrent chromatographic and secondary chemical equilibria as occurring inside each theoretical plate in three successive steps: 1) equilibration between the MP and SP of the species involved in the secondary chemical process; 2) chemical evolution of each species according to the involved kinetic law (frequently, but not necessarily, an irreversible kinetic of first-order) for a time Δt corresponding to that of residence of the MP inside each theoretical plate ($\Delta t = t_0/N_{th}$, being t_0, the dead time); 3) shifting of the MP from the actual plate to the next one, according to the flow direction imposed by the MP through the column. The whole of these three steps have to be repeated a number of times (*n_shift*) sufficient to elute all injected species from the column (in its practical use, this number is calculated by the following equation: $n\_shift = (tr_{max} + 3W_{0.5})/\Delta t$, where tr_{max} is the retention time of the last eluted species and $3W_{0.5}$ is three times the width at half-height of the last eluted peak). On the whole, the TPM model proved to be very effective and relatively easy to implement into computer programs that automatically iterate the simulation of experimental dynamic chromatograms until achieving good agreement. However, because the time of simulation is exponentially related to N_{th}, only separations with not too marked efficiencies (smaller than about 10,000 theoretical plates) can be treated in an iterative way yielding reasonable simulation times. Examples of widely employed computer programs implementing the TPM model are SIMUL (Jung, 1992), ChromWin (Trapp & Schurig, 2001a), Auto DHPLC y2k (Gasparrini et al., 2002; Cabri et al., 2008a; Cirilli et al., 2009), and ChromWin_2D (Trapp et al., 2003; Trapp, 2004a).

The *Classical Stochastic Model* (CSM) or simply the *Stochastic Model* (SM), which was developed in its current formulation by the joint contribution from Keller and Giddings (Keller & Giddings, 1960) and Kramer (Kramer, 1975), has also proved to be very useful. An additional contribution to the approach, called the *Improved Stochastic Model* (SM+), was much more recently given (Trapp & Schurig, 2001b). The CSM model exclusively refers to the instance of mono-equilibrium processes featuring first-order kinetics occurring during a chromatographic separation. In this case, the concurrent chromatographic and secondary chemical equilibria are indirectly taken into account by expressing the profile of the dynamic chromatograms by means of two typologies of time-dependant distribution functions: the first one, $\phi(t)$, is related to the fraction of molecules of I_A and I_B never involved in the secondary chemical equilibrium [$\phi(t) = \phi I_A(t) + \phi I_B(t)$], and the other one, $\varphi(t)$, expressing the plateau zone due to the quote of I_A and I_B molecules that underwent at least once the transformation during the chromatographic separation [$\varphi(t) = \varphi I_A(t) + \varphi I_B(t)$]. While $\phi(t)$ [and then also its parent components $\phi I_A(t)$ and $\phi I_B(t)$] is not related to the rate constants of the secondary process and may be successfully described by a classical Gaussian distribution, func-

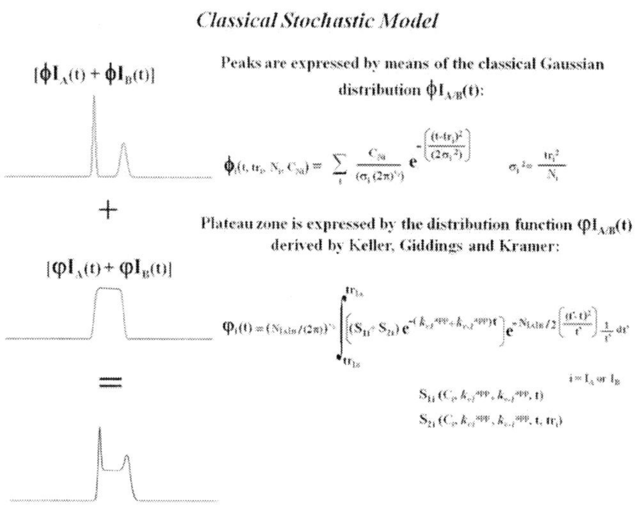

Fig. 3.3. Distribution functions expressing the static and dynamic components of dynamic chromatograms within the Classical Stochastic Model (CSM).

tion $\varphi(t)$ is exponentially related to both kv_1^{app} and kv_{-1}^{app}, as it may be evinced by its mathematical expression derived by Keller, Giddings, and Kramer and shown in Fig. 3.3.

The positive feature of the CSM model is the very short time of simulation, which results are significantly independent of N_{th}. This allows a quick treatment of dynamic chromatograms registered, by resorting to very efficient techniques, such as High Resolution Gas Chromatography (HRGC) and Ultra High-Performance Liquid Chromatography (UHPLC). An example of the relationships between time of simulation (sec) of dynamic chromatograms and chromatographic efficiency (number of theoretical plates) involved by the TPM and CSM models is given in Fig. 3.4.

Two computer programs implementing the CSM model and widely tested on a large variety of first-order processes involving both constitutional and conformational isomerizations (*i.e.*, enantiomerizations, diastereomerizations, tautomerizations, etc.) are ChromWin (Trapp & Schurig, 2001c) and Auto DHPLC y2k (Gasparrini et al., 2002a; Cabri et al., 2008b; Cirilli et al., 2009a). To strongly reduce the computational time that, in the aforementioned iterative procedure of comparison between simulated and experimental chromatograms is not negligible, the Auto DHPLC y2k program modifies kinetic and, if desired, chromatographic parameters, which is automated by the use of an algorithm driven by a simplex procedure. In this way, the user is not busy in this tedious step. Moreover, the same program also implements the possibility of taking tailing effects into account, thus extending the applicability of the procedure (for both TPM and CSM methods) to the frequent occurrence of non-linear sample repartition between MP and SP.

Even faster than the just described model is that based on the derivation of the so-called *Unified Equation of Chromatography* (UEC) (Trapp, 2006),

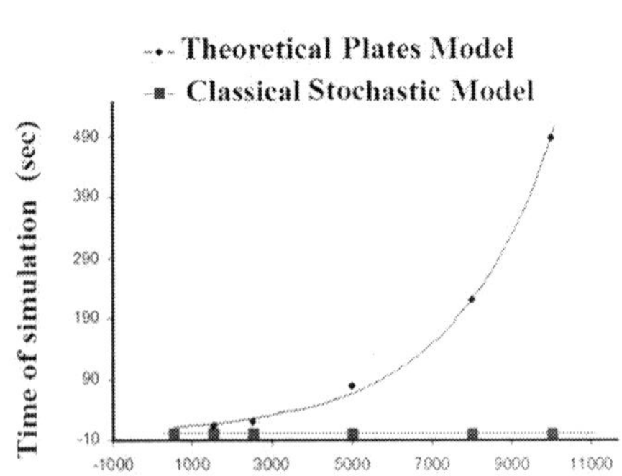

Fig. 3.4. Relationships between time of simulation (sec) of dynamic chromatograms and chromatographic efficiency (number of theoretical plates) involved by the Theoretical Plates (TPM) and Classical Stochastic (CSM) Models.

which overcomes an earlier approximate version (Trapp & Schurig, 2001d) only working in the simplified case of enantiomerization processes. The UEC allows one to directly calculate rate constants of secondary chemical first-order reactions by a few iterative steps, without the need of performing a computationally extensive simulation of elution profiles. The only parameters required to perform the calculus are: i) the retention times of the reacting species, tr_{IA} and tr_{IB}; ii) the peak widths at half-height, $W_{0.5}I_A$ and $W_{0.5}I_B$; iii) the relative height of the plateau, h_p; iv) the initial amounts of the reacting species, I_{A0} and I_{B0}; and v) the equilibrium constant, $K_{eq}=I_B/I_A$. Although very fast, this approach does not assure estimations with acceptable accuracy when the plateau height approaches that of the residual adjacent peaks (Cirilli et al., 2009b) or when the dynamic chromatograms are affected by marked asymmetry (Uray et al., 2010). Furthermore, because any simulation of the experimental dynamic chromatograms is not performed by this model, it is not possible to directly check the reliability of the estimated rate constants by superimposition of simulated and experimental profiles.

A further fairly commonly employed mathematical approach is that often labeled with the generic term of the *Deconvolution Method* (DM), although there are at least three different ways in which this method may practically be addressed. Such an approach, in fact, requires that the dynamic chromatogram is resolved into the components related to the fractions of molecules reacted (area of the plateau zone) and not reacted (area of the residual peaks on either side of the dynamic profile) during the separation. Thus, deconvolution can be performed by: i) the combined use of two or more tools of separation, which give rise to multidimensional hardware systems (Trapp, 2004b; Marriott et al., 2001); ii) the Gaussian or exponentially modified Gaussian functions (Krupcik et al., 2000a; Krupcik et al., 2000b; Oswald et al., 2002a; Oswald et al., 2002b); or iii) the combined use of two or more detectors (one of which must be chiro-optical) as monitoring tools of mono-dimensional chromatographic separations (Mannschreck et al., 1988; Mannschreck & Kiessl, 1989; Wolf et al., 1995; Allenmark & Oxelbark, 1997; Nishikawa et al., 1997; Kusano et al., 1999).

Finally, a novel stochastic approach to the DC has just been proposed in 2010 (Pasti et al., 2010). The novelty of the model is a microscopic point of view of the interconnections existing between repartition and secondary chemical equilibria, so that it might be suitably referred to as the *Microscopic Stochastic Model* (MSM). In its formalism, the shape of a dynamic chromatogram is calculated in the frequency domain when the reaction follows a simple reversible first-order scheme. Then, the derived solutions are expressed in closed form in the Fourier domain. However, at the moment, the model is not implemented in any dedicated standalone software and, because of the quite complex mathematics involved, it is unlikely that such an approach may routinely be used in DC experiments. Nevertheless, the MSM model could acquire specific interest on the basis of its ability to correlate macroscopic classical chromatographic parameters with the behavior properties of individual molecules (Pasti et al., 2005; Felinger et al., 2005).

Calculation of Free Energy Activation Barriers and Their Enthalpic and Entropic Contributions

Stereochemical or constitutional lability of some organic compounds plays a major role in all aspects of chemistry spanning a wide bridge from drug development to supramolecular chemistry. A proper evaluation of the propensity of such compounds to undergo isomerization is commonly given by the associated free energy activation barrier $\Delta G^{\neq}$. This latter is logarithmically correlated to the rate constant of the abovementioned process and promptly calculated by employing the Eyring equation:

$$-\Delta G^{\neq} = RT \ln\left(\frac{hk_{\nu}}{\kappa K_B T}\right)$$

where R is the universal gas constant, T is the absolute temperature, h is Planck's constant, κ is transmission coefficient and K_B is the Boltzmann constant. Moreover, it is possible to split out from a set of $\Delta G^{\neq}$ values the relevant enthalpic ($\Delta H^{\neq}$) and entropic ($\Delta S^{\neq}$) contributions by performing a number of kinetic determinations at different temperatures and carrying out a relevant van't Hoff analysis. In fact, enthalpic and entropic activation quantities may be assessed as the slope and intercept of van't Hoff plots of $\Delta G^{\neq}/T$ versus T^{-1}, respectively (see Fig. 3.5), which are based on the Gibbs equation rewritten in the following form:

$$\frac{\Delta G^{\neq}}{T} = \frac{\Delta H^{\neq}}{T} - \Delta S^{\neq}$$

Interesting information achievable from such a kind of investigation concerns the evaluation of the influence of temperature on a given chemical equilibrium. In fact, appreciable entropic contributions are responsible for marked variations of $\Delta G^{\neq}$ values in response to modest temperature variations. In turn, $\Delta S^{\neq}$ values of non-negligible extent are diagnostic for the occurrence of non-monomolecular transformations (*i.e.*, reaching of the transition state that does not match a first-order process). It is widely known that isomerizations are first-order reactions, but they can be based either on monomolecular or bimolecular mechanisms. Conformational changes are normally based on monomolecular mechanisms and, as such, do not require assistance of a secondary molecular partner. Atropoisomerizations are intramolecular events which fall into this class of reactions, featuring pure first-order kinet-

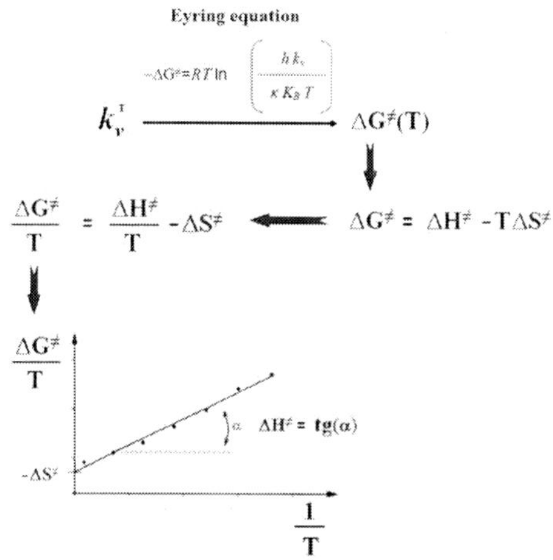

Fig. 3.5. Evaluation of enthalpic and entropic contributions to activation barriers by van't Hoff analysis.

ics (Gasparrini et al., 1997a; Gasparrini et al., 2000a; Gasparrini et al., 2001a; Dell'Erba et al., 2002a; Gasparrini et al., 2002b; Borsato et al., 2004a; Dalla Cort et al., 2005a; Lunazzi et al., 2010a). On the contrary, configurational isomerizations (i.e., processes involving rupture and reformation of chemical bonds) are commonly promoted by species that act as catalysts and therefore do not modify their concentration during interconversion (bimolecular mechanism). Tautomeric equilibria (Ballini et al., 2000; Fontana et al., 2002; Angelini et al., 2007; Siani et al., 2008), enantiomerizations (Gasparrini et al., 2003; Cirilli et al., 2007; Cirilli et al., 2009c; Trapp et al., 2002a) and epimerizations driven by reversible proton abstraction from stereogenic atoms (Cabri et al., 2008c; D'Acquarica et al., 2010a; Cabri et al., 2011) are important examples of configurational isomerizations, featuring pseudo first-order kinetics. Thus, it is attended that conformational isomerizations are barely affected by temperature, while the opposite trend should be predicted for configurational interconversions. For the latter reactions, an activation entropy roughly up to about −40 to −30 e.u. could be theoretically predicted because the loss of translational and rotational degrees of freedom, having values of ~ −50 e.u., should be partially compensated by new low-frequency motions arising in the transition state and quantifiable at about 10–20 e.u. (Page & Jencks, 1971). Also well-known is the case of rather negative activation entropies related to enantiomerization processes of the configurational type, involving the formation of zwitterionic species by heterolytic bond cleavage, as reported for the enantiomerization of aziridines and diaziridines (Trapp & Schurig, 2000a; Shustov et al., 1988; Shustov et al., 1989). According to such considerations, $|\Delta S^{\neq}|$ values always smaller than 17 e.u. have indeed been detected by DC experiments in a wide range of conformational stereochemical isomerizations (prevalently enantiomerizations and epimerizations) (see Table 3.A). In contrast, configurational isomerizations, again studied by DC methods and including enantiomerizations, epimerizations and geometrical diastereomerizations, afforded $-\Delta S^{\neq}$ values always larger than 19 e.u. (Table 3.A).

Application of DC Methods Within Extreme Operating Conditions

The great advantage in using DC techniques is promptly highlighted when they are compared with classical kinetic determinations based on batch-wise procedures. Typically, in the latter case, isomerization rate constants are obtained by monitoring the amount of variations of one of the reacting species as a function of time. Frequently, a chromatographic technique is employed as the monitoring tool, so that the progressive residual amount of reacting species or the increasing amount of the formed product is evaluated by the areas underlying the registered chromatographic peaks. When the studied isomerization involves chiral molecules, as necessarily occurs in enantiomerizations, the variation of enantiomeric excess is monitored as a function of time by off-line enantioselective chromatography. In a typical experiment, pure or enriched samples are allowed to equilibrate into an isolated and thermostated system in the presence of reaction solvent, and the progression of the isomerization is monitored by chromatography under conditions of suppressed interconversion. Although rigorous and of general applicability, batch-wise approaches are usually laborious, time-consuming and expensive in terms of the amount of product to process. Moreover, a preliminary collection of pure or highly enriched reactants at preparative or semi-preparative scale must be accomplished, starting from the equilibrated mixture of the couple of interconverting species to analyze (a racemic mixture, for the case of enantiomerizations). All these drawbacks are completely overcome by using the DC approach. However, two main limitations should be taken into account also in this case: i) the limited range of solvents that can be used, and ii) the perturbing effect of the SP (see the section on Perturbing Effects of Stationary Phases on $\Delta G^{\neq}$ Values Measured by DC Methods).

A great number of enantiomerizations/diastereomerizations of chiral species have been studied by DC approaches in the past two decades. In several cases, the chromatographic results were supported and/or compared to alternative kinetic methods, such as dynamic nuclear magnetic resonance (DNMR) spectroscopy (Gasparrini et al., 1995; Dell'Erba et al., 2002b; Gasparrini et al., 2002c; Dalla Cort et al., 2005b) and Stopped-Flow Gas Chromatography (SFGC) (Wolf, 2005b; Trapp & Schurig, 2000b). It was therefore found out that the DC methods, in the whole of their different forms, can ensure the coverage of a very wide range of activation energies, from the very small value of about 15 kcal mol^{-1} (Gasparrini et al., 2000b) to the much greater value of 38 kcal mol^{-1} (Trapp et al., 2002b). A schematic representation of the optimal

Table 3.A. Activation Entropies ($\Delta S^{\neq}$, e.u.) and Free Energies ($\Delta G^{\neq}$, kcal mol^{-1}) of Conformational and Configurational Isomerizations Calculated by DHPLC and DHRGC.

Conformational isomerizations			Configurational isomerizations		
Compound	$\Delta S^{\neq}$ (e.u.)	$\Delta G^{\neq}$ (kcal mol^{-1})	Compound	$\Delta S^{\neq}$ (e.u.)	$\Delta G^{\neq}$ (kcal mol^{-1})
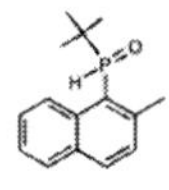	(diastereomerization) DHPLC		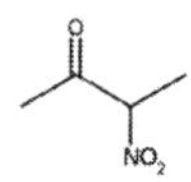	(enantiomerization) DHRGC	
	0.5	14.8 (−68 °C)		−19	29.8 (160 °C)
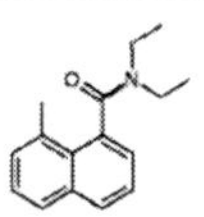	(enantiomerization) DHPLC		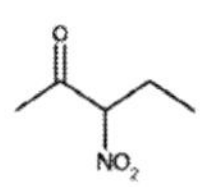	(enantiomerization) DHRGC	
	−9.9	22.7 (50 °C)		−35	29.8 (160 °C)
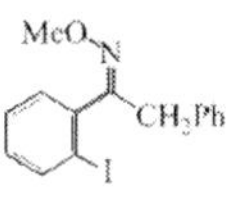	(enantiomerization) DHPLC		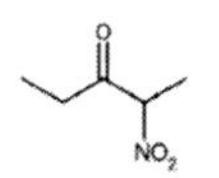	(enantiomerization) DHRGC	
	2.0	15.0 (−70 °C)		−37	29.3 (160 °C)
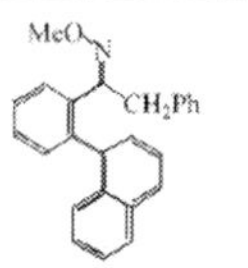	(enantiomerization) DHPLC		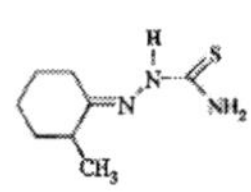	(enantiomerization) DHPLC Second-order process	
	−3.0	19.8 (−10 °C)		−40	22.0 (50 °C)
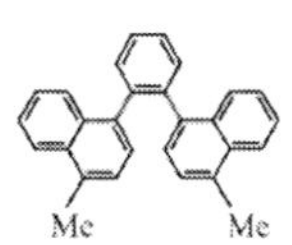	(diastereomerization) DHPLC		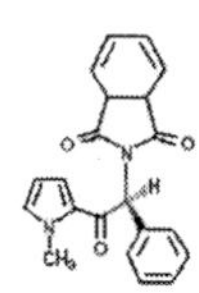	(enantiomerization) DHPLC First-order process	
				−40	24.0 (45 °C)
				Second-order process	
	−4.0	19.5 (−5 °C)		−40	18.5 (−5 °C)
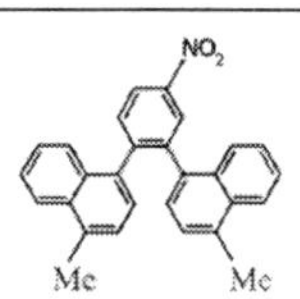	(diastereomerization) DHPLC		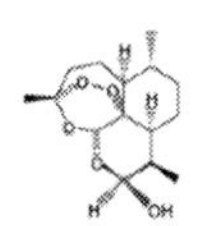	(epimerization) DHPLC First-order process (different pHs)	
	5.0	19.0 (−5 °C)		−31 ÷ −20	22.1 (25 °C)
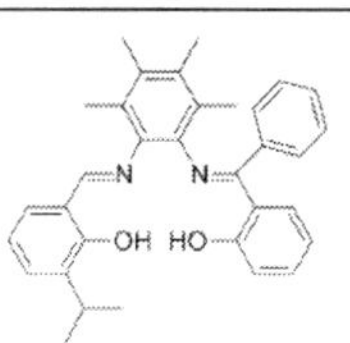	(enantiomerization) DHPLC		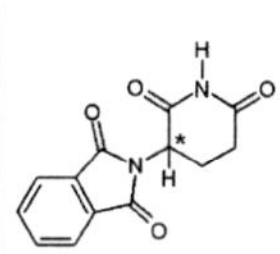	(enantiomerization) DHRGC	
	−9.2	21.8 (30 °C)		−59	38.0 (220 °C)
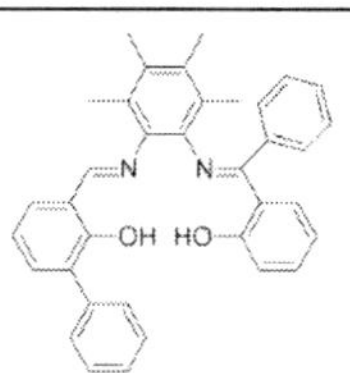	(enantiomerization) DHPLC			(enantiomerization) DHRGC	
	−16.9	21.9 (30 °C)		−55	33.6 (190 °C)

(continued on next page)

Table 3.A. *(continued)*

Conformational isomerizations			Configurational isomerizations		
Compound	$\Delta S^{\neq}$ (e.u.)	$\Delta G^{\neq}$ (kcal mol^{-1})	Compound	$\Delta S^{\neq}$ (e.u.)	$\Delta G^{\neq}$ (kcal mol^{-1})
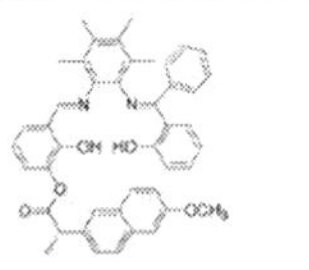	(diastereomerization) DHPLC				
	−2.9	21.9 (30 °C)			
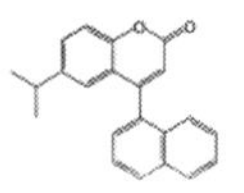	(enantiomerization) DHPLC				
	−2.0	21.7 (30 °C)			
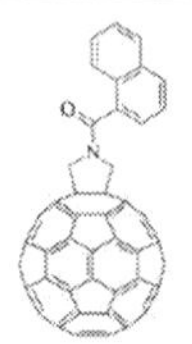	(enantiomerization) DHPLC				
	—	14.3 (−70 °C)			
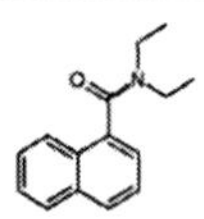	(enantiomerization) DHPLC				
	—	14.9 (−60 °C)			

range of applicability of DC (including GC, HPLC, and UHPLC) and DNMR techniques in the determination of activation barriers for isomerization processes is given in Fig. 3.6.

Table 3.A contains some representative examples of isomerizations whose activation barriers determined by DC methods range between the extreme values of 15 kcal mol^{-1} and 38 kcal mol^{-1}. Both temperature and residence time of the interconverting species inside the column are parameters that can be tuned to approach the upper or lower limits of the above range. The residence time may be effectively modulated by changing the flow-rate, but also by drastically improving the chromatographic efficiency. Very high efficiencies, in fact, allow better resolutions in shorter times, and this strongly reduces the residence time of stereolabile species exhibiting exceptionally low activation barriers to interconversion. A second strategy is to reduce the column temperature, but cryo-chromatography at temperatures ranging from −50° down to −80°C presents some practical and instrumental difficulties, mainly related to the increased viscosity of the mobile phases. Higher viscosities in turn result in a large pressure drop across the column, with limitations in the range of available linear velocities of the eluent. The implementation of a novel and efficient technique, such as Ultra-High Performance Liquid Chromatography (UHPLC), may overcome, or at least limit, the impasse.

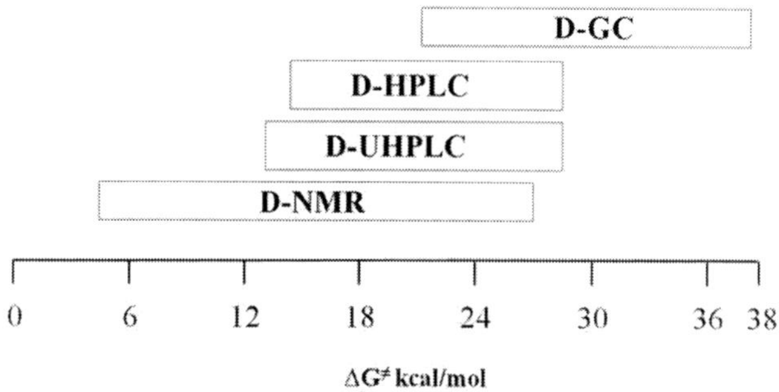

Fig. 3.6. Optimal range of applicability of DC (including GC, HPLC, and UHPLC) and DNMR techniques in the determination of activation barriers for isomerization processes.

Ultra High-Performance Liquid Chromatography (UHPLC)

Analytical LC separations in the field of modern life sciences are faced with formidable challenges, linked to the requirement of a sharp reduction in the overall analysis time and to the ever increasing complexity of samples that must be analyzed. In order to perform fast separations while maintaining acceptable efficiency, resolution and overall chromatographic performance, a compromise is necessary between eluent flow rate, column length, and back pressure.

According to theoretical treatments of the LC chromatographic process, one potential approach to increase column efficiency is to decrease the average size, d_p, of the packing particles in the column (Giddings, 1965; Knox, 1977; Poppe, 1997). Thus, during the last three decades, HPLC has witnessed a smooth evolution of the packing material size from the original 10 micron to 5 micron, and later to 3 micron. A parallel shortening of the standard column length has been observed, moving from the classical 30 or 25 cm to the 10 or 5 cm format, or even to very short 2–3 cm column length for fast (but less efficient) separations in the sub-minute range. In order to retain high efficiency together with reduced analysis time, sub-2 micron spherical porous particles have been proposed as new packing materials in the last few years (Wu et al., 2001; Jerkovich et al., 2003). Columns packed with sub-2 micron particles have very low permeabilities (permeability is proportional to d_p^2) while their optimum flow velocity for maximum efficiency is higher compared to columns packed with larger particles. The combination of reduced particle size with high eluent flow rates results in a drastic increase in the column inlet pressure, which is proportional to the inverse of d_p^2 (Desmet et al., 2006; Neue & Kele, 2007; Gritti & Guiochon, 2008).

Columns packed with sub-2 micron particles move very quickly outside the pressure range of classical HPLC (roughly up to 40 MPa) when they are operated at their optimum flow rates or above, and this poses two problems in terms of mechanical stability of the stationary phase particles and in terms of dedicated instruments, both of which are required to function under operative pressures reaching 100 MPa or higher.

Today, several manufacturers produce analytical LC instruments that are able to deal with pressures higher than 40 MPa, such as UPLC® for Ultra Performance Liquid Chromatography or other variants of the technique (RRLC for Rapid Resolution Liquid Chromatography, UHPLC for Ultra High-Performance Liquid Chromatography, VHPLC for Very High-Pressure Liquid Chromatography). Columns and stationary phases compatible with extreme pressures are available as well. However, the large repertoire of stationary phase chemistries present for HPLC columns is not found for the UHPLC counterpart (Guillarme et al., 2007; Wales et al., 2008; Carr et al., 2009).

Enantioselective LC systems can in principle benefit from a substantial increase in sample throughput by employing smaller particles packed in short columns and using high linear velocities of the eluent. Unfortunately, if the number and types of achiral conventional stationary phases for UHPLC applications are scarce compared to HPLC, the situation is even worse in chiral LC, where no UHPLC-dedicated chiral stationary phase (CSP) is commercially available so far.

Recently, a brush-type CSP for UHPLC applications has been developed as the result of transition from the 5 micron to the sub-2 micron format of the underlying silica particles (Cancelliere et al., 2010a). This UHPLC brush-type CSP combines the use of reduced particle size with an established chiral selector for the generation of advanced materials with high throughput and/or high resolution capabilities. The known diaminocyclohexane—3,5-dinitrobenzoyl (DACH-DNB) CSP (Cancelliere et al., 2006)—was prepared starting from 1.9 micron spherical silica particles using a synthetic strategy that generates the intermediate DACH-CSP in a single step starting from a slurry of bare silica, plus the chiral 1,2-diamine and glycidoxypropyltrimethoxysilane. Subsequent treatment of the intermediate silica with 3,5-dinitrobenzoyl chloride gave the final CSP where π-acidic aromatic fragments are fixed on the diamine framework through amide linkages (Fig. 3.7).

When the efficiency of a stainless-steel (100 × 4.1 mm I.D.) column packed with 1.9 micron DACH-DNB CSP was monitored as a function of the eluent flow-rate (van Deemter analysis), a value of H_{min} = 5.2 micron was found at the optimum linear velocity μ_{opt} = 4.00 mm/s, using methyl benzoate as test solute and 10% chloroform in *n*-hexane as eluent. The van Deemter plot of the column packed with 1.9 micron DACH-DNB CSP showed a flat portion at high linear velocities of the eluent, suggesting a potential high efficiency use of the column in the subminute separation regime (Fig. 3.8).

Indeed, several very fast chiral separations on the 1.9 micron DACH-DNB-CSP have been reported for a range of compounds including alkyl-aryl sulfoxides, secondary phosphine oxides and acylated amines. The combination of high flow-rates, short column length (50 × 4.1 mm I.D.) and large enantioselectivity

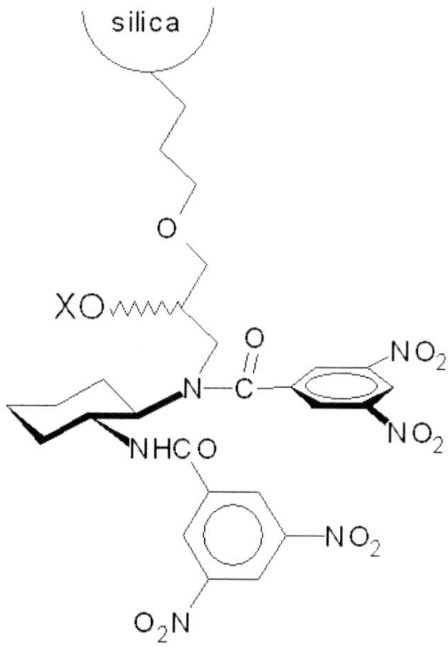

Fig. 3.7. Chemical structure of the DACH-DNB CSP. X = H or 3,5-dinitrobenzoyl.

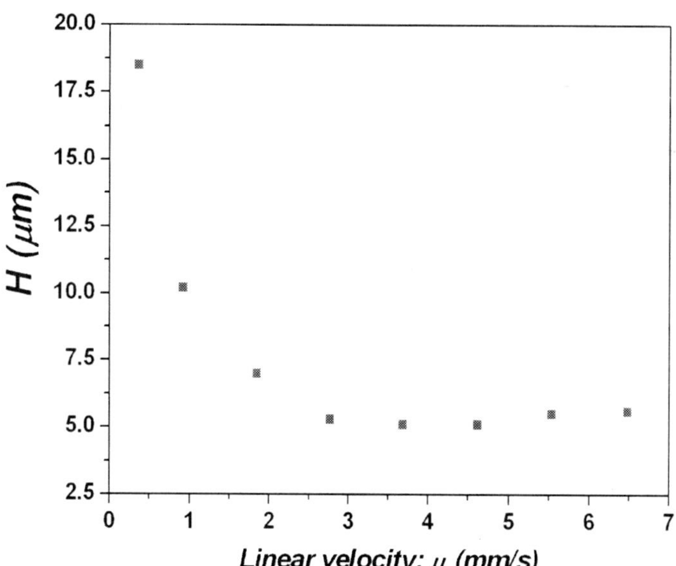

Fig. 3.8. van Deemter plot for the 1.9 micron DACH-DNB CSP packed into a 100 × 4.1 mm I.D. stainless-steel column (not optimized packing procedure), using *n*-hexane/$CHCl_3$ 90/10 (v/v) as an eluent, at column temperature T_{col} = 25°C, and methyl benzoate as a test solute (K^1 = 1.29).

(α = 1.87) resulted in a complete separation of the enantiomers of a chiral 1-naphthamide in less than fifteen seconds (Fig. 3.9).

The chromatographic separation of stereolabile chiral compounds is one particular case in which enantioselective UHPLC columns and hardware find immediate practical applications. Two extreme scenarios can be envisaged when a stereolabile chiral compound, featuring energy barriers to enantiomer interconversion lower than 18 kcal mol^{-1}, must be resolved by chromatography. In one situation, the column temperature can be lowered down to cryogenic temperatures to a point where the half-life times of the interconverting enantiomers is commensurate to the analysis time. With high-viscosity eluents, the column inlet pressure rises rapidly with decreasing temperature, and only UHPLC systems can be used under these extreme experimental conditions. In the other situation, the chromatographic time scale is shifted into the seconds regime by the combined use of short columns packed with sub 2-micron particles and high eluent flow-rates. Under these conditions, the overall analysis time can be cut by a factor of ten, compared to a conventional column, and the time scale of the separation can approach the time scale of the enantiomer interconversion at a given temperature.

With the aim of comparing the UHPLC technique advantages with the well consolidated HPLC, atropoisomeric chiral species have been resolved on a chiral column based on the DACH-DNB selector, and the chromatographic time scale was shifted from the minutes into the seconds range. This is the case of a chiral bis-ketone with two stereogenic axes (Fig. 3.10), which was effectively resolved into its three stereoisomers (a couple of conformational enantiomers and an achiral *meso*-form). In chromatographic runs performed at 10°C by resorting to both HPLC (Gasparrini et al., 1995) and UHPLC methods (Cancelliere et al., 2010b), the very different residence time that the stereoisomers spent inside the column appeared evident. Accord-

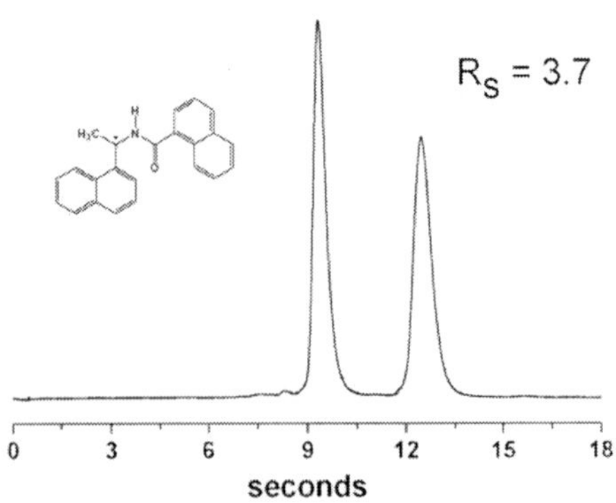

Fig. 3.9. Ultra-fast enantioresolution of *N*-(1-(naphthalen-5-yl)ethyl)-1-naphthamide on the 1.9 micron DACH-DNB CSP packed into a stainless-steel (50 × 4.1 mm I.D.) column. Eluent: *n*-hexane/dichloromethane/1,4-dioxane (30/30/40, v/v/v) + 5 % methanol. Flow-rate: 4.00 mL/min. T_{col} = 25°C.

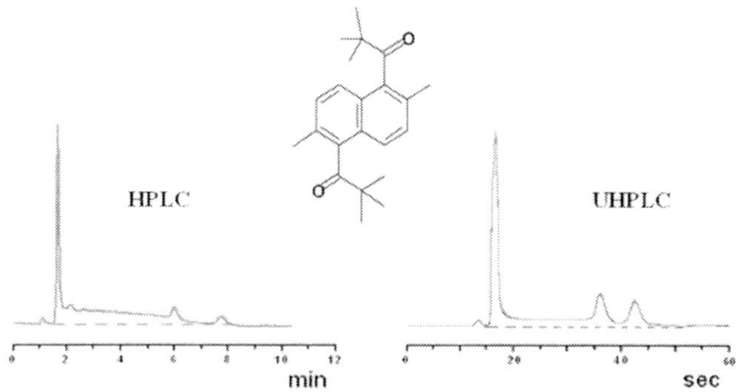

Fig. 3.10. Different residence time and extent of enantiomerization experienced by a stereolabile chiral bis-ketone when subjected to dynamic HPLC and UHPLC (DACH-DNB CSPs packed into stainless-steel (100 × 4 mm I.D.) and (50 × 4 mm I.D.) columns for the DHPLC and DUHPLC experiments, respectively. Eluent: *n*-hexane/2-propanol/methanol 95/4/1 (v/v/v). Flow-rate: 1.00 mL/min for DHPLC and 3.00 mL/min for DUHPLC. T_{col} = 10°C.

ingly, a marked difference in the extent of isomerization is clearly visible in the resulting dynamic chromatograms, the plateau almost lacking in the case of UHPLC (Fig. 3.10). Schematically, if we state that the residence time decreases by a factor of ten when changing from HPLC to UHPLC (a quite low value, relative to optimized conditions), it may be generalized that, to assure the same half-life time ($t_{1/2}$) as the process, the operative temperature has to be increased by a ΔT amount predictable by the following linear equation: $\Delta T = 0.8611 \times \Delta G^{\neq} + 1.0676$.

This means that, to measure an activation energy of 15 kcal mol^{-1}, an operative temperature of −49°C will be required by UHPLC, for a run time of about 1 min, whereas HPLC will require −63°C for a run time of about 10 min. From a different point of view, the same transition of technique (i.e., from UHPLC to HPLC) would allow one to measure $\Delta G^{\neq}$ values lower than about 1 kcal mol^{-1} at any established temperature in the range −100 to +30°C, under the same conditions (Fig. 3.6). However, in general, a much more marked modulation of plateau zones may be obtained by suitable modest column temperature changes, compared to huge changes of the eluent flow-rate (Cancelliere et al., 2010c). Thus, very high activation barriers, not far from 40 kcal mol^{-1}, can be estimated by heating the column up to 200°C, for $t_{1/2}$ of about 30 min. Such temperature values are typical of HRGC, which is suitable only for thermally stable and volatile compounds. A great number of stereoisomerizations (commonly enantiomerizations) have been studied by Dynamic HRGC (DHRGC), featuring $\Delta G^{\neq}$ values almost close to 30 kcal mol^{-1}. Selected examples from the literature are collected in Table 3.A. In the case of thalidomide, an extreme enantiomerization barrier of 38.0 kcal mol^{-1} at 220°C was reported (Trapp et al., 2002c).

Perturbing Effects of Stationary Phases on $\Delta G^{\neq}$ Values Measured by DC Methods

Calculation of activation barriers by DC is affected by additive contributions arising from the particular type of SP used. Such perturbing effects may be either inhibitory or promotional, so that the measured apparent rate constants kv_1^{app} and kv_{-1}^{app} may result in under- or over-estimatation with respect to the related values found in the MP. Such deviations are more marked as the species involved in the interconversion process increase retention. Therefore, when the species under investigation are enantiomers, the forward and backward rate constants found are necessarily different from each other, since the transient adducts formed by interaction with the chiral selector on the SP are diastereoisomers. Thus, apparent enantiomerization rate constants determined by DC techniques (kv_{en}^{app}) are generally expressed as an arithmetic mean of the apparent rate constants for the forward and backward processes [i.e., $kv_{en}^{app} = (kv_1^{app} + kv_{-1}^{app})/2$]. Comparisons of rate constants measured by both DC and DNMR or classic batch-wise approaches stressed again the importance to distinguish among conformational and configurational isomerizations. Generally, in the first case, rate constants are underestimated, and moderate SP perturbations are observed (Gasparrini et al., 1997b; Gasparrini et al., 2000c; Gasparrini et al., 2001b; Dell'Erba et al., 2002c; Gasparrini et al., 2002d; Borsato et al., 2004b; Dalla Cort et al., 2005c; Lunazzi et al., 2010b). A logical explanation may be that, after complexation of the labile species on the SP, the hindered bond rotation relevant to the studied conformational stereomutation may easily find additional physical opposition by the phase, with a consequent increase of the related activation barrier. Such an effect, defined as an Indirect Perturbing Contribution of SP (SP_{IPC}) (Cirilli et al., 2009d), can be reduced, but never suppressed, by minimizing the residence time of the species inside the column. An illustrative example of such evidence is showed in Fig. 3.11, where both classic and dynamic chromatograms are reported for the two couples of residual enantiomers of a hindered secondary aryl phosphine oxide, which is endowed with two asymmetry elements, the phosphorous atom (P) and the axis colinear with the P–Csp^2 bond (Gasparrini et al., 2000d).

While the *R*/*S* inversion of the phosphorous is a strongly disfavored configurational process, not observed at temperatures below 25°C, rotational motion around the stereogenic axis is a conformational modification that occurs very easily. At −65°C, the ^{31}P-DNMR technique afforded an activation barrier of 14.8 kcal mol^{-1} for the interconversion of the more stable (the *R* or *S* synclinal conformer, *sc*) into the less stable species (the *R* or *S* anticlinal conformer, *ac*). The equivalent measure performed by DHPLC (simulation based on TPM) at the same temperature on the separated residual enantiomers [i.e., by DHPLC on the split couples of equilibria $R(+sc) \rightleftharpoons R(-ac)$ and $S(-sc) \rightleftharpoons S(+ac)$] afforded the same value for the couple of diastereomers less retained [from

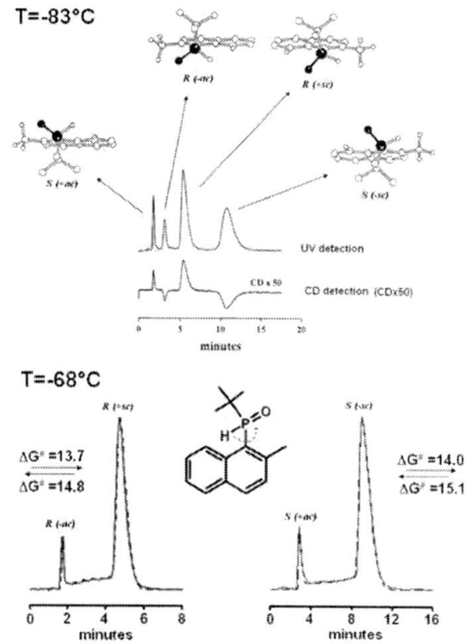

Fig. 3.11. Explicative example of indirect perturbing SP_{IPC} effect displayed by the racemic version of the DACH-DNB stationary phase on the diastereomerization rate constants of a hindered secondary aryl phosphine oxide.

R(+*sc*) to *R*(−*ac*)], whereas a bit greater barrier (+ 0.3 kcal mol^{-1}) was seen for the most retained couple [from *S*(−*sc*) to *S*(+*ac*)]. Thus, it may be generalized that, to obtain kinetic data as close as possible to those achieved by classical batch-wise or DNMR approaches, one should choose operating conditions that allow, at the same time, a suitable chromatographic resolution, a considerable plateau height, and the shortest residence time of the interconverting species inside the column.

In the case of configurational isomerizations (enantiomerizations as well as tautomerizations or acid catalyzed geometrical isomerizations), the SP may affect much more strongly the corresponding apparent pseudo first-order rate constants. This is because both the chiral selector and the chromatographic matrix might not be kinetically inert in principle. The SP can, in fact, act as promoter or inhibitor agent, increasing or decreasing the enantiomerization barrier of the studied chiral samples. This second type of SP effect was referred to as a Direct Perturbing Contribution (SP_{DPC}) (Cirilli et al., 2009e). Thus, while SP_{IPC} arises from a modification of the species involved in the isomerization, which changes its structure to form the proper *SP:substrate* adduct, the SP_{DPC} effect comes as an additive term governed by independent catalytic sites bonded to the SP, which, in most cases, are represented by acid or basic groups acting as promoters. Therefore, from a mathematical point of view, the forward and backward rates of bimolecular processes leading to configurational isomerizations inside a chromatographic tool can be described by the two following general types of kinetic Equations (Cirilli et al., 2009f):

$$ {}^{1}kv^{s} = {}^{2}kv^{s}_{IPCBase} \times [Base] + {}^{2}kv^{s}_{IPCAcid} \times [Acid] + {}^{2}kv^{s}_{DPC} \times [S_{SP}] + {}^{1}k_{0} \quad (1) $$

$$ {}^{1}kv^{m} = {}^{2}kv^{m}_{Base} \times [Base] + {}^{2}kv^{m}_{Acid} \times [Acid] + {}^{1}k_{0} \quad (2) $$

and by Equation 3 the related apparent one

$$ {}^{1}kv^{app} = {}^{2}kv^{app}_{IPCBase} \times [Base] + {}^{2}kv^{app}_{IPCAcid} \times [Acid] + {}^{2}kv^{app}_{DPC} \times [S_{SP}] + {}^{1}k_{0} \quad (3) $$

The superscript 1 or 2 at the left side of the rate constants refers to the kinetic order; superscript *s* (*stationary*) or *m* (*mobile*) at the right side of the rate constants refers to the chromatographic phase in which the constant is considered; Base and Acid represent the catalytic species potentially acting in the MP and SP; the term S_{SP} refers to the sites of SP that can display catalytic effects; finally, ${}^{1}k_{0}$ expresses a first-order rate constant that accounts for any generic contribution not coming from either the SP or explicit catalytic species. According to Equation 1, any kind of SP_{DPC} contribution can be potentially and easily removed by *ad hoc* experiments. Indeed, this was performed for the specific case of a chiral ketone displaying an anti-MAO activity, which enantiomerizes by a keto-enolate equilibrium (Cirilli et al., 2009g). The pseudo first-order enantiomerization rate constants for the process catalyzed by diethylamine (DEA) were measured by both batch-wise and DHPLC approaches at different DEA concentrations and temperatures. The results showed a negligible ${}^{1}k_{0}$ term, so that Equations 1 and 2 may be rewritten in the simplified forms 4 and 5, respectively, which, in turn, afford Equation 6, expressing the related apparent pseudo first-order enantiomerization rate constant ${}^{1}kv^{app}$:

$$ {}^{1}kv^{s} = {}^{2}kv^{s}_{IPC} \times [DEA] + {}^{2}kv^{s}_{DPC} \times [S_{SP}] \quad (4) $$

$$ {}^{1}kv^{m} = {}^{2}kv^{m} \times [DEA] \quad (5) $$

$$ {}^{1}kv^{app} = {}^{2}kv^{app}_{IPC} \times [DEA] + {}^{2}kv^{app}_{DPC} \times [S_{SP}] \quad (6) $$

According to Equation 6, at each temperature the SP_{DPC} contribution expressed by the term ${}^{2}kv^{app}_{DPC}$ $[S_{SP}]$ was split out as the intercept in plots of ${}^{1}kv^{app}$ versus DEA concentration, while, from the same relationships, the slope afforded the apparent second-order enantiomerization rate constants ${}^{2}kv^{app}_{IPC}$. By comparison of these latter ones with the equivalent data coming from the batch-wise determination SP_{IPC} effects of +27% and +20% were quantified on the second-order rate constants at temperatures of 35 and 45°C, respectively (*i.e.*, a reduction of only 0.15 and 0.12 kcal mol^{-1} of the correspondent $\Delta G^{\neq}$ values). Interestingly, in this case, the SP showed a promotion effect, which is normally not found in monomolecular isomerizations (*vide supra*). Presumably, the SP can slightly increase the formal DEA concentration on its surface, and DEA absorption may achieve a saturation level only over a critical catalyst concentration, estimated as 2×10^{-2} M for the specific studied case. Therefore, even minor SP_{IPC} effects could be expected for catalyst concentrations smaller than the critical limit.

DC as a Tool to Quantify Catalytic Sites Bonded on Chromatographic Supports

A severe problem in chromatographic analyses requiring a rigorous quantification of chiral drugs is when the species to resolve may be involved into secondary equilibria during their separation. In fact, both the selector and matrix of the stationary phase may be endowed on their surfaces with chemical groups responsible for possible catalytic effects. Such an event was quite recently specifically addressed in a study showing that even International Pharmacopoeias can contain shortcomings arising from the aforementioned chemical lability (Cabri et al., 2008d; D'Acquarica et al., 2010b). Dihydroartemisinin (DHA) is a powerful antimalarial drug which may easily interconvert between its two epimeric forms α and β. An effective strategy was proposed to set optimized operating conditions (suitable choice of column length, MP flow-rate and temperature) aimed at slowing down the secondary equilibrium. A marked differential SP_{DPC} effect was also stressed, which commercial columns, containing the same stationary phase as a discriminating tool (all RP-C18 columns, in the specific case), can have in promoting the $\beta \rightleftharpoons \alpha$ epimerization. In other words, the concentration of catalytic sites present on the SP surfaces not prepared according to standardized and uniform procedures may significantly differ from one to each other. However, the experimental determination of the ${}^{2}kv^{app}{}_{DPC} \times [S_{SP}]$ term present in Equation 6 of the Perturbing Effects of Stationary Phases on $\Delta G^{\neq}$ Values Measured by DC Methods section may represent a valuable tool to gain quantitative information on the abundance of S_{SP} sites. The results obtained could have direct implication in any analytical investigation aimed at quantitating this family of drugs.

In parallel studies performed on an anti-MAO drug (Cirilli et al., 2009h), within the limit of the adopted approximations (which include the assumption of amino groups as basic S_{SP} sites bonded to the silica-matrix, uncovered by the deposed chiral selector), the $[S_{SP}]$ concentration was estimated as 4.5×10^{-3} M. This, in turn, was assessed to correspond to uncovered amino groups included in a range from 3% to 5% of the whole parent amino groups originally bonded to the matrix. Therefore, such an approach discloses the potential for wide applicability to studies aimed at providing accurate characterization of the surface of chromatographic supports. Nevertheless, for a rigorous application of DC methods to these kinds of determinations further dedicated studies have to be envisaged.

References

Allenmark, S.; Oxelbark, J. Inversion barriers in sulfoxide functions studied by enantioselective liquid chromatography and chiroptical methods. Chirality 1998, 9, 638–642.

Angelini, G.; De Maria, P.; Fontana, A.; Pierini, M.; Siani, G. Ionization and tautomerization of 2-nitrocyclohexanone in aqueous solution. J. Org. Chem. 2007, 72, 4039–4047.

Ballini, R.; Cerritelli, S.; Fontana, A.; De Maria, P.; Pierini, M.; Siani, G. Equilibrium constants for ionisation and enolisation of 3-nitrobutan-2-one. Eur. J. Org. Chem. 2000, 1641–1646.

Borsato, G.; Della Negra, F.; Gasparrini, F.; Misiti, D.; Lucchini, V.; Possamai, G.; Villani, C.; Zambon, A. Internal motions in a fulleropyrrolidine tertiary amide with axial chirality. J. Org. Chem. 2004, 69, 5785–5788.

Bürkle, W.; Karfunkel, H.; Schurig, V. Dynamic phenomena during enantiomer resolution by complexation gas chromatography. A kinetic study of enantiomerization. J. Chromatogr. 1984, 288, 1–14.

Cabri, W.; Ciogli, A.; D'Acquarica, I.; Di Mattia, M.; Galletti, B.; Gasparrini, F.; Giorgi, F.; Lalli, S.; Pierini, M.; Simone, P. On-column epimerization of dihydroartemisinin: an effective analytical approach to overcome the shortcomings of the international pharmacopoeia monograph. J. Chromatogr. B 2008, 875, 180–191.

Cabri, W.; D'Acquarica, I.; Simone, P.; Di Iorio, M.; Di Mattia, M.; Gasparrini, F.; Giorgi, F.; Mazzanti, A.; Pierini, M.; Quaglia, M.; Villani, C. Stereolability of dihydroartemisinin, an ntimalarial drug: a comprehensive thermodynamic investigation ? Part I. J. Org. Chem. 2011, in press.

Cancelliere, G.; D'Acquarica, I.; Gasparrini, F.; Maggini, M.; Misiti, D.; Villani, C. Twenty years of research on silica-based chiral stationary phases. J. Sep. Sci. 2006, 29, 770–781.

Cancelliere G.; Ciogli, A.; D'Acquarica, I.; Gasparrini, F.; Kocergin, J.; Misiti, D.; Pierini, M.; Ritchie, H.; Simone, P.; Villani, C. Transition from enantioselective high performance to ultra high performance liquid chromatography: A case study of a brush-type chiral stationary phase based on sub 5-micron to sub 2-micron silica particles. J. Chromatogr. A 2010, 1217, 990–999.

Carr, P.W.; Wang, X.; Stoll, D.R. Effect of pressure, particle size, and time on optimizing performance in liquid chromatography. Anal. Chem. 2009, 81, 5342–5353.

Cirilli, R.; Ferretti, R.; La Torre, F.; Secci, D.; Bolasco, A.; Carradori, S.; Pierini, M. High-performance liquid chromatographic separation of enantiomers and diastereomers of 2-methylcyclohexanone thiosemicarbazone, and determination of absolute configuration and configurational stability. J. Chromatogr. A 2007, 1172, 160–169.

Cirilli, R.; Costi, R.; Di Santo, R.; La Torre, F.; Pierini, M.; Siani, G. Perturbing effects of chiral stationary phase on enantiomerization second-order rate constants determined by enantioselective dynamic high-performance liquid chromatography: a practical tool to quantify the accessible acid and basic catalytic sites bonded on chromatographic supports. Anal. Chem. 2009, 81, 3560–3570.

Cirilli, R.; Costi, R.; Di Santo, R.; Gasparrini, F.; La Torre, F.; Pierini, M.; Siani, G. A rational approach to predict and modulate stereolability of chiral alpha substituted ketones. Chirality 2009a, 21, 24–34.

D'Acquarica, I.; Gasparrini, F.; Pierini, M.; Villani, C.; Zappia, G. Dynamic HPLC on chiral stationary phases: a powerful tool for the investigation of stereomutation processes. J. Sep. Sci. 2006, 29, 1508–1516.

D'Acquarica, I.; Gasparrini, F.; Kotoni, D.; Pierini, M.; Villani, C.; Cabri, W.; Di Mattia, M.; Giorgi, F. Stereodynamic investigation of labile stereogenic centres in dihydroartemisinin. Molecules 2010, 15, 1309–1323.

Dalla Cort, A.; Gasparrini, F.; Lunazzi, L.; Mandolini, L.; Mazzanti, A.; Pasquini, C.; Pierini, M.; Rompietti, R.; Schiaffino, L. Stereomutations of atropisomers of sterically hindered salophen ligands. J. Org. Chem. 2005, 70, 8877–8883.

Dell'Erba, C.; Gasparrini, F.; Grilli, S.; Lunazzi, L.; Mazzanti, A.; Novi, M.; Pierini, M.; Tavani, C.; Villani, C. Conformational studies by dynamic NMR. 86. structure, stereodynamics, and ryogenic enantioseparation of the stereolabile isomers of O-dinaphthylphenyl derivatives. J. Org. Chem. 2002, 67, 1663–1668.

Desmet, G.; Clicq, D.; Nguyen, D.T.-T.; Guillarme, D.; Rudaz, S.; Veuthey, J.-L.; Vervoort, N.; Torok, G.; Cabooter, D.; Gzil, P. Practical constraints in the kinetic plot representation of chromatographic performance data: theory and application to experimental data. Anal. Chem. 2006, 78, 2150–2162.

Ellis, L.A.; Roberts, D.J. Chromatographic and hyphenated methods for elemental speciation analysis in environmental media. J. Chromatogr. A 1997, 774, 3–19.

Felinger, A.; Pasti, L.; Dondi, F.; van Hulst, M.; Schoenmakers, P.J.; Martin, M. Stochastic theory of size exclusion chromatography: pk Shape analysis on single columns. Anal. Chem. 2005, 77, 3138–3148.

Fontana, A.; De Maria, P.; Pierini, M.; Siani, G.; Cerritelli, S.; Macaluso, G. Ab initio analysis on metal ion catalysis in the enolization reactions of some acetylheterocycles: kinetics of the enolization reactions of 3-acetyl-5-methylisoxazole, 5-acetyl-3-methylisoxazole and 3(5)-acetylpyrazole. J. Phys. Org. Chem. 2002, 15, 247–257.

Gasparrini, F.; Lunazzi, L.; Misiti, D.; Villani, C. Organic stereochemistry and conformational analysis from enantioselective chromatography and dynamic nuclear magnetic resonance measurements. Acc. Chem. Res. 1995, 28, 163–170.

Gasparrini, F.; Misiti, D.; Pierini, M.; Villani, C. Enantiomerization barriers by dynamic HPLC stationary phase effect. Tetrahedron: Asymm. 1997, 8, 2069–2073.

Gasparrini, F.; Lunazzi, L.; Mazzanti, A.; Pierini, M.; Pietrusiewicz, K.M.; Villani, C. Comparison of dynamic HPLC and dynamic NMR in the study of conformational stereodynamics: case of the enantiomers of a hindered secondary phosphine oxide. J. Am. Chem. Soc. 2000, 122, 4776–4780.

Gasparrini, F.; D'Acquarica, I.; Pierini, M.; Villani, C. Chromatographic resolution and enantiomerization barriers of axially chiral 1-naphthamides. J. Sep. Sci. 2001, 24, 941–946.

Gasparrini, F.; Grilli, R.; Leardini, L.; Lunazzi, L.; Mazzanti, A.; Nanni, D.; Pierini, M.; Pinamonti, M. Conformational studies by dynamic NMR. 89. Stereomutation and cryogenic enantioseparation of conformational antipodes of hindered aryl oximes. J. Org. Chem. 2002, 67, 3089–3095.

Gasparrini, F.; Pierini, M.; Villani, C.; De Maria, P.; Fontana A.; Ballini, R. Enantiomerization study of some a-nitroketones by dynamic high-resolution gas chromatography J. Org. Chem. 2003, 68, 3173–3177.
Giddings, J.C. Kinetic processes and zone diffusion in chromatography. J. Chromatogr. 1960, 3, 443–453.
Giddings, J.C. Comparison of theoretical limit of separating speed in gas and liquid chromatography. Anal. Chem. 1965, 37, 60–63.
Gritti, F.; Guiochon, G. Heat exchanges in fast, high-performance liquid chromatography. A complete thermodynamic study. Anal. Chem. 2008, 80, 6488–6499.
Guetens, G.; De Boeck, G.; Wood, M.; Maes, R.A.; Eggermont, A.A.; Highley, M.S.; van Oosterom, A.T.; de Bruijn, E.A.; Tjaden, U.R. Hyphenated techniques in anticancer drug monitoring. I. Capillary gas chromatography-mass spectrometry. J. Chromatogr. A 2002, 976, 229–238.
Guetens, G.; De Boeck, G.; Highley, M.S.; Wood, M.; Maes, R.A.; Eggermont, A.A.; Hanauske, A.; de Bruijn, E.A.; Tjaden, U.R. Hyphenated techniques in anticancer drug monitoring. II. Liquid chromatography-mass spectrometry and capillary electrophoresis-mass spectrometry. J. Chromatogr. A 2002a, 976, 239–247.
Guillarme, D.; Nguyen, D.T.-T.; Rudaz, S.; Veuthey, J.-L. Recent developments in liquid chromatography-impact on qualitative and quantitative performance. J. Chromatogr. A 2007, 1149, 20–29.
Holt, R.M.; Newman, M.J.; Pullen, F.S.; Richards, D.S.; Swanson, A.G. High-performance liquid chromatography/NMR spectrometry/mass spectrometry: further advances in hyphenated technology. J. Mass Spectrom. 1997, 32, 64–70.
Jacobson, J.; Melander, W.R.; Vaisnys, G.; Horváth, C. Kinetic study on cis-trans proline isomerization by high-performance liquid chromatography. J. Phys. Chem. 1984, 88, 4536–4542.
Jerkovich, A.D.; Mellors, J.S.; Jorgenson, J.W. The use of micrometer-sized particles in ultra high pressure liquid chromatography. LC–GC Eur. 2003, 16, 20–23.
Jung, M. Program Simul, No. 620. Quantum chemistry program exchange. QCPE Bull. 1992, 3, 12.
Katsanos, N.A.; Thede, R.; Roubani-Kalantzopoulou, F. Diffusion, adsorption and catalytic studies by gas chromatography. J. Chromatogr. A 1998, 795, 133–184.
Keller, R.A.; Giddings, J.C. Multiple zones and spots in chromatography. J. Chromatogr. 1960, 3, 205–220.
Knox, J.H. Practical aspects of LC theory. J. Chromatogr. Sci. 1977, 15, 352–364.
Kramer, R. Simultan-reaktionsgaschromatographie mit reversibler reaktion erster ordnung. I. J. Chromatogr. 1975, 107, 241–252.
Krupcik, J.; Oswald, P.; Majek; P.; Sandra, P.; Armstrong, D.W. Determination of the interconversion energy barrier of enantiomers by separation methods. J. Chromatogr. A 2003, 1000, 779–800.
Krupcik, J.; Oswald, P.; Spanik, I.; Majek, P.; Bajdichova, M.; Sandra, P.; Armstrong, D.W. On the use of a peak deconvolution procedure for the determination of energy barrier to enantiomerization in dynamic chromatography. Analusis 2000, 28, 859–863.
Krupcik, J.; Oswald, P.; Spanik, I.; Majek, P.; Bajdichova, M.; Sandra, P.; Armstrong, D.W. The use of computerized peak deconvolution for determination of energy

barrier to enantiomerization in dynamic gas chromatography. J. Microcol. Sep. 2000a, 12, 630–636.

Kusano, T.; Tabatabai, M.; Okamoto, Y.; Boehmer, V. The cone-to-cone interconversion of partially O-methylated calix[4]arenes: first experimental values for the energy barriers. J. Am. Chem. Soc. 1999, 121, 3789–3790.

Lunazzi, L.; Mancinelli, M.; Mazzanti, A.; Pierini, M. Stereomutation of axially chiral aryl coumarins. J. Org. Chem. 2010, 75, 5927–5933.

Mannschreck, A.; Andert, D.; Eiglsperger, E.; Gmahl, E.; Buchner, H. Chiroptical detection. Novel possibilities of its application to enantiomers. Chromatographia 1988, 25, 182–188.

Mannschreck, A.; Kiessl, L. Enantiomerization during HPLC on an optically active sorbent. Deconvolution of experimental chromatograms. Chromatographia 1989, 28, 263–266.

Marriott, P.; Trapp, O.; Shellie, R.; Schurig, V. Time-resolved cryogenic modulation reveals isomer interconversion profiles in dynamic chromatography. J. Chromatogr. A 2001, 919, 115–126.

Melander, W.R.; Lin, H.-J.; Jacobson, J.; Horváth, C. Dynamic effect of secondary equilibria in reversed-phase chromatography. J. Phys. Chem. 1984, 88, 4527–4535.

Neue, U.D.; Kele, M. Performance of idealized column structures under high pressure. J. Chromatogr. A 2007, 1149, 236–244.

Nishikawa, T.; Hayashi, Y.; Suzuki, S.; Kubo, H.; Ohtani, H. On-column enantiomerization of 3-hydroxybenzodiazepines during chiral liquid chromatography with optical rotation detection. J. Chromatogr. A 1997, 767, 93–100.

Oswald, P.; Desmet, K.; Sandra, P.; Krupcik, J.; Armstrong, D.W. Evaluation of reversible and irreversible models for the determination of the enantiomerization energy barrier for N-(p-methoxybenzyl)-1,3,2-benzodithiazol-1-oxide by supercritical fluid chromatography. Chirality 2002, 14, 334–339.

Oswald, P.; Desmet, K.; Sandra, P.; Krupcik, J.; Armstrong, D.W. Determination of the enantiomerization energy barrier of some 3-hydroxy-1,4-benzodiazepine drugs by supercritical fluid chromatography. J. Chromatogr. B 2002a, 779, 283– 295.

Page, M.I.; Jencks, W.P. Entropic contributions to rate accelerations in enzymic and intramolecular reactions and the chelate effect. Proc. Natl. Acad. Sci. U.S.A. 1971, 68, 1678–1683.

Pasti, L.; Cavazzini, A.; Felinger, A.; Martin, M.; Dondi, F. Single-molecule observation and chromatography unified by lévy process representation. Anal. Chem. 2005, 77, 2524–2535.

Pasti, L.; Cavazzini, A.; Nassi, M.; Dondi, F. Dynamic chromatography: a stochastic approach. J. Chromatogr. A 2010, 1217, 1000–1009.

Poppe, H. Some reflections on speed and efficiency of modern chromatographic methods. J. Chromatogr. A 1997, 778, 3–21.

Shustov, G.V.; Kadorkina, G.K.; Kostyanovsky, R.G.; Rauk, A. Asymmetric nitrogen. 67. Geminal systems. 41. Chiroptical properties of N-chloro and N-bromo derivatives of three-membered nitrogen heterocycles: aziridines and diaziridines. J. Am. Chem. Soc. 1988, 110, 1719–1726.

Shustov, G.V.; Varlamov, S.V.; Chervin, I.I.; Aliev, A.E.; Kostyanovsky, R.G.; Kim, D.; Rauk, A. Asymmetric nitrogen. 72. Geminal systems. 46. N-chlorooxaziridines:

optical activation, absolute configuration, and chiroptical properties. J. Am. Chem. Soc. 1989, 111, 4210–4215.

Siani, G.; Angelini, G.; De Maria, P.; Fontana, A.; Pierini, M. Solvent effects on the rate of the keto–enol interconversion of 2-nitrocyclohexanone. Org. Biomol. Chem. 2008, 6, 4236–4241.

Trapp, O.; Schurig, V. Stereointegrity of Tröger's base: gas-chromatographic determination of the enantiomerization barrier. J. Am. Chem. Soc. 2000, 122, 1424–1430.

Trapp, O.; Schurig, V. ChromWin - A computer program for the determination of enantiomerization barriers in dynamic chromatography. Comput. Chem. 2001, 25, 187–195.

Trapp, O.; Schurig, V. Approximation function for the direct calculation of rate constants and gibbs activation energies of enantiomerization of racemic mixtures from chromatographic parameters in dynamic chromatography. J. Chromatogr. A 2001a, 911, 167–175.

Trapp, O.; Schoetz, G.; Schurig, V. Stereointegrity of thalidomide: gas-chromatographic determination of the enantiomerization barrier. J. Pharm. Biomed. Anal. 2002, 27, 497–505.

Trapp, O.; Shellie, R.; Marriott, P.; Schurig, V. Simulation of elution profiles for two-dmensional dynamic gas chromatographic experiments. Anal. Chem. 2003, 75, 4452–4461.

Trapp, O. Evaluation and prediction of stereoisomerizations in comprehensive two-dimensional chromatography. J. Chem. Inf. Comput. Sci. 2004, 44, 1671–1679.

Trapp, O. Unified equation for access to rate constants of first-order reactions in dynamic and on-column reaction chromatography. Anal. Chem. 2006, 78, 189–198.

Uray, G.; Jahangir, S.; Fabian, W.M. On- and off-column enantiomerization of 4,4'-bis-quinolin-2-ones: a comparison of auto DHPLC y2k and DCXplorer calculated thermodynamic data generated by dynamic high performance liquid chromatography with theoretically calculated data. J. Chromatogr. A 2010, 1217, 1017–1023.

Wales, T.E.; Fadgen, K.E.; Gerhardt, G.C.; Engen, J.R. High-speed and high-resolution UPLC separation at zero degrees celsius. Anal. Chem. 2008, 80, 6815–6820.

Ward, T.J.; Ward, K.D. Chiral separations: fundamental review 2010. Anal. Chem. 2010, 82, 4712–4722.

Wilkins, C.L. Hyphenated techniques for analysis of complex organic mixtures. Science 1983, 222, 291–296.

Wolf, C.; König, W.A.; Roussel, C. Influence of substituents on the rotational energy barrier of atropisomeric biphenyls—studies by polarimetry and dynamic gas chromatography. Liebigs Ann. 1995, 781–786.

Wolf, C. Stereolabile chiral compounds: analysis by dynamic chromatography and stopped-flow methods. Chem. Soc. Rev. 2005, 34, 595–608.

Wolf, C. Dynamic Stereochemistry of Chiral Compounds. The royal society of chemistry: Cambridge, United Kingdom, 2008.

Wu, N.; Lippert, J.A.; Lee, M.L. Practical aspects of ultrahigh pressure capillary liquid chromatography. J. Chromatogr. A 2001, 911, 1–12.

4

High-temperature Liquid Chromatography

Thorsten Teutenberg

Institut für Energie- und Umwelttechnik e. V. (IUTA), (Institute for Energy and Environmental Technology), Duisburg, Germany

A Brief Definition of High-temperature Liquid Chromatography

What is High-temperature Liquid Chromatography?

Although the question "what is high-temperature liquid chromatography" seems trivial, it is not easy to give an answer. Up to now, a single definition of this technique does not exist, although it has emerged as the topic of many scientific meetings and symposia. Looking through the literature, a range of terms has been used:

- Subcritical water chromatography (Fogwill & Thurbide, 2007; 2008; He & Yang, 2003; Kondo & Yang, 2003; Lamm & Yang, 2003; Miller & Hawthrone, 1997; Teutenberg et al., 2006a; Yang et al., 1999; Yang et al., 2002a; Yang et al., 2002b; Yang, 2007; Yisong et al., 2005)
- Subcritical fluid chromatography (West & Lesellier, 2005)
- Elevated-temperature liquid chromatography (Abbott et al., 1981; Bidlingmeyer & Henderson, 2004; Chen & Horvath, 1993; Coym & Dorsey, 2004b; Hao et al., 2007; Jones & Yang, 2003; Le et al., 1996; Le et al., 1998; Li et al., 1997; Lippert et al., 2007; Neue & Mazzeo, 2001; Sandra & Vanhoenacker, 2007; Sheng et al., 1997; Thompson et al., 2001; Ueki et al., 2006; Vanhoenacker & Sandra, 2006; Xiang et al., 2006; Zhu et al., 2004)
- Superheated water chromatography (Al-Khateeb & Smith, 2008; Chienthavorn & Smith, 1999; Chienthavorn et al., 2004; Chienthavorn et al., 2005; Coym & Dorsey, 2004a; Dugo et al., 2007; Fields et al., 2001; Ingelse et al., 1998; Kephart & Dasgupta, 2002; Nakajima et al., 2003; Smith & Burgess, 1996; 1997; Smith et al., 1999; Smith, 2006; 2008; Tajuddin & Smith, 2002; 2005; Teutenberg et al., 2001; Wilson, 2000; Yarita et al., 2003; Yarita et al., 2005)
- Hot eluent liquid chromatography (Kephart & Dasgupta, 2000)

- (Ultra) High-temperature liquid chromatography (HT-HPLC) (Albert et al., 2005; Djordjevic et al., 1998; Djordjevic et al., 1999; Giegold, 2008; Giegold et al., 2008b; Gika et al., 2008; Gritti & Guiochon, 2007; Guillarme & Heinisch, 2005; Heinisch et al., 2008; Louden et al., 2001; McNeff et al., 2007; Molander et al., 1999; Nguyen et al., 2007; Pereira, 2006; Pereira et al., 2007; Plumb et al., 2007; Riddle & Guiochon, 2006; Sanagi et al., 2004a; Sanagi et al., 2004b; Shen et al., 2006; Stoll et al., 2006; Teutenberg, 2009; Teutenberg et al., 2009b; Teutenberg et al., 2009c; 2009d; Teutenberg, 2010; Trones et al., 1998; Trones et al., 1999; Vanhoenacker & Sandra, 2005; Yamaki et al., 1996; Yang et al., 2005)
- Thermal aqueous liquid chromatography (TALC) (Guillemin et al., 1981)
- And others (Kondo et al., 2002)

It is difficult to select a temperature range and to assign this region to define "high-temperature liquid chromatography", as it will be called throughout this chapter. In my opinion, one must consider the mobile phases used. Since I will exclusively discuss reversed-phase liquid chromatography (RP-HPLC), relevant binary solvent systems that are used in RP-HPLC will be considered and are listed in Table 4.A along with other solvents which might also be used as modifiers.

Whereas water has the highest boiling point among common solvents due to strong hydrogen bonding, the boiling points for the other solvents are much lower. For instance, acetone starts to boil at 56°C. For the much more common solvents methanol and tetrahydrofuran, the normal boiling temperatures are 65°C and 66°C, respectively. This means that from the perspective of the pure components, a lower temperature limit of about 60°C for high-temperature liquid chromatography would be appropriate. Adjusting the temperature above 60°C then requires raising the outlet pressure of the column above atmospheric pressure. Otherwise, a phase transition would be inevitable when a solvent gradient is run from pure water to pure organic solvent. Therefore, increasing the temperature above 60°C means that the domain of high-temperature liquid chromatography is entered in reversed-phase HPLC.

Now that we have defined the lower temperature limit, we can try to define the upper temperature limit. Again, it is helpful to have a look at the data presented in Table 4.A. Besides the normal boiling temperature, I have listed the critical temperatures of these solvents. From a purely thermodynamic standpoint, a liquid turns into a supercritical fluid once it is above the critical temperature. Most of the organic solvents will become a supercritical fluid around 230°C to 240°C, while this temperature is much higher for water. Therefore, increasing the temperature above 374°C means that the domain of supercritical fluid chromatography has been entered and we have completely exited the domain of high-temperature liquid chromatography.

Table 4.A. Compilation of the Most Important Physical Property Data of the Pure Substances.

Solvent	Normal boiling-temperature* [°C]	Critical pressure* p_{cr} [bar]	Critical temperature* T_{cr} [°C]
Water	100.0	220.5	374.2
Acetonitrile	81.9	48.3	274.9
Acetone	56.3	47.0	235.0
Ethanol	78.3	63.8	243.1
Methanol	64.6	81.0	239.5
2-Propanol	82.3	47.6	235.2
Tetrahydrofuran	66.0	51.9	267.2

*Data taken from Teutenberg et al., 2009b.

What Is the Useful Temperature Range?

Even if it is possible to use the complete temperature range for high-temperature HPLC (Kephart & Dasgupta, 2002), the question needs to be addressed of what is the highest temperature which can be used in routine analysis? The requirements to make use of this technique are a stationary phase that is stable at the highest temperature you would like to apply, and a heating system that is able to generate the desired temperature. Most conventional LC heating systems are only capable of raising the temperature to 80°C. Although it is possible with every chromatographic system that is equipped with a column oven to enter the domain of high-temperature liquid chromatography, the region cannot be exploited further. Therefore, some instrument manufacturers have developed special heating systems which have an upper temperature limit of about 200°C (Selerity Technologies Inc, 2010; SIM—Scientific Instruments Manufacturer GmbH, 2010; ZirChrom Separations Inc, 2010). One must consider another very important aspect: the stationary phase. In the last few years, column manufacturers have created silica-based reversed-phase columns with considerably improved stability. From many recent studies, data shows that temperatures as high as 200°C will not lead to an immediate collapse of the column, and some columns can be used over a reasonably long time without total degradation (Teutenberg et al., 2009a).

Although the domain of high-temperature liquid chromatography potentially extends up to 374°C, the useful temperature range for routine analysis that will be considered in this chapter is currently limited to approximately 200°C. This is quite reasonable, because specially designed heating systems, as well as suitable stationary phases, that both generate and withstand these temperatures are now commercially available.

The Heating System

Without a specially designed heating system, the advantages of high eluent temperatures cannot be put into practice.

Preheating of the Mobile Phase

Silica or polystyrene column packings are relatively poor thermal conductors compared to stainless steel. Preventing the formation of axial and radial temperature gradients within the column requires preheating the mobile phase. (Thompson et al., 2001).

First of all, let us consider what happens to the flow profile of the mobile phase inside the column if the eluent is not thermally equilibrated. An illustration has been given by Wolcott and Snyder, which is also used here to highlight the effect of band broadening caused by thermal mismatch (Wolcott et al., 2000). Fig. 4.1 depicts an HPLC column that is heated to 70°C. Fig. 4.1.A illustrates the situation when the mobile and stationary phases are both thermally equilibrated. In this case, there is no band-broadening due to thermal mismatch. The second example (Fig. 4.1.B) describes the problem when the mobile and stationary phases are not thermally equilibrated. Here, the eluent is not preheated to the temperature of the stationary phase, and it enters the column with a low temperature (e.g., 22°C).

Since only the column walls are heated, the solvent near the column wall heats up faster than the solvent in the centre of the column and a radial temperature gradient builds up. Consequently, the viscosity of the eluent along the wall is lower than that at the column centre and hence, the mobile phase moves faster at the column wall than in the centre of the column. This generates a convex flow profile, causing the chromatographic band to broaden, which may

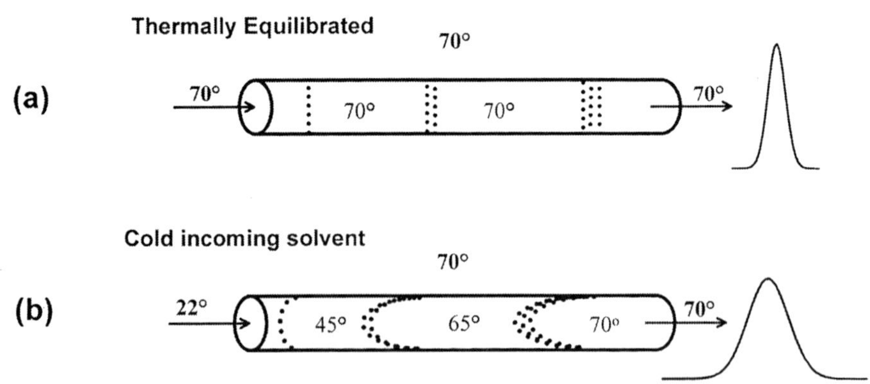

Fig. 4.1. Figure 1 from Wolcott et al., 2000. Band broadening due to thermal effects. (a) Ideal case, no thermal effects; (b) effect of incoming mobile phase that is at a lower temperature than the column. Assume oven temperature of 70°C. Numbers shown inside column suggest plausible solvent temperature at column center. With permission from Elsevier.

even result in split peaks. Thus, the overall separation quality and efficiency are both reduced. In addition to the radial temperature gradient, there is also an axial temperature gradient, because the temperature of the mobile phase gradually approaches the oven temperature at the column outlet.

Fig. 4.2 compares a block heating oven with a preheating unit to an air-circulation oven without a preheater. The chromatograms reveal that for the air-circulation oven, the elution for all compounds is shifted to higher retention times because the mobile phase is not brought to the temperature of the stationary phase. Although the temperature can be accurately controlled with a precision of about +/- 0.1 K for most commercially available heating ovens, this does not automatically mean that there is a homogenous temperature profile in an axial direction within the column.

A much more dramatic example is given in Fig. 4.3. This experiment was performed to visualize the effect of peak distortion at very high eluent temperatures. The upper chromatogram in Fig. 4.3.A resulted when the mobile phase was not preheated before it entered the column. In this case, peaks are broad and distorted. In contrast, the lower chromatogram in Fig. 4.3.B was obtained when the mobile phase was adequately preheated to the temperature of the stationary phase. Here, the peaks have a smaller band width and a Gaussian shape. Moreover, the retention of all analytes except uracil is shifted to Iower times. This means that the temperature within the column is higher

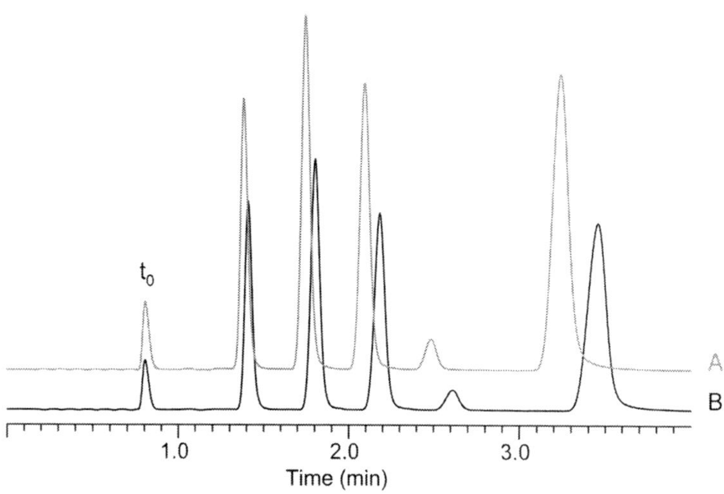

Fig. 4.2. Chromatographic comparison of a contact oven (A) with an air-based oven (B). Separation of five PAH and uracil on a Waters XBridge BEH-C_{18} column (75 mm L × 4.6 mm id, 2.5 µm d_p) with Shimadzu LC-10. Mobile phase: water/methanol, 19/81 (v/v). Detection: UV at 254 nm. Flow rate: 1 mL/min. Injection volume: 1 µL. Column temperature: 80 °C. Column oven: A: SIM HT-HPLC 200 (Scientific Instruments Manufacturer, Oberhausen, Germany); B: Column Thermostat Jetstream 2 (Thermotechnic Products, Langenzersdorf, Austria). Teutenberg et al., 2010, unpublished results.

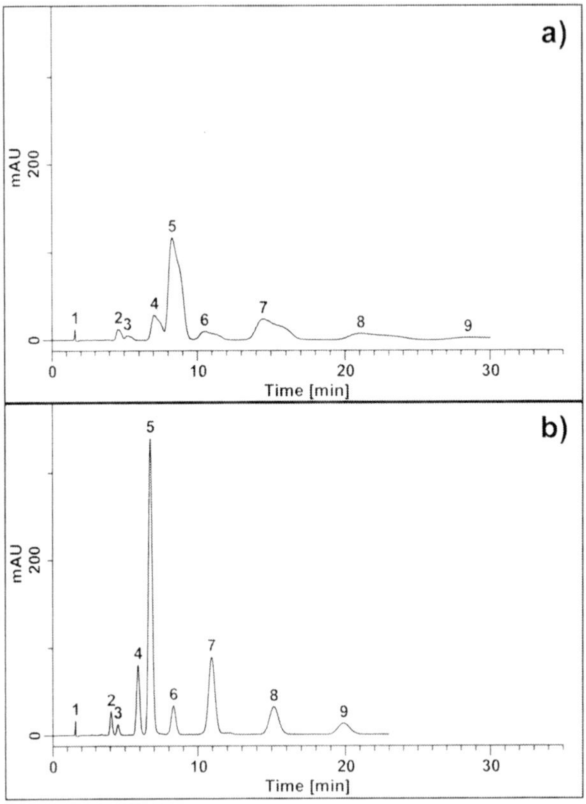

Fig. 4.3. Effect of eluent preheating at very high temperature; a) without preheating; b) with preheating. Separation of eight PAH and uracil on a Waters XBridge BEH-C_{18} column (75 mm L × 4.6 mm id, 2.5 μm d_p) with Shimadzu LC-10 and SIM HT-HPLC 200 column oven. Mobile phase: water/acetonitrile, 50/50 (v/v). Detection: UV at 254 nm. Flow rate: 0.5 mL/min. Injection volume: 3 μL. Column temperature: 150 °C. Peak assignment: 1) uracil; 2) naphthalene; 3) acenaphthylene; 4) fluorene; 5) anthracene; 6) pyrene; 7) chrysene; 8) benzo[k]fluoranthene; 9) indeno[1,2,3-cd]pyrene. Teutenberg & Wiese, 2009, unpublished results.

when the mobile phase is preheated. This example demonstrates that mobile phase preheating is a must, especially when columns with a diameter of 2 mm or greater are used.

Column Heating

Air-bath Ovens

There are different approaches to heat the mobile and stationary phases. The most common heating ovens for liquid chromatography are based on the same principle as GC ovens. Here, air is used to heat the column to the desired temperature. It is important to emphasize that the heat transfer of air is very poor compared to a contact medium like water, silicone oil or a heated metal block. This is because air is a good insulator, especially if the air is not blown against the surface it should be controlling. While air-bath ovens might work well if the column heater is only used for maintaining a constant temperature, rapid temperature changes, e.g., temperature programming, results in long equilibration times. However, long equilibration times become critical not only when the temperature is increased to 200°C, but also when the temperature is only raised slightly above ambient temperature as shown in a recent study by de Villiers and Sandra (de Villiers et al., 2006).

Water-jacket Ovens

A water jacket oven is much more efficient than an air-bath oven when the heat transfer is considered. A decisive advantage is that the fluid medium is in close contact with the column wall, thereby enabling a fast heat transfer. However, there are technical difficulties, which have to be considered such as potential contamination of the fluid medium with the mobile phase. (Teutenberg, 2010).

Block-heating Ovens

A third option to heat the mobile and stationary phases is to use a contact heater. To establish a tight contact between the column and the heating unit, the column must be enclosed between two aluminium shells, which are connected with a heating block (Teutenberg et al., 2006a). To optimize the heat transfer, the aluminium shells must be tailor-made for each column. Otherwise, the formation of small voids can lead to local temperature differences in the radial and axial direction of the column.

Post-column Cooling of the Mobile Phase

Heating the mobile and stationary phases up to a temperature of 200°C means that the eluent must be cooled down prior to detection. This is also required if special hyphenation techniques are applied, which is discussed in the "Special Hyphenation Techniques" section. Although many detectors are equipped with thermostatted cells, the mobile phase will not be cooled down to ambient temperature if the eluent enters the detector with a very high temperature. Modern heating systems should therefore allow for precise eluent cooling before the mobile phase is introduced into the detector.

Temperature Programming

A more demanding approach is to use temperature gradients in liquid chromatography. This technique is explained in the "Method Development Using Temperature as an Active Variable" section, which focuses on method development strategies. Moreover, the "Special Hyphenation Techniques" section will give examples based on new technologies that rely on temperature programming. However, it is useful to first discuss how a temperature gradient works, in contrast to a solvent gradient. Usually, if the temperature is increased during a chromatographic run, the same effect on analyte retention is observed as for solvent gradient mode, which means that the elution strength of the mobile phase is increased (Teutenberg et al., 2009c). The difference between these two gradient modes is that the solvent front always moves from the inlet to the outlet of the column, while temperature immediately affects the whole column. Consider that, depending on the type of column oven, a significant lag may occur between the oven temperature and the programmed gradient (Nikitas et al., 2008). This temperature delay depends significantly on the heating system and should be as small as possible.

System Set-up

Fig. 4.4 depicts the design of a column oven based on contact heating. Here, the transfer capillary leading from the injector to the column is tightly enclosed between two aluminium shells. Due to the close contact between the aluminium block and the stainless steel capillary, fast and precise preheating of the eluent is possible. The precisely defined length of the preheating capillary presents another advantage of this set-up. This guarantees that if the capillary needs to be replaced, the length of the preheating capillary will remain constant. This is a decisive advantage if a method transfer has to be made to a different HPLC system. Provided that the same column oven is used and the connecting capillary is kept constant, no change of retention times will be observed. This differs from conventional air-based column ovens where the length of the connecting capillary is not controlled (Spearman et al., 2004). In this case, the method may have to be adjusted in order to compensate for the shift in retention times.

For eluent cooling, the capillary is also placed between two aluminium shells, and a Peltier element is used to cool the mobile phase prior to detection. While the influence of temperature on the chromatographic parameters is important in method development, the influence of temperature on the signal-to-noise ratio should also be considered. A post column cooling unit may be essential for detectors where temperature has a negative impact on the signal. This is true for the fluorescence detector, where the signal can be quenched if the inlet temperature is increased (Agarwal, 2008). In contrast, a gain in the signal intensity might be achieved for detectors where the liquid mobile phase is converted to a gas during the detection process, which holds true for the evaporative light scattering detector (ELSD), the charged aerosol detector (CAD) and the mass spectrometer (Pereira et al., 2007; Teutenberg, 2010). Therefore, a modular oven concept where the preheating temperature, the column temperature and the post-column cooling temperature can be adjusted individually offers the greatest flexibility to optimize the chromatography and the limit of detection. The column oven depicted in Fig. 4.4 includes all features listed above, and is the most flexible heating system currently commercially available.

Besides the use of a specially designed heating system, only minor modifications need to be made to the HPLC system. Most importantly, the mobile phase must be kept from boiling inside the column. This can be accomplished by using either a back-pressure regulator or a transfer capillary that generates the desired back-pressure. Fig. 4.5 depicts the complete system set-up for high-temperature HPLC.

An understanding of the relationship between vapor pressure of the mobile phase and temperature is mandatory for a proper back-pressure adjustment. The lack of this understanding often prevents the practitioner from using temperature as an active variable in liquid chromatography.

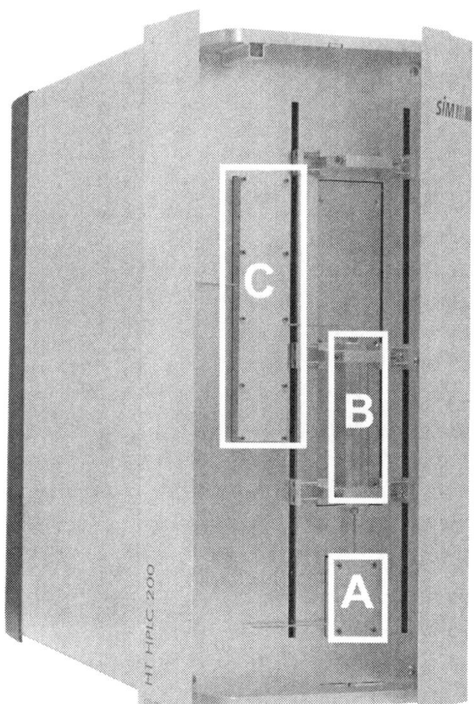

Fig. 4.4. Modular heating oven for temperature programming based on contact heating. A: eluent preheating unit; B: column heating unit; C: eluent post column cooling unit. (See http://www.sim-gmbh.de/index.php?option=com_content&task=view&id=64&Itemid=502&lang=en)

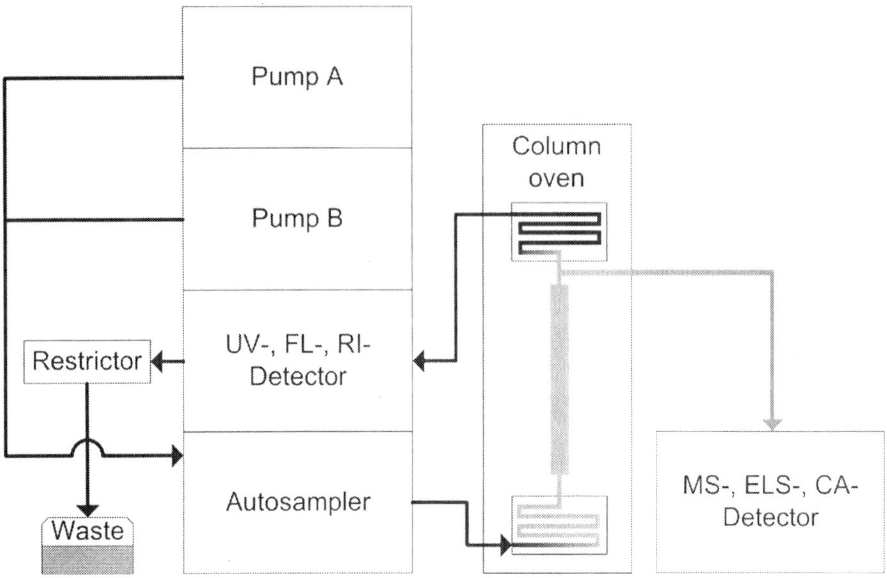

Fig. 4.5. Figure 9 from Teutenberg et al., 2009b. Scheme of a typical HPLC system with a modular heating oven for high-temperature operation. With permission from Elsevier.

Mobile Phase Considerations

The practitioner generally lacks information about the aggregate state of the mobile phase and its dependence on temperature and pressure. It is difficult to search for these data in single publications. In many cases, these data are published in physical chemistry journals, which are mostly overlooked by the chromatographic community. Therefore, we have collected and measured the most important data comprising the vapor pressure, the viscosity and the static permittivity over a large temperature interval from ambient temperature up to 200°C, or even 250°C, for all the binary solvent mixtures given in Table 4.A. These data have been published in three consecutive papers in the Journal of Chromatography A (Teutenberg et al., 2009b; 2009c; 2009d).

Influence of Temperature on Vapor Pressure

In the literature, data are readily found for the vapor pressure of the pure components. However, in reversed-phase liquid chromatography, separations are not usually carried out with a mobile phase consisting of a pure solvent. Instead, the mobile phase consists of water and an organic co-solvent, which is denoted with % A and % B, respectively. Water is the weakest solvent in RP

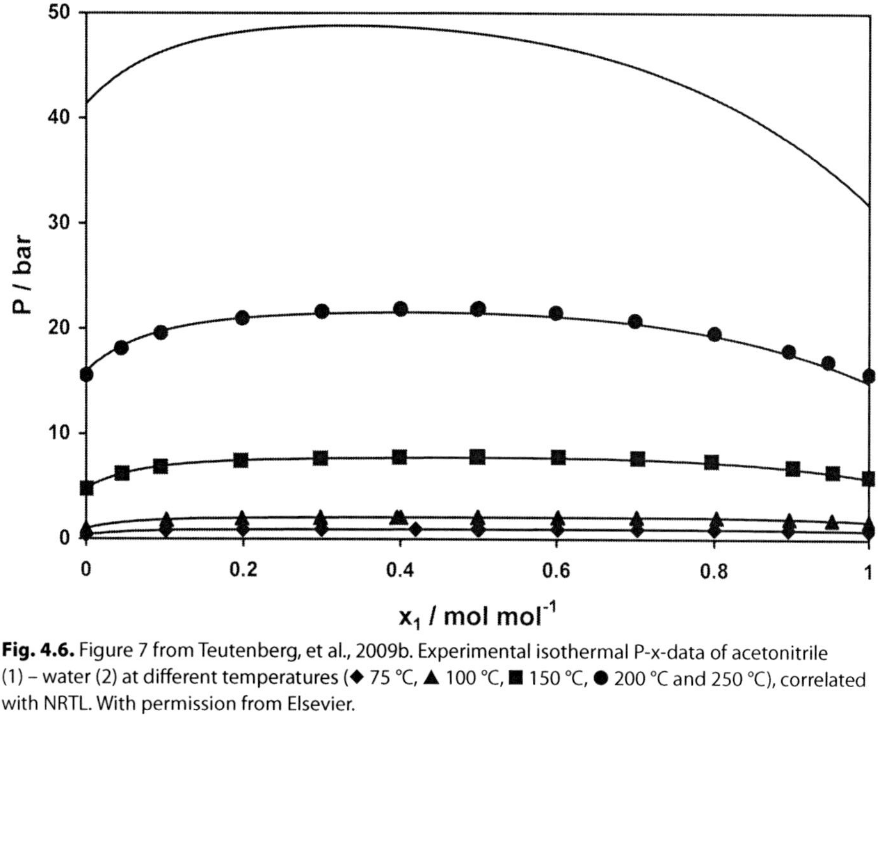

Fig. 4.6. Figure 7 from Teutenberg, et al., 2009b. Experimental isothermal P-x-data of acetonitrile (1) – water (2) at different temperatures (◆ 75 °C, ▲ 100 °C, ■ 150 °C, ● 200 °C and 250 °C), correlated with NRTL. With permission from Elsevier.

separations, and hence the elution strength of the mobile phase is varied by applying a solvent gradient. Usually, the solvent gradient starts with a high water concentration and ends with a high concentration of the organic solvent. Therefore, a description of the bubble point line of binary eluent mixtures is necessary. But how do the vapor pressures of these mixtures change at a defined temperature when the concentration of the organic solvent is increased? Examples are listed for the two solvent systems water–acetonitrile and water–methanol, which are depicted in Figs. 4.6 and 4.7.

For the system water–acetonitrile, the vapour pressure runs through a maximum as the concentration of the organic modifier changes from 0 to 100%, whereas the system water–methanol always exhibits its highest vapor pressure for pure methanol. Moreover, higher temperature results in higher vapor pressure. Therefore, the back pressure must be adjusted to prevent the mobile phase from boiling. The highest back pressure observed at a temperature of 200°C is for the system water–methanol, which is around 40 bar.

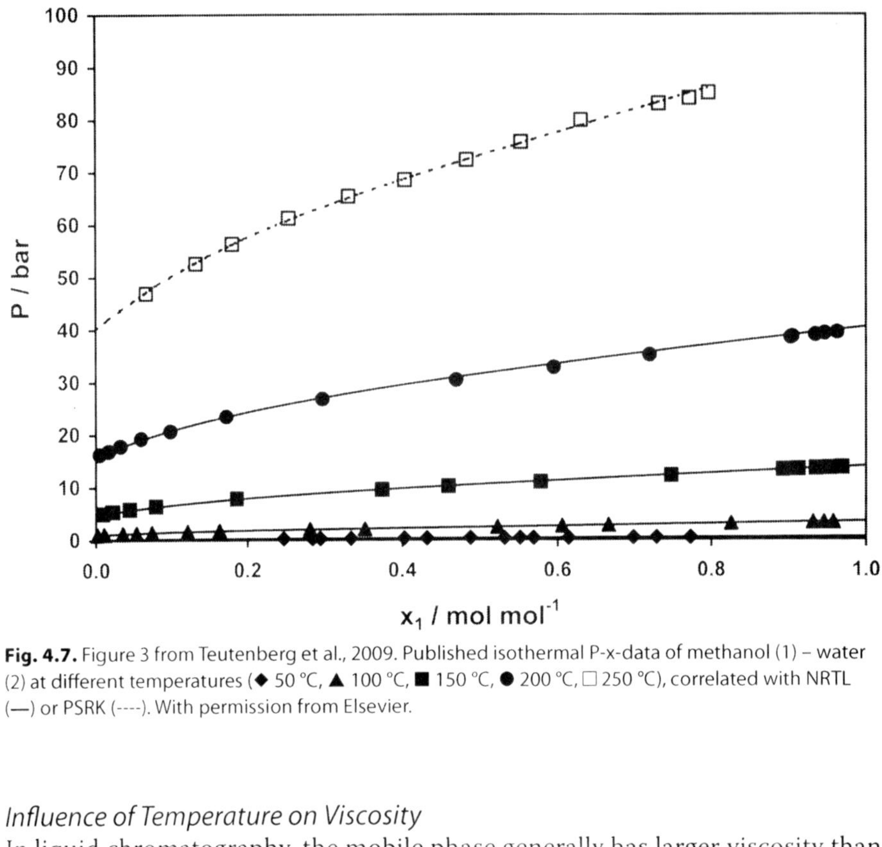

Fig. 4.7. Figure 3 from Teutenberg et al., 2009. Published isothermal P-x-data of methanol (1) – water (2) at different temperatures (◆ 50 °C, ▲ 100 °C, ■ 150 °C, ● 200 °C, □ 250 °C), correlated with NRTL (—) or PSRK (----). With permission from Elsevier.

Influence of Temperature on Viscosity

In liquid chromatography, the mobile phase generally has larger viscosity than in gas chromatography. This is a drawback because the high viscosity causes a low diffusion coefficient and strong flow resistance of the mobile phase resulting in a high back pressure and a high mass transfer resistance.

For a pure compound, e. g. liquid water, the viscosity decreases as temperature increases. When a solvent gradient is applied at ambient temperature, a more or less steep increase in the system pressure is observed. This increase in the system pressure strongly depends on the organic co-solvent. Every practitioner probably knows that a binary mixture of water–methanol is much more troublesome than the respective mixture of water–acetonitrile in solvent gradient mode. The reason is that most liquid chromatographic methods are run at ambient temperature, where the system water–methanol exhibits a huge viscosity maximum. It seems that in many cases this has led to a widespread reluctance to use methanol instead of acetonitrile. Fortunately, as the temperature increases, the huge "excess" viscosity which is experienced at ambient temperature lowers, which is highlighted in Figs. 4.8 and 4.9.

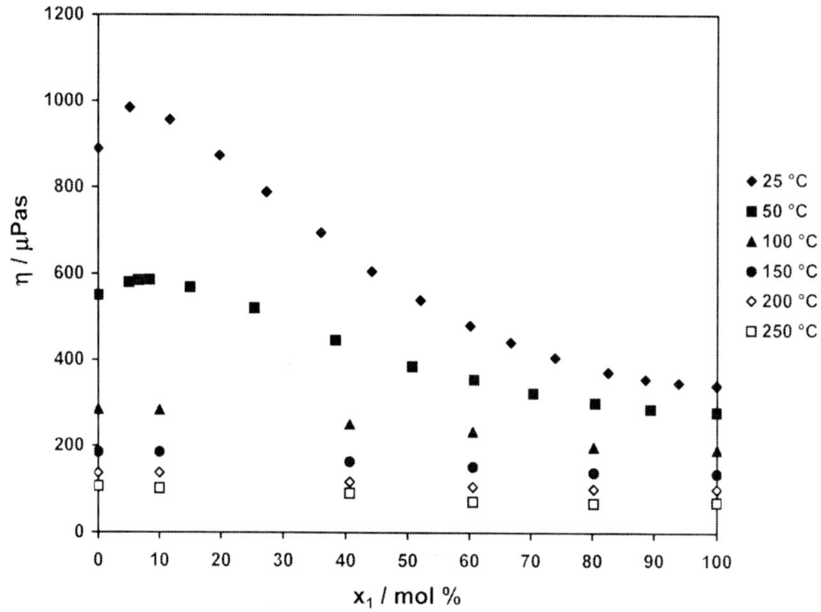

Fig. 4.8. Figure 7 from Teutenberg et al., 2009. Experimentally determined viscosities of the binary mixture acetonitrile (1) – water (2) at different temperatures and 100 bar. With permission from Elsevier.

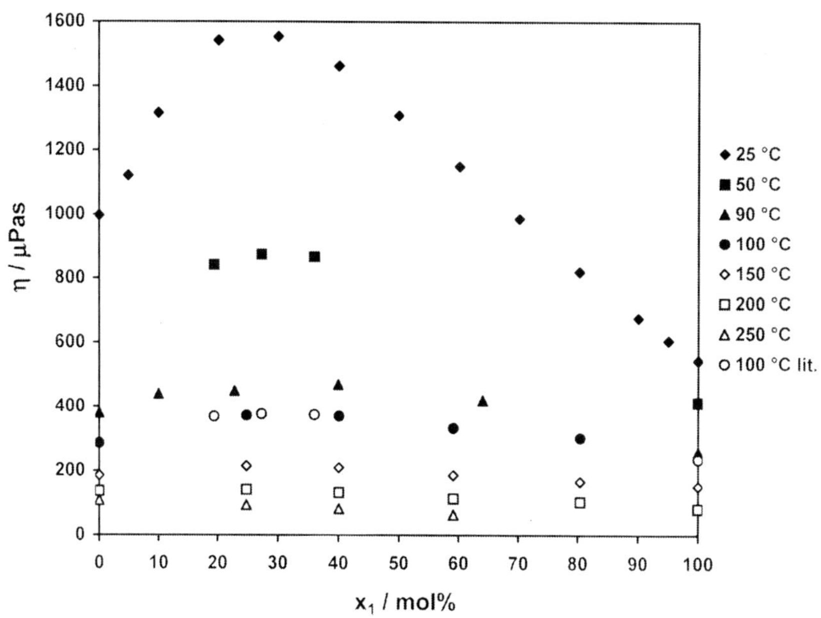

Fig. 4.9. Figure 5 from Teutenberg et al., 2009d. Experimentally determined viscosities of the binary mixture methanol (1) – water (2) at different temperatures and 100 bar. With permission from Elsevier.

What is the practical impact of these results? Since there is hardly a difference between water–acetonitrile and water–methanol mixtures at temperatures above 150°C, both mixtures can be used at the same conditions. This means that for a given flow rate, similar maximum pressures for both mixtures are observed.

Influence of Temperature on Static Permittivity

The static permittivity is a measure of the polarity of a substance and like vapor pressure and viscosity, it is strongly dependent on the temperature. (In literature, often the old term "dielectric constant" instead of "static permittivity" is found. But from a physical point of view this is not acceptable because the static permittivity of a solvent or a mixture is always a function of pressure and temperature.) In general, the static permittivity decreases when the temperature increases. Especially with water, the decrease of static permittivity is responsible for the dissociation of the hydrogen bonds.

The static permittivity of pure water around ambient temperature is much higher than the static permittivity of methanol or acetonitrile. For this reason, water is a weak solvent in reversed-phase HPLC. By increasing the concentration of the organic co-solvent during a solvent gradient, the elution strength of the mobile phase is increased, which means that the time the solute spends in the stationary phase can be greatly reduced. Now it is clear why a temperature gradient has the same effect as a solvent gradient in RP-HPLC. Increasing the temperature reduces the static permittivity of the mobile phase. Hence, the solvent strength of the mobile phase increases even if the concentration of the organic co-solvent is kept constant.

In the scientific literature, most studies on high-temperature HPLC have focused on a pure water mobile phase (Coym & Dorsey, 2004a; Smith et al., 1999; Smith, 2006; 2008; Teutenberg et al., 2001). However, the same rules apply for binary mixtures composed of water and an organic co-solvent, which means that by increasing temperature or applying a solvent gradient the elution strength of the mobile phase will increase. Fig. 4.10 shows the dependence of the static permittivities of the binary solvent mixture water–acetonitrile as a function of temperature and composition.

The interested reader will find more data covering all relevant solvent systems in a recently published article, which also covers the measuring technique for the determination of the static permittivities over a large temperature interval (Teutenberg et al., 2009c).

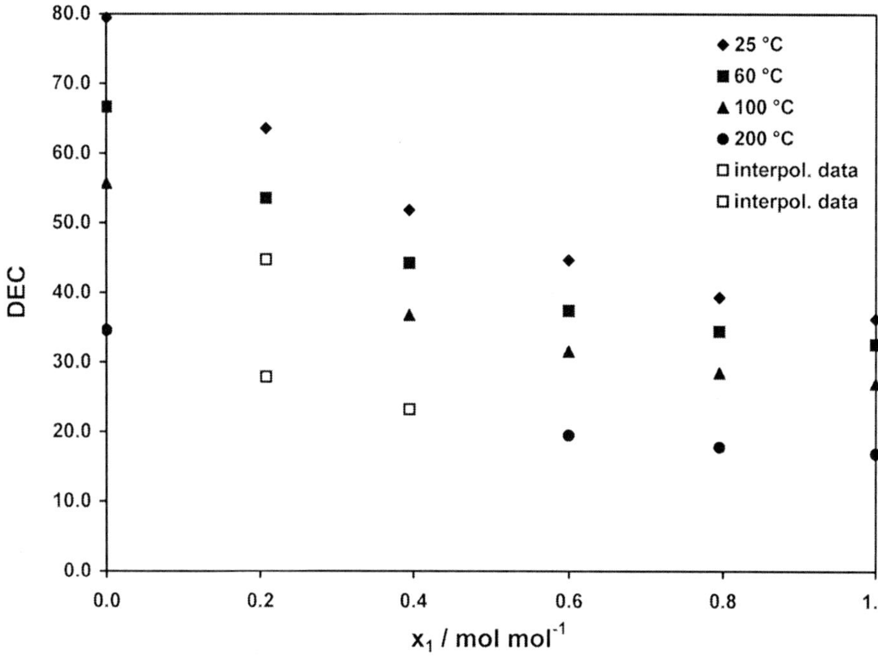

Fig. 4.10. Figure 3 from Teutenberg et al., 2009c. Static permittivities of the binary solvent mixture acetonitrile (1) – water (2) as a function of temperature and composition. With permission from Elsevier.

The System Water-THF

In liquid chromatography, the components of the mobile phase should be fully miscible in any concentration. Otherwise, problems can arise because of the immiscibility of the mobile phase components. At a first glance, all solvent systems which are studied in this chapter are fully miscible in any concentration. However, this is not true for the system water–tetrahydrofuran which deserves special attention. As the temperature increases above 70°C, the two solvents start to separate because the system is characterized by a miscibility gap, which is pressure and temperature dependent. The extent of the miscibility gap is illustrated in Fig. 4.11 at two pressures. As pressure increases, the miscibility gap becomes less pronounced. Above a critical pressure of 247 bar, the miscibility gap is nonexistent. (Please note that in this context, "critical pressure" refers to the pressure above which the miscibility gap is nonexistent.

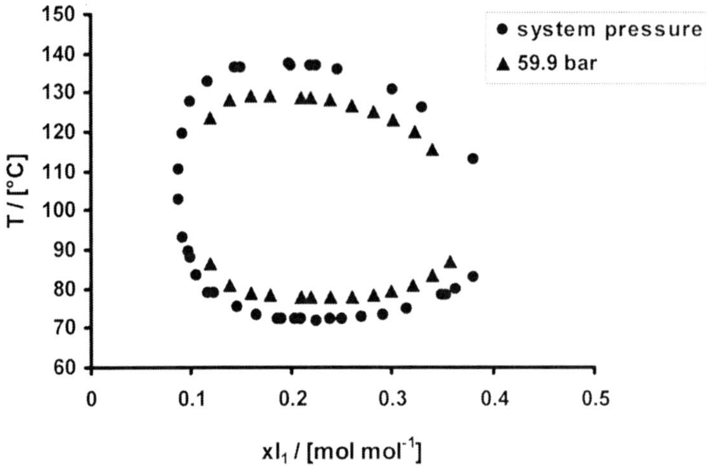

Fig. 4.11. Liquid-liquid phase equilibrium of tetrahydrofuran (1) and water (2). For details, see text.

This has nothing to do with the definition of the critical pressure of a supercritical fluid.)

Although this mixture exhibits a miscibility gap in the temperature range between 70 and 140°C, it can be used as a mobile phase in high-temperature liquid chromatography. However, for the practical application of high-temperature liquid chromatography, the miscibility gap should be avoided. This means that the temperature must be adjusted below 70 or above 140°C when a solvent gradient is applied over the full concentration range. If it is necessary to adjust the temperature between 70°C and 140°C, the concentration of THF should always be higher than 40 mol percent. In this case, the miscibility gap does not form regardless of the pressure and temperature in the column.

Suitable Stationary Phases

Although high-temperature liquid chromatography attracts much interest, data on the stability of common stationary phases at high temperatures is limited. Some authors have published valuable information about the ruggedness of different types of columns at elevated temperatures and extended pH (Claessens et al., 1996; Fonseca et al., 2004; He & Yang, 2003; Teutenberg et al., 2006b; Teutenberg et al., 2007; Teutenberg et al., 2009a; Wilson, 2000; Yang, 2003). However, a standardized protocol for a column aging procedure at high temperatures has not existed until now. As a result, the data generated by different authors are difficult to compare. In the last few years, our laboratory has developed several test procedures to monitor the stability of HPLC columns at high eluent temperature and extreme pH. This section concentrates which columns can be used at extreme conditions.

Undoubtedly, the column is still the most important part of the whole chromatographic system because it provides the greatest amount of control over the selectivity and chemical interaction with the analyte. Moreover, the column must be replaced more often than other hardware components. Therefore, the best column for a separation is not necessarily a column that gives the fastest separation. The batch-to-batch reproducibility of a column is a very strong criterion. If a method has been validated and is run in a regulated environment, it can be devastating to have to re-validate a complete method because the manufacturer made some changes to the packing material. This has led to a widespread use of silica-based stationary phases whose bonding chemistry is well understood and which exhibit an excellent batch-to-batch reproducibility. In the last few years, hybrid materials based on silica gel have emerged as the most rugged stationary phases, when compared to "conventional" silica reversed-phase stationary phases (Al-Khateeb & Smith, 2008; Liu et al., 2005; Shen et al., 2006; Teutenberg et al., 2009a). An example highlighting the extraordinary stability of these phases is given in Fig. 4.12 which compares a silica-based hybrid stationary phase and a polybutadiene coated zirconium dioxide stationary phase, after both columns were exposed for 25 hours to a temperature of 150°C. Obviously, there is hardly any change in the chromatographic properties for both columns.

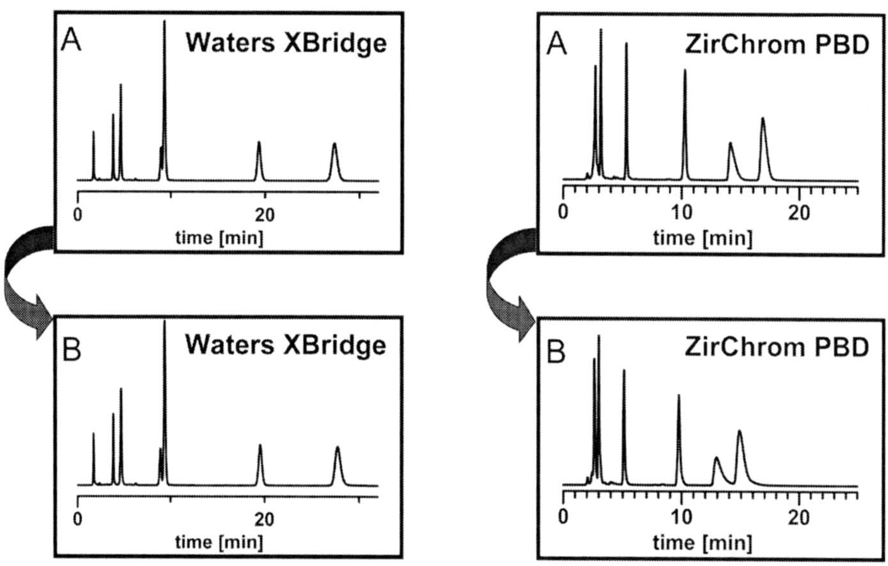

Fig. 4.12. Stability of a Waters XBridge and a ZirChrom PBD column. A: test chromatogram obtained for the brand new columns. B: test chromatogram obtained after the columns were exposed for 25 hours to a mobile phase consisting of water/methanol (90/10, v/v). For detailed test procedure, see Teutenberg et al., 2009a.

Table 4.B lists all columns tested in the last several years, according to a procedure that is published elsewhere (Teutenberg et al., 2009a). When the upper temperature limit is approached, the pH of the mobile phase should lie between 3 and 6. A higher or lower pH might lead to a faster degradation, and thus the maximum temperature should be decreased to prolong the column lifetime. Please note that there might be other columns which also have an enhanced temperature stability. It is likely that column manufacturers will further improve the stability of their materials in the near future.

When it is only necessary to increase the speed of a separation, a temperature around 100°C will, in most cases, lead to a significant acceleration of the elution. In this respect, other materials like the Zorbax StableBond column, the pHidelity or Blaze column can be used. A higher stability is needed when special hyphenation techniques are employed and the mobile phase consists of pure water. The commercial availability of hybrid silica particles with extended pressure, temperature and pH ranges is really a boost for high-temperature HPLC, because silica-based packings are the "gold-standard" in the pharmaceutical industry. Nevertheless, I'm also quite optimistic that alternative materials will be used on a large scale once the methods and techniques I will describe in the "Special Hyphenation Techniques" section become standard procedures in routine laboratories.

Table 4.B. Overview of the Stationary Phases Which Can be Used at Elevated Temperatures.

Manufacturer	Column description	Maximum temperature/°C
Agilent	Zorbax SB C-18	100
Phenomenex	Gemini NX C18	120
Polymer Laboratories	PLRP-S	200
Restek	pHidelity C-18	100
Sachtleben	Sachtopore NP	200[Δ]
Sachtleben	Sachtopore RP	150[▲]
Selerity	Blaze 200 C-18	100
Thermo	Hypercarb	200
Waters	XBridge C-18	150-180
YMC	Triart C-18	120
ZirChrom	Carb	200
ZirChrom	Phase	200[Δ]
ZirChrom	PBD	120-150
ZirChrom	PS	120-150

[Δ]This is a normal phase column without a surface modification.

[▲]This phase should only be used with a pure water mobile phase at temperatures above 80°C.

Method Development Using Temperature as an Active Variable

Influence of Temperature on Resolution

A probable aim of all chromatographic separation procedures is to optimize the resolution while minimizing the analysis time. Equation 4.1 is the well-known master equation of resolution in liquid chromatography. This section discusses the influence of temperature on the resolution as proposed by Knox and Thijssen. [Knox and Thijssen assumed equal peak width ($w_2 = w_1$) and consideration of the retention of the first peak of the critical pair. Other expressions were derived by Purnell or Said. Please note that you will find a lot of other "Master" equations, as they are frequently termed. In principle, they all contain the three terms which are also included in this equation.]:

$$R = \left(\frac{k_1}{k_1 + 1}\right) \cdot (\alpha - 1) \cdot \frac{\sqrt{N}}{4} \tag{4.1}$$

This fundamental equation describes how resolution depends on three terms: retention k, selectivity α and efficiency N. Each of these terms contributes to alter the resolution of the separation and can be influenced by temperature, which will be shown in the following paragraphs.

Influence of Temperature on Retention

Usually, retention will decrease if the temperature is increased in reversed-phase HPLC. The influence of temperature on retention is given by the van't Hoff equation (4.2):

$$\ln k = -\frac{\Delta H}{R} \cdot \frac{1}{T} + \frac{\Delta S}{R} + \ln \beta \tag{4.2}$$

Here, ΔH is the enthalpy of solute transfer from the mobile into the stationary phase, ΔS is the entropy of solute transfer from the mobile into the stationary phase, R is the ideal gas constant and β is the volume phase ratio of the stationary and mobile phases.

Assuming that all analytes strictly obey the van't Hoff equation, a van't Hoff plot can be constructed when the retention of a compound is measured at two different temperatures. Both points can then be connected with a straight line and retention can be predicted at any given temperature. Therefore, measuring the retention factor at two different temperatures is sufficient to describe the peak movement as the temperature is increased. It is the same concept which is incorporated into method development software which assists the practitioner to predict the retention of sample components as the temperature is varied.

Influence of Temperature on Selectivity

The ability of a chromatographic system to discriminate between different analytes is called selectivity (α). Selectivity is determined as the ratio of the retention factors of two analytes and can be written as:

$$\alpha = \frac{k_2}{k_1} \tag{4.3}$$

As shown in the previous paragraph, temperature has a significant influence on retention. Hence, temperature can also be used to control the selectivity of the separation, although in some textbooks it is stated that selectivity is not affected by temperature unless temperature modifies the nature of the analyte. It is also wrong to claim that at higher temperatures, the selectivity will always decrease. Both statements are untrue, and this section will explain why temperature can be used to change the selectivity of the phase system. Usually, increasing the temperature leads to decreasing retention in reversed-phase HPLC as shown before. Regardless of whether the analytes obey the van't Hoff equation or if a curvature is observed, the selectivity will change if the ratio of k_2 and k_1 is not constant with temperature (Teutenberg, 2009). This means that even if all analytes obey the van't Hoff equation, a change in the separation factor may be obtained if the slopes of the van't Hoff plots for the selected analytes differ as demonstrated by Edge et al. (2006).

Influence of Temperature on Efficiency

The discussion on how temperature influences the chromatographic efficiency is controversial. The van Deemter equation, which should be known to every chromatographer, and can be written as:

$$H_u = A + \frac{B}{u} + C \cdot u \tag{4.4}$$

Here, the Height Equivalent to a Theoretical Plate H(u) (HETP) depends on three terms, which are the band broadening due to eddy diffusion (A-term), longitudinal diffusion (B-term) and the resistance to mass transfer between and within the mobile and stationary phases (C-term) and the mobile phase flow rate u. Physically, it is assumed that the A-term does not depend on temperature. However, the remaining B- and C-term are both temperature dependent. This is because the B-term is directly proportional to the diffusion coefficient while the C-term is inversely proportional to the diffusion coefficient D_M, which is temperature-dependent:

$$B \propto D_M \qquad C \propto \frac{1}{D_M} \tag{4.5 and 4.6}$$

From a purely theoretical standpoint, the goal is always to minimize band broadening and thus minimize H by adjusting the flow rate of the mobile phase to the optimum linear velocity. This is highlighted in a plot of the HETP against the linear velocity of the mobile phase in Fig. 4.13.

At velocities higher and lower than the optimum linear velocity there is an increase of the H(u)-curve. However, when the temperature is increased, the profile of this curve changes. The minimum of the H(u)-curve is shifted to higher linear velocities. In addition, there is a much flatter increase of H at flow rates higher than the optimum. Some authors describe this as a "flattening out" of the van Deemter curve. This means that if a separation is carried out at a mobile phase flow rate that is much higher than the optimum flow rate, the loss in efficiency at higher temperatures is less pronounced than at lower temperatures. Note that because the optimum linear velocity is shifted to higher flow rates, working at low velocities below the optimum flow rate can result in a significant loss of efficiency when the temperature is increased. This means that the flow rate should be high enough to suppress the effects of longitudinal diffusion of molecules. The net benefit of operating HPLC columns at higher temperatures therefore is that the operator needs not worry much about the flow rate as long as it is higher than the optimum linear velocity. However, it must be stressed that there is no absolute increase in the

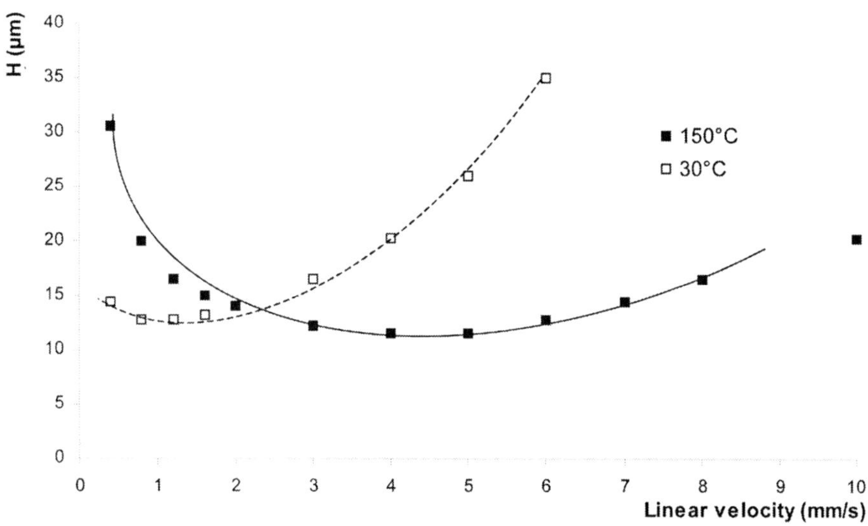

Fig. 4.13. Figure 2 from Vanhoenacker & Sandra, 2006. Plot of plate height (H) vs. linear velocity (u) for propylparaben (k ~ 3.6 at 30°C) on a Blaze 200 column (150 mm L × 4.6 mm id, 5 µm d_p). Mobile phase: water/ACN 60: 40, detection: DAD at 210 nm. Copyright Wiley-VCH Verlag GmbH & Co. KGaA. Reproduced with permission.

efficiency, because it is not possible to lower the minimum of the van Deemter curve. This is often not correctly presented or is misunderstood if people speak of an increase in the efficiency by increasing temperature.

Method Development

This chapter does not give a full description of how method development is carried out in liquid chromatography, instead there are a number of excellent publications and also text books which should be consulted. Dolan and Snyder have contributed significantly to the development of suitable software tools that can be used for method development, and the interested reader should refer to the respective literature (Dolan et al., 1979; Dolan et al., 1987; Dolan et al., 1989; Dolan et al., 1998a; Dolan et al., 1998b; Dolan et al., 1999a; 1999b; Dolan et al., 1999c; Snyder et al., 1979; Snyder et al., 1989; Zhu et al., 1996a; Zhu et al., 1996b; Zhu et al., 1996c; Zhu et al., 1996d). Recently, the authors published an updated version of a book on high performance gradient elution (Snyder & Dolan, 2007). The practitioner will find exhaustive information on how to proceed with structured method development, which is not governed by trial and error. In our laboratory, we also use computer optimization software like DryLab® or ChromSwordAuto which are indispensable tools to develop, optimize and validate liquid chromatographic methods.

The chromatograms that are shown in the following paragraphs are meant to make the practitioner familiar with the basic concept of using temperature for method development. This section will start with isothermal and isocratic separations and then move to more complex method development strategies, including a combination of solvent and temperature gradient programming.

Isothermal and Isocratic Separations

The first application is a separation of steroids on a polybutadiene-coated zirconium dioxide stationary phase at isothermal and isocratic conditions. The column used had a length of 150 mm and an internal diameter of 4.6 mm. The particle diameter was 3 µm. The flow rate was adjusted to 1 mL/min and detection was performed at 200 nm, because most of the target substances did not show a high UV activity at longer wavelengths. The temperature was adjusted to 25°C and the mobile phase consisted of 75/25 (v/v) water-acetonitrile. Since in this case retention cannot be influenced by a pH adjustment, the mobile phase does not contain any additives.

The resulting chromatogram in Fig. 4.14 at a temperature of 25°C looks not very exciting. All the compounds eluted with strong tailing, and the peak widths were excessively broad. Moreover, the analysis time was about 20 minutes. In order to optimize the separation, the temperature was increased to 120°C using pure water as the mobile phase. The separation now takes more than 30 minutes, but the resolution between adjacent peaks is very high so that the temperature can be further increased. As becomes obvious, by

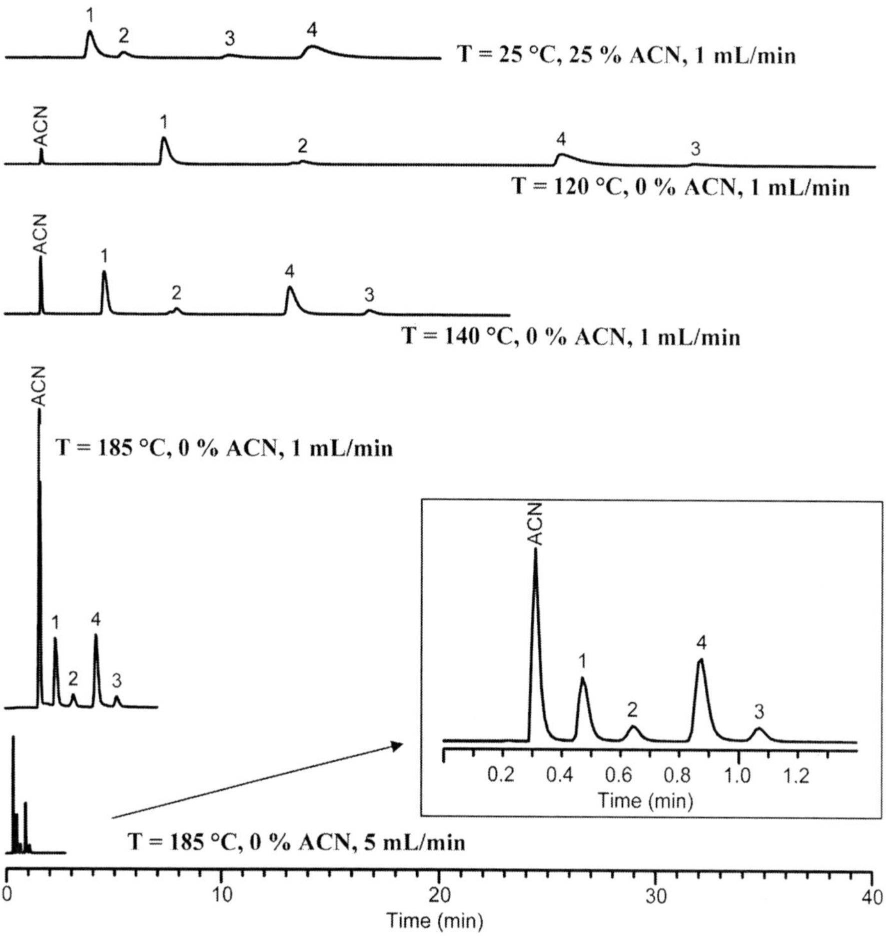

Fig. 4.14. Isocratic and isothermal separation of a mixture containing four steroids on a polybutadiene-coated zirconium dioxide column (ZirChrom-PBD, 4.6 × 150 mm, 3 μm, 300 Å). Peak assignment: (1) estriol, (2) androstadiendione, (3) dehydroepiandrosterone, (4) estrone; detection: UV at 200 nm. Chromatographic conditions: see figure.

increasing the temperature to 140 and 185°C, the analysis time can be reduced to 18 and 5.5 minutes, respectively.

But we can make it even faster. At 185°C, a significant increase of the flow rate due to the lower viscosity of water can be achieved. We were able to increase the flow rate to 5 mL/min without exceeding the recommended pressure limit of this column at 300 bars, which led to an additional five-fold reduction in analysis time. A close inspection of the peaks in the insertion window reveals that the temperature of the mobile phase matched the temperature of the stationary phase, because all the peaks eluted symmetrically and no peak splitting was observed. More importantly, the increase in flow rate led to a proportional decrease of the retention time for all analytes. On doubling the flow rate, retention was halved. In other words: if there had been a so-called thermal mismatch between the eluent entering the column and the stationary phase, radial and axial temperature gradients would have built up in the column as was demonstrated in Fig. 4.1. However, the absence of peak splitting and the respective decrease of the analytes' retention times unambiguously confirm that the mobile phase preheating is adequate and also effective at high flow rates up to 5 mL/min.

Temperature Gradient and Isocratic Separation
Another example refers to the separation of a sulfonamide mixture on a silica C-18 bridged ethyl hybrid stationary phase using water as the mobile phase. This method was developed using the DryLab software, selecting the temperature gradient mode for liquid chromatography. First of all, two basic measurements were performed in a temperature range between 60°C and 180°C. The slopes of the temperature gradients were adjusted to 2 and 6°C/min so that all compounds eluted within the gradient window. Afterwards, the retention data was used to calculate the necessary parameters for the DryLab software. The practitioner is then able to optimize the method at the computer, which means that no further experiments are necessary. Fig. 4.15 shows the optimized method for the separation of the sulfonamide mixture. The run time of the method is about 13 minutes, and the resolution between the critical peak pair sulfadiazine and sulfathiazole is 3.4.

Fig. 4.16 shows the comparison between the simulated chromatogram as predicted by the DryLab software and the experimental run. The highest error is only 1.5% for the predicted retention times. Therefore, a prediction can be made with high precision.

In this example, pure water has been used as the mobile phase. The change of the elution strength was achieved with a temperature gradient instead of a solvent gradient. Contrary to an isocratic separation, in solvent gradient mode, a band compression can be achieved which reduces peak broadening. Why is this? When a solvent gradient is considered, a band migrates through the column in a mobile phase of continuously increasing strength. As a result, the tail of the band always moves in a mobile phase that is slightly stronger than the mobile phase at the front of the band. The tail therefore tends to move faster than the front. The resulting narrowing of the band is referred to as gradient compression, which serves to partly counteract the normal broadening of the band during migration through the column (Gritti & Guiochon, 2008a; 2008b; Neue et al., 2006). Applying a temperature gradient yields a different result. In this case, the temperature immediately affects the whole band. This also means that a compression is not observed when the column is operated isocratically under temperature-programmed

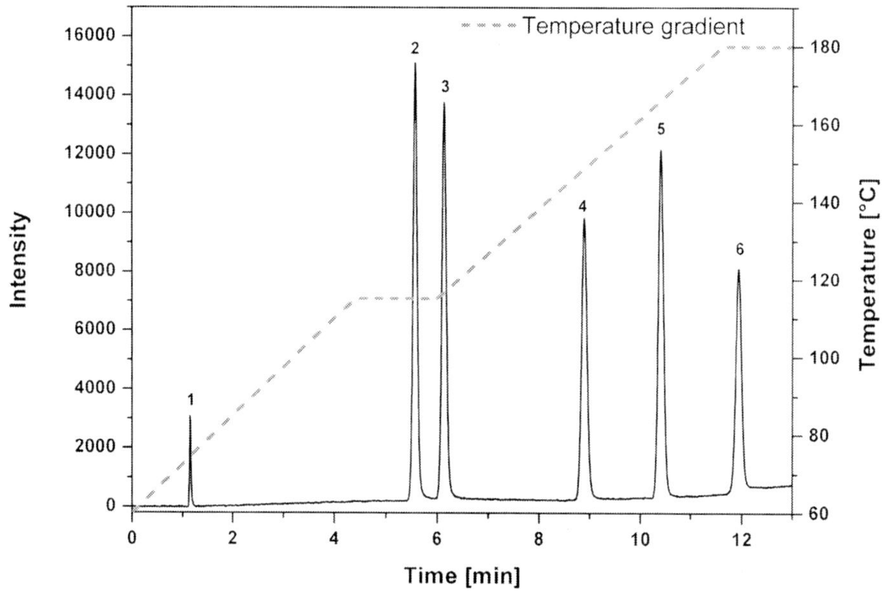

Fig. 4.15. Chromatogram of the separation of five sulfonamides and uracil with a temperature gradient on a Waters XBridge BEH-C_{18} column (75 mm L × 4.6 mm id, 2.5 µm d_p) with Shimadzu LC-10. Mobile Phase: 100% deionized water with 0.1% HCOOH; Detection: UV at 270 nm. Flow rate: 1.0 mL/min. Injection volume: 1 µL. Temperature gradient: see figure. Peak assignment: 1: uracil, 2: sulfadiazine, 3: sulfathiazole, 4: sulfamerazine, 5: sulfamethoxazole, 6: sulfamethazine.

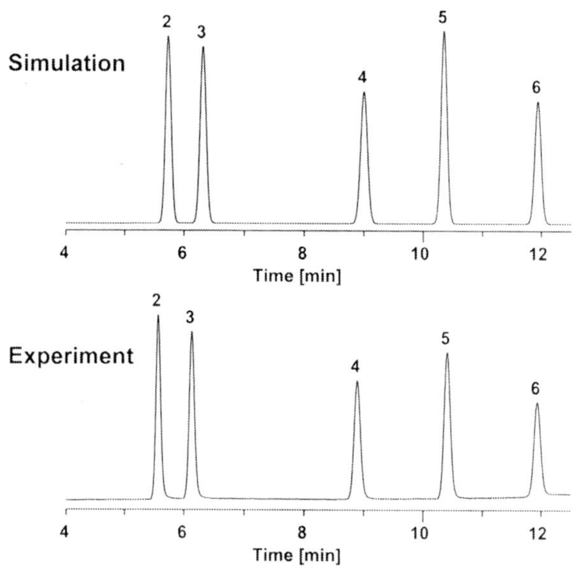

Fig. 4.16. Comparison of the simulated and experimentally obtained chromatograms of the separation depicted in Fig. 4.15.

conditions. Therefore, if the method does not require a pure water mobile phase, I would strongly recommend using a binary mobile phase in high-temperature liquid chromatography. The next paragraph will show how the separation can be significantly accelerated by the combined use of solvent and temperature gradients.

Simultaneous Temperature and Solvent Gradient Separation
The goal was to establish a fast method for the separation of sulfonamides and trimethoprim, where all target analytes were at least baseline resolved. The equipment that was available for method development consisted of an Agilent HPLC system with a maximum operating pressure of 400 bars. In order to achieve a fast elution, the temperature had to be increased to decrease the viscosity of the mobile phase. First, we performed isothermal measurements at 70 and 90°C in solvent gradient mode. Neither the separation at 70°C nor the separation at 90°C was successful. At the lower temperature, the last peak pair was not fully resolved, while at the higher temperature, co-elution of the first peaks could be observed. Hence, a temperature gradient was applied in addition to the solvent gradient. The temperature was linearly increased from 70 to 90°C, because at the lower temperature, a better resolution is achieved for the first peaks while the higher temperature is beneficial for the late-eluting compounds as is depicted in Fig. 4.17.

It was possible to speed up the separation even further by adjusting the solvent and temperature gradients. Both gradients were optimized by applying two-step gradients and the final method is depicted in Fig. 4.18.

It should also be examined if the retention times can be precisely reproduced in consecutive runs or if it was a mere accident that the optimal resolution was achieved with a concomitant temperature gradient. Therefore, Fig. 4.18 shows an overlay of nine consecutive runs in the insertion window, highlighting that a shift in retention times was not observed.

This example highlights the fact that the simultaneous use of solvent and temperature programming can significantly enhance the selectivity of the separation as was also pointed out by Nikitas et al. (2008) and Pappa-Louisi et al. (2009). Even small changes in temperature can influence the peak movement decisively. Of course, this is an advanced technique, and temperature programming should only be performed if the heating system allows for a fast temperature change. Indeed, there are other heating systems which could in principle also be used for temperature programming, but one must keep in mind that the cooling of the mobile phase can take as long as 30 minutes. In this case, speaking of fast chromatography would be ridiculous, and then it makes no practical sense to apply temperature programming. For the method depicted in Fig. 4.18, the re-equilibration of the temperature could be achieved in the same time as for the re-equilibration of the mobile phase,

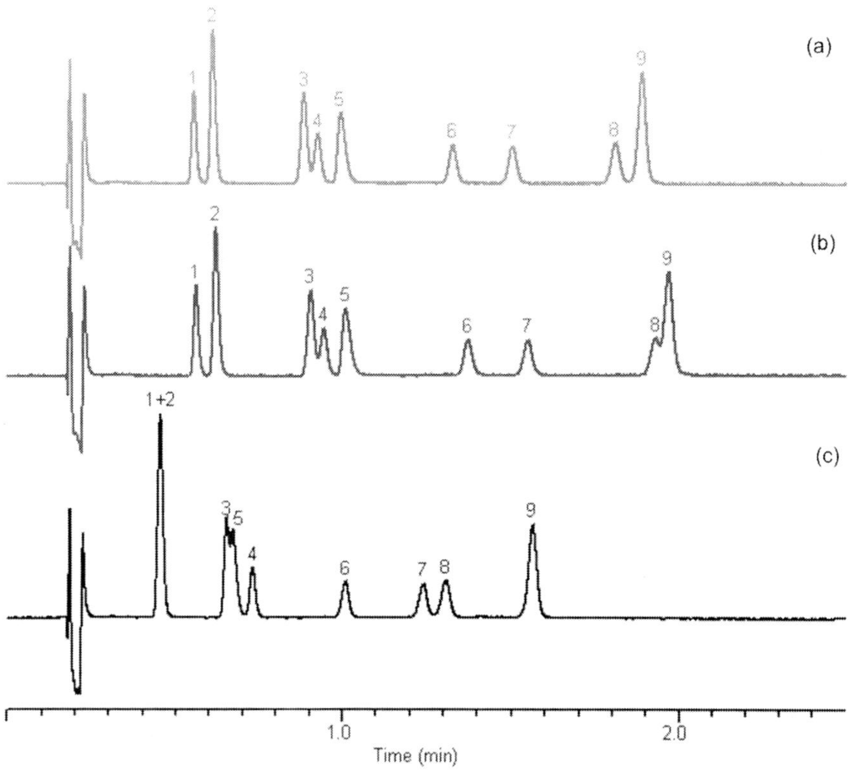

Fig. 4.17. Figure 1 from Giegold et al., 2008b. Copyright Wiley-VCH Verlag GmbH & Co. KGaA. Reproduced with permission. Chromatograms of the separation of eight sulfonamides and trimethoprim with solvent and (a) temperature gradient from 70 to 90°C in 2 min, (b) 70°C isothermal and (c) 90°C isothermal. Stationary phase: Agilent Zorbax Stable Bond C_{18} column (50 mm L × 3.0 mm id, 1.8 µm d_p) with Agilent 1100. Injection volume: 1 µL of a 500 µg/mL sample. Peak assignment: 1: sulfadiazine, 2: sulfathiazole, 3: N^4-acetylsulfadiazine, 4: sulfamerazine, 5: trimethoprim, 6: N^4-acetylsulfamerazine, 7: sulfamethazine; 8: sulfamethoxazole, 9: N^4-acetylsulfamethazine. Flow rate: 1.4 mL/min; solvent gradient: from 7 to 15% acetonitrile within 2.5 min; solvent A: 0.1% formic acid in deionized water, solvent B: 0.1% formic acid in acetonitrile.

because the heating system has a refrigeration compressor for a very fast cooling of the stationary phase.

As was already outlined, whenever possible, computer optimization software should be used to model the retention of the sample compounds. Currently, work is being carried out in our laboratory to implement temperature gradient optimization for method development in high-temperature HPLC. The first results are quite encouraging, as shown in the "Temperature Gradient and Isocratic Separation" section. Although temperature programming is

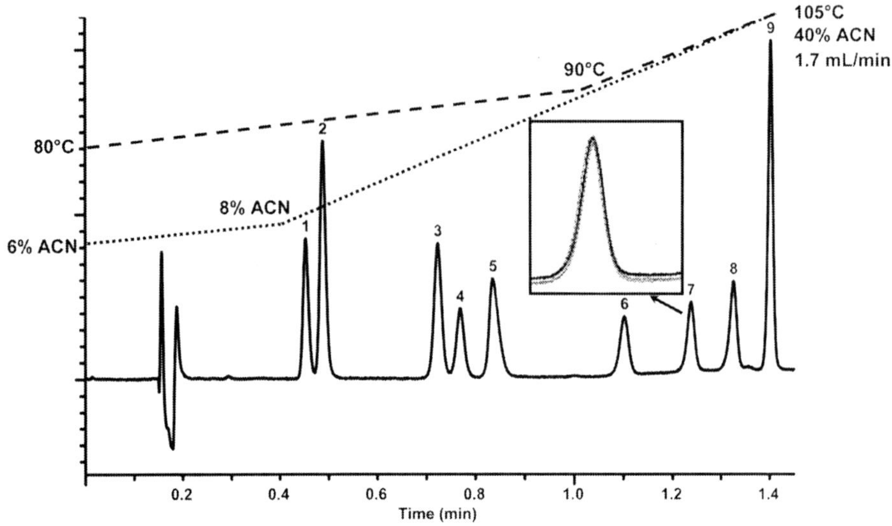

Fig. 4.18. Determination of sulfonamides and trimethoprim using high temperature HPLC with simultaneous temperature and solvent gradient on Agilent Zorbax Stable Bond C_{18} column (50 mm L × 3.0 mm id, 1.8 μm d_p) with Agilent 1100. Insertion window shows overlay of nine chromatograms. Peak assignment: 1: sulfadiazine, 2: sulfathiazole, 3: N^4-acetylsulfadiazine, 4: sulfamerazine, 5: trimethoprim, 6: N^4-acetylsulfamerazine, 7: sulfamethazine; 8: sulfamethoxazole, 9: N^4-acetylsulfamethazine. For chromatographic conditions, see figure.

regarded as too specialized to be of any practical value in most routine laboratories, the example depicted in Fig. 4.17 unambiguously reveals that even small changes in temperature can greatly affect the selectivity. In this respect, the availability of advanced software packages will be of great help for the practitioner, who otherwise will waste precious time without finding the optimum separation conditions.

Analyte Stability

One of the most frequently asked questions is whether a compound may undergo rapid degradation during the time it spends in the heated column. In this context, hydrolysis, oxidation, isomerisation and epimerization are the major types of chemical reactions that affect the stability of analytes and are considered undesirable. This fear has contributed significantly to the widespread reluctance to use high-temperature HPLC in the pharmaceutical industry.

A very easy approach to evaluate analyte stability is to use a UV detector at a fixed wavelength. In liquid chromatography, the peak areas at a fixed wavelength should be constant, provided that the flow rate is not changed. It is the same assumption that is also used in the DryLab software for peak tracking when UV detection is used. Therefore, a simple experiment can be performed where dependence of the peak areas of the target analytes on the temperature are measured. If all other parameters are kept constant, the peak area of the compounds should not change. A plot of the peak area against the temperature for all analytes will reveal if there is a major deviation from this rule. Fig. 4.19 shows a plot that exemplifies the steroid mixture already introduced in the "Isothermal and Isocratic Separations" section. Clearly the peak areas remain constant over the studied temperature range. This means that a degradation of the selected steroids was not observed.

However, a thermally induced degradation may occur even at a much lower temperature, which is highlighted in Fig. 4.20. Here, the same experiment was carried out using different model compounds.

Amoxicillin starts to degrade at 40°C, while the other compounds are stable over the selected temperature range. This means that it is not possible to define a threshold temperature above which a thermal degradation will occur. It depends on the analyte and the chromatographic conditions. Other factors like the stationary phase might also contribute to a degradation as was shown recently by Giegold et al. (2008a). Therefore, analyte stability should always be determined on the column which is used to run the method. Do not use or select a reference column to measure analyte stability and then transfer the method to another column. The result might be that the analyte is stable on one column but will degrade on another column under identical conditions.

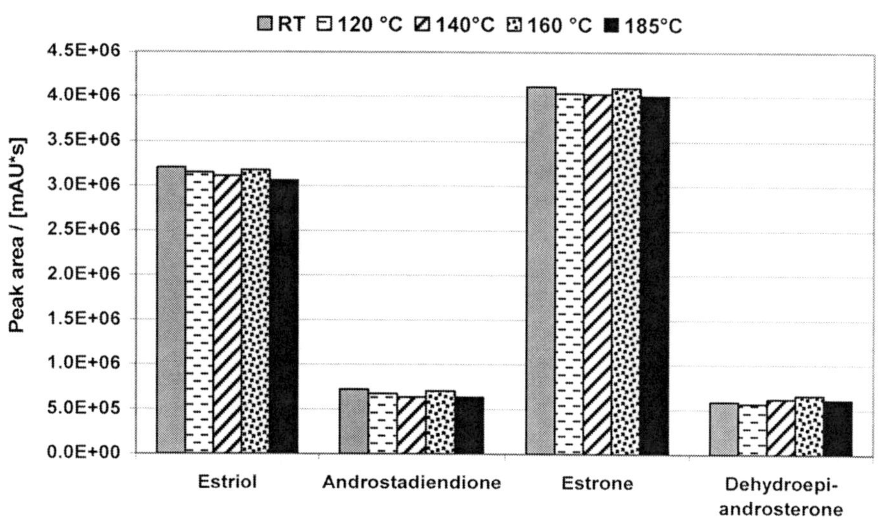

Fig. 4.19. Comparison of the peak areas of selected steroids at different temperatures. Mobile phase at 25°C: 75/25 (v/v) water/acetonitrile; mobile phase at 120, 140, 160 and 185°C: deionized water. Column: ZirChrom-PBD, 4.6 × 150 mm, 3 μm, 300 Å. Flow rate: 1 mL/min.

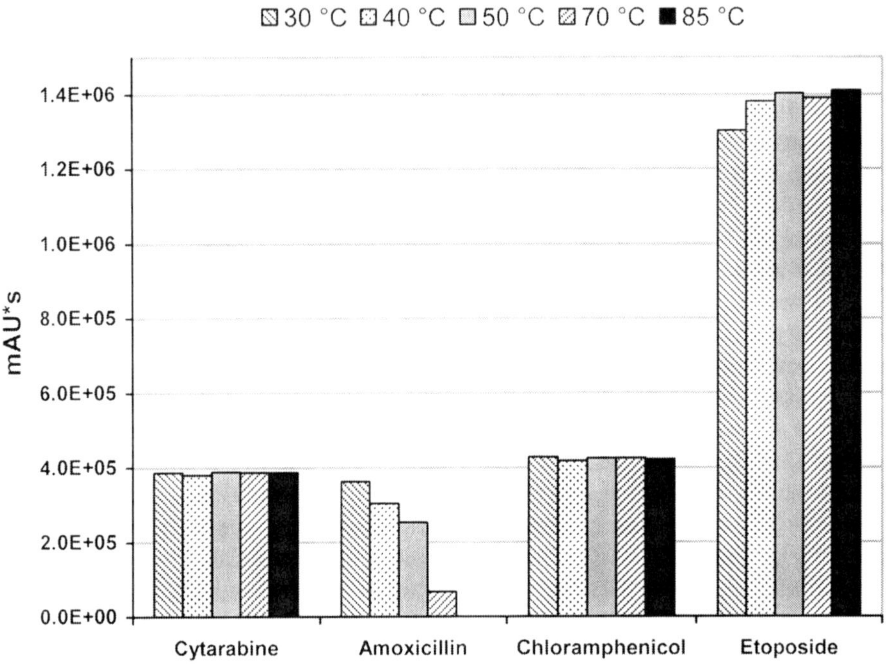

Fig. 4.20. Figure 2 from Teutenberg, 2009. Comparison of the peak areas of selected cytostatic and antibiotic drugs at different temperatures. Mobile phase: 85/15 (v/v) water/acetonitrile + 0.1% formic acid each. Column: ZirChrom-PBD, 4.6 × 150 mm, 3 μm, 300 Å. Flow rate: 1 mL/min. With permission from Elsevier.

Special Hyphenation Techniques

Although I have shown in the "Mobile Phase Considerations" section that all binary solvent systems can be used at high temperatures, the huge potential of high-temperature liquid chromatography lies in the employment of special hyphenation techniques. Nearly all of these techniques rely on a mobile phase consisting of pure water or water with only a minimal portion of an organic co-solvent. Here, the effect is exploited that the static permittivity of water, or a binary mixture of water and an organic co-solvent, significantly decreases if the temperature is increased (Teutenberg et al., 2009c).

LC Taste®

An interesting process that is known as LC Taste®, and makes use of high-temperature HPLC, is the determination of gustatorily active compounds in complex mixtures (Reichelt et al., 2010; Roloff et al., 2006). Flavour compositions contain at least two different sensorially active substances, such as synthetic, natural or nature-identical aromatic substances or plant extracts.

Mostly, however, flavors are complex mixtures of many sensorially active components. Flavoring substances interact with the flavor receptors on the tongue and are responsible for the gustatory impressions. The proportions of the sensorially active substances in a flavor composition can vary enormously, and they naturally have a strong influence on the overall sensory impression of the flavor composition. It is not the absolute amount of a sensorially active component in a flavor that is decisive, but its sensory contribution. Many sensorially important components in foodstuffs are not yet even known, since they are contained in only very small amounts, although they make a significant olfactory and/or gustatory contribution. More meaningful than the amount that is contained in a mixture is therefore the so-called odour or taste activity value, which is defined as the quotient of the concentration of a sensorially active component and its olfactory or gustatory threshold value.

The LC Taste® system uses the advantages of a separation based on high-temperature liquid chromatography and combines it with an in vivo detection of taste active compounds by a sensory tester or sensory panel. Therefore, the analytical and sensory data can be correlated in a way similar to the hyphenation of gas chromatography and olfactometry. Pure water is the preferred eluent because it is not toxic and will not interfere with the detection process of the human being or the technical detector. However, the elution strength of pure water at elevated temperatures is not sufficient for the elution of non-polar components. Therefore, the addition of modifiers that will increase the elution strength of the mobile phase is necessary if these compounds need to be analyzed. Toxic organic solvents like methanol or acetonitrile are strictly forbidden if a human being directly tastes the eluate. However, ethanol is a very convenient co-solvent which—up to a certain concentration—has no negative impact on the sensory impression, and also significantly enhances the elution strength of the mobile phase if solvent gradient elution is applied.

The concept of LC taste® allows for the combined use of temperature and solvent programming. This is a decisive advantage, because the elution strength can be increased by a simultaneous change of temperature and the mobile phase composition. Moreover, there is a far greater chance to elute all compounds in the mixture within one chromatographic run. Also, a band compression due to solvent gradient programming is possible as was already outlined in the "Temperature Gradient and Isocratic Separation" section, and hence the peak capacity can be tremendously increased. However, the ethanol concentration in the mobile phase should preferably be in the range from 5 to 30 percent by weight. Therefore, a concomitant temperature gradient is necessary so that the concentration of ethanol will be as low as possible. This application highlights the fact that besides the optimization of the selectivity, temperature programming in combination with solvent programming exhibits a unique feature to reduce the amount of the organic co-solvent as much as possible.

Another requirement for LC taste® is that the temperature of the mobile phase leaving the column should not be higher than 40°C. Otherwise, the burning recorded by the human operator would not be attributed to the flavour of the separated compound but to the high eluent temperature. Therefore, a heating system with a module for eluent cooling, as described in the "System Set-up" section, is mandatory.

Isotope Ratio Mass Spectrometry

Isotope ratio mass spectrometry (IRMS) can be used to distinguish between compounds that are either faked or are synthesized by different pathways, but which are chemically identical except for their abundance ratio of stable isotopes. More specifically, isotope ratio mass spectrometry is used to very precisely measure the abundance of the heavy to the light isotopes of carbon (Abramson et al., 2001; Mason et al., 1998). The rates at which heavier isotopes participate in chemical and physical processes are slightly different from those for lighter isotopes. The difference in rates leads to a subtle variation in the natural abundance of isotopes, owing to a variety of fractionation processes. The stable isotope composition of compounds is a function of their origin and history. Information about precise isotope ratios is very important in the nuclear, geological, agricultural, environmental and health sectors. Gas chromatography was mainly used in these studies to hyphenate the separation step with an isotope ratio mass spectrometer (GC-IRMS) (Ferchaud et al., 1998; Mason et al., 1998).

Isotope ratios are expressed relative to reference standards, rather than being reported as absolute isotope values. The isotope research community has established the δ-notation, which is the difference in the $^{13}C/^{12}C$ isotope ratio of the sample and an internationally agreed standard normalized by the $^{13}C/^{12}C$ isotope ratio of the standard. The resulting $\delta^{13}C$ value is given in ‰:

$$\delta^{13}C <‰> = <(R_{Sample}/R_{Reference})-1> \times 1000 \tag{4.6}$$

Samples can be directly measured versus a reference gas, which is calibrated against the international reference.

However, the drawback of GC-IRMS analysis is that many analytes of interest cannot be measured without derivatization. This procedure is not only extremely time consuming, but also bears the risk of an isotope fractionation due to the derivatization process (Ferchaud et al., 1998). If it is not possible to correct the measured $\delta^{13}C$ values for this isotope fractionation, the results are suspect. In a recent instrumental development to overcome this limitation, an LC interface for coupling high performance liquid chromatography to isotope ratio mass spectrometry has been introduced (Krummen et al., 2004). In this system, all compounds are quantitatively converted into CO_2

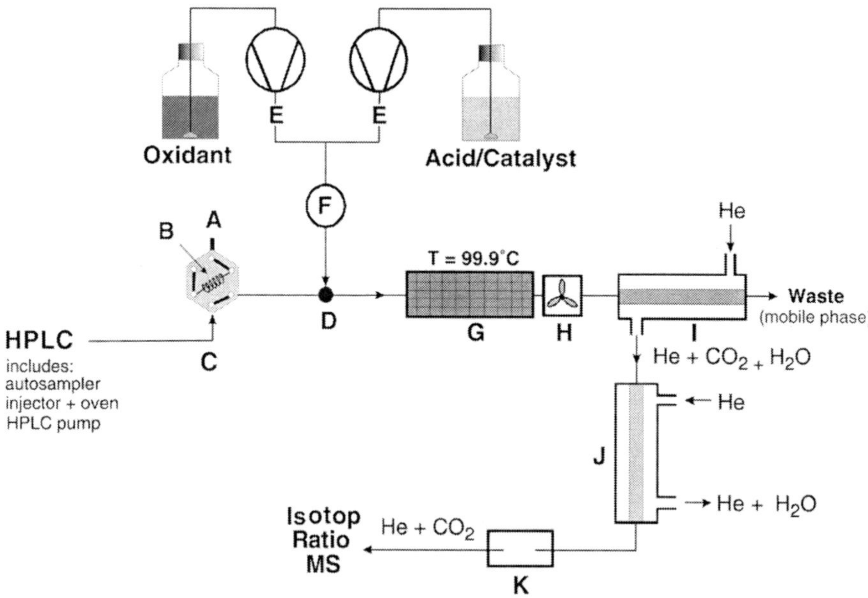

Fig. 4.21. Figure 1 from Krummen et al., 2004. Scheme of the Finnigan™ LC IsoLink. A: Needle port; B: sample loop; C: 6-port valve; D: T-piece; E: two-head-pump; F: pulse damper; G: oxidation reactor; H: cooler; I: CO_2 separation unit; J: gas dryer; K: open split. With permission from Elsevier.

while the analyte is still dissolved in the aqueous liquid phase. The chemical oxidation is typically performed by peroxodisulfate under acidic conditions. The CO_2 is removed from the eluent and entrained into a flow of helium by a miniature separation unit. This helium stream passes a water trap system and is then directed to the ion source of the IRMS via an open split assembly. For a better understanding, Fig. 4.21 depicts the scheme of the so-called LC-IsoLink™.

The work currently being carried out in the author's laboratory is also focused on this topic. For the elution of steroids, a polybutadiene-coated zirconium dioxide column gave the best results. A baseline separation of all analytes for the HPLC-IRMS hyphenation was achieved by linearly increasing the temperature from 90 to 120°C during 40 minutes. The resulting chromatogram is shown in Fig. 4.22.

The flow rate has to be adjusted between 300 and 500 μL/min, because the IsoLink interface is not designed for higher flow rates. This is a drawback, because as was pointed out in the "Influence of Temperature on Efficiency" section, there is a shift of the minimum of the van Deemter curve at higher eluent temperatures towards higher flow rates. Another serious issue is that

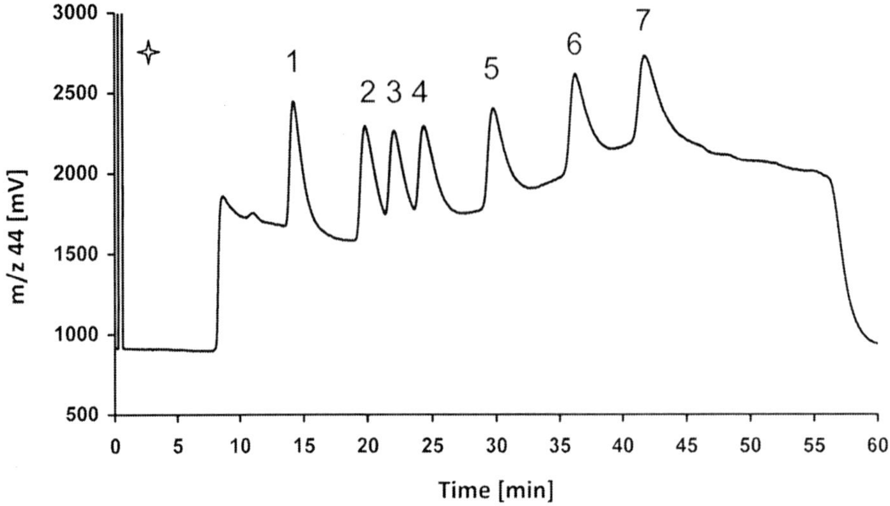

Fig. 4.22. Temperature gradient separation of selected steroids using IRMS-detection. Temperature programme: from 90 to 120°C within 40 minutes. *: CO_2 reference gas; 1: 19-nortestosterone; 2: testosterone; 3: epitestosterone; 4: trans-dehydroandrosterone; 5: etiocholan-3α-ol-17-on; 6: cis-androsterone; 7: 5α-androstane-3α, 17β-diol. Flow rate of mobile phase (water): 500 μL/min.

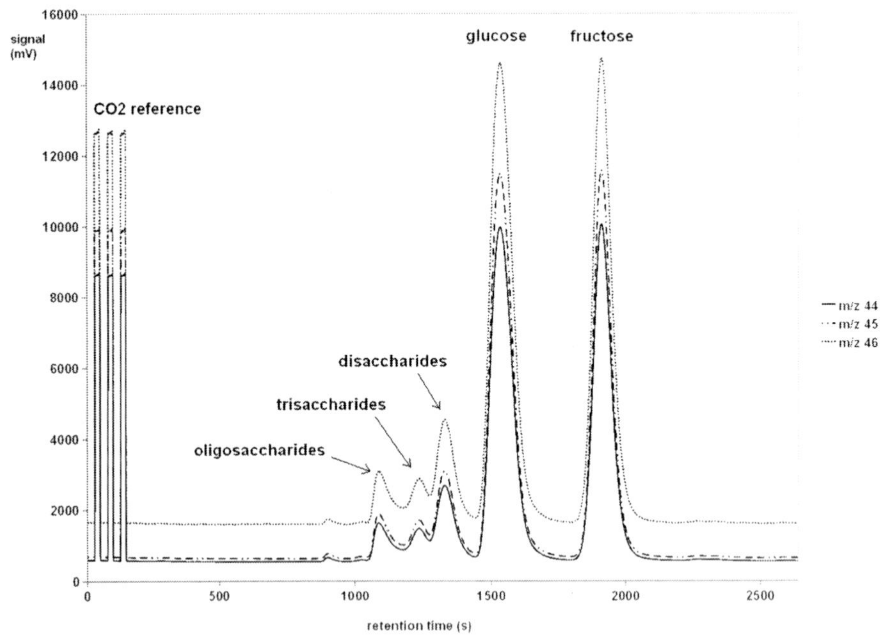

Fig. 4.23. Figure 2 from Elfein & Raezke, 2008. LC-IRMS chromatogram of adulterated polyfloral honey. $\delta^{13}C$ values: fructose −27.4 ‰, glucose −27.0 ‰, disaccharides −26.4 ‰, trisaccharides −24.3 ‰, oligosaccharides −26.7 ‰. Adulteration: ca. 11 ‰ rice syrup. With permission from EDP Sciences.

if the sample cannot be dissolved in pure water, the solvent peak that elutes at the void time of the column can exert a strong influence on the measured delta values. Therefore, the separation of the target analytes from the solvent peak is very important for precise measurements.

HPLC-IRMS is already a routine method. In 2008, Elflein and Raezke published a paper where they described an LC-IRMS method for the detection of honey adulteration (Elflein & Raezke, 2008). A chromatogram of this method is given in Fig. 4.23.

Currently, this is the first and only worldwide accredited method for the detection of adulteration in honey. Please note that the separated compounds were very polar and hence could be eluted at ambient temperature with a water-only mobile phase. In contrast to this, the example given in Fig. 4.22 unambiguously shows that even non-polar compounds can be eluted and separated on a suitable stationary phase without any derivatisation. Hence, this technique is not only applicable for polar analytes like sugars, but also for non-polar sample constituents like steroids.

Drug Screening

The screening of synthetic and natural chemical sources usually is the starting point in drug discovery. High-throughput screening technologies have been developed and implemented that are able to test tens of thousands of compounds or more per day for their activity in various assay types, ranging from receptor binding and enzyme-inhibition to whole cell assays. While such high-throughput techniques are highly efficient for the screening of pure compound samples (libraries), the screening of complex mixtures is more demanding. de Boer and Irth present a different approach, which is based on high-temperature HPLC, and is focused on the determination of biologically active compounds in complex mixtures (de Boer et al., 2005).

Examples of complex mixtures in drug discovery are samples originating from natural products, reaction mixtures from solution-phase combinatorial chemistry, and in vitro or in vivo metabolic profiling. In all cases, non-active sample constituents at widely different concentration ranges are present next to an unknown number of pharmacologically active compounds. The main difficulties encountered are the correlation of biological activity with chemical analysis data for the rapid identification of active compounds and the presence of matrix components that interfere with the assay readout. Identification requires fractionation, mostly performed by off-line liquid chromatography, in combination with fraction collection. The whole process of screening and fractionation must be repeated until the bioactive compound is isolated. It is obvious that this process can be very laborious, time-consuming and error-prone.

In order to overcome these drawbacks, an on-line coupling of a separation step with the biochemical assay would be advantageous. Therefore, de Boer and Irth (2005) described a method in which an HPLC separation was hyphenated to an enzyme assay. The biochemical assay was based on a continuous-flow enzyme-substrate reaction and the subsequent detection of reaction products by electrospray-ionization mass spectrometry (ESI-MS). Inhibition of enzyme activity by compounds eluting from the HPLC column resulted in a temporary change of product concentration, which was determined by ESI-MS as a negative peak in the extracted ion chromatograms (EIC) of the products. A schematic illustration of the experimental design is shown in Fig. 4.24.

However, a limitation of this approach is that the enzyme assay can only be coupled on-line if the concentration of the organic modifier in the mobile phase does not exceed a threshold where the enzymatic activity will be lost. To overcome this problem, high-temperature HPLC was used. Increasing the temperature of the mobile phase offers the possibility of reducing the amount of the organic co-solvent, so that the enzymatic activity will be maintained after the separation step. This drawback could be circumvented by post-column addition of an aqueous solution, but this has the disadvantage of diluting the effluent and requires a more complex analytical system.

The authors showed that biochemical assays could be performed in the presence of organic modifier concentrations up to 15%, as long as the reaction time did not exceed 3 to 5 minutes. Similar results were obtained for cathepsin B, in which the presence of 10% methanol in the enzyme-substrate

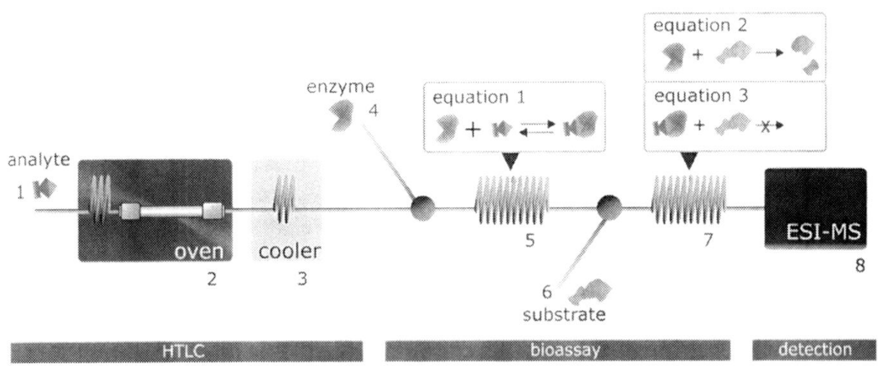

Fig. 4.24. Figure 1 from de Boer et al., 2005. Overview of the continuous-flow system. Key: 1, sample introduction; 2, HTLC oven containing the heating coil and column; 3, cooler; 4, superloop with enzyme solution; 5, bioreactor A; 6, superloop with substrate solution; 7, bioreactor B; 8, ESI-MS. The enzyme continuously converts the substrate into products (eq 2) if no bioactive compounds are eluting from the column. Bioactive compounds present in the effluent bind to the enzyme (eq 1) resulting in a decrease of product turnover (eq 3). The reaction products are continuously detected by ESI-MS. With permission from ACS.

reaction—which corresponded to 20% methanol in the column—leads to an 11% decrease of product formation.

Note that the use of 20% methanol in the mobile phase can be regarded as very high if temperature programming is used in combination with solvent programming. Again, the concomitant use of these two gradient techniques allows for a band compression, leading to higher peak capacities than if the separation is carried out isocratically. So the same conclusions can be drawn as for the concept of LC taste®, which was described in the previous section. A decisive advantage of the technique described here is that the organic co-solvent is not limited to ethanol. In principle, all solvents listed in Table 4.A can be used. As was also noted earlier, complex mixtures usually contain different compounds with a wide range of polarities. Therefore, the right strategy is to find a mobile phase which, at a given maximum temperature, exhibits the lowest polarity. Then, the probability will be high that it is possible to elute all sample components from the column.

Current activities in the author's own laboratory are focused on establishing this concept for the screening of the toxic and allergenic potential of house dust samples. From the discussion in this chapter, it is evident that a lot of basic research is needed in order to find the optimum conditions for this approach. However, once the technical difficulties are solved, this technique could be a real alternative to the time-consuming off-line procedure.

Closing Remarks and Future Prospects

The huge potential of high-temperature liquid chromatography is the hyphenation with special detection techniques, because of the possibility to completely eliminate the organic co-solvent. Another aspect is the hyphenation with mass spectrometry and the combination with miniaturization. Although capillary liquid chromatography is not well established in the pharmaceutical industry, it has many advantages over conventional liquid chromatography. The reduction of the column diameter and the concomitant reduction of the mobile phase flow rate are very beneficial for mass spectrometry. However, for capillary liquid chromatography, it is a prerequisite that the practitioner is very well trained on the analytical instruments. Here, integrating the columns and other parts of the liquid chromatographic system on a chip could be a real boost for miniaturization. I hope that in the near future, lab-on-a-chip devices will be more readily available, because this would also make high-temperature liquid chromatography more attractive for the routine laboratory. However, we should not forget that a good separation can only be achieved if there are stationary phases which can be used over a prolonged period with no degradation or loss of efficiency. As for separations at very high pressure, the application of very high temperatures demands columns with a very low bleed and a high chemical stability. The design and development of rugged materials is expected to continue and the more these techniques described in the "Special

Hyphenation Techniques" section will be used in routine analysis, the more there will be pressure on column manufacturers to develop stable stationary phases.

References

Abbott, S.; Achener, P.; Simpson, R.; Klink, F. Effect of radial thermal gradients in elevated temperature high-performance liquid chromatography. J. Chromatogr. 1981, 218, 123-135.

Abramson, F.P.; Black, G.E.; Lecchi, P. Application of high-performance liquid chromatography with isotope-ratio mass spectrometry for measuring low levels of enrichment of underivatized materials. J. Chromatogr. A 2001, 913, 269-273.

Agarwal, R. Temperature sensitivity and fluorescence detection. J. Sep. Sci. 2008, 31, 128-132.

Al-Khateeb, L.; Smith, R.M. Superheated water chromatography on phenyl bonded hybrid stationary phases. J. Chromatogr. A 2008, 1201, 61-64.

Albert, M.; Cretier, G.; Guillarme, D.; Heinisch, S.; Rocca, J.L. Some advantages of high temperature for the separation of pharmaceutical compounds with mass spectrometry detection. J. Sep. Sci. 2005, 28, 1803-1811.

Bidlingmeyer, B.A.; Henderson, J. Investigation of retention on bare silica using reversed-phase mobile phases at elevated temperatures. J. Chromatogr. A 2004, 1060, 187-193.

Chen, H.; Horvath, C. Rapid separation of proteins by reversed phase HPLC at Elevated Temperatures. Anal. Meth. Instr. 1993, 1, 213-222.

Chienthavorn, O.; Smith, R.M. Buffered Superheated Water as an Eluent for Reversed-Phase High Performance Liquid Chromatography. Chromatographia 1999, 50, 485-489.

Chienthavorn, O.; Smith, R.M.; Saha, S.; Wilson, I.D.; Wright, B.; Taylor, S.D.; Lenz, E.M. Superheated water chromatography-nuclear magnetic resonance spectroscopy and mass spectrometry of vitamins. J. Pharm. Biomed. Anal. 2004, 36, 477-482.

Chienthavorn, O.; Smith, R.M.; Wilson, I.D.; Wright, B.; Lenz, E.M. Superheated water chromatography-nuclear magnetic resonance spectroscopy of kava lactones. Phytochem. Anal. 2005, 16, 217-221.

Claessens, H.A.; van Straten, M.A.; Kirkland, J.J. Effect of buffers on silica-based column stability in reversed-phase high-performance liquid chromatography. J. Chromatogr. A 1996, 728, 259-270.

Coym, J.W.; Dorsey, J.G. Superheated Water Chromatography: A Brief Review of an Emerging Technique. Anal. Lett. 2004a, 37, 1013-1023.

Coym, J.W.; Dorsey, J.G. Reversed-phase retention thermodynamics of pure-water mobile phases at ambient and elevated temperature. J. Chromatogr. A 2004b, 1035, 23-29.

de Boer, A.R.; Alcaide-Hidalgo, J.M.; Krabbe, J.G.; Kolkman, J.; van Emde Boas, C.N.; Niessen, W.M.; Lingeman, H.; Irth, H. High-temperature liquid chromatography coupled on-line to a continuous-flow biochemical screening assay with electrospray ionization mass spectrometric detection. Anal. Chem. 2005, 77, 7894-7900.

de Villiers, A.; Lauer, H.; Szucs, R.; Goodall, S.; Sandra, P. Influence of frictional heating on temperature gradients in ultra-high-pressure liquid chromatography on 2.1mm I.D. columns. J. Chromatogr. A 2006, 1113, 84-91.

Djordjevic, N.M.; Houdiere, F.; Fowler, P. High temperature and temperature programming in capillary HPLC. Biomed. Chromatogr. 1998, 12, 153-154.

Djordjevic, N.M.; Fowler, P.W.J.; Houdiere, F. High Temperature and Temperature Programming in High-Performance Liquid Chromatography: Instrumental Considerations. J. Microcolumn Sep. 1999, 11, 403-413.

Dolan, J.W.; Gant, J.R.; Snyder, L.R. Gradient Elution in High-Performance Liquid-Chromatography .2. Practical Application to Reversed-Phase Systems. J. Chromatogr. 1979, 165, 31-58.

Dolan, J.W.; Snyder, L.R.; Quarry, M.A. Computer-Simulation as a Means of Developing an Optimized Reversed-Phase Gradient-Elution Separation. Chromatographia 1987, 24, 261-276.

Dolan, J.W.; Lommen, D.C.; Snyder, L.R. Drylab Computer-Simulation for High-Performance Liquid-Chromatographic Method Development .2. Gradient Elution. J. Chromatogr. 1989, 485, 91-112.

Dolan, J.W.; Snyder, L.R.; Djordjevic, N.M.; Hill, D.W.; Saunders, D.L.; Van Heukelem, L.; Waeghe, T.J. Simultaneous variation of temperature and gradient steepness for reversed-phase high-performance liquid chromatography method development - I. Application to 14 different samples using computer simulation. J. Chromatogr. A 1998a, 803, 1-31.

Dolan, J.W.; Snyder, L.R.; Saunders, D.L.; Van Heukelem, L. Simultaneous variation of temperature and gradient steepness for reversed-phase high-performance liquid chromatography method development - II. The use of further changes in conditions. J. Chromatogr. A 1998b, 803, 33-50.

Dolan, J.W.; Snyder, L.R.; Djordjevic, N.M.; Hill, D.W.; Waeghe, T.J. Reversed-phase liquid chromatographic separation of complex samples by optimizing temperature and gradient time II. Two-run assay procedures. J. Chromatogr. A 1999a, 857, 21-39.

Dolan, J.W.; Snyder, L.R.; Djordjevic, N.M.; Hill, D.W.; Waeghe, T.J. Reversed-phase liquid chromatographic separation of complex samples by optimizing temperature and gradient time I. Peak capacity limitations. J. Chromatogr. A 1999b, 857, 1-20.

Dolan, J.W.; Snyder, L.R.; Wolcott, R.G.; Haber, P.; Baczek, T.; Kaliszan, R.; Sander, L.C. Reversed-phase liquid chromatographic separation of complex samples by optimizing temperature and gradient time III. Improving the accuracy of computer simulation. J. Chromatogr. A 1999c, 857, 41-68.

Dugo, P.; Buonasera, K.; Crupi, M.L.; Cacciola, F.; Dugo, G.; Mondello, L. Superheated water as chromatographic eluent for parabens separation on octadecyl coated zirconia stationary phase. J. Sep. Sci. 2007, 30, 1125-1130.

Edge, A.M.; Shillingford, S.; Smith, C.; Payne, R.; Wilson, I.D. Temperature as a variable in liquid chromatography: development and application of a model for the separation of model drugs using water as the eluent. J. Chromatogr. A 2006, 1132, 206-210.

Elflein, L.; Raezke, K.P. Improved detection of honey adulteration by measuring differences between C-13/C-12 stable carbon isotope ratios of protein and sugar compounds with a combination of elemental analyzer - isotope ratio mass spectrometry and liquid chromatography - isotope ratio mass spectrometry (delta C-13-EA/LC-IRMS). Apidologie 2008, 39, 574-587.

Ferchaud, V.; Le Bizec, B.; Monteau, F.; Andre, F. Determination of the exogenous character of testosterone in bovine urine by gas chromatography-combustion-isotope ratio mass spectrometry. Analyst 1998, 123, 2617-2620.

Fields, S.M.; Ye, C.Q.; Zhang, D.D.; Branch, B.R.; Zhang, X.J.; Okafo, N. Superheated water as eluent in high-temperature high-performance liquid chromatographic separations of steroids on a polymer-coated zirconia column. J. Chromatogr. A 2001, 913, 197-204.

Fogwill, M.O.; Thurbide, K.B. Rapid column heating method for subcritical water chromatography. J. Chromatogr. 2007, 1139, 199-205.

Fogwill, M.O.; Thurbide, K.B. Carbon dioxide modified subcritical water chromatography. J. Chromatogr. A 2008, 1200, 49-54.

Fonseca, D.A.; Gutierrez, H.R.; Collins, K.E.; Collins, C.H. Rapid method for evaluating reversed-phase high-performance liquid chromatography column stability. J. Chromatogr. A 2004, 1030, 149-155.

Giegold, S. Application of high temperature - high performance liquid chromatography (HT-HPLC): About the influence of temperature on the analysis time, selectivity and different detection systems (Dissertation); Department 8 - Chemistry, University of Siegen: Siegen, Germany, 2008.

Giegold, S.; Holzhauser, M.; Kiffmeyer, T.; Tuerk, J.; Teutenberg, T.; Rosenhagen, M.; Hennies, D.; Hoppe-Tichy, T.; Wenclawiak, B. Influence of the stationary phase on the stability of thalidomide and comparison of different methods for the quantification of thalidomide in tablets using high-temperature liquid chromatography. J. Pharm. Biomed. Anal. 2008a, 46, 625-630.

Giegold, S.; Teutenberg, T.; Tuerk, J.; Kiffmeyer, T.; Wenclawiak, B. Determination of sulfonamides and trimethoprim using high temperature HPLC with simultaneous temperature and solvent gradient. J. Sep. Sci. 2008b, 31, 3497-3502.

Gika, H.G.; Theodoridis, G.; Extance, J.; Edge, A.M.; Wilson, I.D. High temperature-ultra performance liquid chromatography-mass spectrometry for the metabonomic analysis of Zucker rat urine. J. Chromatogr. B 2008, 871, 279-287.

Gritti, F.; Guiochon, G. Unusual behavior of the height equivalent to a theoretical plate of a new poroshell stationary phase at high temperatures. J. Chromatogr. A 2007, 1169, 125-138.

Gritti, F.; Guiochon, G. Exact peak compression factor in linear gradient elution I. Theory. J. Chromatogr. A 2008a, 1212, 35-40.

Gritti, F.; Guiochon, G. The ultimate band compression factor in gradient elution chromatography. J. Chromatogr. A 2008b, 1178, 79-91.

Guillarme, D.; Heinisch, S. Detection Modes with High Temperature Liquid Chromatography- A Review. Sep. Purif. Technol. 2005, 34, 181-216.

Guillemin, C.L.; Millet, J.L.; Dubois, J. Thermal Aqueous Liquid Chromatography—The TALC Technique. J. High Resolut. Chromatogr. Chromatogr. Commun. 1981, 4, 280-286.

Hao, Z.; Lu, C.Y.; Xiao, B.; Weng, N.; Parker, B.; Knapp, M.; Ho, C.T. Separation of amino acids, peptides and corresponding Amadori compounds on a silica column at elevated temperature. J. Chromatogr. A 2007, 1147, 165-171.

He, P.; Yang, Y. Studies on the long-term thermal stability of stationary phases in subcritical water chromatography. J. Chromatogr. A 2003, 989, 55-63.

Heinisch, S.; Desmet, G.; Clicq, D.; Rocca, J.L. Kinetic plot equations for evaluating the real performance of the combined use of high temperature and ultra-high pressure in liquid chromatography—Application to commercial instruments and 2.1 and 1 mm ID columns. J. Chromatogr. A 2008, 1203, 124-136.

Ingelse, B.A.; Janssen, H.-G.; Cramers, C.A. HPLC-FID with Superheated Water as the Eluent: Improved Methods and Instrumentation. J. High Resolut. Chromatogr. 1998, 21, 613-616.

Jones, A.; Yang, Y. Separation of nonpolar analytes using methanol-water mixtures at elevated temperatures. Anal. Chim. Acta 2003, 485, 51-55.

Kephart, T.S.; Dasgupta, P.K. Hot eluent capillary liquid chromatography using zirconia and titania based stationary phases. Anal. Chim. Acta 2000, 414, 71-78.

Kephart, T.S.; Dasgupta, P.K. Superheated water eluent capillary liquid chromatography. Talanta 2002, 56, 977-987.

Kondo, T.; Yang, Y.; Lamm, L. Separation of polar and non-polar analytes using dimethyl sulfoxide-modified subcritical water. Anal. Chim. Acta 2002, 460, 185-191.

Kondo, T.; Yang, Y. Comparison of elution strength, column efficiency, and peak symmetry in subcritical water chromatography and traditional reversed-phase liquid chromatography. Anal. Chim. Acta 2003, 494, 157-166.

Krummen, M.; Hilkert, A.W.; Juchelka, D.; Duhr, A.; Schluter, H.J.; Pesch, R. A new concept for isotope ratio monitoring liquid chromatography/mass spectrometry. Rapid Commun. Mass Spectrom. 2004, 18, 2260-2266.

Lamm, L.J.; Yang, Y. Off-line coupling of subcritical water extraction with subcritical water chromatography via a sorbent trap and thermal desorption. Anal. Chem. 2003, 75, 2237-2242.

Le, X.C.; Ma, M.; Wong, N.A. Speciation of Arsenic Compounds Using High-Performance Liquid Chromatography at Elevated Temperature and Selective Hydride Generation Atomic Fluorescence Detection. Anal. Chem. 1996, 68, 4501-4506.

Le, X.C.; Li, X.-F.; Lai, V.; Ma, M.; Yalcin, S.; Feldmann, J. Simultaneous speciation of selenium and arsenic using elevated temperature liquid chromatography separation with inductively coupled plasma mass spectrometry detection. Spectrochim. Acta B 1998, 53, 899-909.

Li, J.W.; Hu, Y.; Carr, P.W. Fast separations at elevated temperatures on polybutadiene-coated zirconia reversed-phase material. Anal. Chem. 1997, 69, 3884-3888.

Lippert, J.A.; Johnson, T.M.; Lloyd, J.B.; Smith, J.P.; Johnson, B.T.; Furlow, J.; Proctor, A.; Marin, S.J. Effects of elevated temperature and mobile phase composition on a novel C18 silica column. J. Sep. Sci. 2007, 30, 1141-1149.

Liu, Y.; Grinberg, N.; Thompson, K.C.; Wenslow, R.M.; Neue, U.D.; Morrison, D.; Walter, T.H.; Gara, J.E.O.; Wyndham, K.D. Evaluation of a C 18 hybrid stationary phase using high-temperature chromatography. Anal. Chim. Acta 2005, 554, 144-151.

Louden, D.; Handley, A.; Taylor, S.; Sinclair, I.; Lenz, E.; Wilson, I.D. High temperature reversed-phase HPLC using deuterium oxide as a mobile phase for the separation of model pharmaceuticals with multiple on-line spectroscopic analysis (UV, IR, 1H-NMR and MS). Analyst 2001, 126, 1625-1629.

Mason, P.M.; Hall, S.E.; Gilmour, I.; Houghton, E.; Pillinger, C.; Seymour, M.A. The use of stable carbon isotope analysis to detect the abuse of testosterone in cattle. Analyst 1998, 123, 2405-2408.

McNeff, C.V.; Yan, B.; Stoll, D.R.; Henry, R.A. Practice and theory of high temperature liquid chromatography. J. Sep. Sci. 2007, 30, 1672-1685.

Miller, D.J.; Hawthrone, S.B. Subcritical Water Chromatography with Flame Ionization Detection. Anal. Chem. 1997, 69, 623-627.

Molander, P.; Trones, R.; Haugland, K.; Greibrokk, T. Aspects and applications of non-aqueous high temperature packed capillary liquid chromatography. Analyst 1999, 124, 1137-1141.

Nakajima, R.; Yarita, T.; Shibukawa, M. Analysis of alcohols by superheated water chromatography with flame ionization detection. Bunseki Kagaku 2003, 52, 305-309.

Neue, U.D.; Mazzeo, J.R. A theoretical study of the optimization of gradients at elevated temperature. J. Sep. Sci. 2001, 24, 921-929.

Neue, U.D.; Marchand, D.H.; Snyder, L.R. Peak compression in reversed-phase gradient elution. J. Chromatogr. A 2006, 1111, 32-39.

Nguyen, D.T.T.; Guillarme, D.; Heinisch, S.; Barrioulet, M.P.; Rocca, J.L.; Rudaz, S.; Veuthey, J.L. High throughput liquid chromatography with sub-2 mu m particles at high pressure and high temperature. J. Chromatogr. A 2007, 1167, 76-84.

Nikitas, P.; Pappa-Louisi, A.; Papachristos, K.; Zisi, C. Theory and application of the two-mode gradient elution in liquid chromatography involving simultaneous changes in temperature and mobile-phase composition. Anal. Chem. 2008, 80, 5508-5514.

Pappa-Louisi, A.; Nikitas, P.; Papachristos, K.; Balkatzopoulou, P. Multimode Gradient Elution in Reversed-Phase Liquid Chromatography: Application to Retention Prediction and Separation Optimization. of a Set of Amino Acids in Gradient Runs Involving Simultaneous Variations of Mobile-Phase Composition, Flow Rate, and Temperature. Anal. Chem. 2009, 81, 1217-1223.

Pereira, L. Ultra high temperature in HPLC with a porous graphite carbon column. LCGC North Am. - Appl. Noteb. 2006, June 2006, 75-77.

Pereira, L.; Aspey, S.; Ritchie, H. High temperature to increase throughput in liquid chromatography and liquid chromatography-mass spectrometry with a porous graphitic carbon stationary phase. J. Sep. Sci. 2007, 30, 1115-1124.

Plumb, R.; Mazzeo, J.R.; Grumbach, E.S.; Rainville, P.; Jones, M.; Wheat, T.; Neue, U.D.; Smith, B.; Johnson, K.A. The application of small porous particles, high temperatures, and high pressures to generate very high resolution LC and LC/MS separations. J. Sep. Sci. 2007, 30, 1158-1166.

Reichelt, K.V.; Peter, R.; Paetz, S.; Roloff, M.; Ley, J.P.; Krammer, G.E.; Engel, K.H. Characterization of Flavor Modulating Effects in Complex Mixtures via High Temperature Liquid Chromatography. J. Agric. Food Chem. 2010, 58, 458-464.

Riddle, L.A.; Guiochon, G. Separation of free sterols by high temperature liquid chromatography. J. Chromatogr. A 2006, 1137, 173-179.

Roloff, M.; Erfurt, H.; Kindel, G.; Schmidt, C.-O.; Krammer, G. Process for the separation and sensory evaluation of flavours (Patent); WO 2006/111476 A1, World Intellectual Property Organization; Symrise GmbH & Co. KG, Holzminden, Germany, 2006.

Sanagi, M.M.; Heng, S.H.; Ibrahim, W.A.W.; Naim, A.A. High Temperature High Performance Liquid Chromatography of Triazole Fungicides on Carbon-Clad Zirconia Stationary Phase. Malaysian J. Chem. 2004a, 6, 55-66.

Sanagi, M.M.; See, H.H.; Ibrahim, W.A.; Naim, A.A. High temperature liquid chromatography of triazole fungicides on polybutadiene-coated zirconia stationary phase. J. Chromatogr. A 2004b, 1059, 95-101.

Sandra, P.; Vanhoenacker, G. Elevated temperature-extended column length conventional liquid chromatography to increase peak capacity for the analysis of tryptic digests. J. Sep. Sci. 2007, 30, 241-244.

Selerity Technologies Inc. : Products : Series 9000 Total Temperature Controller. http://www.selerity.com/main/main_products_hplc_9000.html (July 2010).

Shen, S.; Lee, H.; McCaffrey, J.; Yee, N.; Senanayake, C.; Grinberg, N. High Temperature High Performance Liquid Chromatography of Substituted Anilines using a C18 Hybrid Stationary Phase. J. Liq. Chromatogr. Relat. Technol. 2006, 29, 2823-2834.

Sheng, G.C.; Shen, Y.F.; Lee, M.L. Elevated temperature liquid chromatography using reversed-phase packed capillary columns. J. Microcolumn Sep. 1997, 9, 63-72.

SIM - Scientific Instruments Manufacturer GmbH. SIM HT Column Oven. http://www.sim-gmbh.de/index.php?option=com_content&task=view&id=64&Itemid=502&lang=en (July 2010).

Smith, R.M.; Burgess, R.J. Superheated Water-a Clean Eluent for Reversed-phase High-performance Liquid Chromatography. Anal. Commun. 1996, 33, 327-329.

Smith, R.M.; Burgess, R.J. Superheated water as an eluent for reversed-phase high-performance liquid chromatography. J. Chromatogr. A 1997, 785, 49-55.

Smith, R.M.; Burgess, R.J.; Chienthavorn, O.; Bone, J.R. Superheated water: a new look at chromatographic eluents for reversed-phase HPLC. LCGC Int. 1999, 17, 938, 940, 942, 944-945.

Smith, R.M. Superheated water: the ultimate green solvent for separation science. Anal. Bioanal. Chem. 2006, 385, 419-421.

Smith, R.M. Superheated water chromatography—A green technology for the future. J. Chromatogr. A 2008, 1184, 441-455.

Snyder, L.R.; Dolan, J.W.; Gant, J.R. Gradient Elution in High-Performance Liquid-Chromatography .1. Theoretical Basis for Reversed-Phase Systems. J. Chromatogr. 1979, 165, 3-30.

Snyder, L.R.; Dolan, J.W.; Lommen, D.C. Drylab Computer-Simulation for High-Performance Liquid-Chromatographic Method Development .1. Isocratic Elution. J. Chromatogr. 1989, 485, 65-89.

Snyder, L.R.; Dolan, J.W. High-Performance Gradient Elution—The Practical Application of the Linear-Solvent-Strength Model. Wiley-Interscience, John Wiley & Sons: Hoboken, NJ, 2007.

Spearman, L.; Smith, R.M.; Dube, S. Monitoring effective column temperature by using shape selectivity and hydrophobicity and the effects of mobile phase temperature. J. Chromatogr. A 2004, 1060, 147-151.

Stoll, D.R.; Cohen, J.D.; Carr, P.W. Fast, comprehensive online two-dimensional high performance liquid chromatography through the use of high temperature ultra-fast gradient elution reversed-phase liquid chromatography. J. Chromatogr. A 2006, 1122, 123-137.

Tajuddin, R.; Smith, R.M. On-line coupled superheated water extraction (SWE) and superheated water chromatography (SWC). Analyst 2002, 127, 883-885.

Tajuddin, R.; Smith, R.M. On-line coupled extraction and separation using superheated water for the analysis of triazine herbicides in spiked compost samples. J. Chromatogr. A 2005, 1084, 194-200.

Teutenberg, T.; Lerch, O.; Götze, H.J.; Zinn, P. Separation of selected anticancer drugs using superheated water as the mobile phase. Anal. Chem. 2001, 73, 3896-3899.

Teutenberg, T.; Goetze, H.J.; Tuerk, J.; Ploeger, J.; Kiffmeyer, T.K.; Schmidt, K.G.; Kohorst, W.G.; Rohe, T.; Jansen, H.D.; Weber, H. Development and application of a specially designed heating system for temperature-programmed high-performance liquid chromatography using subcritical water as the mobile phase. J. Chromatogr. A 2006a, 1114, 89-96.

Teutenberg, T.; Tuerk, J.; Holzhauser, M.; Kiffmeyer, T.K. Evaluation of column bleed by using an ultraviolet and a charged aerosol detector coupled to a high-temperature liquid chromatographic system. J. Chromatogr. A 2006b, 1119, 197-201.

Teutenberg, T.; Tuerk, J.; Hozhauser, M.; Giegold, S. Temperature stability of reversed phase and normal phase stationary phases under aqueous conditions. J. Sep. Sci. 2007, 30, 1101-1114.

Teutenberg, T. Potential of high temperature liquid chromatography for the improvement of separation efficiency—A review. Anal. Chim. Acta 2009, 643, 1-12.

Teutenberg, T.; Hollebekkers, K.; Wiese, S.; Boergers, A. Temperature and pH-stability of commercial stationary phases. J. Sep. Sci. 2009a, 32, 1262-1274.

Teutenberg, T.; Wagner, P.; Gmehling, J. High-temperature liquid chromatography. Part I. Determination of the vapour pressures of binary solvent mixtures—Implications for liquid chromatographic separations. J. Chromatogr. A 2009b, 1216, 6471-6480.

Teutenberg, T.; Wiese, S.; Wagner, P.; Gmehling, J. High-temperature liquid chromatography. Part III: Determination of the static permittivities of pure solvents and binary solvent mixtures—Implications for liquid chromatographic separations. J. Chromatogr. A 2009c, 1216, 8480-8487.

Teutenberg, T.; Wiese, S.; Wagner, P.; Gmehling, J. High-temperature liquid chromatography. Part II: Determination of the viscosities of binary solvent mixtures—Implications for liquid chromatographic separations. J. Chromatogr. A 2009d, 1216, 8470-8479.

Teutenberg, T. High-Temperature Liquid Chromatography—A User's Guide for Method Development; R.M. Smith, Series Ed., RSC Chromatography Monographs, Vol. 13.; Royal Society of Chemistry: Cambridge, 2010.

Thompson, J.D.; Brown, J.S.; Carr, P.W. Dependence of thermal mismatch broadening on column diameter in high-speed liquid chromatography at elevated temperatures. Anal. Chem. 2001, 73, 3340-3347.

Trones, R.; Andersen, T.; Hunnes, I.; Greibrokk, T. Modified laser light-scattering detector for use in high temperature micro liquid chromatography. J. Chromatogr. A 1998, 814, 55-61.

Trones, R.; Andersen, T.; Greibrokk, T. Improved modification of a laser light-scattering detector for use in packed capillary high temperature liquid chromatography. J. High Resolut. Chromatogr. 1999, 22, 283-286.

Ueki, Y.; Umemura, T.; Iwashita, Y.; Odake, T.; Haraguchi, H.; Tsunoda, K. Preparation of low flow-resistant methacrylate-based monolithic stationary phases of different hydrophobicity and the application to rapid reversed-phase liquid chromatographic separation of alkylbenzenes at high flow rate and elevated temperature. J. Chromatogr. A 2006, 1106, 106-111.

Vanhoenacker, G.; Sandra, P. High temperature liquid chromatography and liquid chromatography-mass spectroscopy analysis of octylphenol ethoxylates on different stationary phases. J. Chromatogr. A 2005, 1082, 193-202.

Vanhoenacker, G.; Sandra, P. Elevated temperature and temperature programming in conventional liquid chromatography—fundamentals and applications. J. Sep. Sci. 2006, 29, 1822-1835.

West, C.; Lesellier, E. Separation of substituted aromatic isomers with porous graphitic carbon in subcritical fluid chromatography. J. Chromatogr. A 2005, 1099, 175-184.

Wilson, I.D. Investigation of a Range of Stationary Phases for the Separation of Model Drugs by HPLC Using Superheated Water as the Mobile Phase. Chromatographia 2000, 52, 28-34.

Wolcott, R.G.; Dolan, J.W.; Snyder, L.R.; Bakalyar, S.R.; Arnold, M.A.; Nichols, J.A. Control of column temperature in reversed-phase liquid chromatography. J. Chromatogr. A 2000, 869, 211-230.

Xiang, Y.; Liu, Y.; Lee, M.L. Ultrahigh pressure liquid chromatography using elevated temperature. J. Chromatogr. A 2006, 1104, 198-202.

Yamaki, S.; Isobe, T.; Okuyama, T.; Shinoda, T. Reversed-phase liquid chromatography on a microspherical carbon column at high temperature. J. Chromatogr. A 1996, 728, 189-194.

Yang, X.Q.; Ma, L.J.; Carr, P.W. High temperature fast chromatography of proteins using a silica-based stationary phase with greatly enhanced low pH stability. J. Chromatogr. A 2005, 1079, 213-220.

Yang, Y.; Jones, A.D.; Eaton, C.D. Retention Behavior of Phenols, Anlines, and Alkylbezenes in Liquid Chromatographic Separations Using Subcritical Water as the Mobile Phase. Anal. Chem. 1999, 71, 3808-3813.

Yang, Y.; Jones, A.D.; Mathis, J.A.; Francis, M.A. Flame ionization detection after splitting the water effluent in subcritical water chromatography. J. Chromatogr. A 2002a, 942, 231-236.

Yang, Y.; Lamm, L.J.; He, P.; Kondo, T. Temperature effect on peak width and column efficiency in subcritical water chromatography. J. Chromatogr. Sci. 2002b, 40, 107-112.

Yang, Y. Stationary Phases for High-Temperature Liquid Chromatography. LCGC Eur. 2003, 16, 37-41.

Yang, Y. Subcritical water chromatography: A green approach to high-temperature liquid chromatography. J. Sep. Sci. 2007, 30, 1131-1140.

Yarita, T.; Nakajima, R.; Shibukawa, M. Superheated water chromatography of phenols using poly(styrene-divinylbenzene) packings as a stationary phase. Anal. Sci. 2003, 19, 269-272.

Yarita, T.; Nakajima, R.; Shimada, K.; Kinugasa, S.; Shibukawa, M. Superheated water chromatography of low molecular weight polyethylene glycols with ultraviolet detection. Anal. Sci. 2005, 21, 1001-1003.

Yisong, S.; Jen-Fon, J.; Weibing, Z. The Method Development of Subcritical Water Chromatography. Chin. J. Chromatogr. 2005, 23, 238-242.

Zhu, C.R.; Goodall, D.M.; Wren, S.A.C. Elevated temperature HPLC: Principles and applications to small molecules and biomolecules. LCGC Eur. 2004, 17, 530-540.

Zhu, P.L.; Dolan, J.W.; Snyder, L.R. Combined use of temperature and solvent strength in reversed-phase gradient elution .2. Comparing selectivity for different samples and systems. J. Chromatogr. A 1996a, 756, 41-50.

Zhu, P.L.; Dolan, J.W.; Snyder, L.R.; Djordjevic, N.M.; Hill, D.W.; Lin, J.T.; Sander, L.C.; Van Heukelem, L. Combined use of temperature and solvent strength in reversed-phase gradient elution .4. Selectivity for neutral (non-ionized) samples as a function of sample type and other separation conditions. J. Chromatogr. A 1996b, 756, 63-72.

Zhu, P.L.; Dolan, J.W.; Snyder, L.R.; Hill, D.W.; Van Heukelem, L.; Waeghe, T.J. Combined use of temperature and solvent strength in reversed-phase gradient elution .3. Selectivity for ionizable samples as a function of sample type and pH. J. Chromatogr. A 1996c, 756, 51-62.

Zhu, P.L.; Snyder, L.R.; Dolan, J.W.; Djordjevic, N.M.; Hill, D.W.; Sander, L.C.; Waeghe, T.J. Combined use of temperature and solvent strength in reversed-phase gradient elution .1. Predicting separation as a function of temperature and gradient conditions. J. Chromatogr. A 1996d, 756, 21-39.

ZirChrom Separations Inc. Metalox 200-C Column Heater. http://www.zirchrom.com/metalox.asp (July 2010).

5

High-efficiency Liquid Chromatography Separations Achieved by Monolithic Silica Columns

Kosuke Miyamoto[1], Tohru Ikegami[1], Takeshi Hara[2], Yasushi Ishihama[3], and Nobuo Tanaka[1,4]

[1]Kyoto Institute of Technology, Department of Biomolecular Engineering, Matsugasaki, Sakyo-ku, Kyoto 606-8585, Japan; [2]Physical Chemistry Institute, Justus Liebig University Giessen, 58 Heinrich Buff Ring, D-35392 Giessen, Germany; [3]Kyoto University, Graduate School of Pharmaceutical Sciences, Yoshida, Sakyo-ku, Kyoto 606-8501, Japan; and [4]GL Sciences, 237-2 Sayamagahara, Iruma 358-0032, Japan

Introduction

Ultrahigh column efficiency has been the focus of many chromatographers studying theories, columns, packing materials, or the separation of complex mixtures. Increasing the plate number (N) of a separation medium is the simplest way to increase separation capability, as resolution of two adjacent peaks (R_s) is proportional to $\sqrt{N}$. Such motivation could also be a driving force for a new technique and the advances of separation science. The possibility of generating a large number (e. g., 100,000–1,000,000) of theoretical plates has been studied since the beginning of high-performance liquid chromatography (HPLC), especially in liquid chromatography (LC) using particle-packed columns, where it is hard to achieve because of the relatively slow molecular diffusion of a solute in liquid phase. Guiochon recently discussed in detail the limit in performance of LC (Guiochon, 2006). Here we briefly review some studies that explore extremely high column efficiency. We also discuss possibilities of practical uses of monolithic silica columns capable of generating 100,000–1,000,000 theoretical plates.

In LC, the efficiency (N, number of theoretical plates, equation 1) of a column (length L), is related to the plate height (H, equation 2), the particle size (d_p) of packing materials, and the variance (σ^2) of a Gaussian-shaped solute band acquired by the solute during its migration through a column. A reduced plate height (h) measures about 2 in a conventional system, which comprises pressure-driven operation of a packed column at optimum linear velocity. N can be calculated from a retention time of a solute (t_R) and the base peak width ($t_W=4\sigma$, σ is a standard deviation) for a symmetrical peak. The potential performance one can achieve using particles of a certain size (d_p) can be described by H. H depends on mobile phase linear velocity (u) and d_p, as shown in equation 3 (Giddings, 1965; Bristow & Knox, 1977), where C_E, C_D, and C_S are coefficients indicating the contributions of the eddy diffusion, molecular diffusion, and slow mass transfer, respectively, to band broadening. Although N under optimized conditions is nearly proportional to $1/d_p$, the pressure drop (ΔP) (equation 4, where η is solvent viscosity, ϕ a flow resistance factor; t_0, column dead time, and $u=L/t_0$) of a column (length L) is proportional to $1/d_p^2$. High permeability of a column (K), (equation 5, $u_F=\varepsilon u$, ε: total column porosity), leads to small ΔP. The smaller d_p yields higher efficiency by the faster mass transfer based on the shorter diffusion path length inside and outside the particles, whereas the small d_p results in a high ΔP due to the small interstitial flow paths. This leads to a limit in N with certain d_p at certain ΔP. High efficiency (large N) and low pressure (small ΔP) are desirable for high resolution and high speed. Small particles allow the use of a shorter column to effect similar separation than do large particles, thus leading to a shorter separation time at the expense of a higher ΔP. Separation time dictated by L is an important factor for the selection of d_p and L, especially for ultrahigh-efficiency separations.

$$N = L^2/\sigma^2 = 16t_R^2/t_W^2 \tag{1}$$

$$H = \sigma^2/L = L/N = hd_p \tag{2}$$

$$H = C_E d_p + C_D/u + C_S d_p^2 u \tag{3}$$

$$\Delta P = \phi \eta u L/d_p^2 \tag{4}$$

$$K = u_F \eta L/\Delta P \tag{5}$$

High Efficiency Separations Achieved by Various Chromatographic Methods

The presence of a limit in N for certain d_p at certain ΔP in each chromatographic system is illustrated in Fig. 5.1. Fig. 5.1.A shows the plots of a plate time, t_0/N, against a plate number, N, in HPLC for particle sizes, 1.4–7 μm (Poppe, 1997), calculated for the pressure limit of 40 MPa. The plots predict that it will be easier to generate very large N by using large d_p, but at the expense of the time required for elution. The plots for several particle sizes shown in Fig. 5.1.A form a dashed line in Fig. 5.1.B as an envelope for the curves; this is known as the Knox-Saleem limit in pressure-driven operation of a packed column (PD-PC) (Knox & Gilbert, 1979). Under no circumstances does a packed column give a plot beyond this limit at certain ΔP.

Table 5.A summarizes the results of previous successful attempts related to high-efficiency separations, reflecting the progress of LC. The first application of LC involved a long column packed with particles. The use of a long column packed with relatively large particles is a straightforward approach to increasing the N of an HPLC column (Scott & Kucera, 1979; Menet et al., 1984; Karlsson & Novotny, 1988). Scott and Kucera demonstrated an efficiency of 650,000 plates using a 14-m-long column packed with 5 μm

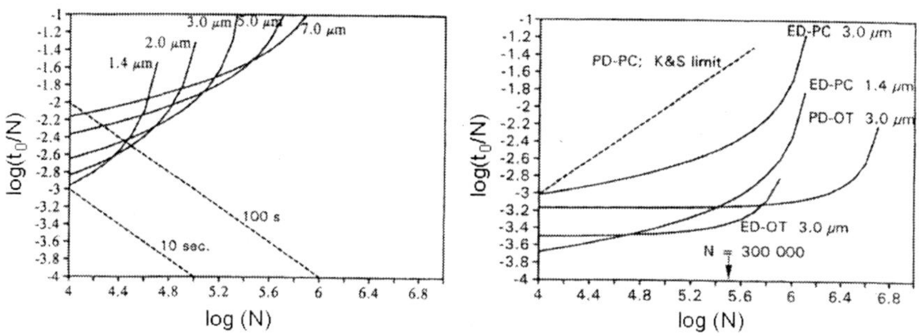

Fig. 5.1. (a) Plot of plate time, t_0/N, against required plate number, N, in high-performance liquid chromatography with particle sizes,of 1.4 μm, 2 μm, 3 μm, 5 μm, and 7 μm. Assumed parameters: maximum pressure ΔP=40 MPa; viscosity η=0.001 Pa s; flow resistance factor φ=1000; diffusion coefficient 1×10^{-9} m^2 s^{-1}; and Knox equation; $h=1.0v^{1/3}+1.5/v+0.05v$. (b) Knox-Saleem limit for a packed column in pressure-driven operation (PD-PC) shown in Figure 1a (dashed line). Electrodriven operation of a particle-packed column (ED-PC) assumes $h=0.75v^{1/3}+1.5/v+0.05v$, maximum voltage 40 kV, and electroosmotic mobility $\mu_{eo}=5\times10^{-8}$ m^2 V^{-1} s^{-1}. Pressure-driven operation of an open tubular column assumes φ =32, h=2/v+Cv, where $C=(1/96)(1+6k+11k^2)/(1+k)^2$ with k=3. Electrodriven operation of an open tubular column (ED-OT) assumes maximum voltage 40 kV, h=2/v+Cv, where $C=(1/16)(k/(1+k))^2$ with k=3, and electroosmotic mobility $\mu_{eo}=5\times10^{-8}$ m^2 V^{-1} s^{-1}. Reproduced with permission from (Poppe, 1997). Copyright 2006 Elsevier.

silica particles with a t_0 of about 500 min (solute=alkylbenzenes, mobile phase=tetrahydrofuran) (Scott & Kucera, 1979). Using a 22-m column packed with 7- to 8-μm silica particles, Menet et al. achieved an N of more than 900,000 theoretical plates with a t_0 as long as 18 h (solute=toluene, mobile phase=dichloromethane) (Menet et al., 1984). Karlsson and Novotny reported high-efficiency separation (N=226,000) in a relatively short time using a 1.95-m capillary column packed with 5-μm octadecylsilylated (ODS) silica particles with t_0=33 min (solute=methoxynaphthalene, mobile phase=acetonitrile) (Karlsson & Novotny, 1988). The high efficiencies reported in these studies, however, were obtained for solutes with very small retention factors. Such high efficiencies have not been reported for well-retained solutes. These early attempts were performed employing columns packed with particles measuring 5–8 μm.

Ultrahigh pressure LC (UHPLC), a technique developed by Jorgenson and coworkers, utilized very high (up to 500 MPa) pressure and columns packed with particles smaller than 2 μm (MacNair et al., 1997; MacNair et al., 1999; Mellors & Jorgenson, 2004). A tenfold increase in pressure to 400 MPa for the calculation would shift the whole plot to the right by one logarithm unit (Fig. 5.1). Using a 66-cm column packed with 1.5-μm silica-ODS particles (solute=hydroquinone, mobile phase=10% acetonitrile) a separation with up to 330,000 theoretical plates was demonstrated (MacNair et al., 1997). Similarly, N=230,000 was obtained using a 46-cm column packed with 1-μm

Table 5.A. High-efficiency Separations that had been Reported for Pressure-driven Liquid Chromatography.

Mode	Column size: d_p[a]	Solute	N	t_0	Ref.
Particulate column	14 m × 1 µm, 5 µm	Benzene	650,000	8.5 h	Scott & Kucera, 1979
	22 m × 1 µm, 7-8 µm	Toluene	920,000	18 h	Menet et. al., 1984
	16.2 m × 44 µm, 10 µm	ethylbenzene	500,000	4.8 h	Tanaka et al., 1985
	1.95 m × 44 µm, 5 µm	methoxy-naphthalene	226,000	33 min	Karlsson & Novotny, 1988
	66 cm × 33 µm, 1.5 µm	hydroquinone	330,000	12 min	MacNair et al., 1997
	46 cm × 33 µm, 1 µm	hydroquinone	230,000	3 min	Macnair et al., 1999
Open tubular column	27.5 m × 32 µm	toluene	2.800,000	6.3 h	Tijssen et al., 1981
	105 m × 39 µm	benzene	1,170,000	8 h	Kucera & Guiochon, 1984
	19.6 m × 51 µm	acetophenone	1,143,000	50 min	Liu et al., 1992
	5 m × 11 µm	Anthracene	1,010,000	50 min	Swart et al., 1995
Monolithic silica	130 cm × 50 µm, 10 µm	hexylbenzene	106,000	25 min	Ishizuka et al., 2002
	140 cm × 4.6 µm, 3.5 µm	amylbenzene	108,000	19 min	Ikegami et al., 2004

[a]Abbreviations: d_p, particle size for a particulate column: d_d, domain size for a monolithic silica column: N, plate number; t_0, column dead time.

particles with t_0 of less than 3 min at 500 MPa (MacNair et al., 1999). UHPLC using small particles is a sure and attractive means of achieving a large number of theoretical plates per unit time. It requires equipment compatible with the high pressure, and is primarily valued for reducing separation time, not for gaining an ultrahigh number of theoretical plates.

Fig. 5.1.B compares the limits of performance of a packed column (PC) and an open-tubular column under pressure-driven (PD) and electrodriven (ED) conditions under a limiting pressure drop (40 MPa) or an electric field (40 kV) (Poppe, 1997). For a certain N value, the smaller t_0/N value corresponds to the faster elution. Open-tubular column (OT) and/or ED operation can produce column efficiency beyond the Knox-Saleem limit in PD-PC.

As the N one can generate by PD-PC is restricted by limitations in pump pressure, high-efficiency separations have also been studied by using columns with the highest permeability, open-tubular columns (Tsuda et al., 1978; Tsuda & Novotny, 1978). As indicated in Fig. 5.1.B, PD-OT yields much faster separation and gives much greater N than PD-PC under the same ΔP (Poppe,

1997; Guiochon, 1981). In order to generate a large N, the column size should be very small so as to avoid slow mass transfer of a solute in a column (Poppe, 1997; Knox & Gilbert, 1979; Tsuda & Nakagawa, 1983), as indicated in Fig. 5.1.B, where the results of calculation for a 3-μm capillary were shown (Poppe, 1997). On the other hand, band broadening associated with a detector was unavoidable for a small-sized open tube capillary column, as described by Knox and Gilbert (Knox & Gilbert, 1979). These attempts at ultrahigh efficiency were first carried out using capillary columns measuring 30–50-μm in diameter.

Tijssen and colleagues reported 2,800,000 theoretical plates using a 32-μm inside diameter (I.D.) and 27.5-m long open-tubular column of etched soft glass (solute=toluene and N-propylaniline, mobile phase=isooctane) (Tijssen et al., 1981). Kucera and Guiochon achieved 1,100,000 theoretical plates for unretained benzene using a 105-m etched glass capillary column measuring 39 μm I.D. The authors also obtained 400,000 plates for anisole with a retention factor (k) of 2 (mobile phase=heptane) (Kucera & Guiochon, 1984). A marked decrease in column efficiency was observed with an increase in solute retention. Erni and coworkers used a shorter column (L=19.6 m, I.D.=51 μm) than the examples shown above but employed high temperatures of up to 200 °C; they reported column efficiencies greater than 1,000,000 theoretical plates for retained solutes with k values of up to 0.1 (solute=acetophenone and benzene, mobile phase=methanol or aqueous acetonitrile) (Liu et al., 1992).

Poppe and coworkers used a 5-m long, 11.4-μm-I.D. capillary coated with polyacrylate to form a column with a flow path of approximately 8.4 μm. The column generated more than 1,000,000 theoretical plates for anthracene with a k of approximately 0.1 and t_0 of about 50 min and resulted in the separation of six anthracene derivatives (mobile phase=acetonitrile) (Swart et al., 1995). A decrease in N was observed with the increase in retention, as demonstrated in equation 6, which describes the dependence of plate height on retention of solutes in such a system. The column efficiency experimentally obtained was much higher than that achieved by PD-PC (mentioned above), but it was much lower than the results of calculation shown by the plots in Fig. 5.1.B for PD-OT assuming k=3, although the comparison may include the difference in the pressure utilized.

$$h=2/\nu+C\nu \tag{6a}$$

$$C= (1/96)(1+6k+11k^2)/(1+k)^2 \tag{6b}$$

Investigators have also sought an increase in column permeability by employing a column packed loosely (Tsuda et al., 1982). A glass column with an I.D. of 1 mm was dry-packed with 10-μm silica particles and was subsequently drawn to 44 μm I.D. to form a loosely-packed column of 16.2-m in length. The column generated 500,000 theoretical plates after octadecylsilylation for ethylbenzene with a k=0.11 in methanol at 10 MPa with t_0 of 27 min (Tanaka et al., 1985). The plot of the results is located outside the Knox-Saleem limit (Fig. 5.1.A). This column contained small amount of packing materials attached to the glass wall which resulted in high permeability, although an ordinary particulate column requires dense packing of particles for the stability of the packed bed.

Capillary electrochromatography using a packed column (ED-PC) can generate a large N with a short t_0 by using smaller particles than PD-PC (Dittman & Rozing, 1996). In this case, small particles can be used, because the solvent is driven by electro osmosis. The plots in Fig. 5.1.B for ED-PC assumed $h=0.75\nu^{1/3}+1.5/\nu+0.05\nu$, reflecting the reduction in the first term (A term) in equation 3 with a plug flow in ED rather than a parabolic flow in PD operation. This is also an excellent method to achieve a large N per unit time, but it is not necessarily suitable for generating extremely large N, as the plots in Fig. 5.1.B suggest. Problems encountered in practical operations include bubble formation, inadequate stability of the packed bed, necessity of an ultrahigh electric field, or inadequate electro osmotic flow for a long capillary column. The electric field available with commercial equipment was not adequate to generate 1,000,000 theoretical plates within a reasonable period of time by ED-PC (Eeltink et al., 2006; Stol et al., 2003).

It is possible to achieve an increase in permeability by using a monolithic column consisting of a network structure of silica skeletons (Svec, 2004; Minakuchi et al., 1996; Guiochon, 2007). A monolithic silica column that can be prepared with various external porosities, or varying throughpore size:skeleton size ratios, possesses the potential for ultrahigh-efficiency separations. Small-sized silica skeletons can contribute to fast mass transfer in the stationary phase. The greater porosity and the relatively large flowthrough pores allow higher permeability than conventional packed columns showing similar efficiencies (Leinweber & Tallarek, 2003; Miyabe et al., 2003; Ishizuka et al., 2002). Desmet and coworkers showed that monolithic silica columns could provide faster separations than particulate columns packed with particles of any size, over a range greater than 80,000 theoretical plates at ΔP=40 MPa (Motokawa et al., 2002; Desmet et al., 2005). More than 100,000 theoretical plates can be generated using a 1.3m monolithic silica capillary column (MSC) in 80% acetonitrile with t_0=25 min with extremely low ΔP (less than 1 MPa). About 108,000 theoretical plates can also be generated by using commercially available monolithic silica C_{18} columns with an I.D. of 4.6 mm; this was demonstrated when 14 columns were connected to form a 140-cm column system (solute=alkylbenzene, 80% acetonitrile, t_0=19 min) (Ikegami et al., 2004). The use of monolithic silica capillary columns seems to be a promising approach to generating a very large N, based on the high permeability and relatively high efficiency allowed by the silica network structure (Guiochon, 2006). In the following sections, we describe generation of 100,000–1,000,000 theoretical plates by using monolithic silica columns, while most HPLC applications are carried out by employing columns that provide 10,000–30,000 theoretical plates.

Structural Features and Chromatographic Properties of Monolithic Silica Columns

Structures of Monolithic Silica Columns

Fig. 5.2.A shows a scanning electron microscopy (SEM) image of a monolithic silica column prepared from tetramethoxysilane (TMOS) in a test tube, and Fig. 5.2.B–D shows those prepared from a mixture of TMOS and methyltrimethoxysilane (MTMS) in a fused silica capillary of 50–200 μm I.D. Monolithic silica columns prepared in a test tube (monolithic silica rods: MSRs) present smooth and evenly distributed skeletons compared to monolithic silica capillary columns (MSCs). Although shrinkage of skeletons is allowed during the formation of the co-continuous structures in a test tube, the skeletons

Fig. 5.2. Scanning electron micrographs of a monolithic silica column. (a) Monolithic silica prepared from tetramethoxysilane (TMOS) in a test tube. Monolithic silica columns prepared from a mixture of TMOS and methyltrimethoxysilane (MTMS) in a 50-μm fused-silica capillary (b), in a 100-μm fused-silica capillary (c), and in a 200-μm fused silica capillary (d). The arrows in panel a indicate the through-pore size and the skeleton size. Reproduced with permission from reference (Motokawa, 2002). Copyright 2002 Elsevier.

must be covalently attached to the tube wall for preparation in a capillary to avoid shrinking of the whole structure. Otherwise, the shrinkage tends to pull the skeletons away from the wall resulting in a void between the silica skeleton and the tube wall (Ishizuka et al., 2002; Nakanishi, 1997). The structures of monolithic silica materials prepared in a capillary, shown in Fig. 5.2.B–D, were less homogeneous than that shown in Fig. 5.2.A, but they are relatively homogeneous compared to those prepared from TMOS in a capillary (Motokawa et al., 2002). It is not easy to synchronize phase separation and gelation for the preparation in a capillary needed for forming homogeneous structures. See (Nakanishi, 1997) for details of the preparation method.

Fig. 5.3 shows the plots of through-pore size versus skeleton size for MSRs and for MSCs in several batches, as well as for particle-packed columns (Motokawa et al., 2002; Minakuchi et al., 1997; Minakuchi et al., 1998; Hara et al., 2006; Nakanishi & Tanaka, 2007). The size of through-pores and silica skeletons range from 1.3 to 3.5 μm and from 0.9 to 2.3 μm, respectively. A through-pore size/skeleton size ratio of about 1.5 was found for many monolithic silica columns, while the ratio was found to be about 1.0 for some materials prepared to have a greater phase ratio or a lower porosity.

The monolithic silica shown in Fig. 5.2.B (MS-H(50)-II in reference 31), possesses a through-pore size of about 2 μm and skeleton size of about 1.5 μm, whereas monolithic silica F in Fig. 5.3 (MS(100)-F in [Hana, et al., 2006]) possesses smaller skeletons (0.9 μm) and through-pores (1.3 μm). This material was produced with a higher phase ratio (i.e., there was a greater amount of

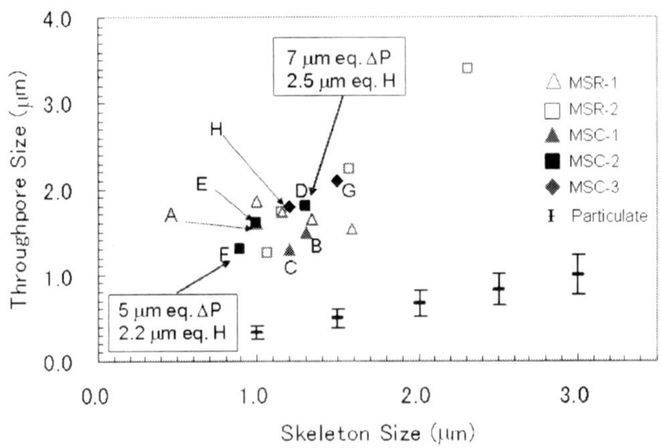

Fig. 5.3. Plots of through-pore size versus skeleton size for monolithic silica columns prepared in a test tube (MSRs), monolithic silica columns prepared in a capillary (MSCs), and for particle-packed columns. Throughpore-size/skeleton-size ratios in a range 0.25–0.4 are indicated as bars for the particle-packed columns. Reproduced with permission from reference (Nakanishi & Tanaka, 2007). Copyright 2007 American Chemical Society.

silica in the column compared to the previous preparations), resulting in the more homogeneous structures shown in Fig. 5.4.B, compared to MS-H(50)-II shown in Fig. 5.4.A. The former possesses co-continuous structures, whereas the latter can be considered as agglomerated spheres. It has been predicted that a decrease in porosity and an increase in homogeneity would improve the performance of monolithic columns (Gzil et al., 2004). A similar increase in structural homogeneity and efficiency in high speed separations was achieved recently for a hybrid monolithic silica capillary column (MS-H) by a similar approach of increasing the phase ratio (Hara et al., 2010).

Throughpore-size/skeleton-size ratios ranging from 1 to 2 obtained for the monolithic silica columns (Fig. 5.3) are much larger than those of packed columns at 0.25–0.4 (the range indicated as bars in Fig. 5.3). The particle size dictates the size of interstitial voids for PCs, while the structures can be designed for monolithic columns. The greater throughpore-size/skeleton-size ratio means that the external porosity (the porosity outside the skeletons) of a monolithic silica column is greater than that of a particle-packed column that

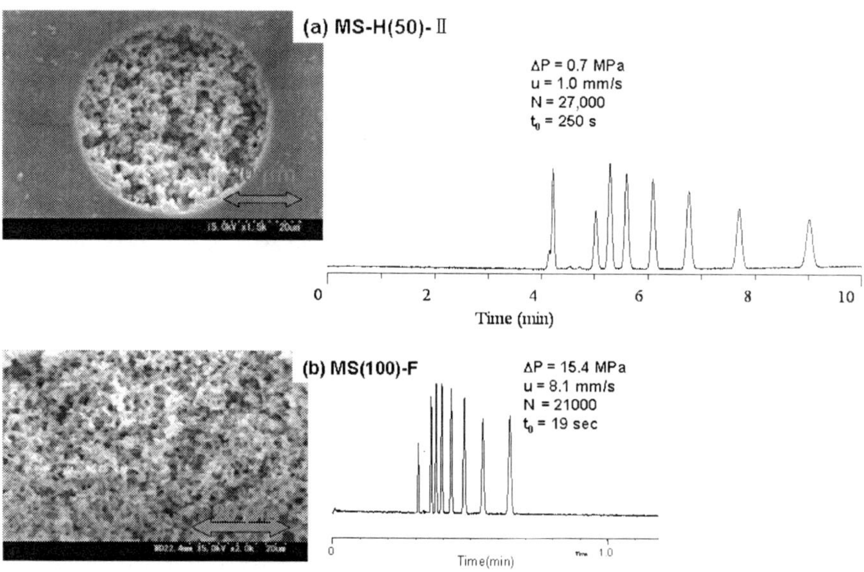

Fig. 5.4. Scanning electron microscopy images and chromatograms obtained for monolithic silica capillary columns. Chromatograms were obtained for uracil (the first peak) and alkylbenzenes ($C_6H_5(CH_2)nH$, n=0-6). (a) MS-H(50)-II prepared from a mixture of TMOS and MTMS 100 μm ID, 20-cm long. Mobile phase: acetonitrile/water = 80/20. Temperature: 30 °C. (b) MS(100)-F (shown in Fig. 5.3) prepared from TMOS, 100 μm ID, 20-cm long. Mobile phase: acetonitrile/water = 80/20. Temperature: 30 °C. Panel (a) reproduced with permission from reference (Motokawa et al., 2002). Copyright 2002 Elsevier. Panel (b) reproduced with permission from reference (Hara et al., 2006). Copyright 2006 American Chemical Society.

is commonly about 0.4. External porosity of about 0.6 was found for monolithic silica rod columns, and external porosity of about 0.8 for monolithic silica capillary columns; total porosities of about 0.8 and those greater than 0.9 were observed, respectively.

Chromatographic Properties of Monolithic Silica Columns

Commercially available silica rod columns, Chromolith columns, (4.6-mm diameter, 1- to 10-cm long) modified by octadecylsilylation and coated with polyetheretherketone (PEEK) resin, possessing 2-μm through-pores, 1 to1.5-μm skeletons, and with mesopores measuring around 13 nm result in back pressure equivalent to 7- to 10-μm particles and a column efficiency equivalent to 3.5- to 4-μm particles (Leinweber & Tallarek, 2003; Miyabe et al., 2003; Cabrera et al., 2000). The performance in terms of column efficiency and pressure drop was higher than that of commonly used 5-μm particles; this was made possible by the large-sized through-pores and the small-sized skeletons. The thin skeletons lead to high efficiency based on fast equilibration of solutes between a mobile liquid phase and a stationary solid phase on the mesopore walls, and the large macropores contribute to high permeability. Typically, permeability (K, equation 5) $K= 8\times10^{-14}\ m^2{\sim}20\times10^{-14}\ m^2$ was observed for monolithic silica rod columns compared to $2\text{-}4\times10^{-14}\ m^2$ for a column packed with 5-μm particles. The flow resistance parameter, ϕ, is estimated to be 50-150 for monolithic silica columns having a domain size of 3.5 μm (2-μm through-pores and 1.5-μm skeletons) compared to 700–1000 for a particulate column.

Another type of monolithic silica rod column has become commercially available recently. MonoClad-C18-HS columns possess smaller skeletons and through-pores (1.2 μm) with larger mesopores (18 nm) than Chromolith RP-18, showed permeability $K= 2.4\times10^{-14}\ m^2$ which is similar to that of a column packed with 5-μm particles, while producing column efficiency similar to that of 2.5-μm particles with minimum H as small as 5.5-μm (Miyazaki et al., 2010). Total porosity and external porosity are ca. 75% and 55%, respectively, for a MonoClad column, which are smaller than those of a Chromolith column. The monolithic silica rod (3 mm diameter) was octadecylsilylated and covered with two layers of polymers in a stainless-steel tube (8 mm OD, 2 mm thickness). The structure allowed the fabrication of 25-cm long columns, longer than Chromolith columns, allowing full use of their high permeability, with a suggested pressure limit of 30 MPa. Such a column can produce as much as 45,000 theoretical plates with a t_0 of 1.8 min or less with back pressure less than 10 MPa.

It was possible to obtain 80,000 and 120,000 theoretical plates by coupling two or three 25-cm MonoClad columns, respectively, in a pressure range of 15–25 MPa, and a t_0 of 4–7 min. The results can be compared favourably with those reported with columns packed with 1.7 μm totally porous particles or

2.7 μm superficially porous particles at much higher pressures (Cabooter et al., 2008), although only the latter was examined for different solutes and mobile phases. Although the separation on the connected monolithic silica columns requires a longer time for generating similar numbers of theoretical plates based on the use of a longer column system, similar column efficiency was obtained at much lower pressure, 15–25 MPa, which can be delivered with conventional HPLC equipment.

Monolithic silica capillary columns can also provide 10,000–20,000 theoretical plates at low back pressures. The chromatograms presented in Fig. 5.4 show that a 25-cm MS-H(50)-II column (d_d=3.5 μm, d_d: domain size) yielded N=27,000 after octadecylsilylation, with very low ΔP, whereas the 20-cm column MS(100)-F (d_d=2.2 μm) generated N=21,000 with t_0=19 s. The separation impedance values ($E=(t_0/N)(\Delta P/N)(1/\eta)$) of these columns were between 500 and 1000, which are much smaller than the values found for a PC.

Fig. 5.5 compares the performance of reversed-phase monolithic silica columns with particulate columns using a kinetic plot analysis at a 40-MPa pressure limit. The plots of $\log(t_0/N^2)$ versus log(N) allow an effective comparison of the column performance in terms of column efficiency and separation time, as proposed by Desmet et al. (2005). In the case of particulates, there is an optimum particle size for each desired N, where a faster separation (smaller t_0) is possible than with other particle sizes. In Fig. 5.5, the plots for the second generation monolithic silica capillary column with the small domains (plot F in Fig. 5.3) and those for MonoClad C18-HS are shown below the curves representing the plots for particle-packed columns (d_p=1.4 – 5 μm) in a region where N is greater than 30,000 and 50,000, respectively, thus indicating superior performance for producing N in this range. The optimum performance of these monolithic silica columns having small domains is seen at a range of N of approximately 200,000–500,000.

For high efficiency separations (N=300,000–1,000,000), the optimum particle size for a particle-packed column is in the range of 5- to 10-μm under the 40-MPa pressure limit, as shown in Fig. 5.6 (Miyamoto et al., 2008). A particle size of 5 to 10 μm is suitable for a long column to be used with conventional HPLC equipment, as has been examined previously. Fig. 5.6 also shows similar kinetic plots for monolithic silica capillary columns prepared under the conditions similar to MS-H(50)-II. The monolithic columns are shown to provide smaller $\log(t_0/N^2)$ values than particulate columns for a range of N greater than 80,000 with a 40 MPa pressure limit. The plots indicate that 100,000 theoretical plates can be generated with a t_0 of about 250 s, and the monolithic columns shown in Fig. 5.6 will show optimum performance at around 1,000,000 theoretical plates when used at 40 MPa. A typical monolithic column can provide 300,000 plates faster by a factor of three than a particulate column packed with 5-μm particles, and 1,000,000 plates faster by a factor of about five than a particulate column packed with 10-μm particles or

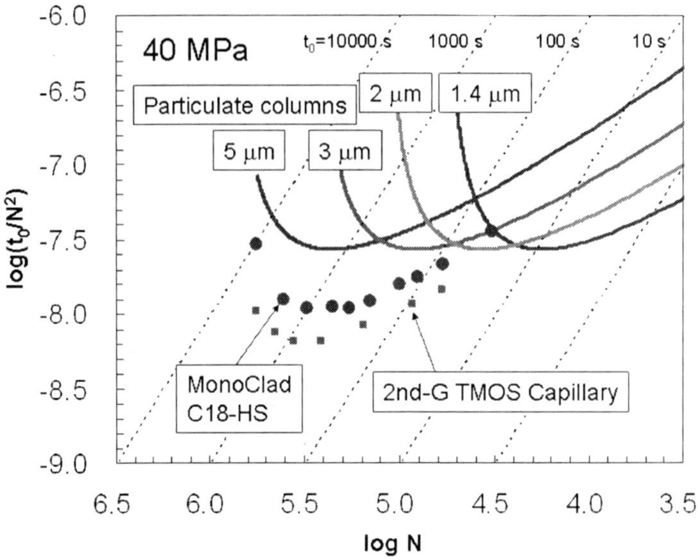

Fig. 5.5. Plots of log(t_0/N^2) versus log(N) for monolithic silica C_{18} columns and particle-packed columns. The curves for particle-packed columns were obtained by assuming the following parameters: η=0.00046 Pa s, φ = 700, D_m=2.22x10^{-9} m^2/s, and Knox equation, h=0.65$v^{1/3}$+2/v+0.08v. Maximum pressure: 40 MPa. The particle diameters for the particle-packed columns were: 1.4 μm, 2 μm, 3 μm, and 5 μm. Symbols: The performance of MS(100)-F (X) in Figure 3 and MonoClad C18-HS columns obtained experimentally are also indicated.

particles of any size; monolithic columns thus show a definite advantage over particulate columns in achieving ultrahigh efficiency. Fig. 5.6 suggests that when one million theoretical plates are desired, a t_0 value of 5,000 to 10,000 s will be needed at this ΔP. With the columns shown in Fig. 5.6, it is expected that 1,000,000 theoretical plates could be accomplished with an inlet pressure of 40 MPa using a 10-μm column.

Examples of High-efficiency Separations

It is possible to prepare an octadecylsilylated monolithic silica capillary column of up to 500 cm in length. Such a column commonly produced 400,000–500,000 theoretical plates under optimum conditions. Fig. 5.7 shows a separation of alkylbenzenes in 80% acetonitrile on a column system of 1140 cm consisting of three monolithic capillary columns connected in series (448+448+244 cm) (Miyamoto et al., 2008). At about 47 MPa and at 30°C, the column system produced ca. 1,000,000 theoretical plates for the alkylbenzenes with k values of up to 2.1. Smaller numbers of theoretical plates were obtained for alkylbenzenes retained longer. High column efficiencies were also observed for polynuclear aromatic hydrocarbons (PAHs), which are Envi-

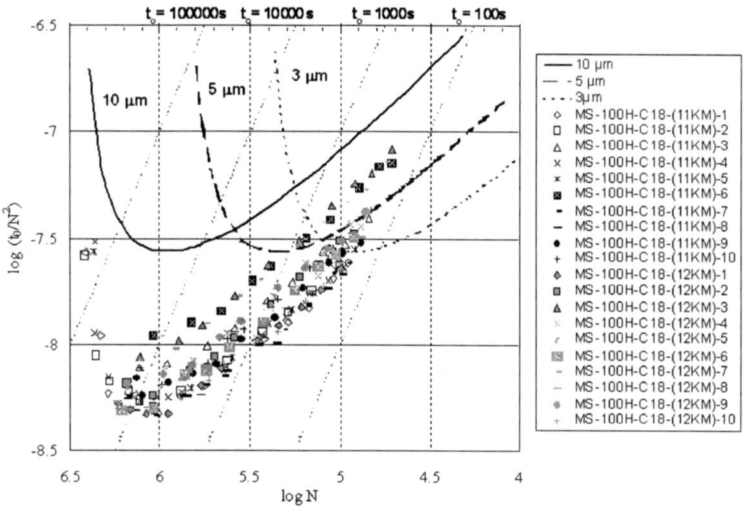

Fig. 5.6. Plots of log(t_0/N_2) versus log(N) for monolithic silica C18 capillary columns prepared from a mixture of TMOS and MTMS. The curves for particle-packed columns were obtained as in Fig. 5.5. Maximum pressure: 40 MPa. The particle diameters for the particle-packed columns are 3 µm, 5 µm, and 10 µm, respectively. The symbols represent the experimental values obtained for twenty monolithic silica C18 capillary columns prepared in two batches. Reproduced with permission from reference (Miyamoto et al., 2008). Copyright 2008 American Chemical Society.

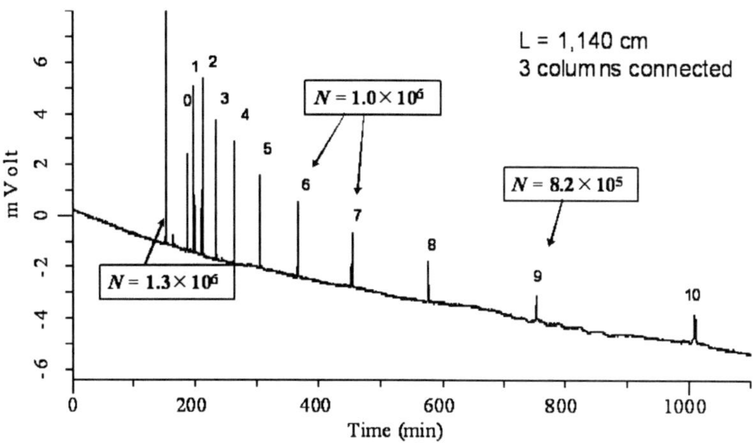

Fig. 5.7. Separation of alkylbenzenes. Mobile phase: 80% CH_3CN. Column: three monolithic silica C18 columns connected in series, effective length 1140 cm. ΔP = 35.4 MPa. u = 1.24 mm/s. Detection: 210 nm. Temperature: 30oC. Sample: thiourea and alkylbenzenes, $C_6H_5C_NH2_N$+1 (n=0-10). The number of theoretical plates for the peak is indicated. Reproduced with permission from reference (Miyamoto et al., 2008). Copyright 2008 American Chemical Society.

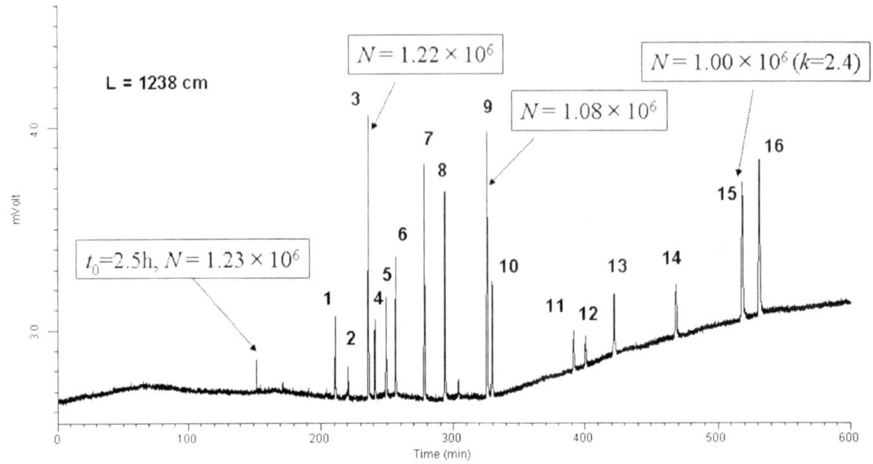

Fig. 5.8. Separation of polynuclear aromatic hydrocarbons. Mobile phase: 80% CH_3CN. Column: three monolithic silica C18 columns connected in series, effective length 1238 cm (total length 1244 cm). ΔP = 46.6 MPa. u = 1.31 mm/s. Detection: 210 nm. Temperature: 30°C. Sample: 16-PAHs primary pollutants designated by EPA. Peak numbers; 1, naphthalene; 2, acenaphthylene; 3, fluorine; 4, acenaphthene; 5, phenanthrene; 6, anthracene; 7, fluoranthene; 8, pyrene; 9, chrysene; 10, benz(a) anthracene; 11, benzo(b)fluoranthene; 12, benzo(k)fluoranthene; 13, benzo(a)pyrene; 14, dibenz(a,h) anthracene; 15, indeno(1,2,3-cd)pyrene; and 16, benzo(g,h,i)perylene. Reproduced with permission from reference (Miyamoto et al., 2008). Copyright 2008 American Chemical Society.

ronmental Protection Agency-designated priority pollutants under similar conditions. Fig. 5.8 shows the separation of 16 PAHs using a 1238-cm column system (448+448+342 cm), which produced 1,230,000 theoretical plates for an unretained peak, and 1,000,000 theoretical plates for the PAHs with retention factors (k values) of up to 2.4.

High column efficiencies of around 1,000,000 theoretical plates were obtained with the monolithic silica capillary columns for solutes with greater retention factors in a much shorter time than in previous attempts. The reduction in column efficiency for retained solutes was shown to be less when compared to the results reported for open tubular columns mentioned earlier. The through-pore size of about 2 μm is much smaller than the capillary diameter of the open tubular columns employed in the previous studies, which can contribute to the smaller contribution of the slow mass transfer in the mobile phase to the band broadening for the monolithic columns. In fact, 1,000,000 theoretical plates were obtained with a t_0 of about 2.5 h; this figure agrees with the prediction based on Fig. 5.6.

Monolithic silica capillary columns of 70–90 cm were very useful for a LC-mass spectrometry (MS) system extensively used in metabolomic and proteomic studies, resulting in high peak capacity by gradient elution (Tolstikov et al., 2003; Luo et al., 2005). Fig. 5.9 shows chromatograms of gradient

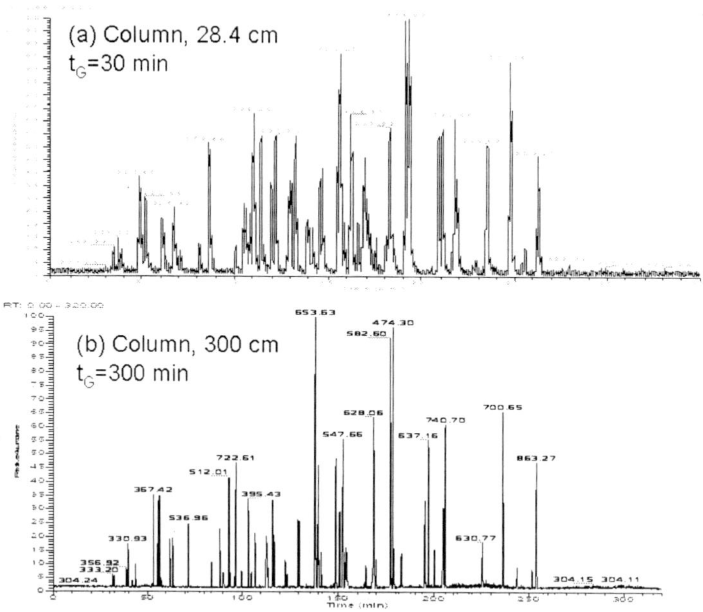

Fig. 5.9. Gradient separation of a BSA digest. Column: MS-100H-C18, (a) 28.4 cm and (b) 300 cm. The mobile phases consisted of (A) 0.1% formic acid in water and (B) 0.1% formic acid in acetonitrile. A linear gradient of 5% to 40% B in 30 min for the short column and in 300 min for the long column was employed. A BSA digest sample dissolved in water (100 nL, 1 nmol/200 μL) was injected at the start of the gradient with a flow rate of 1.2 μL/min for the 28.4-cm column and 0.67 μL/min for the 300-cm column. Reproduced with permission from reference (Miyamoto et al., 2008). Copyright 2008 American Chemical Society.

elution of a tryptic digest of bovine serum albumin on a 28.4-cm octadecylsilylated (ODS) column (a) and 300-cm MSC-ODS column (b) with gradient times proportional to the column length (Miyamoto et al., 2008). The shorter column and the longer one produced peak capacities of about 125 in 23 min and 380 in 215 min, respectively, for the elution range of the peptides between the two peaks with mass numbers of 331 and 863. Peak capacity obtainable per unit time is nearly three times greater for the shorter column, but the real peak capacity of the longer column for the sample mixture is greater than that of the shorter column by a factor of three. Long monolithic silica capillary columns will be very effective in increasing resolution of a complex mixture.

Recently a 350-cm monolithic silica capillary column was applied for a reversed-phase gradient separation of peptides for a proteomic study. An extremely long gradient time of longer than 40 h was applied in the one-dimensional LC separation coupled with tandem mass spectrometry resulting in identification of more than 16,000 peptides and 2,200 proteins with one run (Fig. 5.10), and more than 22,000 peptides and 2,600 proteins in the Escherichia coli proteome to effect nearly complete coverage with triplicate

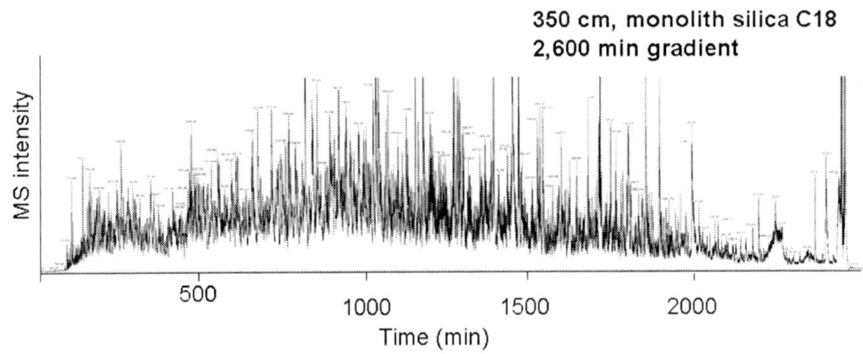

Fig. 5.10. Base peak chromatogram of E. coli whole proteome analysis using a monolithic silica C18 column (L=350 cm, 100 μm ID) with the injection of 4 μg E. coli whole cell lysate. Mobile phase; (A) 0.5% acetic acid and (B) 0.5% acetic acid and 80% acetonitrile with a two-step linear gradient of 5% to 40% B in 2470 min, and 40% to 100% B in 5 min, then 100% B for 10 min. Flow rate; 500 nL/min. The MS scan range; 300-1500. Reproduced with permission from reference (Iwasaki et al., 2010). Copyright 2010 American Chemical Society.

runs (Iwasaki et al., 2010). More than 90% of membrane proteins were identified with this approach. The small sample amount required, simple instrumentation, and the simple operation of the system are among the advantages of one-dimensional capillary LC system for proteomics.

Applications of Long Monolithic Silica Capillary Columns for Separations Based on Small Separation Factors

Isotope separation usually requires a large number of theoretical plates, because of the small separation factors (Tanaka & Thornton, 1977; Turowski et al., 2003; Kimata et al., 1996). For the separation of deuterated compounds from protiated compounds, one can expect a single isotope effect (the separation factor provided by one deuterium substitution) of up to 1.010 per deuterium atom on the retention factors, (k_H/k_D) (Turowski et al., 2003). Fig. 5.11 shows an example of the isotopic separation of benzene, benzene-d, benzene-1,3,5-d_3, benzene-d_6, toluene, toluene-α,α,α-d_3, and toluene-d_8 performed on a 440-cm monolithic silica capillary column. Toluene and toluene-α,α,α-d_3 were separated with a resolution greater than unity based on a separation factor of 1.0166. The single isotope effects observed with benzene and the aromatic C-H/C-D of toluene were calculated to be 0.70–0.72%. The single isotope effect observed for the aliphatic C-H/C-D of toluene was ca. 0.55%. The chromatogram in Fig. 5.11 inset (b) was obtained for isotopic benzenes, using an 850-cm column system consisting of two columns connected in series, 500 cm and 350 cm. All isotopologues from benzene-d_6 to benzene were separated in this chromatogram by the difference of one deuterium, although the sample was supposed to contain only four isotopologues.

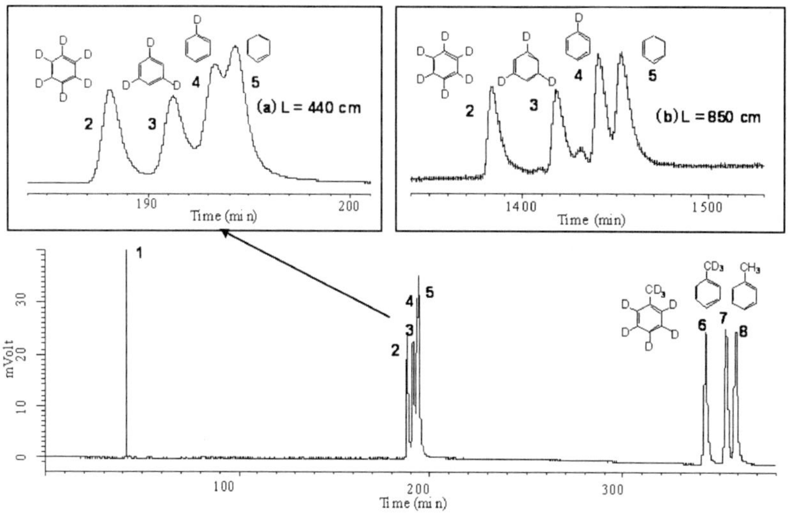

Fig. 5.11. Separation of benzene and toluene isotopologues. Mobile phase: 30% CH_3CN. Column: monolithic silica C18 column, 440 cm. Detection: 210 nm. Temperature: 30°C. u = 1.76 mm/s. ΔP = 39.6 MPa. Samples: 1, thiourea; 2, benzene-d_6; 3, benzene-1,3,5-d_3; 4, benzene-d; 5, benzene; 6, toluene-d_8; 7, toluene-α,α,α-d_3; 8, toluene. Inset a: Magnification of part of the chromatogram at 185-200 min. Inset b: Chromatogram for the benzene isotopologues obtained with two monolithic silica C18 columns connected in series, total length 850 cm in a ternary mobile phase, acetonitrile-methanol-water (10/5/85). Detection: 210 nm. Temperature: 30°C. u = 1.02 mm/s. ΔP = 34 MPa. Reproduced with permission from reference (Miyamoto et al., 2008). Copyright 2008 American Chemical Society.

This example shows the utility of long monolithic silica capillary columns for the separation of closely related compounds of structural similarity. According to calculations, it is possible to produce Rs of unity for a pair of compounds with a separation factor as small as 1.0056 using 1,000,000 theoretical plates at a k of 2.5. Such a small separation factor was found for diastereomers based on isotopic chirality caused by the presence of a phenyl and a deuterated phenyl group (Kimata et al., 1996). A long column would be useful for the separation of complex mixtures in a relatively narrow range of a chromatogram with little effort required for method development for isocratic elution as well as for a complex mixture of a wide range of compounds with long gradient elution.

Further Possibility of High-efficiency Separations Using Monolithic Silica Columns

In addition to the above-described methods, high-efficiency capillary columns can also be prepared by polymerizing alkyl methacrylate (Núñez et al., 2007), acrylic acid (Horie et al., 2007), or acrylamide (Ikegami et al., 2006) in monolithic silica capillary columns that have anchor groups on the silica surface. Facile preparation of polar- and nonpolar-modified surfaces (Ikegami et al., 2008; Núñez et al., 2008) as well as those with ion-exchange functionality (Watanabe et al., 2009), and stability against the high flow rate required for multidimensional applications (Tanaka et al., 2004) make monolithic silica attractive, despite the disadvantages related to the silica preparation and the chemical modification in individual columns. Hydrophilic interaction liquid chromatography (HILIC) (Alpert et al., 1994; Yoshida, 1997; Hemström et al., 2006) will become an important tool for the separation of highly polar substances that are biologically important, but are not well retained in a reversed-phase system. High-efficiency columns for HILIC and ion exchange LC will be useful for life sciences-oriented applications.

Long capillary columns can be prepared easily, as it is not necessary to pack a column after chemical modification of silica surfaces, and the columns are free from frits. Monolith modification possesses additional advantages. It is possible to control the amount of stationary phase and to maintain the performance of silica support after modification, unless the stationary phase hinders the mass transfer of solutes in the stationary phase and the solvent flow in a column. As suggested earlier in the comparison of various forms of chromatography, monolithic silica columns may be better able to provide means for practical applications of high-efficiency separations using commonly available equipment. Further increase in N to several million theoretical plates in reversed-phase mode will require a monolithic silica column of ca. 5-μm d_d and 1-2 days of t_0 at ΔP=50-100 MPa.

Conclusions

We have described structural features and the performance of long monolithic silica columns. A recent experiment using a MSC of 90–500 cm long reported results of around 100,000–500,000 theoretical plates. When three columns were connected in series to form an 11–12 m column system, more than 1,000,000 theoretical plates were generated for solutes with k up to 2.4 with a pressure drop of 40–50 MPa (Miyamoto et al., 2008). The features of monolithic silica columns, (i) high permeability based on large throughpores and high porosity leading to high-speed operation, (ii) high efficiency based on small-sized skeletons, (iii) facile preparation of polar and nonpolar stationary phases by simple polymerization reactions of functional monomers in the monolithic silica column having anchor groups, and (iv) facile preparation of a long capillary column from monomer solutions, may make them a practical tool for very high efficiency chromatographic separations.

Acknowledgment

This work was supported in part by Grants-in-Aid for Scientific Research funded by the Ministry of Education, Sports, Culture, Science and Technology of Japan, No. 20350036. We thank Hirotaka Takubo, Shohei Miwa, Chun Yang, Shota Miyazaki, and Masayoshi Ohira for their assistance.

References

Alpert, A.J.; Shukla, M.; Shukla, A.K.; Zieske, L.R.; Yuen, S.W.; Ferguson, M.A.J.; Mehlert, A.; Pauly, M.; Orlando, R. Hydrophilic-interaction chromatography of complex carbohydrates. J. Chromatogr. A 1994, 676, 191–202.

Bristow, P.A.; Knox, J.H. Standardization of test conditions for high-performance liquid chromatography columns. Chromatographia 1977, 10, 279–289.

Cabooter, D.; Lestremau, F.; Lynen, F.; Sandra, P.; Desmet, G. Kinetic plot method as a tool to design coupled column systems producing 100,000 theoretical plates in the shortest possible time. J. Chromatogr. A 2008, 1212, 23–34.

Cabrera, K.; Lubda, D.; Eggenweiler, H.-M.; Minakuchi, H.; Nakanishi, K. A new monolithic-type HPLC column for fast separations. J. High Resolut. Chromatogr. 2000, 23, 93–99.

Desmet, G.; Clicq, D.; Gzil, P. Geometry-independent plate height representation methods for the direct comparison of the kinetic performance of LC supports with a different size or morphology. Anal. Chem. 2005, 77, 4058–4070.

Dittmann, M.M.; Rozing, G.P. Capillary electrochromatography: a high-efficiency microseparation technique. J. Chromatogr. A 1996, 744, 63–74.

Eeltink, S.; Desmet, G.; Vivó-Truyols, G.; Rozing, G.P.; Schoenmakers, P. J.; Kok, W.T. Performance limits of monolithic and packed capillary columns in high-performance liquid chromatography and capillary electrochromatography. J. Chromatogr. A 2006, 1104, 256–262.

Giddings, J.C. Dynamics of Chromatography: Principles and Theory, Marcel Dekker: New York 1965.

Guiochon, G. Conventional packed columns vs packed or open tubular microcolumns in liquid chromatography. Anal. Chem. 1981, 53, 1318–1325.

Guiochon, G. The limits of the separation power of unidimensional column liquid chromatography. J. Chromatogr. A 2006,1126, 6–49.

Guiochon, G. Monolithic columns in high-performance liquid chromatography. J. Chromatogr. A 2007, 1168, 101–168.

Gzil, P.; Vervoort, N.; Baron, G.V.; Desmet, G. General rules for the optimal external porosity of LC supports. Anal. Chem. 2004, 76, 6707–6718.

Hara, T.; Kobayashi, H.; Ikegami, T.; Nakanishi, K.; Tanaka, N. Performance of monolithic silica capillary columns with increased phase ratios and small-sized domains. Anal. Chem. 2006, 78, 7632–7642.

Hara, T.; Makino, S.; Watanabe, Y.; Ikegami, T.; Cabrera, K.; Smarsly, B.; Tanaka, N. The performance of hybrid monolithic silica capillary columns prepared by changing feed ratios of tetramethoxysilane and methyltrimethoxysilane. J. Chromatogr. A, 2010, 1217, 89–98.

Hemström, P.; Irgum, K. Hydrophilic interaction chromatography. J. Sep. Sci. 2006, 29, 1784–1821.

Horie, K.; Ikegami, T.; Hosoya, K.; Saad, N.; Fiehn, O.; Tanaka, N. Highly efficient monolithic silica capillary columns modified with poly(acrylic acid) for hydrophilic interaction chromatography. J. Chromatogr. A 2007, 1164, 198–205.

Ikegami, T.; Dicks, E.; Kobayashi, H.; Morisaka, H.; Tokuda, D.; Cabrera, K.; Hosoya, K.; Tanaka, N. How to utilize the true performance of monolithic silica columns. J. Sep. Sci. 2004, 27, 1292–1302.

Ikegami, T.; Fujita, H.; Horie, K.; Hosoya, K.; Tanaka, N. HILIC mode separation of polar compounds by monolithic silica capillary columns coated with polyacrylamide. Anal. BioAnal. Chem. 2006, 386, 578–585.

Ikegami, T.; Tomomatsu, K.; Takubo, H.; Horie, K.; Tanaka, N. Separation efficiencies in hydrophilic interaction chromatography. J. Chromatogr. A 2008, 1184, 474–503.

Ishizuka, N.; Kobayashi, H.; Minakuchi, H.; Nakanishi, K.; Hirao, K.; Hosoya, K.; Ikegami, T.; Tanaka, N. Monolithic silica columns for high-efficiency separations by high-performance liquid chromatography. J. Chromatogr. A 2002, 960, 85–96.

Iwasaki, M.; Miwa, S.; Ikegami, T.; Tomita, M.; Tanaka, N.; Ishihama, Y. One-dimensional capillary liquid chromatographic separation coupled with tandem mass spectrometry unveils the Escherichia coli proteome on a microarray scale. Anal. Chem. 2010, 82, 2616–2620.

Karlsson, K.-E.; Novotny, M.V. Separation efficiency of slurry-packed liquid chromatography microcolumns with very small inner diameters. Anal. Chem. 1988, 60, 1662–1665.

Kimata, K.; Kobayashi, M.; Hosoya, K.; Araki, T.; Tanaka, N. Chromatographic separation based on isotopic chirality. J. Am. Chem. Soc. 1996, 118, 759–762.

Knox, J.H.; Gilbert, M.T. Kinetic optimization of straight open-tubular liquid chromatography. J. Chromatogr. 1979, 186, 405–418.

Kucera, P.; Guiochon, G. Use of open-tubular columns in liquid chromatography. J. Chromatogr. 1984, 283, 1–20.

Leinweber, F.C.; Tallarek, U. Chromatographic performance of monolithic and particulate stationary phases: hydrodynamics and adsorption capacity. J. Chromatogr. A 2003, 1006, 207–228.

Liu, G.; Djordjevic, N.M.; Erni, F. High-temperature open-tubular capillary column liquid chromatography. J. Chromatogr. 1992, 592, 239–247.

Luo, Q.Z.; Shen, Y.F.; Hixson, K.K.; Zhao, R.; Yang, F.; Moore, R.; Heather J.; Mottaz, M.; Smith, R.D. Preparation of 20 mm I.D. silica-based monolithic columns and their performance for proteomics analyses. Anal. Chem. 2005, 77, 5028–5035.

MacNair, J.E.; Lewis, K.C.; Jorgenson, J.W. Ultrahigh-pressure reversed-phase liquid chromatography in packed capillary columns. Anal. Chem. 1997, 69, 983–989.

MacNair, J.E.; Patel, K.D.; Jorgenson, J.W. Ultrahigh-pressure reversed-phase capillary liquid chromatography: isocratic and gradient elution using columns packed with 1.0-mm particles. Anal. Chem. 1999, 71, 700–708.

Mellors, J.S.; Jorgenson, J.W. Use of 1.5-mm porous ethyl-bridged hybrid particles as a stationary-phase support for reversed-phase ultrahigh-pressure liquid chromatography. Anal. Chem. 2004, 76, 5441–5450.

Menet, H.G.; Gareil, P.C.; Rosset, R.H. Experimental achievement of one million theoretical plates with microbore liquid chromatographic columns. Anal. Chem. 1984, 56, 1770–1773.

Minakuchi, H.; Nakanishi, K.; Soga, N.; Ishizuka, N.; Tanaka, N. Effect of domain size on the performance of octadecylsilylated continuous porous silica columns in reversed-phase liquid chromatography. J. Chromatogr. A 1998, 797, 121–131.

Minakuchi, H.; Nakanishi, K.; Soga, N.; Ishizuka, N.; Tanaka, N. Effect of skeleton size on the performance of octadecylsilylated continuous porous silica columns in reversed-phase liquid chromatography. J. Chromatogr. A 1997, 762, 135–146.

Minakuchi, H.; Nakanishi, K.; Soga, N.; Ishizuka, N.; Tanaka, N. Octadecylsilylated porous silica rods as separation media for reversed-phase liquid chromatography. Anal. Chem. 1996, 68, 3498–3501.

Miyabe, K.; Cavazzini, A.; Gritti, F.; Kele, M.; Guiochon, G. Moment analysis of mass-transfer kinetics in C18-silica monolithic columns. Anal. Chem. 2003, 75, 6975–6986.

Miyamoto, K.; Hara, T.; Kobayashi, H.; Morisaka, H.; Tokuda, D.; Horie, K.; Koduki, K.; Makino, S.; Núñez, O.; Yang, C.; et al. High-efficiency liquid chromatographic separation utilizing long monolithic silica capillary columns. Anal. Chem. 2008, 80, 8741–8750.

Miyazaki, S.; Takahashi, M.; Ohira, M.; Terashima, H.; Morisato, K.; Nakanishi, K.; Ikegami, T.; Miyabe, K.; Tanaka, N. Monolithic silica rod columns for high-efficiency reversed-phase liquid chromatography. J. Chromatogr. A 2010.

Motokawa, M.; Kobayashi, H.; Ishizuka, N.; Minakuchi, H.; Nakanishi, K.; Jinnai, H.; Hosoya, K.; Ikegami, T.; Tanaka, N. Monolithic silica columns with various skeleton sizes and through-pore sizes for capillary liquid chromatography. J. Chromatogr. A 2002, 961, 53–63.

Nakanishi, K. Pore structure control of silica gels based on phase separation. J. Porous Mater. 1997, 4, 67–112.

Nakanishi, K.; Tanaka, N. Sol-Gel with Phase Separation. Hierarchically porous materials optimized for HPLC separations. Acc. Chem. Res. 2007, 40, 863–873.

Núñez, O.; Ikegami, T.; Kajiwara, W.; Miyamoto, K.; Horie, K.; Tanaka, N. Preparation of high efficiency and highly retentive monolithic silica capillary columns for reversed-phase chromatography by chemical modification by polymerization of octadecyl methacrylate. J. Chromatogr. A 2007, 1156, 35–44.

Núñez, O.; Nakanishi, K.; Tanaka, N. Preparation of monolithic silica columns for high-performance liquid chromatography. J. Chromatogr. A. 2008, 1191, 231–252.

Poppe, H. Some reflections on speed and efficiency of modern chromatographic methods. J. Chromatogr. A 1997, 778, 3–21.

Scott, R.P.W.; Kucera, P. Mode of operation and performance characteristics of microbore columns for use in liquid chromatography. J. Chromatogr. 1979, 169, 51–72.

Stol, R.; Poppe, H.; Kok, W.T. Optimization of particle dimensions for high efficiency in capillary electrochromatography. Anal. Chem. 2003, 75, 5246–5253.

Swart, R.; Kraak, J.C.; Poppe, H. Performance of an ethoxyethylacrylate stationary phase for open-tubular liquid chromatography. J. Chromatogr. A 1995, 689, 177–187.

Svec, F. Organic polymer monoliths as stationary phases for capillary HPLC. J. Sep. Sci. 2004, 27, 1419–1430.

Tanaka, N.; Kimura, H.; Tokuda, D.; Hosoya, K.; Ikegami, T.; Ishizuka, N.; Minakuchi, H.; Nakanishi, K.; Shintani, Y.; Furuno, M.; et al. Simple and comprehensive two-dimensional reversed-phase HPLC using monolithic silica columns. Anal. Chem. 2004, 76, 1273–1281.

Tanaka, N.; Kinoshita, H.; Araki, M.; Tsuda, T. On-column preparation of chemically bonded stationary phase with maximum surface coverage and high reproducibility, and its application to packed microcapillary columns. J. Chromatogr. 1985, 332, 57–69.

Tanaka, N.; Thornton, E.R. Structural and isotopic effects in hydrophobic binding measured by high-pressure liquid chromatography, a stable and highly precise model for hydrophobic interactions in biomembranes. J. Am. Chem. Soc. 1977, 99, 7300–7307.

Tijssen, R.; Bleumer, J.P.A.; Smith, A.L.C.; Van Kreveld, M.E. Microcapillary liquid chromatography in open tubular columns with diameters of 10--50 μm: potential application to chemical ionization mass spectrometric detection. J. Chromatogr. 1981, 218, 137–165.

Tolstikov, V.V.; Lommen, A.; Nakanishi, K.; Tanaka, N.; Fiehn, O. Monolithic silica-based capillary reversed-phase liquid chromatography/electrospray mass spectrometry for plant metabolomics. Anal. Chem. 2003, 75, 6737–6740.

Tsuda, T.; Hibi, K.; Nakanishi, T.; Takeuchi, T.; Ishii, D. Studies of open-tubular microcapillary liquid chromatography. II. Chemically bonded octadecylsilane stationary phase. J. Chromatogr. A 1978, 158, 227–232.

Tsuda, T.; Nakagawa, G. Open-tubular liquid chromatography with 5–10-μm I.D. columns. J. Chromatogr. 1983, 268, 369–374.

Tsuda, T.; Novotny, M. Band-broadening phenomena in microcapillary tubes under the conditions of liquid chromatography. Anal. Chem. 1978, 50, 632–634.

Tsuda, T.; Tanaka, I.; Nakagawa, G. Packed microcapillary liquid chromatography with reduced I.D. columns. J. Chromatogr. 1982, 239, 50–513.

Turowski, M.; Yamakawa, N.; Meller, J.; Kimata, K.; Ikegami, T.; Hosoya, K.; Tanaka, N.; Thornton, E.R. Deuterium isotope effects on hydrophobic interactions: the importance of dispersion interactions in the hydrophobic phase. J. Am. Chem. Soc. 2003, 125, 13836–13849.

Watanabe, Y.; Ikegami, T.; Horie, K.; Hara, T.; Jaafar, J.; Tanaka, N. Improvement of separation efficiencies of anion-exchange chromatography using monolithic silica capillary columns modified with polyacrylates and polymethacrylates containing tertiary amino or quaternary ammonium groups. J. Chromatogr. A, 2009, 1216, 7394–7401.

Yoshida, T. Peptide separation in normal phase liquid chromatography. Anal. Chem. 1997, 69, 3038–3043.

6

Silver-ion High-performance Liquid Chromatography - Mass Spectrometry in the Analysis of Lipids

Michal Holčapek and Miroslav Lísa
University of Pardubice, Faculty of Chemical Technology, Department of Analytical Chemistry, Pardubice, Czech Republic

Introduction

The ability of silver-ion and some other metal-ions to interact with double bonds (DBs) has been a well known property for a long time (Dobson et al., 1995; Guha & Janák, 1972; Devries, 1962; Morris, 1962). Silver-ion chromatography, sometimes called argentation chromatography, is a separation technique based on the formation of weak reversible charge-transfer complexes between silver-ion and DBs of unsaturated organic molecules. Silver-ion chromatography can be performed either in a planar arrangement (silver-ion thin layer chromatography, Ag-TLC) or on column [used mainly for preparative purposes or silver-ion high-performance liquid chromatography (Ag-HPLC) used for high-resolution separations]. Ag-TLC is a well-established technique in the lipid analysis because it is cheap, simple to use and applicable both in analytical and preparative scale. Disadvantages of Ag-TLC are mainly the contamination of fractions in the preparative isolation by silver and the limited resolution of TLC technique. Ag-TLC has been reviewed in several works (Dobson et al., 1995; Momchilova & Nikolova-Damyanova, 2003), therefore it is not discussed in this chapter.

The retention in the Ag-HPLC is governed by the DB number. More DBs mean stronger interactions and therefore higher retention times. In practice, the situation is not as simple because other factors have to be taken into account, such as the DB geometry (*cis* vs. *trans*), distances among individual DBs (from the conjugation to remote DBs), overall molecular structures, the type of stationary phase, the composition of mobile phase, the gradient steepness, separation temperature, etc. The vast majority of Ag-HPLC applications is in the area of lipids, but this technique can be used for other compounds containing DBs as well (Guha & Janák, 1972; Williams & Mander, 2001). This chapter focuses solely on lipids, mainly triacylglycerols (TGs), fatty acids (FAs) and their simple derivatives such as fatty acid methyl esters (FAMEs). In principle, other non-polar lipids (e.g., sterols, wax esters, cholesterol esters, vitamins, carotenoids) can be analyzed as well, but there is a lack of published papers in the literature (Joh et al., 1995; Kakela et al., 2002; Shan & Wilson, 2002).

Mechanism of Silver-ion Interaction with Double Bonds

Zeise prepared the first organo-metal complex in 1820s [i.e., Zeise's salt $K[PtCl_3(C_2H_4)]\cdot H_2O$ (Zeise, 1831)], which stimulated more scientific research in the field of organometallic chemistry. Today, organo-metal complexes are of great interest due to their application in bond activation processes and catalysis. In the field of chromatography, organo-metal complexes are used for the separation of unsaturated organic compounds based on their interactions with metals immobilized in the stationary phase (Guha & Janák, 1972), especially silver-ion chromatography uses the capability of unsaturated compounds to form organo-metal complexes containing silver(I) ions. In silver-ion chromatography, unsaturated compounds form weak reversible complexes of different strength with silver-ions immobilized in the stationary phase during their elution through the chromatographic column. It is a dynamic system with continuous establishmen of equilibrium among complexed and free components with a high equilibrium constant. Complexes are of the charge-transfer type, where unsaturated compounds donate electrons to the silver-ion (acceptor). The description of complexation bonding between DB and silver(I) ion by the Dewar-Chatt-Duncanson model (Dewar, 1951; Chatt & Duncanson, 1953) is now widely accepted. This model describes the stabilization of complexes as a combination of σ-donation and π-back-bonding interactions between DBs and metal, i.e., donation of π-electrons from the occupied 2*p* bonding orbital of the olefinic DB into vacant 5*s* and 5*p* orbitals of the silver-ion (σ-type bond, Fig. 6.1.A) and the back-donation of *d*-electrons from occupied *4d* orbitals of the silver-ion into unoccupied π*-2*p* antibonding orbitals of the olefinic DB (π-type bond, Fig. 6.1.B). This model describes only bonding between DB and silver-ion, but the stability constant of complex also depends on steric and polar effects. A number of experimental and theoretical studies have been done for complexes of silver-ion with short olefins, since these complexes are

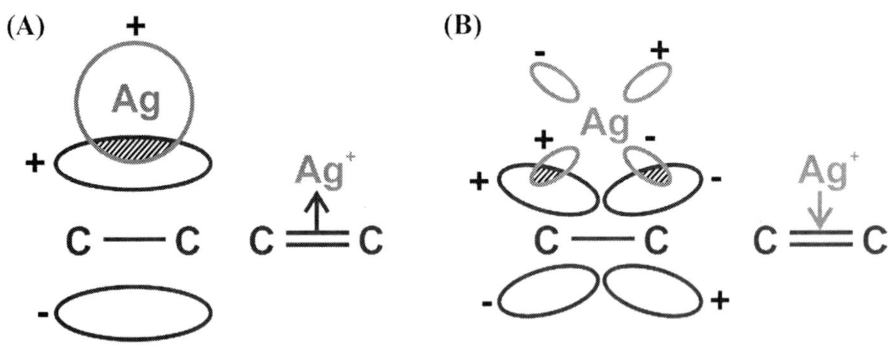

Fig. 6.1. Description of complex bonding between silver-ions and DBs by the Dewar-Chatt-Duncanson model: (A) σ-donation and (B) π-back-bonding interactions between the metal and DBs. Redrawn with permission from (Dewar, 1951).

important in organometallic chemistry. Early studies of stability constants of organo-silver complexes used distribution methods based on the distribution of organic compounds between an organic phase and silver nitrate aqueous solution (Lucas et al., 1943; Winstein & Lucas, 1938). These results have been confirmed later by various analytical techniques, i.e., UV, infrared and Raman spectroscopy based on shifts in absorption maxima between complex and free unsaturated compounds (Hosoya & Nagakura, 1964), X-ray study of organo-silver monocrystals (Gmelin, 1975; Bressan et al., 1967), electron spin resonance (Kasai et al., 1980), etc. General conclusions of complex stability affected by structural factors are as follows. The stability of complexes containing *cis*-DBs are higher than with *trans*-DBs (Lucas et al., 1943; Morris, 1966), complexes of methylene-interrupted DBs are stronger than conjugated ones and the stability of complex increases with increasing distance of DBs (Winstein & Lucas, 1938), the stability of complexes decreases with increasing chain length (Conacher, 1976) and the stability of complexes increases by substitution of hydrogen with deuterium atoms (Cvetanović et al., 1965).

In silver-ion chromatographic process, there are not only simple interactions between DB of unsaturated compound and silver-ion. Electron spin resonance shows the interaction of silver-ion with two molecules (Kasai et al., 1980) and X-ray of monocrystal shows the coordination of one silver-ion with two DBs from different molecules (Gmelin, 1975; Bressan et al., 1967). The interaction of silver-ion with carboxylic oxygen of unsaturated compounds has also been shown (Winstein & Lucas, 1938). In addition to many interactions of silver-ion and unsaturated compounds, the retention is also influenced by the quality of column packing, i.e., density and accessability of silver-ions on the surface of stationary phase. Nowadays, most Ag-HPLC columns are based on the silica matrix chemically modified with alkylphenylsulfonic bonding of silver-ions by ionic bond. Free silanol groups may interact with unsaturated molecules during the chromatographic process and molecules are separated based on the mixed retention mechanism. The separation of glycerolipids is more complex due to a number of combinations of interactions among silver-ions, DBs and carboxylic oxygens. Electrostatic forces within and between long acyl chains of molecule could also influence their retention behavior, but little is known about these forces so far.

Parameters Affecting Silver-ion HPLC

Types of Silver-ion Systems

Three potential ways for embedding silver-ions in the HPLC system have been described (Momchilova & Nikolova-Damyanova, 2003; Nikolova-Damyanova, 2009; Nikolova-Damyanova, 2005): 1) adsorption of silver-ions on the stationary phase, 2) silver-ions are embedded in the stationary phase via ionic bond, 3) addition of silver-ions into the mobile phase.

1) Silver complexation column (Jeffrey, 1991; Schuyl et al., 1998) in past, silver-ion columns for HPLC were prepared by the adsorption of silver-ions (typically in the form of silver nitrate) on silica columns, then washed carefully with different solvents and the final washing with the mobile phase. A certain part of silver-ions is adsorbed on the silica gel. Some papers present good separations with this type of column, but at present they almost disappear due to serious drawbacks, such as the leakage of silver-ion into the mobile phase (critical for mass spectrometry coupling), poor reproducibility and also technical skills required for the reproducible preparation of columns.
2) Strong cation-exchanger modified with silver-ions (Emken et al., 1964; Christie, 1987; Toschi et al., 1993) silver-ions replace initial protons in the $-SO_3H$ functional group [estimated silver content is 50–80 mg (Christie, 2011)] and then $-SO_3Ag$ interacts with DBs during the chromatographic process. This ionic bond is rather stable and no leakage of silver-ions is detected even for long-time use in high-performance liquid chromatography - mass spectrometry (HPLC/MS) experiments. Nowadays, practically all Ag-HPLC/MS papers are based on the ion-exchanger type of silver-ion columns. Some researchers use commercial ChromSpher Lipids column from Varian as the only manufacturer on the market so far, while others rely on their laboratory-made silver-ion columns with comparable performance according to the procedure published by Christie (1987). The manufacturer recommends the use of eluents like dichloromethane, dichloroethane, acetone, acetonitrile, toluene and ethylacetate. Aqueous solvents are discouraged since they permanently alter the column properties, especially small anions that can possibly cause silver precipitation are prohibited. Solvents should be free of peroxides or any reducing agents that can cause the reduction of silver(I) to the metal state. No acids should be used due to the back replacement of silver-ions to protons.
3) The addition of silver-ions into the mobile phase using C18 column (Correa et al., 1999; Nikolova-Damyanova et al., 1993). This approach is not applicable for MS coupling, because non-volatile inorganic salts cause ion suppression and also the contamination of the ion source. Moreover, the retention mechanism in this arrangement is not pure Ag-HPLC as for previous two approaches, but only a combination of silver-ion and reversed-phase modes.

Mobile Phase Composition

The mobile phase optimization is a crucial step in Ag-HPLC because the proper optimization of solvent composition and gradient steepness can significantly improve the separation including the regioisomeric resolution.

Two types of mobile phases are most frequently used in Ag-HPLC. The first type is based on chlorinated solvents, such as dichlormethane or dichloroethane with the addition of other polar modifiers at low concentration, typically acetonitrile, acetone, methanol (Christie 1988; Juaneda et al., 1994; Laakso & Voutilainen, 1996; Nikolova-Damyanova et al., 1992; Nikolova-Damyanova et al., 1995). The second type is hexane based mobile phase with addition of acetonitrile as polar modifier (Adlof & List, 2004; Adlof, 1994; Adlof, 1995; Dugo et al., 2004; Dugo et al., 2006a–c; Mondello et al., 2005). In addition to these two main types, some other solvent combinations have been also reported, for example toluene, hexane and ethylacetate (Schuyl et al., 1998), methanol and acetonitrile (van der Klift et al., 2008), 13 different eluents system containing hexane, heptane or isooctane with the addition of acetonitrile, propionitrile or butyronitrile (Muller et al., 2006), heptane with acetonitrile or acetone (Macher & Holmqvist, 2001), acetone with acetonitrile (Nikolova-Damyanova et al., 1995).

Hexane-acetonitrile mobile phases have a unique property of possible regioisomeric resolution of unsaturated TGs, which has not been reported for chlorinated mobile phases. The disadvantage of hexane-acetonitrile system is a low solubility of acetonitrile in hexane, which is only about 1–1.5% at ambient temperature (Adlof & List, 2004). The miscibility problem strongly limits the range of applicable chromatographic conditions and significantly contributes to the reproducibility problem, which can be partly solved by using continuous magnetic stirring. Two alternative approaches have been tested how to solve the miscibility limitation while keeping an excellent chromatographic resolution including the resolution of positional isomers. When acetonitrile is replaced by propionitrile, then the miscibility is much better and the regioisomeric resolution is almost the same (Muller et al., 2006; Lísa et al., 2009), but a serious health hazard arises due to the toxicity of propionitrile. The second approach relies on the addition of the third solvent with good mutual miscibility with both acetonitrile and hexane. The ideal combination is hexane - 2-propanol - acetonitrile (Lísa et al., 2009; Cvačka et al., 2006; Han et al., 1999; Holčapek et al., 2010; Holčapek et al., 2009), where remarkable improvements in the reproducibility of retention times are observed in comparison with traditional hexane - acetonitrile binary mixture (Muller et al., 2006; Lísa et al., 2009). Standard deviations of retention times for three selected peaks (PLP, PLL and LLL) in hexane - 2-propanol - acetonitrile are 0.4%, 1.0% and 0.7% for one-day measurements compared to 7.4%, 6.8% and 5.2% for hexane - acetonitrile mobile phase (Lísa et al., 2009). Some shifts in retention times can occur on a longer time scale, but they can be efficiently eliminated by the use of the relative retention $r = (t_{R,TG} - t_M)/(t_{R,std} - t_M)$. There are some important issues concerning the mobile phase preparation, which should be followed to obtain good reproducibility. Mobile phases should be prepared fresh every day using solvents dried with molecular sieves and kept in tightly closed containers to avoid evaporation. A low percentage of additives in hexane should be premixed in solvent containers (Dugo et al., 2006; Dugo et al., 2006; Lísa et al., 2009). The degassing of mobile phase by automatic degasser is preferred over continuous stripping with the stream of helium. Columns are conditioned using the low flow rate of initial gradient composition (50 μL/min) overnight and the standard flow rate for one hour before the analysis (Lísa et al., 2009).

Temperature

Temperature has an important role in the optimization of chromatographic separation of lipids, which is not limited to Ag-HPLC, but it is valid for other separation modes as well, such as non-aqueous reversed-phase (NARP) (Holčapek et al., 2005; Lísa & Holčapek, 2008; Holčapek et al., 1999; Holčapek et al., 2003). Increased temperature can result in the loss of resolution for critical TG pairs in NARP mode, where the retention time depends inversely on the temperature. In case of Ag-HPLC, the correlation among retention times, temperature and chromatographic resolution is more complex. The basic rule for the temperature dependence in Ag-HPLC mode is that higher temperature means higher retention times (Adlof & List, 2004; Lísa et al., 2009; Adlof, 2007), which is rather unusual behavior not known for other HPLC and GC modes, where just the opposite behavior is common. The magnitude of this effect is directly related to the DB number and is more evident with *cis*-DBs (Adlof & List, 2004). A possible explanation for this strange behavior has been proposed by Adlof (Adlof & List, 2004) based on the different stability of the acetonitrile complex with silver-ions, which is probably exothermic and thereby less stable at higher temperatures, which allows an increased number of interactions for the analyte with silver-ions at higher separation temperatures, resulting in higher retention.

A temperature decrease, within a certain range, causes lower retention in hexane-acetonitrile mobile phases, as demonstrated for FAMEs (Adlof, 2007) and TGs (Adlof & List, 2004; Adlof, 2007). Fig. 6.2 shows that the retention of FAME standards decreases with decreasing temperature from 20 to -10°C, however it significantly rises with further decrease in temperature (-20°C). A similar trend is observed for TGs, but the retention change is at 0°C (Fig. 6.3). This behavior is not observed in chlorinated solvent systems (Adlof & List, 2004). The sudden increase in retention times at very low temperature can be explained by several temperature-related factors (Adlof, 2007): A) solubility of sample in the mobile phase, B) solubility of acetonitrile in hexane, C) changes in flexibility/3D configuration of analyte or stationary phase. Another possible explanation is that the number of unsaturated molecules coordinated in the complex with silver-ion depends on the temperature, whereas only one unsaturated molecule forms the complex at 25°C in comparison to two coordinated molecules at 0°C (Winstein & Lucas, 1938). Temperature gra-

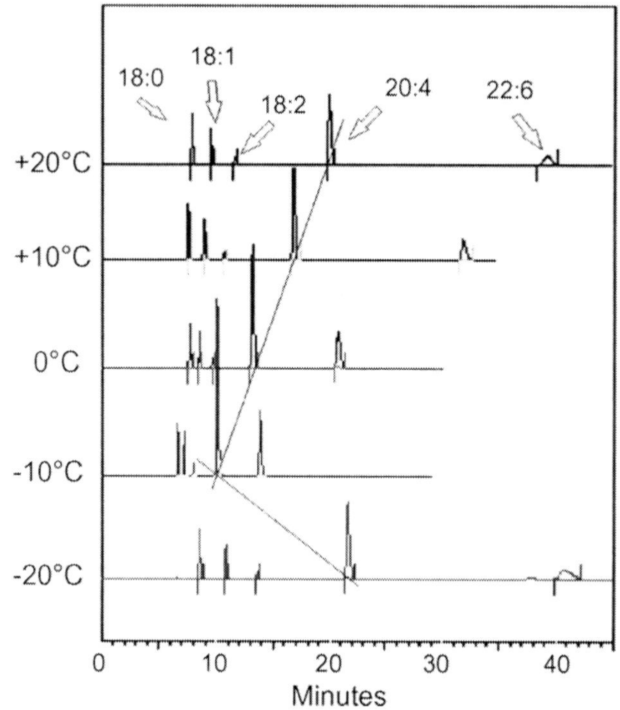

Fig. 6.2. Effect of separation temperature on the retention of FAMEs. Conditions: dual-column Ag-HPLC, 1.0 mL/min, 0.5% acetonitrile in hexane, evaporative light-scattering detection. Reprinted with permission from (Adlof, 2007).

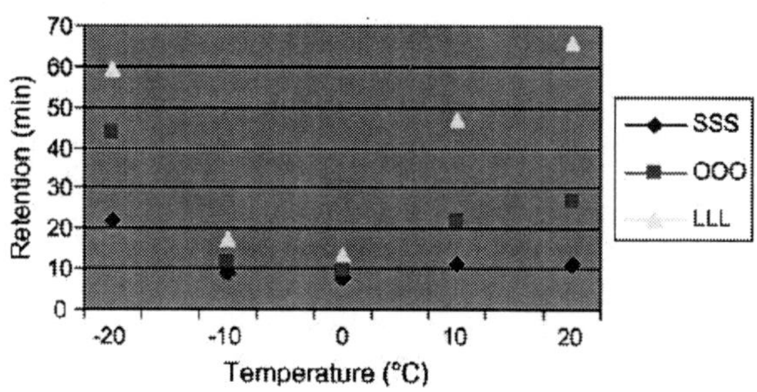

Fig. 6.3. Effect of separation temperature on the retention of TGs. Conditions: dual-column Ag-HPLC, 1.0 mL/min, 0.7% acetonitrile in hexane, evaporative light-scattering detection. Reprinted with permission from (Adlof, 2007).

dients (Adlof, 2007) could be used for the optimization of chromatographic resolution of complex FAMEs or TGs instead of common solvent composition gradient, but in our best knowledge this idea has not yet been used yet in any published paper.

Other Approaches

If the resolution is not sufficient on one silver-ion column, then more columns can be coupled in series, as demonstrated in several works (Adlof & List, 2004; Adlof, 1994; Adlof, 1995; Lísa et al., 2009; Holčapek et al., 2010; Holčapek et al., 2009). An increased length of chromatographic column improves the resolution of critical pairs, for example regioisomeric doublets (Lísa et al., 2009; Holčapek et al., 2010; Holčapek et al., 2009) or DB positional isomers of FA derivatives (Juaneda, et al., 1994). Unlike in NARP systems, the back pressure is not a limiting factor here, because the mobile phases typically consist of low viscosity organic solvents (e.g., hexane, dichloromethane, dichloroethane) with a low percentage of polar modifier. Limiting factors are mainly long retention times associated with the extended column length and peak broadening effects for the multiple column coupling.

Another approach for increasing peak capacity is the use of two-dimensional (2D) HPLC, where the arrangement may be either off-line (Dugo et al., 2004; Dugo et al., 2006; Holčapek et al., 2009; Dugo et al., 2005; Laakso & Christie, 1991) or on-line (Dugo et al., 2006; Mondello et al., 2005; van der Klift et al., 2008). NARP is mostly used in the first dimension for off-line coupling (Dugo et al., 2004; Dugo et al., 2006; Holčapek et al., 2009; Dugo et al., 2005), while the reversed arrangement is more typical for on-line comprehensive 2D (Dugo et al., 2006; Mondello et al., 2005; van der Klift et al., 2008). The absence of any paper with Ag-HPLC mode used in the second dimension for on-line separation indicates that Ag-HPLC is not convenient for fast one minute analysis required in such setup. Van Beek et al. (2008) has reported a focusing effect in the second dimension, because the second dimension solvent (methanol - methyl-*tert*-butyl ether, 70:30) is stronger than the first dimension solvent (acetonitrile - methanol mixture). On-line separation enables the fast separation of TGs in two chromatographic modes without the intervention of the operator and allows possible automation, while an off-line setup requires the fraction collection in the first dimension with their subsequent analysis in the second dimension. An off-line approach is more laborious, but the resolution in both modes can be fully optimized without the limitation in 2D dictated by the fraction collection time (Fig. 6.4). By the way, regioisomeric separations have been reported only in off-line arrangement (Dugo et al., 2004; Dugo et al., 2006; Holčapek et al., 2009).

Recently, on-line and off-line Ag-SFC and NARP-HPLC coupling have been applied for the analysis of TGs in fish oil, where off-line mode gives much better performance due to the sample complexity (Francois et al., 2010). In addition to HPLC, other separation techniques can be performed in the silver-ion mode as well, such as gas-liquid chromatography (GLC) (Guha & Janak, 1972) and supercritical fluid chromatography (SFC) (Demirbuker & Blomberg, 1991; Demirbukeret al., 1992; Dermaux et al., 1999).

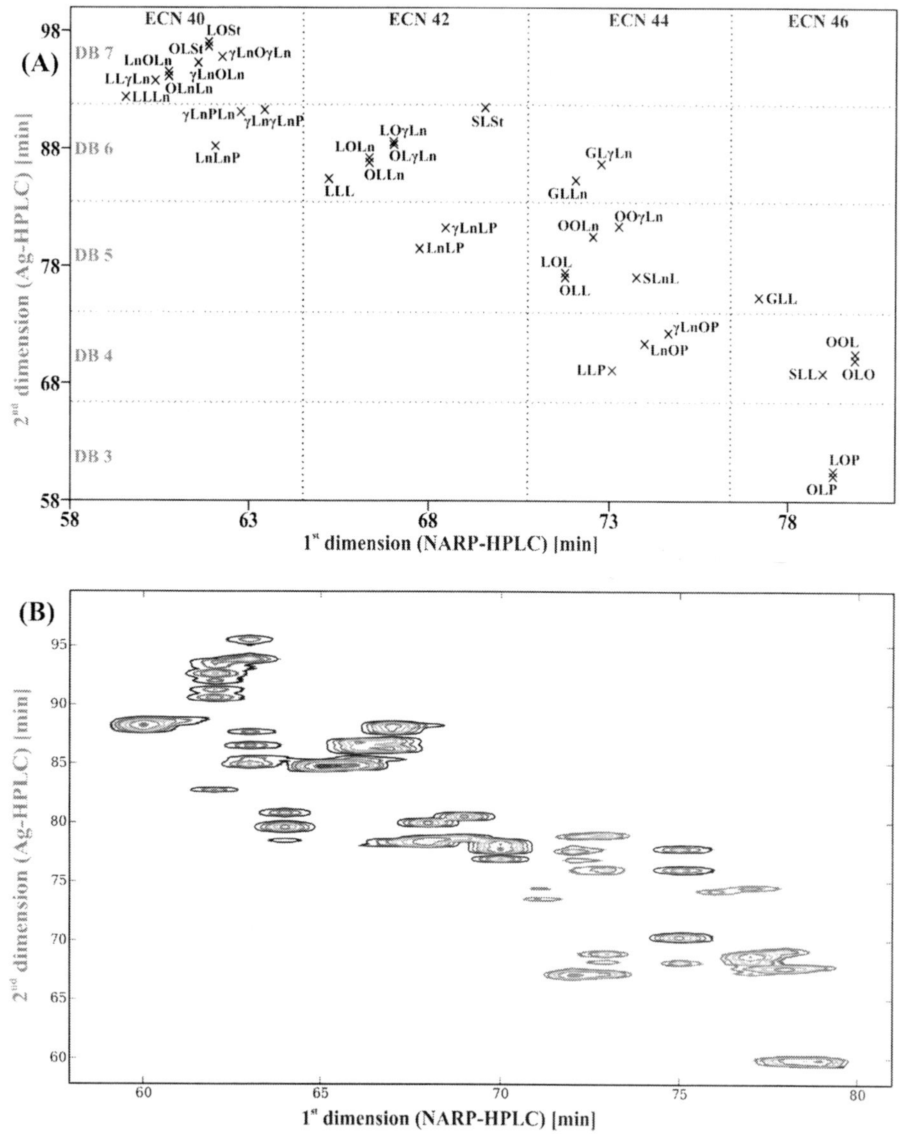

Fig. 6.4. Off-line 2D chromatograms of blackcurrant oil using NARP in the first dimension and silver-ion mode in the second dimension after fraction collection each minute: (A) dot plot with the peak annotation and DB and equivalent carbon number (ECN) labeling, (B) contour plot showing peak intensities. Reprinted with permission from (Holčapek et al., 2009).

Retention Behavior

Fatty Acids and Their Derivatives

FAs are an important lipid class due to their functions in a human body. They are the main constituents of complex lipids, governing their retention behavior in Ag-HPLC because their DBs form complexes with silver-ions. The separation of FAs in their free form using Ag-HPLC is feasible, but in most cases they are separated as FAMEs or other derivatives (Nikolova-Damyanova, 2009). In general, the retention times of FAs in Ag-HPLC are determined by the DB number, their *cis-trans-* configuration and positions of DBs in acyl chains. There is no direct proportionality between retention times of FAs and DB number, but in general, retention times of FAs increase with increasing number of DBs. For example, Δ9,12-18:2 is retained about 3 times longer than a corresponding Δ9-18:1 homolog, Δ9,12,15-18:3 is retained 2.5 times longer than Δ9,12-18:1, and Δ5,8,11,14,17-20:5 is retained 1.5 times longer than Δ5,8,11,14-20:4 (Nikolova-Damyanova et al., 1992).

The complex of silver-ions with *cis*-DBs is stronger than with *trans*-isomers, therefore *cis*-FA isomers are retained more strongly than *trans*-isomers. The retention behavior of polyunsaturated FAs containing *trans*-DBs is given by the number and positions of *trans*-DBs in the acyl chain. Retention times of FAs decrease with increasing number of *trans*-DBs (Toschi et al., 1993; Adlof, 1994; Adlof & Lamm, 1998; Christie & Breckenridge, 1989; Phillips et al., 1997). The retention order of 18:3 FA isomers as methyl- (Adlof, 1994) and phenacyl esters (Juaneda et al., 1994) is the following (Fig. 6.5):

$$\Delta 9t,12t,15t \ll \Delta 9t,12c,15t < \Delta 9c,12t,15t < \Delta 9t,12t,15c \ll \Delta 9c,12t,15c < \Delta 9t,12c,15c < \Delta 9c,12c,15t \ll \Delta 9c,12c,15c.$$

The stability of FA complexes with silver-ions during the chromatographic process is also influenced by their position(s) in the acyl chain and the number of methylene units between two DBs. Conjugated dienes with various configurations and positions of DBs in the acyl chain are eluted always prior to *cis*-monoenes (Adlof & Lamm, 1998; Momchilova & Nikolova-Damyanova, 2000). FAs with separated DBs are retained more strongly than FAs with methylene-interrupted DBs (Nikolova-Damyanova et al., 1992). The series of positional isomers of C18:1 and C18:2 (Nikolova-Damyanova et al., 1992) or conjugated C18:2 (Delmonte et al., 2005) have been studied and their relative retention factors (k) were plotted against DB positions. The retention behavior of all series of isomers is similar, giving sinusoidal curves of k values of individual isomers. In general, FAs with the first DB at C_5 to C_7 are most strongly retained.

Although the acyl chain length is supposed to have no effect in Ag-HPLC, FAs with longer chains are eluted slightly earlier than FAs with short chains (Juaneda et al., 1994). This phenomenon can be ascribed either to the normal-

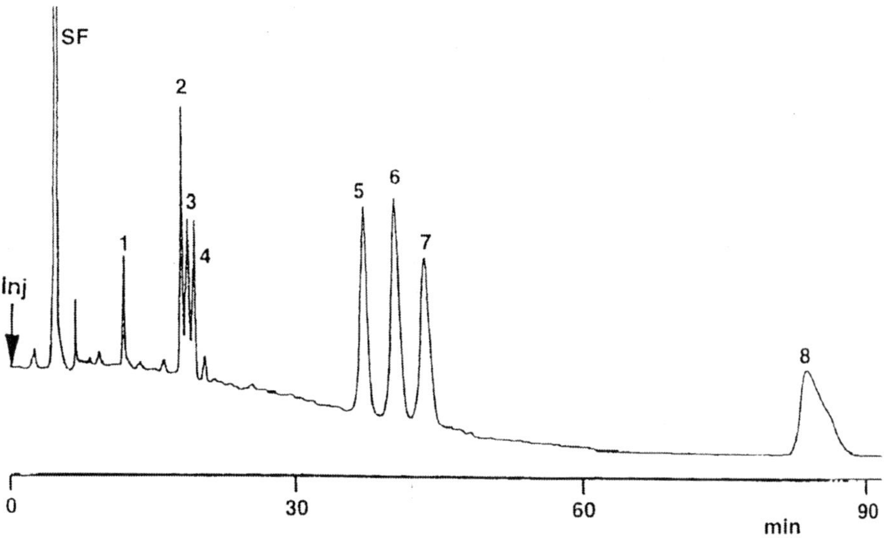

Fig. 6.5. Separation of FA phenacyl esters of C18:3 *cis-trans-* geometrical isomers by Ag-HPLC. Peak annotation: SF - solvent front, 1 - Δ9*t*,12*t*,15*t*, 2 - Δ9*t*,12*c*,15*t*, 3 - Δ9*c*,12*t*,15*t*, 4 - Δ9*t*,12*t*,15*c*, 5 - Δ9*c*,12*t*,15*c*, 6 - Δ9*t*,12*c*,15*c*, 7 - Δ9*c*,12*c*,15*t*, 8 - Δ9*c*,12*c*,15*c*. Reprinted with permission from (Juaneda et al., 1994).

phase effect occurring in silica-based silver loaded columns (Adlof, 1997) or to the lower complex stability for longer alkyl chains (Lucas et al., 1943; Winstein & Lucas, 1938). Similar behavior is reported for TGs (Lísa et al., 2009; Holčapek et al., 2010). The retention of various FA derivatives has been tested, *i.e.*, short chain alkenyl esters (Nikolova-Damyanova et al., 1995), benzyl and phenacyl esters (Nikolova-Damyanova et al., 1996), phenethyl, phenacyl and *p*-methoxyphenacyl esters (Momchilova et al., 1998), 2-naphthacyl, 9-anthrylmethyl and 2-naphthylmethyl esters (Momchilova & Nikolova-Damyanova, 2000). The elution order is the same for all derivatives, but the type of derivative affects the selectivity, mainly for positional isomers of FAs. *p*-Methoxyphenacyl esters provide the best selectivity for positional isomers of Fas, enabling the separation according to the position of DBs and the chain length. Changed composition of mobile phase can reverse the retention order, as demonstrated by the example of *p*-methoxyphenacyl esters, when 2-propanol was replaced by acetonitrile in a dichloromethane mobile phase (Momchilova & Nikolova-Damyanova, 2000).

Triacylglycerols

When retention rules described for FAs and their simple derivatives are compared with TGs, the complexity significantly increases due to the fact that three FAs are present in a TG molecule, and they may occupy three different stereochemical positions, which results in an enormous number of combinations described by an equation $N = 1/2\ (n^3 + n^2)$, counting all isomers except for enantiomers. In case of a relatively simple natural lipidome, we can find at least 10 different FAs, which results in 550 combinations for TGs. Of course, not all combinations are present because only certain routes are preferred in the biosynthesis, but nevertheless it is a challenging task for an analytical chemist. The retention in Ag-HPLC is governed dominantly by a single molecular property, the nature of its unsaturation, *i.e.*, the number, geometry (*cis* vs. *trans*) (Adlof & List, 2004; Holčapek et al., 2010; Holčapek et al., 2009; Adlof et al., 2002) and positions of DBs (Laakso & Voutilainen, 1996; Lísa et al., 2009; Holčapek et al., 2010; Holčapek et al., 2009). The steric availability (*i.e.*, positional isomerism) of DBs (Schuyl et al., 1998; Adlof & List, 2004; Adlof, 1995; Lísa et al., 2009; Holčapek et al., 2010; Holčapek et al., 2009) and the lengths of acyl chains (Lísa et al., 2009) are minor factors affecting the retention as well. The retention order of TGs has been summarized for chlorinated mobile phases (Nikolova-Damyanova, 2009; Christie, 1988; Nikolova-Damyanova et al., 1995) as follows:

SaSaSa (DB=0) < SaSaMo (1) < SaMoMo (2) < SaSaD (2) < MoMoMo (3) < SaMoD (3) < MoMoD (4) < SaDD (4) ≤ SaSaT (3) < SaMoT (4) ≤ MoDD (5) < MoMoT (5) < SaDT (5) ≤ DDD (6) < MoDT (6) ≤ SaTT (6) < DDT (7) < MoTT (7) < DTT (8) < TTT (9),

where symbols are defined as Sa for saturated, Mo for monounsaturated, D for diunsaturated, T for triunsaturated and positional isomers are not distinguished. To avoid the confusion of using the same symbol in two meanings (*i.e.*, S for both stearic acid and saturated, M for both myristic and monounsaturated), we use the symbol Sa for saturated and Mo for monounsaturated unlike the annotation used in most previous works. The retention behavior observed in our experiments in hexane - 2-propanol - acetonitrile mobile phases (Lísa et al., 2009; Holčapek et al., 2010; Holčapek et al., 2009) is mostly in good agreement with published data with few exceptions. In mobile phases consisting of hexane - acetonitrile (Adlof, 1995; Dugo et al., 2004), heptane - acetonitrile (Macher & Holmqvist, 2001), or hexane - 2-propanol - acetonitrile (Lísa et al., 2009; Holčapek et al., 2010; Holčapek et al., 2009), the retention order of all TGs follows the rule that an increasing number of DBs means longer retention, while one exception is reported above for DMoMo (4) < DDSa (4) ≤ SaSaT (3). The explanation of changed retention patterns probably lies in the different mobile phase composition, which affects the retention mechanism. The alkyl chain length has a small effect on the retention, but it does not change the retention order. Fig. 6.6 highlights this effect for TGs with saturated FAs from C7:0 to C22:0, where TGs with longer chains elute earlier (Holčapek et al., 2010). For TGs of the LLSa type found in sunflower

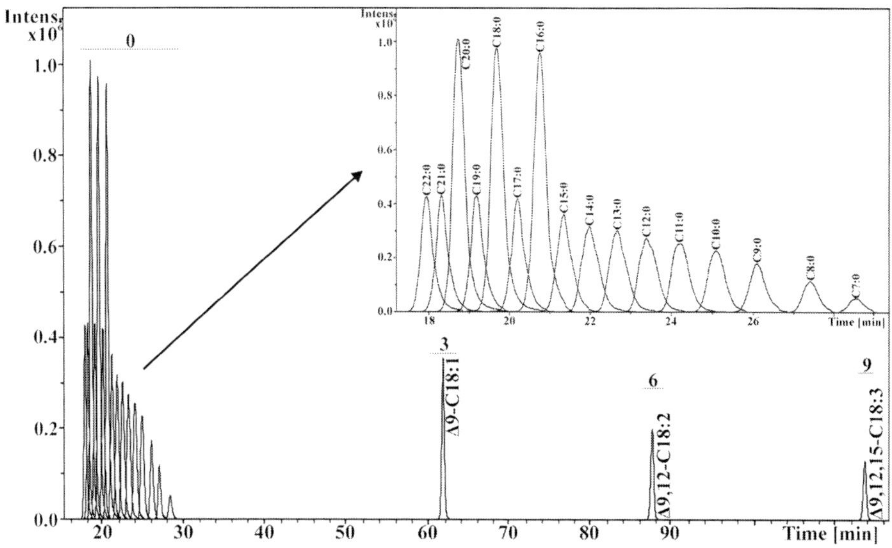

Fig. 6.6. Ag-HPLC/APCI-MS chromatogram of mono-acid TG standards. Numbers correspond to the DB number. Inset zoom shows the resolution of saturated TGs from C7:0 to C22:0. Concentrations of all standards are identical except for C16:0, C18:0 and C20:0 with doubled concentrations. Reprinted with permission from (Holčapek et al., 2010).

oil, the retention decrease in the LLSa series is approximately 0.4 minutes per two methylene units (Lísa et al., 2009). The analysis of a randomized mixture containing six different TGs (PPP, SSS, OOO, LLL, LnLnLn and AAA) and of numerous plant oils (Lísa et al., 2009; Holčapek et al., 2010; Holčapek et al., 2009) enabled the generalization of the retention order of TGs, including positional isomers, valid in hexane - 2-propanol - acetonitrile mobile phases (Fig. 6.7):

SaSaSa (DB=0) << SaMoSa (1) < SaSaMo (1) << SaDSa (2) < SaSaD (2) ≤ SaMoMo (2) < MoSaMo (2) << SaTSa (3) ≤ SaDMo (3) ≤ SaMoD (3) < SaSaT (3) ≤ MoSaD (3) ≤ MoMoMo (3) << SaDD (4) < SaTMo (4) ≤ MoDMo (4) ≤ DSaD (4) ≤ MoMoD (4) ≤ SaMoT (4) < MoSaT (4) << SaTD (5) ≤ MoDD (5) ≤ DMoD (5) ≤ SaDT (5) ≤ MoTMo (5) ≤ MoMoT (5) ≤ DSaT (5) << DDD (6) < MoTD (6) ≤ MoDT (6) ≤ DMoT (6) ≤ SaTT (6) < TSaT (6) << DTD (7) ≤ DDT (7) < MoTT (7) < TMoT (7) << DTT (8) ≤ TDT (8) << TTT (9).

We believe that this order is fully applicable for similar mobile phases containing acetonitrile (or propionitrile) in hexane (or heptane) with or without the addition of 2-propanol.

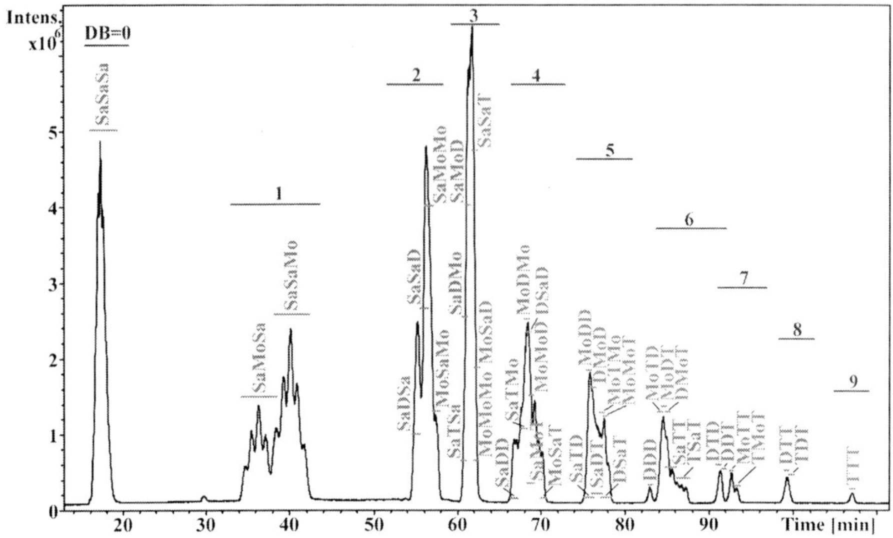

Fig. 6.7. Ag-HPLC/APCI-MS chromatogram of the randomized mixture of PPP, SSS, OOO, LLL, LnLnLn and AAA. Peak annotation: Sa - saturated, Mo - monounsaturated, D - diunsaturated, T - triunsaturated, DB - double bond number. Redrawn with permission from (Holčapek et al., 2009).

The randomized mixture containing mono-acid TGs differing only in the geometry of their DBs (oleic acid, Δ9*c*-C18:1,O *vs.* elaidic acid, Δ9*t*-C18:1, E) shows that TGs differing in the total number of *trans*-DBs can be partly separated in NARP mode (Fig. 6.8.A) and very well in Ag-HPLC mode (Fig. 6.8.B), which provides also the partial separation of regioisomers:

EEE << EOE < OEE << OOE < OEO << OOO.

This retention order is identical to that already published (Dobson et al., 1995), but with the extension for regioisomers. Physico-chemical properties of *trans*-FAs are closer to saturated FAs, due to the linear arrangement of alkyl chains containing *trans*-DBs. Therefore the retention of TGs containing *trans*-FAs is slightly higher in NARP mode, but significantly lower in Ag-HPLC, compared to *cis*-FAs. This effect is more pronounced for outer *sn*-1/3 positions, due to a better steric availability.

The effect of DB position in TGs can be illustrated by the example of linolenic (Ln, Δ9,12,15-C18:3) *vs.* gamma-linolenic (γLn, Δ6,9,12-C18:3) acids. LnLnLn has higher retention than γLnγLnγLn in most mobile phases [acetone - acetonitrile (Leskinen et al., 2008), chlorinated solvents (Laakso & Voutilainen, 1996)], while the reverse order is reported for the hexane - 2-propanol - acetonitrile mobile phase (Holčapek et al., 2010). This example shows that

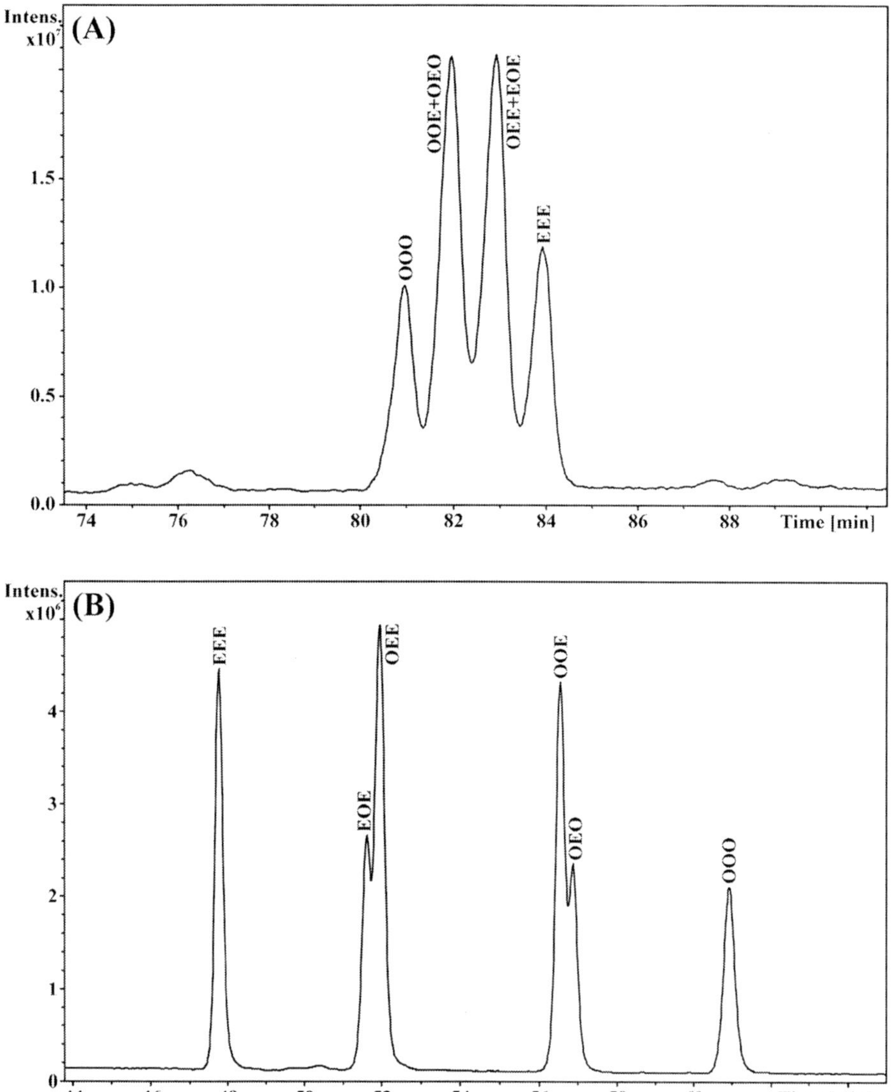

Fig. 6.8. HPLC/APCI-MS chromatograms of the randomized mixture of OOO (Δ9-C18:1) and EEE (Δ9*t*-C18:1): (A) NARP-HPLC using two Nova-Pak C_{18} columns with a total length of 45 cm, flow rate 1 mL/min, column temperature 25°C, gradient of acetonitrile - 2-propanol, (B) Ag-HPLC - three ChromSpher Lipids columns with a total length of 75 cm, flow rate 1 mL/min, column temperature 25°C, gradient of hexane - 2-propanol - acetonitrile. Reprinted with permission from (Holčapek et al., 2009).

the retention mechanism in Ag-HPLC is rather complex and not fully understood so far. If only one FA in a TG molecule differs in the DB position, then only peak broadening is observed due to small differences in retention times of isomers differing in the DB position only. Unlike FAs and their derivatives, the systematic study on the retention behavior of TGs containing unusual DB positions is still missing. Neff *et al.* (1994) has shown that TGs containing a triple bond (crepenynic, Δ9,12triple-C18:2) elute later in comparison to analogous TGs containing DBs at the same position (linoleic, Δ9,12-C18:2).

Other Lipid Classes

Applications of Ag-HPLC/MS for TGs, FAMEs and other simple FA derivatives are well documented, but this technique is easily applicable for other non-polar lipids. However, only few applications are reported in the literature, for example the analysis of underivatized sterols using ternary gradient elution with acetone, hexane and acetonitrile (Shan & Wilson, 2002), wax esters in fish oils (Joh et al., 1995), and vitamins in tissues of freshwater fish-feeding mammals (Kakela et al., 2002). The analysis of phospholipids (PLs) by Ag-HPLC would be highly desirable, but no acceptable compromise has been found so far between the mobile phase composition and the efficient separation of intact PLs, unlike Ag-TLC (Stoll, 1996). Ag-HPLC has been used only for PLs after derivation, when the polar functional group is masked, as applied for sphingomyelins (Smith et al., 1981), phosphatidylethanolamines (Yeung et al., 1977), phosphatidylserines (Yeung et al., 1977) and phosphatidylinositols (Holub & Kuksis, 1971). The other possibility is the hydrolysis of PLs followed by esterification and then the analysis of FA derivatives, but no real Ag-HPLC application for intact PLs has been reported so far. Additional problems can arise from the presence of oxidized PLs in samples because functional groups formed during the oxidation of lipids can lead to irreversible retention on the silver-ion column, which can alter the retention characteristics of this column and reduce the column life-time.

Regioisomeric Determination of Triacylglycerols

The position of FAs on the glycerol skeleton is very important from a nutritional point of view, because enzymes in the human or animal body selectively hydrolyze preferred positions which may result in different biological availability of FAs depending on their stereochemical positions. This differentiation may be especially important in the case of essential FAs, which cannot be synthesized in the organism. The analysis of TG regioisomers is quite challenging, because regioisomers have the identical elemental composition, identical ions in mass spectra and in general similar physico-chemical properties. Ag-HPLC has the ability to resolve TG regioisomers. In our opinion, the following techniques have the highest potential for regioisomeric analysis:

1) Ag-HPLC (Schuyl et al., 1998; Adlof & List, 2004; Adlof, 1995; Dugo et al., 2004; Dugo et al., 2006; Lísa et al., 2009; Holčapek et al., 2010; Holčapek et al., 2009; Adlof, 2007), 2) MS (Holčapek et al., 2009; Holčapek et al., 2005; Lísa & Holčapek, 2008; Holčapek et al., 2003; Leskinen et al., 2008; Byrdwell, 2001; Byrdwell, 2005; Byrdwell et al., 1996; Fauconnot et al., 2004; Jakab et al., 2003; Lísa et al., 2009; Mottram et al., 2001; Mottram & Evershed, 1996; Mottram et al., 1997), 3) enzymatic reactions (e.g., pancreatic lipase, phospholipase A_2) (Janssen et al., 2006) followed by some analytical technique (e.g., Ag-HPLC), and 4) nuclear magnetic resonance (NMR) spectroscopy (Standal et al., 2009). Each technique has certain limitations. NMR provides an absolute answer based on small differences in chemical shifts, but the practical applicability is severely limited by the need of pure compounds in milligram amounts, which is hard to achieve for most biological analyses. Enzyme lipases with various levels of selectivity towards stereochemical positions of particular FAs are known, but their application in a quantitative manner is not straightforward because exact values of enzymatic stereoselectivity may not be known and they may also depend on experimental conditions, alkyl chain length, DB number and positions. Furthermore, another analytical technique (e.g., HPLC) is required to detect the products of enzymatic hydrolysis. In mass spectra of regioisomers, different ratios of $[M+H-R_iCOOH]^+$ ions are observed for *sn*-1/3 *vs.* *sn*-2 positions, but regioisomeric standards should be measured first, for reliable quantitation. Chromatographic peak areas can be directly used for the quantitation, because the peak areas of regioisomers directly correspond to regioisomeric ratios (Lísa et al., 2009; Holčapek et al., 2009). The regioisomeric separation is also feasible in NARP mode using the multiple column coupling and very long retention times about 100–300 minutes, depending on the DB number (Momchilova et al., 2006; Momchilova et al., 2004). Of course, this approach is not yet applicable for routine use, but new developments in reversed-phase column technology (for example sub-2 μm or porous shell particles) could enable such analysis within an acceptable retention window.

Standards of Regioisomers

The severe problem in the regioisomeric analysis of TGs is the lack of standards. A limited range of standards is commercially available mainly from Larodan (Malmö, Sweden) and some other suppliers. Their range is limited and prices are rather high for small amounts probably due to difficult preparative isolation. For this reason, Evans et al. (Li & Evans, 2005) have developed a synthetic procedure for regioisomeric TGs starting from regioisomerically pure 1,3-diacylglycerols or 2-monoacylglycerols dissolved in chloroform, then 4-dimethylaminopyridine and selected fatty acid chlorides are added dropwise with continuous stirring for 30 minutes, which yields pure TG regioisomers.

This procedure has been used for a synthesis of a wide range of regioisomeric series OXO/YOY (Li & Evans, 2005), PXP/YPY (Li et al., 2006), LXL/YLY and AXA/YAY (Gakwaya et al., 2007), where X and Y are saturated FAs from C4:0 to C24:0 and unsaturated FAs containing one to six DBs. They report trends in fragment ions ratios of regioisomers depending on the alkyl chain length (Fig. 6.9) and the DB number (Fig. 6.10). The disadvantage of this approach is that the preparation of a wider series of regioisomers is laborious.

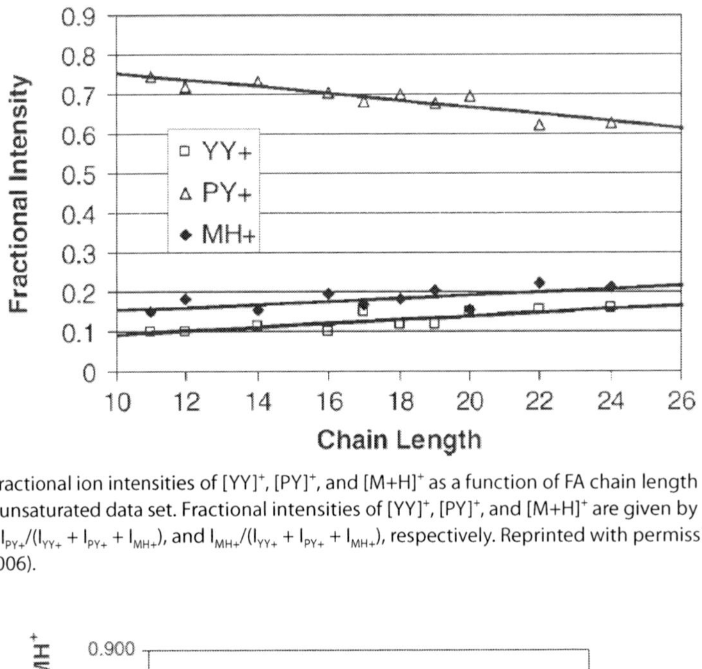

Fig. 6.9. Fractional ion intensities of $[YY]^+$, $[PY]^+$, and $[M+H]^+$ as a function of FA chain length for the YPY monounsaturated data set. Fractional intensities of $[YY]^+$, $[PY]^+$, and $[M+H]^+$ are given by $I_{YY+}/(I_{YY+} + I_{PY+} + I_{MH+})$, $I_{PY+}/(I_{YY+} + I_{PY+} + I_{MH+})$, and $I_{MH+}/(I_{YY+} + I_{PY+} + I_{MH+})$, respectively. Reprinted with permission from (Li et al., 2006).

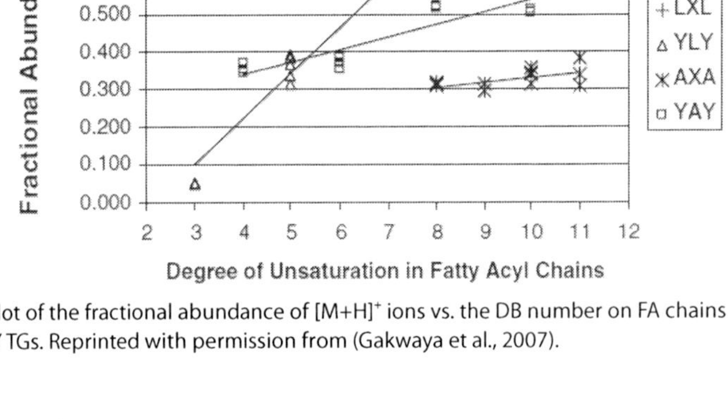

Fig. 6.10. Plot of the fractional abundance of $[M+H]^+$ ions vs. the DB number on FA chains of LXL, AXA, YLY, and YRY TGs. Reprinted with permission from (Gakwaya et al., 2007).

Randomization (i.e., chemical interesterification) is a common industrial process used for the modification of physico-chemical properties of TG mixtures in food and cosmetic products (Akoh & Min, 2002). In our previous work (Lísa et al., 2009), we have modified this procedure for the micro-scale range (low milligram amounts), which overcomes the lack of pure regioisomeric standards by the synthesis of regioisomeric mixtures having a well defined composition. Theoretically, the randomization of binary mixtures of equal amounts of single-acid TGs $R_1R_1R_1$ and $R_2R_2R_2$ should provide 8 combinations of TGs at identical concentrations: $R_1R_1R_1$, $R_1R_2R_1$, $R_1R_1R_2$, $R_2R_1R_1$, $R_1R_2R_2$, $R_2R_2R_1$, $R_2R_1R_2$ and $R_2R_2R_2$. In practice, enantiomers $R_1R_1R_2$ *vs.* $R_2R_1R_1$ and $R_1R_2R_2$ *vs.* $R_2R_2R_1$ cannot be resolved in non-chiral environment, therefore we obtain the following concentration ratios (in parenthesis) for initial TGs and two regioisomeric pairs: $R_1R_1R_1$ (1), $R_1R_2R_1$ (1), $R_1R_1R_2$ + $R_2R_1R_1$ (2), $R_1R_2R_2$ + $R_2R_2R_1$ (2), $R_2R_1R_2$ (1), $R_2R_2R_2$ (1), as illustrated in Fig. 6.11.A. Data obtained for a range of randomization reactions have proven that the reaction provides really a random distribution of statistically expected regioisomers (Lísa et al., 2009; Holčapek et al., 2010; Holčapek et al., 2009). Fig. 6.11.B shows the chromatogram of ternary randomization mixture of OOO, LnLnLn and PPP providing six regioisomeric doublets at concentration ratios 2:1 and one triplet OLnP / LnOP / OPLn with identical concentrations. Standards of all common and relatively rare mono-acid TGs are commercially available, therefore tailored randomization can produce desired regioisomeric combinations. The randomization reaction can be performed also with FAs, if the TG standard is not available.

Silver-ion HPLC of Regioisomers

The strength of interaction between silver-ions and DBs also depends on the spatial availability or hindrance of DBs on the molecular skeleton, which can result in the separation of regioisomers with Ag-HPLC. The effect of a polar organic modifier in hexane on the regioisomeric resolution is absolutely critical. Based on our study of effects of different solvents (Lísa et al., 2009) on the regioisomeric resolution and also a comprehensive literature search, the following statements can be made. Chlorinated solvents have been used in numerous works reporting excellent overall separation, but we have not found any papers that report regioisomeric resolution. Almost all papers (Adlof & List, 2004; Adlof, 1995; Dugo et al., 2004; Dugo et al., 2006; Macher & Holmqvist, 2001; Lísa et al., 2009; Holčapek et al., 2010; Holčapek et al., 2009; Adlof, 2007; Smith et al., 1981) published on the regioisomeric resolution of TGs are based on mobile phases containing acetonitrile in hexane (or heptane), except for one work measured on a silver complexation column in toluene, hexane and ethylacetate (Schuyl et al., 1998). When acetonitrile is replaced by other solvents (e.g., 2-propanol, ethanol, and ethylacetate) (Muller et al., 2006; Lísa et al., 2009), peak shapes and the theoretical number of plates

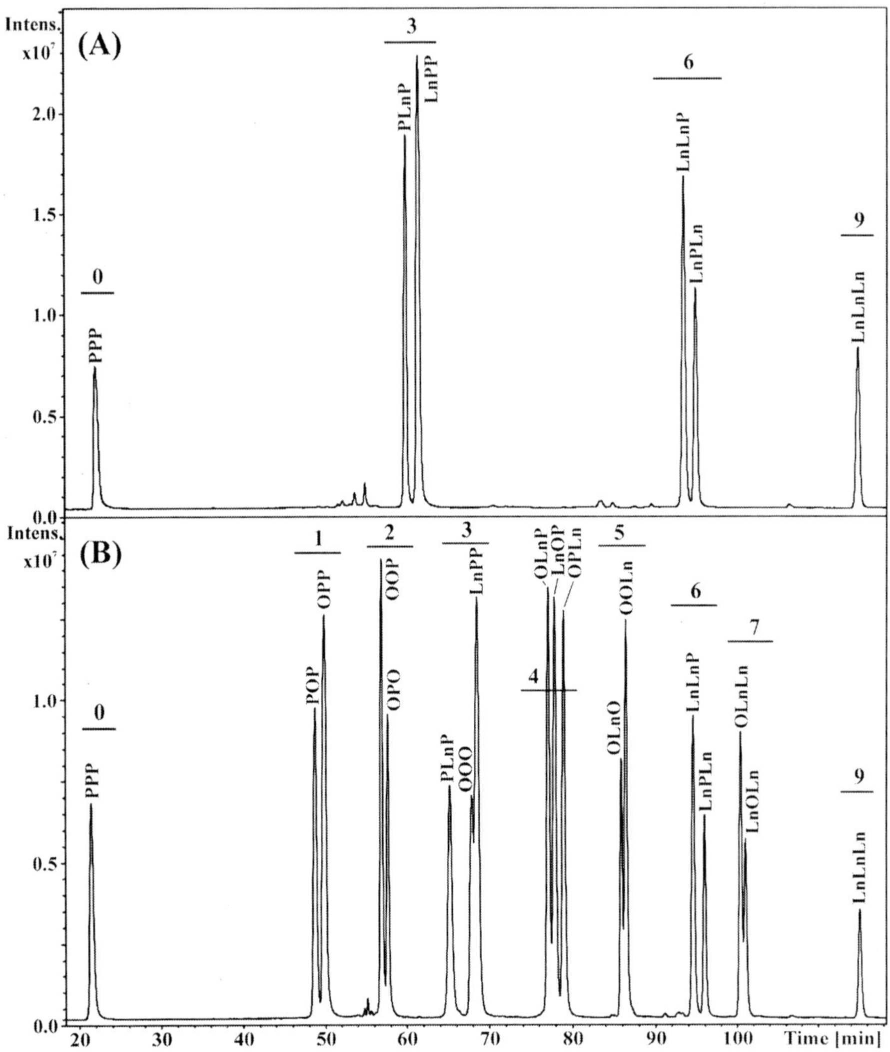

Fig. 6.11. Ag-HPLC/APCI-MS analysis of randomized mixtures: (A) PPP / LnLnLn, (B) PPP / OOO / LnLnLn. Reprinted with permission from (Lísa et al., 2009).

may even be increased for certain compositions, but the regioisomeric resolution is lost or at least dramatically reduced. We believe that acetonitrile plays an essential role in the formation of the complex with silver-ions. This hypothesis is supported by experiments with propionitrile (Muller et al., 2006; Lísa et al., 2009), where similar complex formation can be expected as well. No regioisomeric resolution is achieved for fully saturated TGs (Lísa et al., 2009; Holčapek et al., 2009).

For our studies of retention behavior of regioisomeric TGs, randomized mixtures of mono-acid TGs have been used, as described in the Standards of Regioisomers section. The baseline separation for regioisomers containing up to 3 DBs and at least partial separation for TGs with 4–8 DBs can be achieved (Adlof, 1995; Lísa et al., 2009; Holčapek et al., 2010). As a rule, the bigger the difference in the DB number of FAs means better separation of corresponding TG regioisomers, e.g., the baseline separation of P/L and P/Ln regioisomers is easily achieved (Fig. 6.11.B). The example of partial resolution of LnOLn/OLnLn regioisomers (Fig. 6.11.B) containing 7 DBs shows that the number of DBs is probably not a limiting factor, but the critical requirement for the successful separation of highly polyunsaturated regioisomeric TGs is the difference in the DB number between the two FAs. Polyunsaturated TGs containing FAs differing only by one DB are critical pairs, e.g., O/L (peak splitting) and L/Ln (only peak shoulder). FAs differing by two and more DBs are well separated for TGs containing up to 4 DBs and partially for 5 and more DBs. Some improvement may be expected with further extension of column length or a temperature increase, at expense of longer retention times and also higher costs of multiple columns.

The retention order of regioisomers follows the rule that more DBs in the *sn*-1/3 positions mean a stronger interaction with silver-ions embedded in the stationary phase resulting in higher retention compared with regioisomers having the same DB number in the middle *sn*-2 position. Regioisomers of XYO/YXO and XYL/YXL types (X and Y are saturated FAs) are not resolved. A highly complex retention pattern is observed in the case of TGs containing DB positional isomers, such as linolenic (Δ9,12,15-C18:3) and γ-linolenic (Δ6,9,12-C18:3) acids. When the only difference between two TGs is in DB positions for *sn*-2 FA (OLnO *vs.* OγLnO), then no visible chromatographic separation is observed (Holčapek et al., 2010). On the other hand, the same difference occurring in *sn*-1/3 positions (OOLn *vs.* OOγLn) leads to a very good resolution with the difference of retention times about 1.6 minutes.

The generally accepted opinion is that unsaturated FAs (mainly linoleic) preferentially occupy the *sn*-2 position in plant oils, while it is just the opposite for animal fats, where unsaturated FAs are found mainly in *sn*-1/3 positions. We have applied (Lísa et al., 2009) our method for the quantitation of *sn*-2 occupation of a representative plant oil (sunflower) and animal fat (lard), as compared in Table 6.A, with other literature data for TGs composed of palmitic, oleic and linoleic acids. The positional preference is very strong, because ratios of regioisomeric doublets are close to 100% in accordance with other papers (Leskinen et al., 2007; Leskinen, 2010; Herrera et al., 2010). An interesting example is TGs composed of P, O and L, where the *sn*-2 occupation preference for sunflower oil is in the order of increasing DB number (OLP / LOP / OPL = 63 / 36 / 1) in contrast to the lard sample (OLP / LOP / OPL = 3 / 12 / 85). This clear preference shows the potential of regioisomeric analysis in the determination of food authenticity (plant *vs.* animal sources).

Table 6.A. Regioisomeric Occupation of the *sn*-2 Position for Saturated (Palmitic), Monounsaturated (Oleic) and Diunsaturated (Linoleic) FAs in TGs of Plant (Sunflower, Olive and Palm) Oils and Animal Fat (Lard).

	Sunflower oil			Olive oil	Palm oil	Lard	
Regioisomers	(Lísa et al., 2009)	(Leskinen, 2010)	(Herrera et al., 2010)	(Herrera et al., 2010)	(Leskinen, 2010)	(Lísa et al., 2009)	(Leskinen et al., 2007)
POP / OPP	100 / 0	—	99/1	97/3	86/14	8 / 92	0/100
OOP / OPO	98 / 2	91/9	100/0	98/2	95/5	12 / 88	4/96
PLP / LPP	100 / 0	—	—	—	—	1 / 99	—
LLP / LPL	97 / 3	—	—	—	—	9 / 91	—
OLP / LOP / OPL	63 / 36 / 1	—	—	—	—	3 / 12 / 85	—

Mass Spectrometric Identification and Quantitation of Regioisomers

Mass spectra of TGs provide two basic types of ions sufficient for the structural assignment (Byrdwell et al., 1996; Mottram et al., 1997; Byrdwell & Emken, 1995): 1) molecular adducts with protons $[M+H]^+$ or other small cations depending on the mobile phase composition, typically $[M+Na]^+$ or $[M+NH_4]^+$, and 2) fragment ions formed by the neutral loss (NL) of FA $[M+H-R_iCOOH]^+$. In case of conventional electron ionization (EI), the molecular radical cation $M^{+\cdot}$ provides fragment ions $[M-R_iCOO]^+$ by NL from all three stereochemical positions, but the NL from *sn*-2 is less preferred, as shown for the first time in the pioneering work published in 1964 (Barber & Merren, 1964). By the way, ion annotations $[M+H-R_iCOOH]^+$ and $[M-R_iCOO]^+$ are numerically identical. The same behavior has been reported later for soft ionization techniques, such as atmospheric pressure chemical ionization (APCI) (Holčapek et al., 2003; Fauconnot et al., 2004; Mottram et al., 2001; Mottram & Evershed, 1996; Mottram et al., 1997), electrospray ionization (ESI) (Li & Evans, 2005; Li et al., 2006; Gakwaya et al., 2007; Herrera et al., 2010; Malone & Evans, 2004; Leveque et al., 2010) and matrix-assisted laser desorption/ionization (MALDI) (Pittenauer & Allmaier, 2009). The relative ratio of fragment ions $[M+H-R_iCOOH]^+$ enables one to differentiate FAs in the side *sn*-1/3 positions *vs.* the middle *sn*-2 position, because the NL from the *sn*-2 position, providing an $[M+H-R_2COOH]^+$ ion, has a lower relative abundance in comparison to $[M+H-R_1COOH]^+$ and $[M+H-R_3COOH]^+$ fragment ions (Schuyl et al., 1998; Cvačka et al., 2006; Holčapek et al., 2005; Holčapek et al., 2003; Leskinen et al., 2008; Byrdwell, 2001; Jakab et al., 2003; Lísa et al., 2009; Mottram et al., 2001; Mottram & Evershed, 1996; Mottram et al., 1997; Lísa et al., 2007), as illustrated in Fig. 6.12. The simple approach just determines the prevailing FA in the middle *sn*-2 position based on the lower relative abundance of corresponding $[M+H-R_iCOOH]^+$ fragment ion (Dugo et al., 2004; Lísa & Holčapek, 2008; Holčapek et al., 2003; Mottram et al., 2001; Mottram & Evershed, 1996; Mottram et al.,

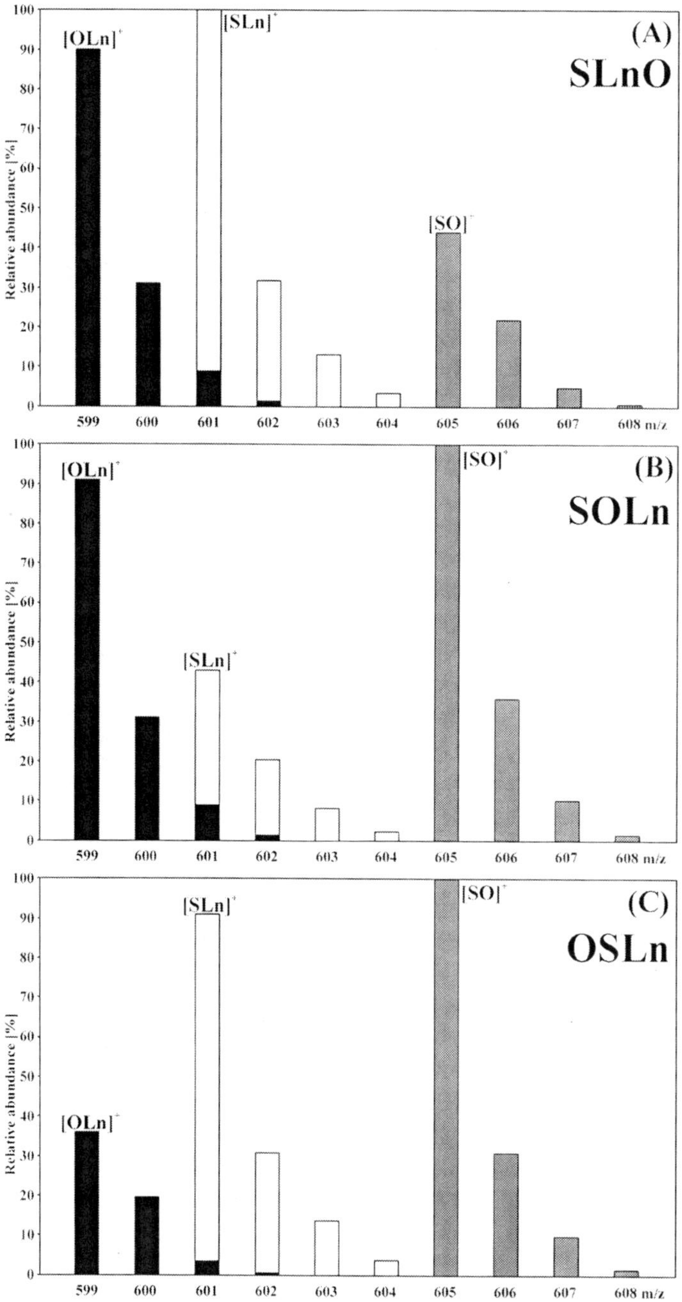

Fig. 6.12. Normalized relative abundances of [M+H-R$_i$COOH]$^+$ fragment ions in APCI mass spectra of the regioisomeric triplet SLnO (A), SOLn (B) and OSLn (C). Reprinted with permission from (Holčapek et al., 2010).

1997), but the type of FAs (mainly DB number and positions) (Holčapek et al., 2005; Li & Evans, 2005; Li et al., 2006; Gakwaya et al., 2007) also affects relative abundances of corresponding $[M+H-R_iCOOH]^+$ fragment ions. An interesting example of the strong effect of DB position is the ratio of $[R_1R_1]^+$ / $[R_1R_2]^+$ fragment ions, which is completely reversed in the case of OOγLn unlike all other regioisomers (Holčapek et al., 2010). A more exact approach is based on the construction of calibration curves using identical standards of regioisomeric pairs mixed at different ratios (Holčapek et al., 2003; Leskinen et al., 2008; Fauconnot et al., 2004; Jakab et al., 2003; Leskinen et al., 2007; Leskinen, 2010; Malone & Evans, 2004; Leskinen et al., 2010). In practice, this approach is time-consuming and also expensive, due to prices of positional isomers and their limited availability. The literature data (Holčapek et al., 2010) measured on five different instruments can be used for the determination of the prevailing FA in *sn*-2 position and also for semi-quantitative purposes.

The type of mass spectrometer has a relatively low effect on $[M+H-R_iCOOH]^+$ relative ratios for tested instruments (single quadrupole LC/MSD from Agilent Technologies, triple quadrupole API 3000 from AB SCIEX, ion trap Esquire 3000 from Bruker Daltonics, quadrupole time-of-flight micrOTOF-Q from Bruker Daltonics and LTQ Orbitrap XL Thermo Fisher Scientific), while the distribution of DBs in individual FA chains is the governing factor for $[M+H-R_iCOOH]^+$ ratios (Fig. 6.13). The main goal of our inter-instrument comparison of APCI mass spectra of TG regioisomers has been to study to what extent are published ratios of $[M+H-R_iCOOH]^+$ ions applicable for different instruments and laboratories. Ratios of $[M+H-R_iCOOH]^+$ fragment ions strongly depend on the DB number and position in individual FAs, while the DB geometry and FA length have only minor influences (Fig. 6.14). The data measured on the quadrupole analyzers are grouped according to the unsaturation level of individual FAs. Good correlation is observed among values in individual groups, especially considering the fact that differences in *cis/trans* isomerism (*e.g.*, PEP *vs.* POP, EPP *vs.* OPP) or high differences in alkyl chain lengths (*e.g.*, AOA *vs.* POP differs by eight carbon atoms) are negligible. On the other hand, positions of DBs have a much stronger effect on $[M+H-R_iCOOH]^+$ ratios, as illustrated by the difference between LnOLn and γLnOγLn (32 *vs.* 12%) (Holčapek et al., 2010) in agreement with previous works (Leskinen et al., 2008; Lísa et al., 2007). Fig. 6.14 suggests that the data for TGs with a given unsaturation pattern can be generalized for other TGs with the same unsaturation pattern, which can solve the lack of unusual standards but still provide accurate regioisomeric determination.

Conclusions and Future Perspectives

Ag-HPLC is certainly a powerful liquid-phase separation technique for the characterization of TGs, FAMEs and other non-polar lipids differing in the

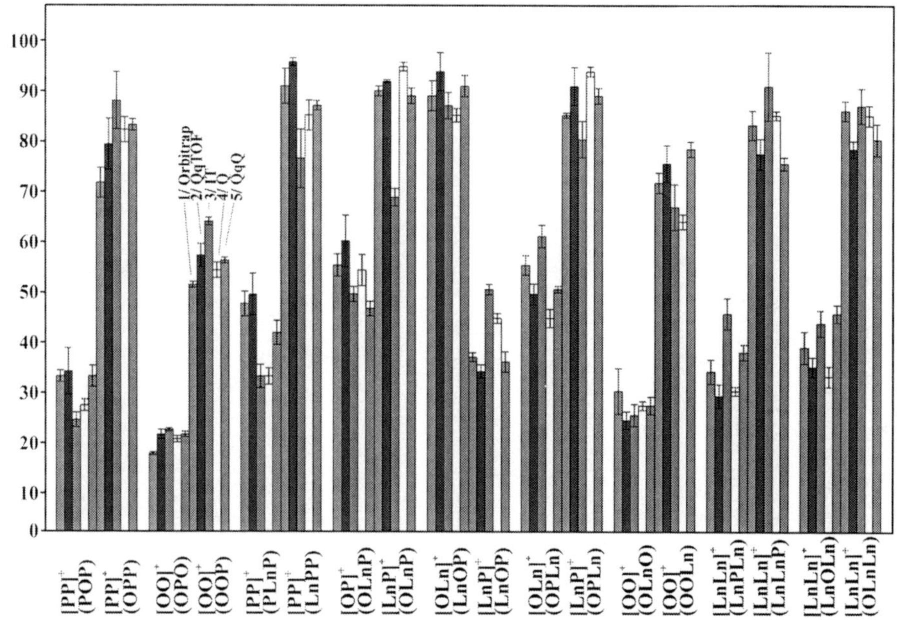

Fig. 6.13. Plots of relative abundances of $[M+H-R_2COOH]^+$ related to $[M+H-R_{1,3}COOH]^+$ (100%) with their standard deviations for the randomized mixture PPP/OOO/LnLnLn measured on five different instruments: 1) LTQ Orbitrap XL (Thermo Fisher Scientific), 2) quadrupole time-of-flight micrOTOF-Q (Bruker Daltonics), 3) ion trap Esquire 3000 (Bruker Daltonics), 4) single quadrupole LC/MSD (Agilent Technologies), 5) triple quadrupole API 3000 (AB SCIEX). Reprinted with permission from (Holčapek et al., 2010).

number, geometry and positions of DBs. The introduction of hexane - acetonitrile mobile phases brings additional dimension in terms of the regioisomeric separation. Ag-HPLC is the most reliable technique for regioisomeric quantitation. Reproducibility problems can be overcome by the addition of a third solvent miscible with both hexane and acetonitrile, for example 2-propanol. Unfortunately, only one supplier on the market offers commercial Ag-HPLC columns. More competition could be a driving force in the innovation and possible introduction of silver-ion columns applicable for ultrahigh-performance liquid chromatography (UHPLC), which is desirable due to the relatively long retention times for well optimized separations. An even more important aspect for a wider acceptance of Ag-HPLC would be the extension of this technique to underivatized PLs, but the principal limitation of mobile phase composition has to be solved. A highly promising trend in extreme chromatographic techniques is the use of 2D chromatography. Ag-HPLC has a very good orthogonality with NARP-HPLC, therefore 2D coupling of these separation modes is a logical choice with high peak capacity.

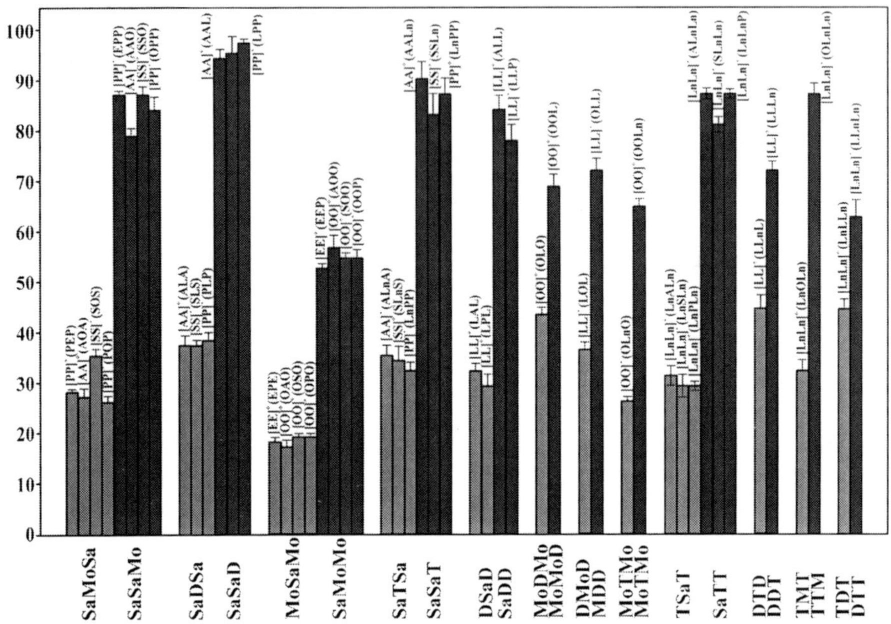

Fig. 6.14. Plots of relative abundances of $[M+H-R_2COOH]^+$ related to $[M+H-R_{1,3}COOH]^+$ (100%) with their standard deviations for randomized mixtures measured on the single quadrupole mass spectrometer; annotation of x axis: Sa - saturated, Mo - monounsaturated, D - diunsaturated, T - triunsaturated; annotation of particular columns correspond to individual $[M+H-R_2COOH]^+$ ions fragmented from TGs written in parentheses. Reprinted with permission from (Holčapek et al., 2010).

The governing factor for the retention order in Ag-HPLC is the total number of DBs, which causes the separation of complex samples into distinct groups differing in the DB number. The second parameter with comparable importance is the DB geometry, where the retention is significantly lower for *trans*-isomers. The positions of DBs have a smaller effect and the smallest difference is for regioisomers, where DBs in outer *sn*-1/3 positions cause slightly stronger retention compared to *sn*-2 regioisomers.

The role of the mobile phase in Ag-HPLC is very strong (illustrated in several examples here), but complete understanding of the retention mechanism needs further systematic studies. Moreover, separation temperature also shows unusual effects. In our opinion, these effects can be attributed, at least in part, to different lipid conformations, depending on the solvent composition and temperature, and subsequently the various extent of weak London forces among individual FA chains. We believe that London forces contribute to the retention mechanism of lipids in Ag-HPLC and other separation modes as well. Hopefully, the ongoing progress in computational chemistry could help to evaluate this hypothesis.

Acknowledgments

This work was supported by the grant project No. MSM0021627502 sponsored by the Ministry of Education, Youth and Sports of the Czech Republic and the project No. 203/09/0139 sponsored by the Czech Science Foundation. We would like to acknowledge the contribution of Dr. Miroslav Polášek (J. Heyrovský Institute of Physical Chemistry, Prague) into the discussion about the potential of London forces in the lipid retention mechanism.

References

Adlof, R. Analysis of triacylglycerol and fatty acid isomers by low-temperature silver-ion high performance liquid chromatography with acetonitrile in hexane as solvent: Limitations of the methodology. J. Chromatogr. A 2007, 1148, 256-259.

Adlof, R.; Lamm, T. Fractionation of cis- and trans-oleic, linoleic, and conjugated linoleic fatty acid methyl esters by silver ion high-performance liquid chromatography. J. Chromatogr. A 1998, 799, 329-332.

Adlof, R.; List, G. Analysis of triglyceride isomers by silver-ion high-performance liquid chromatography - Effect of column temperature on retention times. J. Chromatogr. A 2004, 1046, 109-113.

Adlof, R.O. Normal-phase separation effects with lipids on a silver ion high-performance liquid chromatography column. J. Chromatogr. A 1997, 764, 337-340.

Adlof, R.O. Separation fo cis and trans unsaturated fatty-acid methyl-esters by silver ion high-performance liquid chromatography. J. Chromatogr. A 1994, 659, 95-99.

Adlof, R.O.; Menzel, A.; Dorovska-Taran, V. Analysis of conjugated linoleic acid-enriched triacylglycerol mixtures by isocratic silver-ion high-performance liquid Chromatogr. J. Chromatogr. A 2002, 953, 293-297.

Adlof, R.O. Analysis of triacylglycerol positional isomers by silver ion high-performance liquid-chromatography. Hrc-J. High Res. Chromatog. 1995, 18, 105-107.

Akoh, C.C.; Min, D.B. Food Lipids: Chemistry, Nutrition, and Biotechnology. Marcel Dekker, New York, 2002.

Barber, M.; Merren, T.O. The mass spectrometry of large molecules .1. the triglycerides of straight chain fatty acids. Tetrahedron Letters 1964, 1063-1067.

Bressan, G.; Broggi, R.; Lachi, M.P.; Segre, A.L. Complexes of silver perchlorate with alpha omega-diolefins. J.Organomet. Chem. 1967, 9, 355-360.

Byrdwell, W.C. Atmospheric pressure chemical ionization mass spectrometry for analysis of lipids. Lipids 2001, 36, 327-346.

Byrdwell, W.C. The bottom-up solution to the triacylglycerol lipidome using atmospheric pressure chemical ionization mass spectrometry. Lipids 2005, 40, 383-417.

Byrdwell, W.C.; Emken, E.A.; Neff, W.E.; Adlof, R.O. Quantitative analysis of triglycerides using atmospheric pressure chemical ionization mass spectrometry. Lipids 1996, 31, 919-935.

Byrdwell, W.C.; Emken, E.A. Analysis of triglycerides using atmospheric-pressure chemical lonization mass-spectrometry. Lipids 1995, 30, 173-175.

Chatt, J.; Duncanson, L.A. Olefin co-ordination compounds . 3. Infra-red spectra and structure — attempted perparation of acetylene complexes. J. Chem. Soc. 1953, 2939-2947.

Christie, W.W. A Stable silver-loaded column for the separation of lipids by high-performance liquid-chromatography. J. High Res. Chromatog. 1987, 10, 148-150.

Christie, W.W. http://lipidlibrary.aocs.org/index.html. (accessed 25.01.2011 2011)

Christie, W.W. Separation of molecular-species of triacylglycerols by high-performance liquid- with a silver ion column. J. chromatography. 1988, 454, 273-284.

Christie, W.W.; Breckenridge, G.H.M. Separation of cis-isomers and trans-isomers of unsaturated fatty-acids by high-performance liquid-chromatography in the silver ion mode. J. Chromatogr. 1989, 469, 261-269.

Conacher, H.B.S. Chromatographic determination of cis-trans monoethylenic unsaturation in fats and oils—review. J. Chromatogr. Sci. 1976, 14, 405-411.

Correa, R.A.; Ferraz, V.; Medvedovici, A.; Sandra, P.; Cerne, K.; David, F. Positional and configurational separation of fatty acid isomers by micro reversed-phase liquid chroomatography with an Ag+-containing mobile phase. J. Chromatogr. A 1999, 848, 83-93.

Cvačka, J.; Hovorka, O.; Jiroš, P.; Kindl, J.; Stranský, K.; Valterová, I. Analysis of triacylglycerols in fat body of bumblebees by chromatographic methods. J. Chromatogr. A 2006, 1101, 226-237.

Cvetanović, R.J.; Duncan, F.J.; Falconer, W.E.; Irwin, R.S. Secondary deuterium isotope effects on stability of silver ion-olefin complexes. J. Amer. Chem. Soc. 1965, 87, 1827-1832.

Delmonte, P.; Kataoka, A.; Corl, B.A.; Bauman, D.E.; Yurawecz, M.P. Relative retention order of all isomers of cis/trans conjugated linoleic acid FAME from the 6,8- to 13,15-positions using silver ion HPLC with two elution systems. Lipids 2005, 40, 509-514.

Demirbuker, M.; Blomberg, L.G. Separation of triacylglycerols by supercritical-fluid argentation chromatography. J. Chromatogr. 1991, 550, 765-774.

Demirbuker, M.; Hagglund, I.; Blomberg, L.G. Separation of unsaturated fatty-acid methyl-esters by packed capillary supercrical fluid chromatography—comparison of different column packings. J. Chromatogr. 1992, 605, 263-267.

Dermaux, A.; Medvedovici, A.; Ksir, M.; Van Hove, E.; Talbi, M.; Sandra, P. Elucidation of the triglycerides in fish oil by packed-column supercritical fluid chromatography fractionation followed by capillary electrochromatography and electrospray mass spectrometry. J. Microcolumn Sep. 1999, 11, 451-459.

Devries, B. Quantitative separations of lipid materials by column chromatography on silica impregnated with silver nitrate. Chem. Ind. 1962, 1049-1050.

Dewar, J.S. A review of the pi-complex theory. Bulletin De La Societe Chimique De France 1951, 18, C71-C79.

Dobson, G.; Christie, W.W.; Nikolova-Damyanova, B. Silver ion chromatography of lipids and fatty-acids. J. Chromatogr. B. 1995, 671, 197-222.

Dugo, P.; Favoino, O.; Tranchida, P.Q.; Dugo, G.; Mondello, L. Off-line coupling of non-aqueous reversed-phase and silver ion high-performance liquid chromatography-mass spectrometry for the characterization of rice oil triacylglycerol positional isomers. J. Chromatogr. A 2004, 1041, 135-142.

Dugo, P.; Kumm, T.; Crupi, M.L.; Cotroneo, A.; Mondello, L. Comprehensive two-dimensional liquid chromatography combined with mass spectrometric detection in the analyses of triacylglycerols in natural lipidic matrixes. J. Chromatogr. A 2006a, 1112, 269-275.

Dugo, P.; Kumm, T.; Fazio, A.; Dugo, G.; Mondello, L. Determination of beef tallow in lard through a multidimensional off-line non-aqueous reversed phase-argentation LC method coupled to mass spectrometry. J. Sep. Sci. 2006b, 29, 567-575.

Dugo, P.; Kumm, T.; Chiofalo, B.; Cotroneo, A.; Mondello, L. Separation of triacylglycerols in a complex lipidic matrix by using comprehensive two-dimensional liquid chromatography coupled with atmospheric pressure chemical ionization mass spectrometric detection. J. Sep. Sci. 2006c, 29, 1146-1154.

Dugo, P.; Kumm, T.; Lo Presti, M.; Chiofalo, B.; Salimei, E.; Fazio, A.; Cotroneo, A.; Mondello, L. Determination of triacylglycerols in donkey milk by using high performance liquid chromatography coupled with atmospheric pressure chemical ionization mass spectrometry. J. Sep. Sci. 2005, 28, 1023-1030.

Emken, E.A.; Dutton, H.J.; Scholfield, C.R. Chromatographic separation of cis and trans fatty esters by argentation with macroreticular exchange resin. J. Amer. Oil Chem. Soc. 1964, 41, 388-390.

Francois, I.; Pereira, A.D.; Sandra, P. Considerations on comprehensive and off-line supercritical fluid chromatography x reversed-phase liquid chromatography for the analysis of triacylglycerols in fish oil. J. Sep. Sci. 2010, 33, 1504-1512.

Fauconnot, L.; Hau, J.; Aeschlimann, J.M.; Fay, L.B.; Dionisi, F. Quantitative analysis of triacylglycerol regioisomers in fats and oils using reversed-phase high-performance liquid chromatography and atmospheric pressure chemical ionization mass spectrometry. Rapid Commun. Mass Spectrom. 2004, 18, 218-224.

Gakwaya, R.; Li, X.W.; Wong, Y.L.; Chivukula, S.; Collins, E.J.; Evans, J.J. Examining the collision-induced decomposition spectra of ammoniated triglycerides. III. The linoleate and arachidonate series. Rapid Commun. Mass Spectrom. 2007, 21, 3262-3268.

Gmelin, L. Gmelin Handbuch der anorganischen Chemie. Springer, Berlin, 1975.

Guha, O.K.; Janák, J. Charge-transfer complexes of metals in chromatographic separation of organic compounds. J. Chromatogr. 1972, 68, 325-343.

Han, J.J.; Iwasaki, Y.; Yamane, T. Use of isopropanol as a modifier in a hexane-acetonitrile based mobile phase for the silver ion HPLC separation of positional isomers of triacylglycerols containing long chain polyunsaturated fatty acids. Hrc-J. High Res. Chromatog. 1999, 22, 357-361.

Herrera, L.C.; Potvin, M.A.; Melanson, J.E. Quantitative analysis of positional isomers of triacylglycerols via electrospray ionization tandem mass spectrometry of sodiated adducts. Rapid Commun. Mass Spectrom. 2010, 24, 2745-2752.

Holčapek, M.; Dvořáková, H.; Lísa, M.; Girón, A.J.; Sandra, P.; Cvačka, J. Regioisomeric analysis of triacylglycerols using silver-ion liquid chromatography-atmospheric pressure chemical ionization mass spectrometry: Comparison of five different mass analyzers. J. Chromatogr. A 2010, 1217, 8186-8194.

Holčapek, M.; Velínská, H.; Lísa, M.; Česla, P. Orthogonality of silver-ion and non-aqueous reversed-phase HPLC/MS in the analysis of complex natural mixtures of triacylglycerols. J. Sep. Sci. 2009, 32, 3672-3680.

Holčapek, M.; Lísa, M.; Jandera, P.; Kabátová, N. Quantitation of triacylglycerols in plant oils using HPLC with APCI-MS, evaporative light-scattering, and UV detection. J. Sep. Sci. 2005, 28, 1315-1333.

Holčapek, M.; Jandera, P.; Fischer, J.; Prokeš, B. Analytical monitoring of the production of biodiesel by high-performance liquid chromatography with various detection methods. J. Chromatogr. A 1999, 858, 13-31.

Holčapek, M.; Jandera, P.; Zderadička, P.; Hrubá, L. Characterization of triacylglycerol and diacylglycerol composition of plant oils using high-performance liquid chromatography-atmospheric pressure chemical ionization mass spectrometry. J. Chromatogr. A 2003, 1010, 195-215.

Holub, B.J.; Kuksis, A. Resolution of intact phosphatidylinositols by argentation thin-layer chromatography. J. Lipid Res. 1971, 12, 510-512.

Hosoya, H.; Nagakura, S. Spectroscopic and theoretical studies of charge-transfer type molecular complexes are complexes between monoolefins and metal ions—selective complex formation abilities of copper (I), silver (I) and mercury (II) ions. Bull. Chem. Soc. Jpn. 1964, 37, 249-265.

Jakab, A.; Jablonkai, I.; Forgacs, E. Quantification of the ratio of positional isomer dilinoleoyl-oleoyl glycerols in vegetable oils. Rapid Commun. Mass Spectrom. 2003, 17, 2295-2302.

Janssen, H.G.; Hrncirik, K.; Szoradi, A.; Leijten, M. An improved method for sn-2 position analysis of triacylglycerols in edible oils and fats based on immobilised lipase D (Rhizopus delemar). J. Chromatogr. A 2006, 1112, 141-147.

Jeffrey, B.S.J. Silver-complexation liquid chromatography for fast, high-resolution separations of triacylglycerols. J. Amer. Oil Chem. Soc. 1991, 68, 289-293.

Joh, Y.G.; Brechany, E.Y.; Christie, W.W. Characterization of wax esters in the roe oil of amber fish, seriola-aureovittata. J. Amer. Oil Chem. Soc. 1995, 72, 707-713.

Juaneda, P.; Sebedio, J.L.; Christie, W.W. Complete separation of the geometrical-isomers of linolenic acid by high-performance liquid-Chromatography with a silver ion column. hrc-J.High Res. Chromatog. 1994, 17, 321-324.

Kakela, A.; Kakela, R.; Hyvarinen, H. Analysis of hepatic vitamins A(1), A(2), their fatty acyl esters, and vitamin E for biomonitoring mammals feeding on freshwater fish. Envir. Tox. Chem. 2002, 21, 390-396.

Kasai, P.H.; Mcleod, D.; Watanabe, T. Acetylene and ethylene complexes of copper and silver atoms-matrix-isolation electron-spin-resonance study. J. Amer. Chem. Soc. 1980, 102, 179-190.

Laakso, P.; Voutilainen, P. Analysis of triacylglycerols by silver-ion high-performance liquid chromatography- Atmospheric pressure chemical ionization mass spectrometry. Lipids 1996, 31, 1311-1322.

Laakso, P.; Christie, W.W. Combination of silver ion and reversed-phase high-performance liquid-chromatography in the fractionation of herring oil triacylglycerols. J. Amer. Oil Chem. Soc. 1991, 68, 213-223.

Leskinen, H.M. Quantification of triacylglycerol regioisomers by ultra-high-performance liquid chromatography and ammonia negative ion atmospheric pressure chemical ionization tandem mass spectrometry. Rapid Commun. Mass Spectrom. 2010, 24, 1-5.

Leskinen, H.M.; Suomela, J.P.; Yang, B.; Kallio, H.P. Regioisomer compositions of vaccenic and oleic acid containing triacylglycerols in sea buckthorn (hippophae rhamnoides) pulp oils: Influence of origin and weather conditions. J. Agr. Food Chem. 2010, 58, 537-545.

Leskinen, H.; Suomela, J.P.; Kallio, H. Quantification of triacylglycerol regioisomers in oils and fat using different mass spectrometric and liquid chromatographic methods. Rapid Commun. Mass Spectrom. 2007, 21, 2361-2373.

Leskinen, H.; Suomela, J.P.; Pinta, J.; Kallio, H. Regioisomeric structure determination of alpha- and beta-linolenoyldilinoleoylglycerol in blackcurrant seed oil by silver ion high-performance liquid chromatography and mass spectrometry. Anal. Chem. 2008, 80, 5788-5793.

Leveque, N.L.; Heron, S.; Tchapla, A. Regioisomer characterization of triacylglycerols by non-aqueous reversed-phase liquid chromatography/electrospray ionization mass spectrometry using silver nitrate as a postcolumn reagent. J. Mass Spectrom. 2010, 45, 284-296.

Li, X.W.; Evans, J.J. Examining the collision-induced decomposition spectra of ammoniated triglycerides as a function of fatty acid chain length and degree of unsaturation. I. The OXO/YOY series. Rapid Commun. in Mass Spectrom. 2005, 19, 2528-2538.

Li, X.W.; Collins, E.J.; Evans, J.J. Examining the collision-induced decomposition spectra of ammoniated triglycerides as a function of fatty acid chain length and degree of unsaturation. II. The PXP/YPY series. Rapid Commun. Mass Spectrom. 2006, 20, 171-177.

Lísa, M.; Holčapek, M.; Řezanka, T.; Kabátová, N. High-performance liquid chromatography-atmospheric pressure chemical ionization mass spectrometry and gas chromatography-flame ionization detection characterization of Δ5-polyenoic fatty acids in triacylglycerols from conifer seed oils. J. Chromatogr. A 2007, 1146, 67-77.

Lísa, M.; Holčapek, M.; Boháč, M. Statistical evaluation of triacylglycerol composition in plant oils based on high-performance liquid chromatography-Atmospheric pressure chemical ionization mass spectrometry data. J. Agr. Food Chem. 2009, 57, 6888-6898.

Lísa, M.; Holčapek, M. Triacylglycerols profiling in plant oils important in food industry, dietetics and cosmetics using high-performance liquid chromatography-atmospheric pressure chemical ionization mass spectrometry. J. Chromatogr. A 2008, 1198, 115-130.

Lísa, M.; Lynen, F.; Holčapek, M.; Sandra, P. Quantitation of triacylglycerols from plant oils using charged aerosol detection with gradient compensation. J. Chromatogr. A 2007, 1176, 135-142.

Lísa, M.; Velínská, H.; Holčapek, M. Regioisomeric characterization of triacylglycerols using silver-ion HPLC/MS and randomization synthesis of standards. Anal. Chem. 2009, 81, 3903-3910.

Lucas, H.J.; Moore, R.S.; Pressman, D. The coordination of silver ion with unsaturated compounds II cis- and trans-2-pentene. J. Amer. Chem. Soc. 1943, 65, 227-229.

Macher, M.B.; Holmqvist, A. Triacylglycerol analysis of partially hydrogenated vegetable oils by silver ion HPLC. J. Sep. Sci. 2001, 24, 179-185.

Malone, M.; Evans, J.J. Determining the relative amounts of positional isomers in complex mixtures of triglycerides using reversed-phase high-performance liquid chromatography-tandem mass spectrometry. Lipids 2004, 39, 273-284.

Momchilova, S.; Nikolova-Damyanova, B. Stationary phases for silver ion chromatography of lipids: Preparation and properties. J. Sep. Sci. 2003, 26, 261-270.

Momchilova, S.; Nikolova-Damyanova, B. Silver ion HPLC of p-methoxyphenacyl derivatives of unsaturated fatty acids. II. Chain length vs. double bond position. J. Liq. Chromatogr. R. T. 2000, 23, 2317-2325.

Momchilova, S.; Nikolova-Damyanova, B.; Christie, W.W. Silver ion high-performance liquid chromatography of isomeric cis- and trans-octadecenoic acids - Effect of the ester moiety and mobile phase composition. J. Chromatogr. A 1998, 793, 275-282.

Momchilova, S.; Nikolova-Damyanova, B. Silver ion high performance liquid chromatography of polynuclear aromatic derivatives of positionally isomeric octadecenoic fatty acids. J. Liq. Chromatogr. R. T. 2000, 23, 1319-1330.

Momchilova, S.; Nikolova-Damyanova, B. Silver ion HPLC of p-methoxyphenacyl derivatives of unsaturated fatty acids. I. Mobile phase effects. J. Liq. Chromatogr. R. T. 2000, 23, 2303-2316.

Momchilova, S.; Itabashi, Y.; Nikolova-Damyanova, B.; Kuksis, A. Regioselective separation of isomeric triacylglycerols by reversed-phase high-performaance liquid chromatography: Stationary phase and mobile phase effects. J. Sep. Sci. 2006, 29, 2578-2583.

Momchilova, S.; Tsuji, K.; Itabashi, Y.; Nikolova-Damyanova, B.; Kuksis, A. Resolution of triacylglycerol positional isomers by reversed-phase high-performance liquid chromatography. J. Sep. Sci. 2004, 27, 1033-1036.

Mondello, L.; Tranchida, P.Q.; Staněk, V.; Jandera, P.; Dugo, G.; Dugo, P. Silver-ion reversed-phase comprehensive two-dimensional liquid chromatography combined with mass spectrometric detection in lipidic food analysis. J. Chromatogr. A 2005, 1086, 91-98.

Morris, L.J. Separation of higher fatty acid isomers and vinylogues by thin-layer chromatography. Chem. Ind. 1962, 1238-1240.

Morris, L.J. Separations of lipids by silver ion chromatography. J. Lipid Res. 1966, 7, 717-732.

Mottram, H.R.; Crossman, Z.M.; Evershed, R.P. Regiospecific characterisation of the triacylglycerols in animal fats using high performance liquid chromatography-atmospheric pressure chemical ionisation mass spectrometry. Analyst 2001, 126, 1018-1024.

Mottram, H.R.; Evershed, R.P. Structure analysis of triacylglycerol positional isomers using atmospheric pressure chemical ionisation mass spectrometry. Tetrahedron Letters 1996, 37, 8593-8596.

Mottram, H.R.; Woodbury, S.E.; Evershed, R.P. Identification of triacylglycerol positional isomers present in vegetable oils by high performance liquid chromatography atmospheric pressure chemical ionization mass spectrometry. Rapid Commun. Mass Spectrom. 1997, 11, 1240-1252.

Muller, A.; Dusterloh, K.; Ringseis, R.; Eder, K.; Steinhart, H. Development of an alternative eluent system for Ag+-HPLC analysis of conjugated linoleic acid isomers. J. Sep. Sci. 2006, 29, 358-365.

Neff, W.E.; Adlof, R.O.; Elagaimy, M. Silver ion high-performance liquid chromatography of the triacylglycerols of crepis-alpina seed oil. J. Amer. Oil Chem. Soc. 1994, 71, 853-855.

Nikolova-Damyanova, B. Retention of lipids in silver ion high-performance liquid chromatography: Facts and assumptions. J. Chromatogr. A 2009, 1216, 1815-1824.

Nikolova-Damyanova, B. Silver-ion HPLC of fatty acids and triacylglycerols; J.-T. Lin, T.A. McKeon Eds; HPLC of acyl lipids. HNB Publishing: New York, 2005 221-268.

Nikolova-Damyanova, B.; Christie, W.W.; Herslof, B.G. High-performance liquid chromatography for fatty-acid derivatives in the combined silver ion and reversed-phase modes. J. Chromatogr. A 1993, 653, 15-23.

Nikolova-Damyanova, B.; Herslof, B.G.; Christie, W.W. Silver ion high-performance liquid chromatography of derivatives of isomeric fatty-acids. J. Chromatogr. 1992, 609, 133-140.

Nikolova-Damyanova, B.; Christie, W.W.; Herslof, B.G. Retention properties of triacylglycerols on silver ion high-performance liquid-chromatography. J. Chromatogr. A 1995, 694, 375-380.

Nikolova-Damyanova, B.; Christie, W.W.; Herslof, B. Silver ion high-performance liquid-chromatography of esters of isomeric octadecenoic fatty-acids with short-chain monounsaturated alcohols. J. Chromatogr. A 1995, 693, 235-239.

Nikolova-Damyanova, B.; Christie, W.W.; Herslof, B. Mechanistic aspects of fatty acid retention in silver ion chromatography. J. Chromatogr. A 1996, 749, 47-54.

Phillips, K.M.; Lugogo, R.D.; Harris, R.F.; Tarragotrani, M.T.; Bailey, J.A.; Stewart, K.K. Direct determination of trans-octadecenoic acid in diet composites using a combination of silver ion high performance liquid chromatography and capillary gas chromatography. J. Food Lipids 1997, 4, 173-188.

Pittenauer, E.; Allmaier, G. The Renaissance of high-energy CID for structural elucidation of complex lipids: MALDI-TOF/RTOF-MS of alkali cationized triacylglycerols. J. Amer. Soc. Mass Spectr. 2009, 20, 1037-1047.

Schuyl, P.J.W.; De Joode, T.; Vasconcellos, M.A.; Duchateau, G. Silver-phase high-performance liquid chromatography electrospray mass spectrometry of triacylglycerols. J. Chromatogr. A 1998, 810, 53-61.

Shan, H.; Wilson, W.K. Ternary gradient elution markedly improves silver-ion high performance liquid Chromatogr. of unsaturated sterols. Steroid. 2002, 67, 917-923.

Smith, M.; Monchamp, P.; Jungalwala, F.B. Separation of molecular-species of sphingomyelin and ceramide by argentation and reversed-phase HPLC. J. Lipid Res. 1981, 22, 714-719.

Standal, I.B.; Axelson, D.E.; Aursand, M. Differentiation of fish oils according to species by C-13-NMR regiospecific analyses of triacyglycerols. J. Amer. Oil Chem. Soc. 2009, 86, 401-407.

Stoll, U. Identification of molecular species of phospholipids from pig heart by combining chromatographic techniques. Fett-Lipid 1996, 98, 178-182.

Toschi, T.G.; Capella, P.; Holt, C.; Christie, W.W. A comparison of silver ion HPLC plus with fourier-transform IR spectroscopy for the determination of trans-double bonds in unsaturated fatty-acids. J. Sci. Food Agr. 1993, 61, 261-266.

Van Der Klift, E.J.C.; Vivó-Truyols, G.; Claassen, F.W.; Van Holthoon, F.L.; Van Beek, T.A. Comprehensive two-dimensional liquid chromatography with ultraviolet, evaporative light scattering and mass spectrometric detection of triacylglycerols in corn oil. J. Chromatogr. A 2008, 1178, 43-55.

Williams, C.M.; Mander, L.N. Chromatography with silver nitrate. Tetrahedron 2001,

57, 425-447.
Winstein, S.; Lucas, H.J. The coordination of silver ion with unsaturated compounds. J. Amer. Chem. Soc. 1938, 60, 836-847.
Yeung, S.K.F.; Kuksis, A.; Marai, L.; Myher, J.J. Resolution of molecular-species on intact serine and ethanolamine phosphatides by argentation chromatography of their trifluoroacetamides. Lipids 1977, 12, 529-537.
Zeise, W.C. Von der Wirkung zwischen Platinchlorid und Alkohol, und von den dabei entstehenden neuen Substanzen. Annalen der Physik 1831, 97, 497–541.

7

Recent Advances in Comprehensive Two-dimensional Liquid Chromatography for the Analysis of Natural Products

Francesco Cacciola[1], Paola Donato[2], Luigi Mondello[1,2], and Paola Dugo[1,2]
[1]Dipartimento Farmaco-chimico, Facoltà di Farmacia, Università di Messina, viale Annunziata, 98168 Messina, Italy and [2]Università Campus Bio-Medico, Via Álvaro del Portillo 21, 00128 - Roma, Italy

Introduction

The analysis of real matrices, namely biological, food and environmental samples, poses a high demand for HPLC methods enabling determination and identification of a large number of compounds occurring in the samples.

However, a single column does not often have sufficient separation power for the baseline separation of all components in complex samples, and the extent of peak overlap is enhanced as the number of compounds increases.

The development of comprehensive two-dimensional high performance liquid chromatography (2D-LC) has been relatively slow until recently, even though the advantages of the technique were already described in the late 1970s and early 1980s (Karger et al., 1973; Erni & Frei, 1978; Giddings, 1984). In the 1990s the application of LCxLC was limited to proteomic applications, but in the last decade the utility of this technique in the separation of a variety of other complex mixtures has been demonstrated. The general motivation which lies behind this research area is linked to its potential for substantially more resolving (separating) power in comparison with the conventional one-dimensional (1D) counterparts.

The most general set-up of a comprehensive LC system consists of two pumps, two columns, injector, interface and detector. The interface is in general a high pressure switching valve, and this device is often referred to as a modulator or sampling device.

The suitability of any LC separation system for resolving complex samples can be expressed by theoretical peak capacity, n_c, which determines the maximum number of peaks that can be separated side by side, into the separation space at a desired degree of resolution, e.g., $R_S = 1$ ($n_c = t/w$). The separation time can be specified as the retention time interval $\Delta t_R = t_{R,Z} - t_{R,1}$, between the first (1) and the last (z) sample peaks while w can be estimated from the average bandwidth, w_{av}.

Two-dimensional LC methods improve the likelihood for separation of a higher number of sample components, since the peak capacity of the 2D

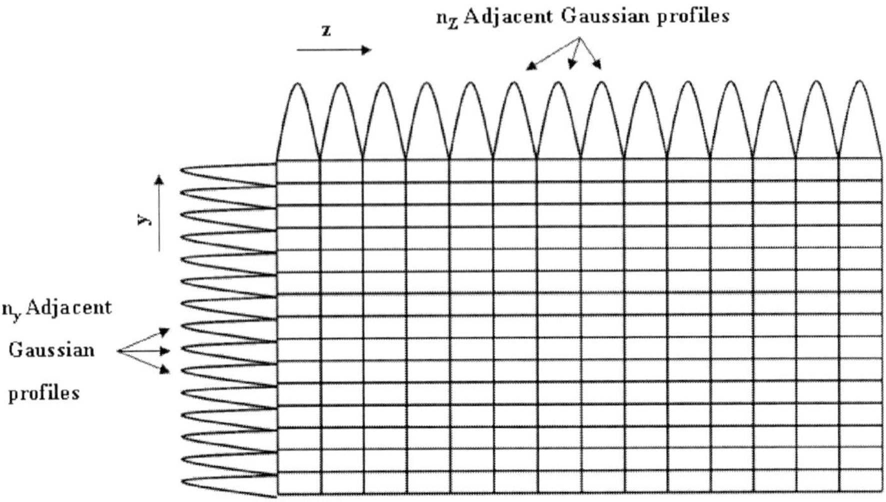

Fig. 7.1. Peak capacity in a two-dimensional system.

separation is multiplicative, which derives from the product of the peak capacities of the constituent 1D separations of the 2D system ($n_{2D} = n_{D1} \times n_{D2}$) (Karger et al., 1973; Giddings, 1984; Giddings, 1987). This finding is illustrated in Fig. 7.1. The peak capacities of the two dimensions are shown as the number of adjacent Gaussian profiles that can be accommodated into the space along with the respective separation coordinates. The resolution is represented by rectangular boxes that divide the separation plane. The total peak capacity, n_{2D}, is therefore approximately equal to the number of such boxes (Giddings, 1987).

In this regard, by careful selection of orthogonal separation systems both in the first and the second dimensions, the peak capacity of the overall systems can dramatically increase. This makes group separations of various samples often possible so that various structurally related classes of compounds can be distinguished and located in different areas of the two-dimensional retention plane, which helps in the interpretation of the 2D data.

Due to the tremendously higher separation power of comprehensive LC compared to its 1D counterpart, the technique has been employed in several fields and has been the subject of various reviews (Liu & Lee, 2000; Evans & Jorgenson, 2004; Guttman et al. 2004; Stroink et al., 2005; Dixon et al., 2006; Jandera, 2006; Shalliker & Gray, 2006; Shellie & Haddad, 2006; Stoll et al., 2007; Dugo et al., 2008a; Dugo et al., 2008b; Guiochon et al. 2008; Pol & Hyotylainen, 2008; François et al., 2009; Herrero et al., 2009; Sandra et al., 2009; Berek, 2010). Comprehensive LC is currently used for the separation of polymers, pharmaceuticals and biological mixtures, natural products and complex environmental and petrochemical samples.

Advances in Theory

In recent years, most theoretical work in comprehensive 2D-LC has been basically devoted to the optimization of 2D-LC methods. In addition to the factors that affect resolution and performance in conventional 1D methods, including retention, selectivity, and column efficiency, comprehensive 2D methods imply other factors as the 2D system must be optimized as a complete system. The most striking example of the dependence of the two dimensions on each other is related to the analysis speed in the second dimension. To this regard, many practitioners have recognized the effect of the sampling rate on the peak capacity of the overall 2D separation, but very few reports are effectively dealing with this issue. In their prominent work, Murphy et al. (1998) investigated the effect of the sampling rate on the effective first dimension peak width and modeled a Gaussian peak as a histogram profile of the average concentration within every sampling period. The outcome of the research was that approximately three to four modulations per each first dimension peak (8σ) are needed in order to avoid serious loss of performance of the 2D system as a result of "under-sampling" of first-dimension peaks. However, this number is only valid when the sampling is "in-phase", which implies that the sampling should start exactly at the beginning of the peak. In experimental practice, this cannot be guaranteed, so a minimum of four samplings is preferable. A few years later, Seeley et al. published another report on this topic investigating the sampling frequency for modulators with various duty cycles (2002). On the one hand, when using an interface consisting of a two-position/ten- or eight-port switching valve provided with two storage loops, the primary effluent is collected throughout the entire sampling period and the duty cycle is equal to unity. On the other hand, when using devices which do not integrally sample the first dimension effluent, the duty cycle is less than 1 as in the case of a two-position/six-port switching valve equipped with only one sampling loop. Seeley defined the (average) peak broadening factor σ^* by introducing parameters as sampling period τ_Z and sampling phase Φ. τ_Z is determined by the modulation period or sampling time t_s and ${}^1\sigma$ of the first dimension peak, while the latter represents the way in which the primary peak is cut into aliquots. The sampling phase Φ is the time difference between the center of the sampling cycle nearest to the peak maximum and the peak maximum, divided by the sampling period. Peak broadening was less significant when the peak maximum was centered in one of the sampling periods and for low duty cycles. For systems with a duty cycle of 1, as is the case for most comprehensive separations, the peak broadening was practically independent of Φ. Basically, Seeley drew the same conclusions as Murphy, extending his investigations to duty cycles less than 1. More recently, Horie at al. nailed down that in general, modulation periods equal to 2.2–2.4 are sufficient to minimize "peak broadening" (2007). On the one hand, shorter modulation periods offer a higher number of cuts across the first dimension peaks, but the secondary column fails to provide sufficient separation power. On the other hand, longer sampling periods in combination with a longer column leads to a gain in resolution in the second dimension, but the overall resolution is in this case jeopardized due to the deleterious effect of the under-sampling. In addition, they introduced a new parameter, e_M (Nobuo factor), which is defined as the ratio of the apparent peak capacity of the first column derived from σ^* to its actual peak capacity in the first dimension. A useful "protocol" for optimizing 2D-LC separations based on proper sampling has been described by Schoenmakers and co-workers (2006). According to this approach, separations in the second dimension of the system have to be isocratic, thus providing some guidelines for choosing the dimensions of columns used in the first and second dimensions under these conditions. Further theoretical work has been carried out by Blumberg and co-workers (Blumberg et al., 2008) comparing 1D and 2D-GC. They demonstrated that there is the potential for tenfold higher peak capacities in 2D-GC compared with 1D-GC separations executed in the same amount of analysis time, but that significant physical limitations affect fully realizing this potential in practice.

More recently, Schure (2009) also provided a list of factors that should be considered first in developing successful 2D-LC separations.

Great effort towards understanding and describing the effect of under-sampling has been devoted by Carr and his co-workers. In their reports, in contrast to all previous studies where the results were based on just a single pair of peaks, the investigation was extended to the overall chromatogram (Davis et al., 2008; Davis et al., 2009; Li et al., 2009). They introduced an average first dimension peak broadening factor, β, as a function of the sampling time (t_s) and the standard deviation of the first dimension peaks prior to sampling (${}^1\sigma$) which can be used together with the product rule for the calculation of the practical peak capacity (Davis et al., 2008). For the determination of this factor, 2D statistical overlap theory, along with the number of observed peaks in simulated comprehensive 2D separations, was used (Davis et al., 2005). In comparison with previous studies (Murphy, 1998; Seeley, 2002; Horie et al., 2007) the peak broadening factors determined in this work were 2–35% larger; as a result, a major effect on the overall peak capacity can be observed. Furthermore, the theory applies to high and low orthogonal separations and also to non-Gaussian peaks.

In order to alleviate the under-sampling problem, several different experimental approaches have been employed aimed at increasing the speed of second-dimension separations in 2D-LC systems. Among these are the use of high pressure conditions (e.g., >400 bar) (Huidobro et al., 2008), high temperature conditions (e.g., >100°C) (Cacciola et al., 2007; Stoll et al. 2008), columns packed with partially porous particles (Dugo et al., 2008; Hajek et al., 2008; Dugo et al., 2009a; Dugo et al., 2009b; Alexander & Ma, 2009; François et al., 2009; Mondello et al., 2010), short monolithic columns (Tanaka et al.,

2004; Ikegami et al., 2006) and multiple second-dimension columns operated in parallel (Opiteck et al., 1997; Alexander & Ma, 2009; François et al., 2009). In particular, the benefits derived from the use of multiple second dimension columns have been recently shown in two theoretical studies (Fairchild et al., 2009; Potts et al., 2010). The idea that lies behind this approach is related to the fact that the sampling rate of the first dimension effluent can be increased because the multiple second dimension columns can simultaneously separate multiple fractions of first dimension effluent. Guiochon and co-workers reported the primary benefits of this approach, viz. increased peak capacity of the 2D-LC system and the possibility to reduce the analysis time required to produce a specified amount of peak capacity, compared with a 2D-LC system employing only a single second-dimension column (Fairchild et al., 2009). The main drawback can be traced back to the strict requirements for the similarity of the multiple second dimension columns. Carr and co-workers evaluated the effects of the first dimension peak capacity and gradient time and the second dimension cycle times on the overall peak capacity of the 2D-LC system (Potts et al., 2010). In addition, they examined the effects of re-equilibration time on under-sampling as measured by the under-sampling factor and the influence of molecular type (peptide vs. small molecule) on peak capacity.

Advances in Practice

The most significant development in the practice of 2D-LC in recent years has been the introduction of second dimension small (ca. 3 μm diameter) partially porous particles (also referred to as "fused-core"). Monolithic columns have been widely used as the second dimension of comprehensive LC systems over the last decade. The choice of such columns was dictated by several reasons; in particular these columns allow the use of high flow rates without loss of resolution, due to the better mass transfer properties of a monolithic skeleton over particle packed columns, thus reducing analysis time. Moreover, because of the higher permeability of these columns, successive cycles of gradient can be performed with very brief re-equilibration times and significantly lower backpressures (Lubda et al., 2001). However, recent works have demonstrated the possibility to use conventional-particle packed beds and wide pore columns to obtain very fast second dimension gradient elution separation (Stoll et al., 2006; Stoll et al., 2007). In 2007, a new particle technology (Fused-Core™) has been developed in order to deliver hyper-fast chromatographic separations while avoiding the reliability issues often associated with fast HPLC (Way, 2007a; Way, 2007b). These new HPLC columns consist of a 1.7 μm solid core with a 0.5 μm porous silica shell surrounding it (dp = 2.7 μm). The major benefit of this particle is therefore the small diffusion path, compared to the approximately 1.5 μm diffusion path of a 3 μm totally porous particle, which reduces axial dispersion of solutes and minimizes peak broadening. The gain in performance can be best demonstrated via the van

Deemter equation. In fact, smaller particle diameter pores reduce the distance a solute travels within the stationary phase, minimizing axial diffusion and the contribution of diffusion in the stagnant mobile phase (Giddings, 1991). Recent studies have already demonstrated the unique performance of these particle columns, offering the advantages of higher efficiencies and rugged design, capable of high pressure operation (Gritti et al., 2007a; Gritti et al., 2007b; Marchetti et al., 2007; Cavazzini et al., 2007; Cunliffe & Maloney, 2007; Donato et al., 2009). Because of these attributes, these columns are well suited to use in the second dimension of 2D-LC systems, and several groups have demonstrated their use in this application. Dugo and co-workers (2008; 2009a; 2009b) and Hajek and co-workers (2008) have both used partially porous materials in the second dimension of 2D-LC systems for analysis of mixtures of antioxidants. In two of the above-mentioned references (Dugo et al., 2008; Hajek et al., 2008) the performance of a partially porous short column was compared to that of a monolithic column. The results demonstrated the possibility to use partially porous columns in the second dimension of 2D-LC systems to obtain fast analyses, using high flow rates, under repetitive gradient conditions with very brief reconditioning times. Alexander and Ma (2009) used two columns packed with partially porous materials in parallel in a 2D-LC system for analysis of impurities present in a pharmaceutically active ingredient. Very fast gradient separations (30 s total cycle time) were achieved at ambient temperature without excessive backpressure and without compromising optimal 1st dimension sampling rates. Francois et al. (2009) and Mondello et al. (2010) exploited the employment of partially porous columns in the first and second dimension of 2D-LC systems, respectively, for analysis of tryptic digests. In both applications the resolving power of the first dimension was extremely high through the serial column coupling. In particular, Francois et al. described the first on-line comprehensive RP-LC × 2RP-LC separation of complex peptide mixtures employing considerably different pH in the two dimensions, which ensured selectivity differences and hence orthogonality (2009). The first dimension comprised of four serially coupled columns packed with fused-core octadecylsilica (ODS) particles and operated at pH 1.8, while fast second dimension separations are carried out on two short ODS columns at pH 10 placed in parallel. However, in the field of proteomics, it is more prevalent to operate the second dimension at low pH, since this typically allows more sensitive MS analysis in the positive ion mode and gives rise to MS/MS spectra compatible with database searching (Toll et al., 2005). This aspect has been successfully investigated by Mondello et al., who optimized a 2D-LC system that performed coupled column chromatography at high pH in the first dimension, with a fast highly efficient second dimension separation at low pH. This held promise for coupling to MS to allow searching of databases for protein identification (2010).

In order to develop suitable 2D-LC methods, several parameters, e.g., selection of stationary and mobile phases, column dimensions and sample transfer modulation, have to be taken into account. Among them, the choice of either isocratic or gradient elution conditions in both dimensions is of utmost importance. To this regard neither mode is clearly the best in all situations, and uses of both modes have been reported extensively in the literature. An interesting approach has been proposed by the group of Jandera and co-workers who have extensively contributed to the progress in this area (Cacciola et al., 2007a; 2007b; 2008; Cesla et al., 2009). Cacciola at al. (2007b) explored various combinations of PEG-silica, phenyl-silica and C18 columns in a single-column or serial (tandem) arrangement in the first dimension and a monolithic column in the second dimension for comprehensive 2D-LC separation of phenolic and flavone natural antioxidants. The novel aspect of this work is related to the use of gradients with matching profiles running in parallel in the two dimensions over the whole 2D separation time range. This approach allowed them to improve system orthogonality and bandwidths suppression, obtaining more regular band distribution over the whole 2D retention plane for samples with partially correlated elution times in the two dimensions, and increasing peak capacity in the different 2D setups. In addition, it eliminates the necessity of second-dimension column re-equilibration after the independent gradient runs for each fraction. Fig. 7.2 shows the profiles of the parallel gradients of acetonitrile run simultaneously in the first and in the second dimension for each set-up tested. In this work, two short trapping columns in the 10-port switching valve sample transfer interface were employed in place of the two conventional loops. This contributed to significant suppression of the band broadening associated with the sample transfer between the two dimensions, since the sample compounds were less strongly retained on the first-dimension column than on the trapping column. Thus, the narrow sample zone adsorbed on the top of the trapping column could be back-flushed to the second dimension column in a small volume of the mobile phase. The sample fraction modulation technique that was employed allowed using a longer monolithic column in the second dimension, resulting in narrower sample bands (spots) in the comprehensive 2D chromatogram. The parallel gradient profile approach has been further investigated in two recent works coming from the same research group (Jandera et al., 2008; Cesla et al., 2009). Jandera et al. (2008) compared several types of bonded C18, amide, phenyl, pentafluorophenyl and poly(ethylene glycol) columns, using the linear free energy relationship method to find suitable column combinations with large differences in separation selectivity and low correlation of retention of representative standards. The phase systems were optimized step-by-step to find the mobile phases and gradients in both dimensions providing best separation selectivity for phenolic compounds. In their most recent work

(Cesla et al., 2009), a 2D system with a bonded polyethylene glycol Discovery HS PEG micro-column in the first dimension and a partially porous C18 in the second dimension were tested. The employment of a second dimension C18 partially porous fused-core column showed improved performance in terms of peak capacity, corresponding to the number of average bandwidths and reduced overall separation time at high temperature (60°C). In addition, for optimization of the parallel linear gradient profiles in both the first and the second dimensions, a resolution mapping window diagram strategy has been investigated (Jandera, 1985; Schoenmakers, 1986). For this task, a new software program was developed that enabled automated attribution of the peaks in subsequent fractions, and compensation for the shifts in the elution times caused by the transfer of subsequent fractions from the first dimension (in a mobile phase with gradually increasing elution strength) into a continuous second dimension gradient. All these examples highlight the usefulness of the parallel matching gradient method which makes a good use of the 2D separation space. The main drawback can be attributed to the fact that the elution scheme must be adapted to each new sample. Nevertheless, this approach will likely be a fruitful area of research as soon as this type of optimization strategy is implemented for analysis of more sample types.

2D-LC Separation of Natural Products

The two most commonly used configurations of 2D-LC systems for the analysis of natural products are represented by normal-phase LC combined with reversed-phase LC (NP×RP), and reversed-phase in both dimensions (RP×RP). On the one hand, for NP×RP separation of carotenoids, either silica or cyanopropyl (CN) stationary phases have been investigated (Dugo et al., 2006a; Dugo et al., 2008d; Dugo et al., 2008e; Dugo et al., 2009c) whereas for triacylglycerol separations, an Ag×RP combination turned out to be useful for this task (Mondello et al., 2005; Dugo et al., 2006b; Dugo et al., 2006c; Van der Klift et al., 2008). On the other hand, for RP×RP separation of natural antioxidants, the effect of several stationary phase chemistries [C18, PEG, strong cation exchange (SCX), CN, NH_2, Phenyl] on the separation selectivity and resolution has been investigated (Cacciola et al., 2007a; Cacciola et al., 2007b; Kivilompolo et al., 2007; Dugo et al., 2008c; Jandera et al., 2008; Pol et al., 2008; Kivilompolo et al., 2008a; Kivilompolo et al., 2008b; Dugo et al., 2009a; Dugo et al., 2009b; Cesla et al., 2009).

LC×LC Chromatography of Carotenoids

Carotenoids are the most broadly distributed pigments in nature (Mendes-Pinto et al., 2005; Herrero et al., 2006). *Citrus* species, particularly orange, are commonly regarded as the most complex natural source of these compounds. Some carotenoids are reported to show different functional properties, such

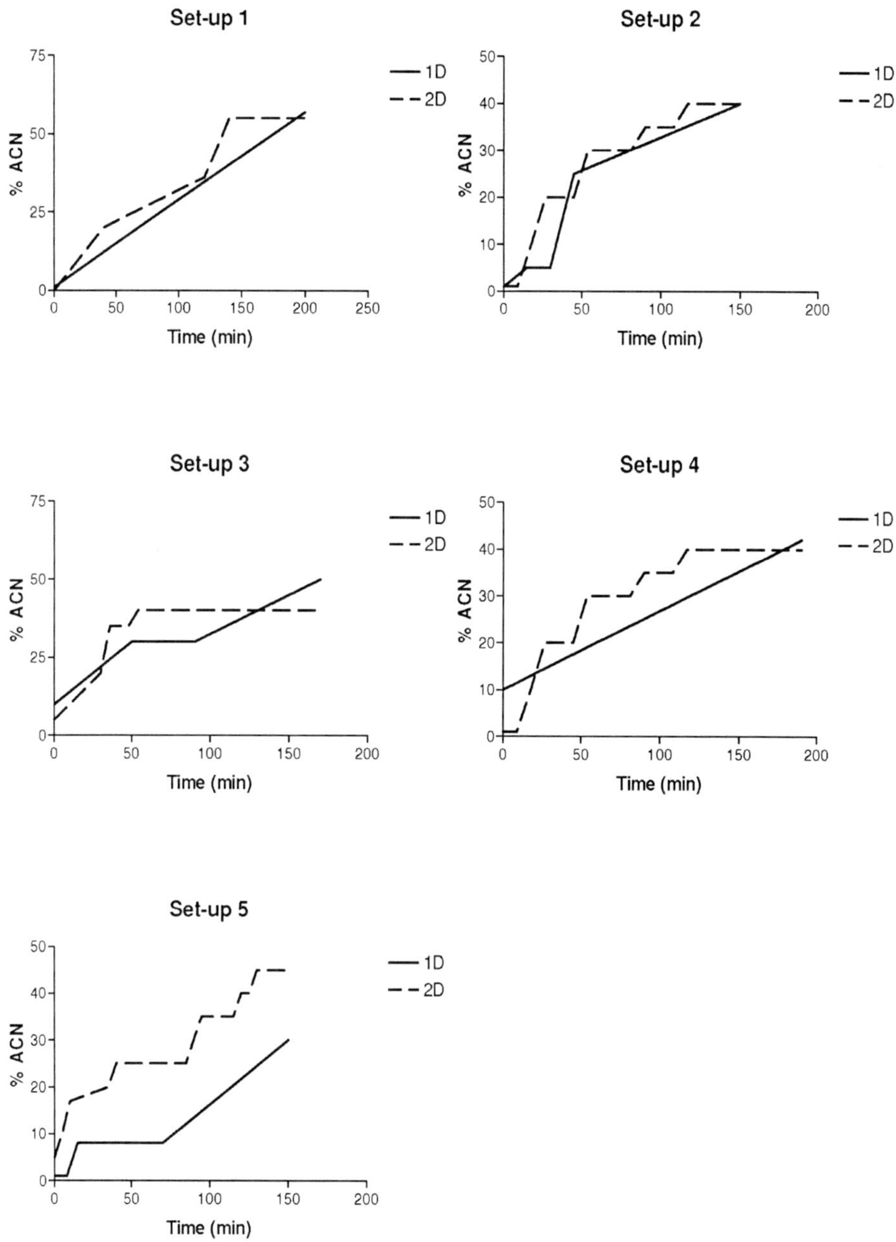

Fig. 7.2. Profiles of the parallel gradients of acetonitrile in the first (full lines) and second (dashed lines) dimensions in the five setups tested. Reprinted with permission from Cacciola et al., 2007.

as antioxidant activity (Beutner et al., 2001; Leach et al., 1998), prevention of cancer (Nishino et al., 1999), cardiovascular diseases (Arab & Steck, 2000; Rao & Rao, 2007) and macular degeneration (Snodderly, 1995). Carotenoids are a very complex group of compounds based on a C40-tetraterpenoid skeleton and are usually classified into two main groups: hydrocarbon carotenoids, known as carotenes (e.g., β-carotene and lycopene), and oxygenated carotenoids, known as xanthophylls (e.g., β-cryptoxanthin and lutein). The elucidation of carotenoid patterns is particularly challenging, because of the complex composition of carotenoids in natural matrices, their great structural diversity and their extreme instability. Dugo et al. exploited a 2D-LC NP×RP platform for the separation and identification of these compounds in intact and saponified citrus samples (Dugo et al., 2006; Dugo et al., 2008d; Dugo et al., 2008e; Dugo et al., 2009c). The dual gradient elution 2D-LC system was comprised of either a silica or a cyano column in the first dimension and a C18 column in the first and second dimension. For identification purposes, PDA and APCI/MS detection were employed. In the earlier work, a microbore silica column in NP mode as the first dimension and a monolithic C18 column in RP mode in the second dimension was employed for the elucidation of the carotenoid pattern in saponified orange essential oil and juice samples (Dugo et al., 2006). In the two subsequent studies aiming to evaluate the native carotenoid composition of mandarin and red orange essential oils, the D1 microbore silica column was replaced with a cyanopropyl column for the separation of free carotenoids and carotenoid esters (Dugo et al., 2008d; Dugo et al., 2008e). Regarding detection, the carotenoid identification was carried out by combining the information provided by the PDA and MS systems and the peaks' relative positions in the two-dimensional chromatograms. In the most recent work, the native carotenoid patterns of seven different monovarietal orange juices were studied by LC×LC-DAD/APCI-IT-TOF-MS for the first time (Dugo et al., 2009c). Specifically, as an estimate of the age and freshness of an orange juice, special attention was devoted to the epoxycarotenoid components. In this sense, it has been observed that the relative proportions of 5,6- and 5,8-epoxycarotenoids could give a hint about the juice freshness. Fig. 7.3 shows the APCI$^+$-IT-TOF-MS spectra obtained corresponding to *cis*-violaxanthin-laureate-myristate (C12:0, C14:0) (A), *all trans*-violaxanthin-laureate (C12:0) (B) and *all trans*-luteoxanthin-di-laureate (C12:0, C12:0) (C) separated in one of the commercial orange oils tested (Valencia). Thanks to the combination of the information coming from the PDA (UV/Vis maxima, spectral fine structures, cis peak, Q-ratio) and the IT-TOF-MS (detection in rapidly alternating positive and negative ionization modes), it was possible to achieve the identification of several *cis*-violaxanthin and *all trans*-luteoxanthin diesters as well as some *all trans*-violaxanthin monoesters, among other minor compounds. To this regard, as previously explained, the diverse composition

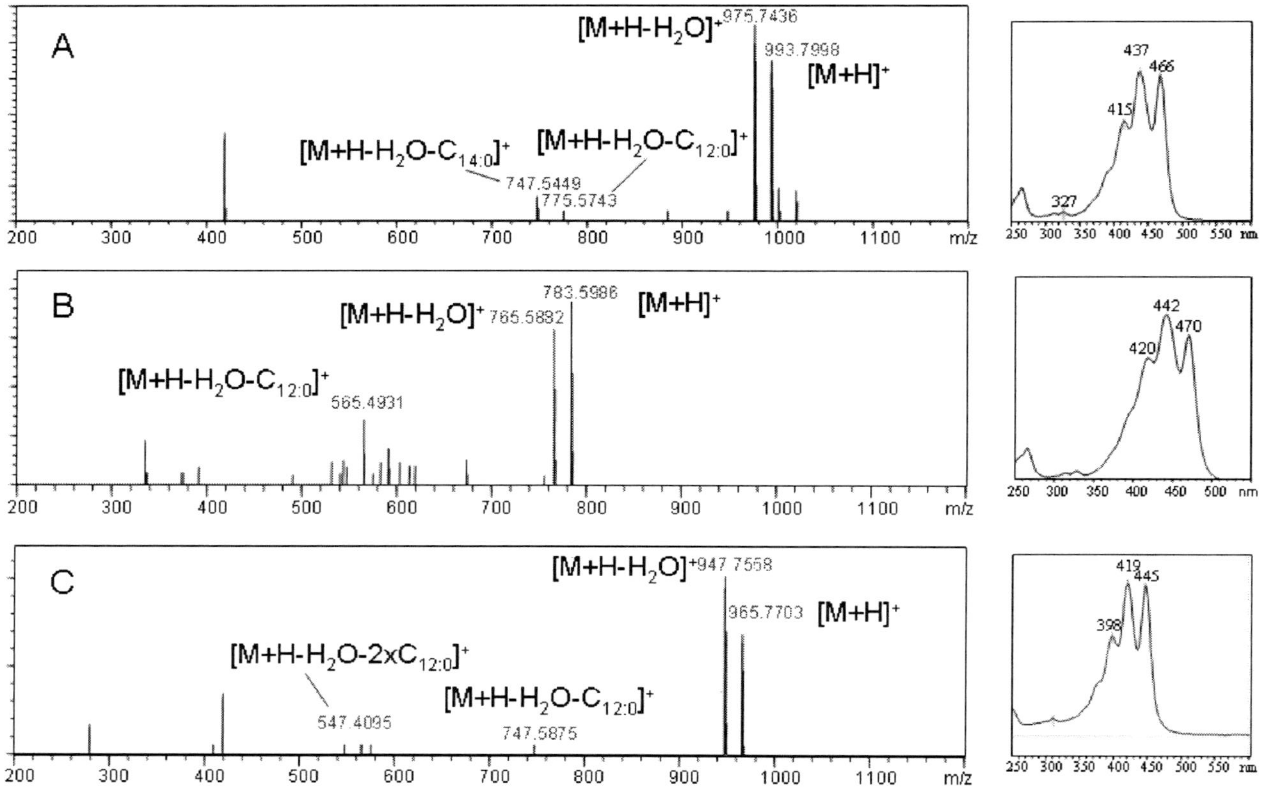

Fig. 7.3. APCI$^+$-IT-TOF-MS spectra obtained corresponding to cis-violaxanthin-laureate-myristate (C12:0, C14:0) (A), all trans-violaxanthin-laureate (C12:0) (B) and all trans-luteoxanthin-dilaureate (C12:0, C12:0) (C) separated in Valencia orange oil.

of the carotenoid fraction can be used as a tool to estimate the relative age and the freshness of the juice. For this reason, the developed method might be used to prevent or detect adulterations or to establish ripeness indexes.

LC×LC Chromatography of Triacylglycerols

Various lipidic matrixes such as donkey milk samples (Dugo et al., 2006b) and plant oils (Mondello et al., 2005; Dugo et al., 2006c; Van der Klift et al., 2008) have been investigated using a comprehensive 2D-LC mode employing silver ion HPLC in one dimension and non-aqueous RP HPLC in the other one. The comprehensive approach was based on the employment of a D1 microbore Ag column and of a D2 octadecylsilyl column connected via a 10-port switching valve. In all cases, the use of APCI-MS greatly increased the potential for component identification thanks to the absence of peak coelution and to the restricted number of components that can be assigned to a spot located in a well-defined position of the 2D plane, making peak identification easier even when commercial standards are not available.

A thorough elucidation of a rice oil TAG profile was reported by Mondello et al. using a three-dimensional Ag-LC × RP-LC-APCI-MS method (2005). A primary Ag microbore column, operated under isocratic elution conditions, was connected to a monolithic column, operated under gradient elution conditions, via a 10-port valve equipped with two 20-μL loops. Due to the highest LC×LC resolving power and MS spectral information, 11 triacylglycerols were positively identified, thus exploring a group of components that are of fundamental importance in food chemistry. A nice application to the analysis of TAGs in donkey milk, using a similar set-up, has been reported by Dugo et al. (2006b). With respect to the previous work, the main difference consisted in the use of gradient elution also in the first dimension, enabling the attainment of peaks with comparable widths in a reasonable analytical time. By using this approach nearly 60 TAGs were separated and identified revealing that the investigated system was an effective tool for the characterization of the TAG fraction of such a complex matrix. The use of a gradient program in both LC dimensions was also applied to the analysis of different plant oils containing TAGs in a wide range of partition numbers (PN) (36–52) and double bond (DB) numbers (1–9) such as linseed and soybean (Dugo et al., 2006c). In addition, the separation of some positional isomers was achieved. In all cases, TAGs were separated under normal phase conditions in the first dimension and reversed phase conditions in the second dimension. However, an interesting modification to the above described platforms was provided by Van der Klift et al. who clearly demonstrated that TAGs could be efficiently separated with a methanol-based solvent on a Ag coated ion exchanger medium, thus avoiding the use of hexane (2008). This innovative comprehensive LC×LC system was successfully employed for the detection of 44 triacylglycerols in corn oil. Fig. 7.4 illustrates the contour plots of corn

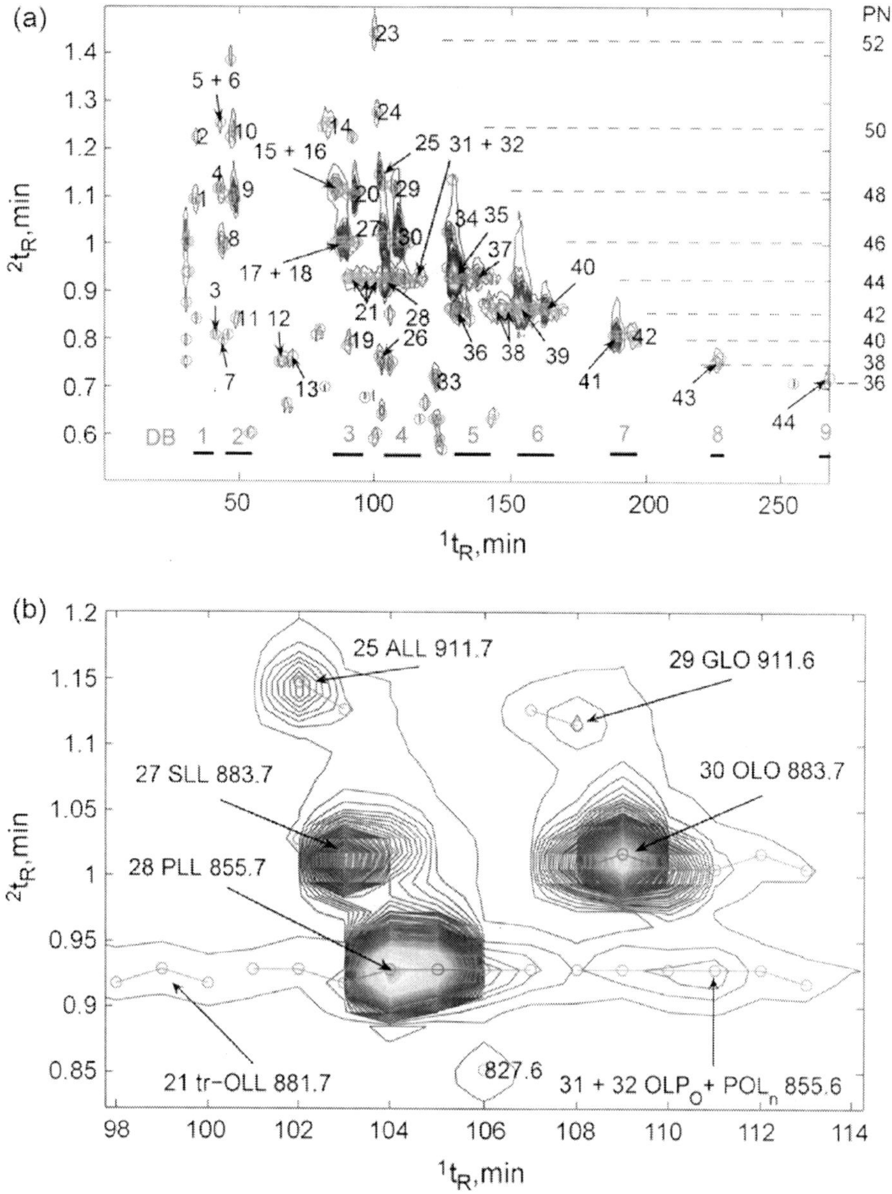

Fig. 7.4. Contour plots of corn oil constructed on the basis of the TIC chromatogram. Dots represent peaks detected automatically by the software. (a) Total contour plot. Dots remaining unlabelled correspond to peaks that could not be identified by manual inspection of the MS. Dashed lines refer to the PN, numbers at the bottom to the DB number; (b) expansion of contour plot in (a). First-dimension from 97.5–115 min, second-dimension from 0.84–1.18 min. Lines linking dots indicate peaks that were pooled by the merging algorithm and are considered to originate from the same compound. Numbers after the name of the TAG correspond with the molecular weight of the TAG. Reprinted with permission from Van der Klift et al., 2008.

oil constructed by means of the TIC chromatogram. The potential of off-line 2D-LC coupling of NARP in the first dimension and silver-ion in the second dimension was successfully demonstrated by Holčapek et al. (2009). This experimental setup provided the highest separation capacity of TAGs in complex natural samples of plant oils and animal fats resulting in the highest number of identified TAGs ever reported for studied samples. The only drawback of the approach is related to the long analysis time; therefore, such method is not useful for routine quality control of TAG mixtures of low complexity.

LC×LC Chromatography of Phenolic Antioxidants

Phenolic antioxidants represent a large family of naturally occurring compounds in fruits, vegetables, and beverages originating from the plants used in production, such as barley, hop, and hop products used in the beer brewing process. It has been demonstrated that some of them show strong antioxidant properties, with beneficial, physiological, and anticarcinogenic properties for human health (Hertog, 1993; Tsuchiya, 1997; Dalluge, 1998). Because of higher peak capacity, selectivity, and resolution, 2D-LC×LC systems have been investigated for the separation of these compounds, taking advantage of the different selectivity of various RP columns. In most works reported in the literature, various combinations of columns have been employed for the separation of phenolic and flavonoid antioxidants in herb, beer and wine samples (Cacciola et al., 2007a; Cacciola et al., 2007b; Kivilompolo et al., 2007; Pol et al., 2007; Kivilompolo et al., 2008a; Kivilompolo et al., 2008b; Dugo et al., 2008c; Jandera et al., 2008; Hajek et al., 2008; Cesla et al., 2009; Dugo et al., 2009a; Dugo et al., 2009b; Cacciola et al., 2010). In most applications, PEG-silica columns (due to the high polarity of the bonded phase) have been used as the first dimension of the comprehensive LC systems, providing complementary selectivity to either ZR-Carbon (Cacciola et al., 2007a) or C18 columns (Hajek et al., 2008; Cacciola et al., 2007b; Jandera et al., 2008; Cesla et al., 2009) employed in the second dimension of the comprehensive 2D system. Phenyl-silica and RP-Amide stationary phases have also been employed for the first dimension separation (Cacciola et al., 2007b; Dugo et al., 2008c; Dugo et al., 2009a; Dugo et al., 2009b). In two recent works, Dugo et al. employed a partially porous C18 column in the second dimension of the 2D-LC set-up (2008c; 2009a). In the first work, Dugo et al. (2008c) compared the performance of the partially porous short column (3.0 cm) to that of a monolithic column of comparable dimensions, for the separation of polyphenolic components in a red wine sample. In the second work by Dugo et al. (2009a), two dimensional liquid chromatography and conventional liquid chromatography were compared for the analysis of polyphenols in red wine samples, in terms of linearity, sensitivity, detection and quantification limits, and separation performance. A good accordance between quantitative results in conventional and comprehensive LC was obtained. On the one hand, due to the higher degree of separation obtained in comprehensive LC, components co-eluted in conventional LC were easily quantified separately. On the other hand, the resultant sensitivity was lower in comprehensive LC, due to the dilution caused by the high second dimension flow rate; but higher resolution allowed the researchers to obtain clearer spectra for positive peak identification. Fig. 7.5 shows a two-dimensional plot corresponding to the separation of a red wine sample, at 278 nm (Fig. 7.5.A) and 354 nm (Fig. 7.5.B).

Two comprehensive two-dimensional liquid chromatography systems connected on-line to electrospray ionization time-of-flight mass spectrometry (ESI-TOF-MS) have been investigated for analysis of *Stevia rebaudiana* and Lamiaceae herbs (Pol et al., 2007; Kivilompolo et al., 2007; Cacciola et

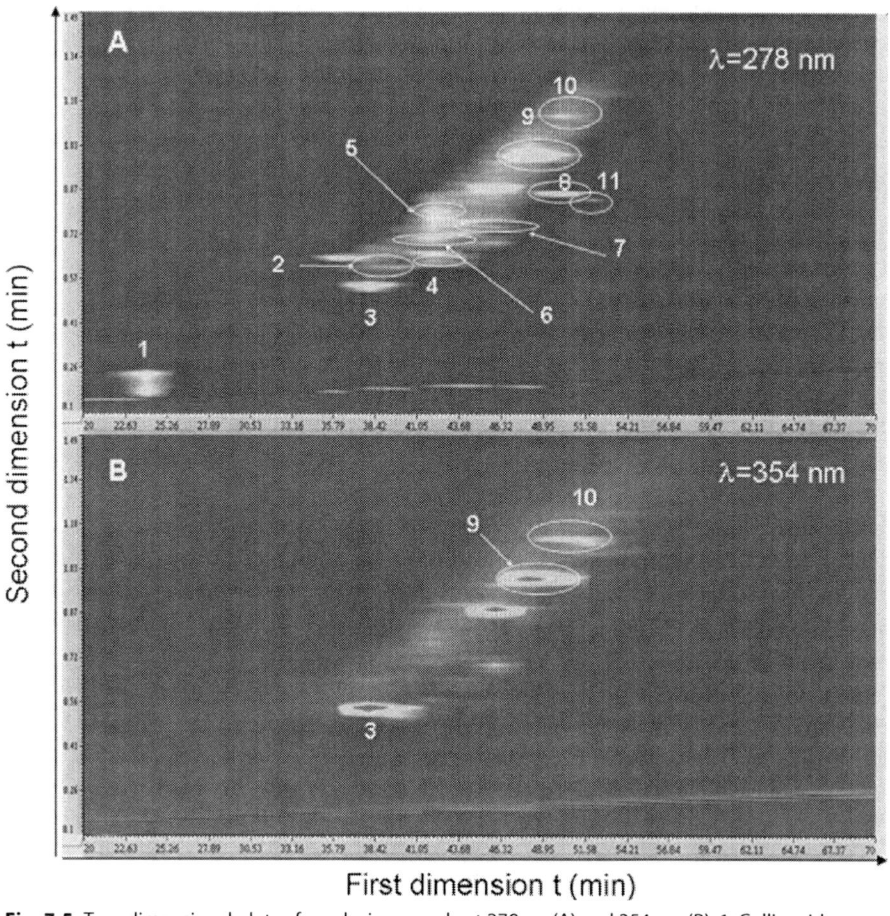

Fig. 7.5. Two-dimensional plots of a red wine sample at 278nm (A) and 354 nm (B). 1. Gallic acid; 2. Procyanidin; 3. Caftaric acid; 4: Tyrosol; 5: Epicatechin; 6: Catechin; 7: Caffeic acid; 8: Ethylgallate; 9: Rutin. Reprinted with permission from Dugo et al., 2009.

al., 2010). Pol et al. (2007) tested different combinations of cation-exchange, amino and octadecylsilane stationary phases tested in comprehensive LC mode for the separation of glycosides contained in an aqueous extract of *Stevia*, as alternative to synthetic sweeteners for treatment of diabetes, hypertension, and for prevention of caries (cavities). On the same topic, Cacciola et al. (2010) has recently reported an innovative 2D-LC platform based on the different selectivity of a polyamine column operated in normal phase mode in the first dimension and a UHPLC C18 column operated in reversed phase mode in the second dimension. To the best of the authors' knowledge, this is the first work that reports the use of UHPLC conditions in the second dimension performed on a short octadecylsilane column with sub 2-μm particles for separation of *Stevia* polyphenolic and stevioside compounds. Kivilompolo et al. (2007) developed a dynamic sonication assisted extraction method and compared it to traditional extraction methods for the recovery of phenolic acids from basil, oregano, rosemary, sage, spearmint and thyme of the *Lamiaceae* family. Selected antioxidative phenolic acids and flavonoids have also been determined from wines and juices by using a combination of two C18 columns, the latter with an ion-pair reagent (tetrapentylammoniumbromide) (2008a). In a recent work, Kivilompolo and Hyotilainen compared the separation power of ultra performance liquid chromatography and comprehensive two-dimensional liquid chromatography for the separation of phenolic compounds in wine and juices (2008b). In the LC×LC approach, the first dimension separation was based on RPLC and the second dimension was performed with ion-pair chromatography by employing three different columns, viz. two short columns packed with either 2.5 or 1.7 μm particles and a monolithic column. The best results were achieved when the monolithic column was employed for the second dimension separation, yielding a n_c of 616.

Data Analysis

Data analysis is a complicated task and requires dedicated software for the elaboration of two-dimensional or three-dimensional plots. As a consequence, analysis of data resulting from 2D-LC experiments is still a challenge, and great effort is required towards the development of effective algorithms for quantitative and qualitative purposes. Usually, data is visualized in 2D plots or contour plots, where the retention times in the first and the second dimensions are plotted along the *x*- and *y*-axes, respectively, while the color of the spots is a measure of intensity. Most developments have been carried out for software applied to gas-chromatography-based instrumentation rather than liquid-chromatography-based systems. One of the main reasons for this situation is related to the much more complicated background signal in 2D-LC rather than those in 2D-GC, at least for UV absorbance detection compared with flame-ionization detection (Reichenbach et al., 2005; Amandor-Munoz & Marriott, 2008; Harynuk et al., 2008; Pierce et al., 2008; Adcock et al., 2009).

As a consequence, the LC×LC methodology does suffer from limited availability of dedicated software essential for both instrumental control and data visualization.

Due to the lack of appropriate programs, the instrumentation in most laboratories is controlled by in-house written software capable of controlling the LC pumps, signal acquisition and switching of the interface valve. To our knowledge, only Shimadzu Corporation (Shimadzu, Kyoto, Japan) is providing a complete solution for comprehensive LC×LC. The main features of the acquisition software (LCsolution) from Shimadzu is not only the control of the valve but also the function that allows the user to write only two times the gradient of the second dimension and repeat it as many times as necessary by means of a loop function.

Quantification in comprehensive LC represents an additional problem and has been rarely reported. To this regard, Mondello and co-workers (2008; 2009a) reported the quantification of a series of standards, as well as real-world sample compounds, by using a dedicated laboratory-constructed LC×LC software program. The results obtained were compared with those observed in conventional LC, without showing statistically significant differences. The only drawback was the loss in sensitivity observed when using the LC×LC approach, compared to conventional LC, because of the sample dilution that occurs when using the former method.

So far, at least two commercially available software packages, namely ChromSquare® (Shimadzu, Kyoto, Japan) and LC Image (Zoex Corporation, Houston, Texas, USA), specially designed for analyzing data from LC×LC analyses, are available on the market. Additional work in this field is considered essential due to the increasing number of 2D-LC separations, especially if they are used more for quantitative rather than qualitative purposes.

Future Perspectives

Comprehensive LC has proven so far its viability as a powerful tool for separation of complex samples. An increasing number of new applications for various sample types is expected as soon as the new generation of instrumentation for 2D-LC becomes commercially available. Great effort will be devoted to the practical selection of separation systems with low selectivity correlations in the two dimensions and the development in this direction will progress with future introduction of new stationary phases. The commercialization within the last two years of HPLC pumping systems with small gradient delay volumes and small partially porous particles were very exciting developments. To this regard, progress in column technology will result in the commercialization of new types of columns with high efficiency, temperature stability, and permeability, enabling wider use of high-pressure and high-temperature fast second-dimension operation, aimed at increasing the overall peak capacity and speed in comprehensive LC. Along with the development of new 2D-LC methods, more robust and flexible software packages are considered necessary in order to handle the complexity of resulting 2D data for both qualitative and quantitative purposes.

References

Adcock J.L.; Adams, M.; Mitrevski, B.S.; Marriott, P.J. Peak modeling approach to accurate assignment of first-dimension retention times in comprehensive two-dimensional chromatography. Anal. Chem. 2009, 81, 6797-6804.

Alexander A.J.; Ma, L. Comprehensive two-dimensional liquid chromatography separations of pharmaceutical samples using dual Fused-Core columns in the 2nd dimension. J. Chromatogr A 2009, 1216, 1338–1345.

Amador-Munoz, O.; Marriot, P. Quantification in comprehensive two-dimensional gas chromatography and a model of quantification based on selected summed modulated peaks. J. Chromatogr. A 2008, 1184, 323-340.

Arab, L.; Steck, S. Lycopene and cardiovascular disease. Am. J. Clin. Nutr. 2000, 71, 1691S-1695S.

Berek, D. Two-dimensional liquid chromatography of synthetic polymers. Anal. Bioanal. Chem. 2010, 396, 421-441.

Beutner, S.; Bloedorn, B.; Frixel, S.; Hernandez Blanco, I.; Hoffman, T.;. Martin, H.-D; Mayer, B.; Noack, P.; Ruck, C.; Schmidt, M.; et al. Quantitative assessment of antioxidant properties of natural colorants and phytochemicals: carotenoids, flavonoids, phenols and indigoids. The role of β-carotene in antioxidant functions. J. Sci. Food Agric. 2001, 81, 559-568.

Blumberg L.M.; David, F.; Klee, M.S.; Sandra, P. Comparison of one-dimensional and comprehensive two-dimensional separations by gas chromatography. J. Chromatogr. A 2008, 1188, 2-16.

Cacciola, F.; Delmonte, P.; Jaworska, K.; Dugo, P.; Mondello, L.; Rader, J. Employing ultra high pressure liquid chromatography as the second dimension in a comprehensive two-dimensional system for analysis of Stevia rebaudiana extracts. J. Chromatogr. A 10.1016/j.chroma.2010.08.081 2010.

Cacciola F.; Jandera, P.; Mondello, L. Temperature effects on separations of phenolic antioxidants. J. Sep. Sci. 2007a, 30, 462–474.

Cacciola, F.; Jandera, P.; Hajdu, Z.; Cesla, P.; Mondello, L. Comprehensive two-dimensional liquid chromatography with parallel gradients for separation of phenolic and flavone antioxidants. J. Chromatogr. A 2007b, 1149, 73-87.

Cavazzini, A.; Gritti, F.; Kaczmarski, K.; Marchetti, N.; Guiochon, G. Mass-Transfer Kinetics in a Shell Packing Material for Chromatography. Anal. Chem. 2007, 79, 5972-5979.

Cesla, P.; Hájek T.; Jandera, P. Optimization of two-dimensional gradient liquid chromatography separations. J. Chromatogr A 2009, 1216, 3443–3457.

Cunliffe, J.M.; Maloney, T.D. Fused-core particle technology as an alternative to sub-2-μm particles to achieve high separation efficiency with low backpressure. J. Sep. Sci. 2007, 30, 3104-3109.

Dalluge, J.J.; Nelson, B.C.; Brown Thomas, J.; Sander, L.C. Selection of column and gradient elution system for the separation of catechins in green tea using high-performance liquid chromatography. J. Chromatogr. A 1998, 793, 265-274.

Davis, J.M. Statistical-overlap theory for elliptical zones of high aspect ratio in comprehensive two-dimensional separations. J. Sep. Sci. 2005, 28, 347-359.

Davis, J.M.; Stoll, D.R.; Carr, P.W. Effect of first-dimension undersampling on effective peak capacity in comprehensive two-dimensional separations. Anal. Chem. 2008a, 80, 461–473.

Davis, J.M.; Stoll, D.R.; Carr, P.W. Dependence of effective peak capacity in comprehensive two-dimensional separations on the distribution of peak capacity between the two dimensions. Anal. Chem. 2008b, 80, 8122-8134.

Dixon, S.P.; Pitfield, I.D.; Perrett, D. Comprehensive multi-dimensional liquid chromatographic separation in biomedical and pharmaceutical analysis: a review. Biomed. Chromatogr. 2006, 20, 508-529.

Donato, P.; Dugo, P.; Cacciola, F.; Dugo, G.; Mondello, L. High peak capacity separation of peptides through the serial connection of LC shell-packed columns. J. Sep. Sci. 2009, 32, 1129-1136.

Dugo, P.; Skerikova, V.; Kumm, T.; Trozzi, A.; Jandera, P.; Mondello, L. Elucidation of carotenoid patterns in citrus products by means of comprehensive normal-phase and reversed-phase liquid chromatography. Anal. Chem. 2006a, 78, 7743-7750.

Dugo, P.; Kumm, T.; Chiofalo, B.; Cotroneo, A.; Mondello, L. Separation of triacylglycerols in a complex lipidic matrix by using comprehensive two dimensional liquid chromatography coupled with atmospheric pressure chemical ionization mass spectrometric detection. J. Sep. Sci. 2006b, 29, 1146–1154.

Dugo, P.; Kumm, T.; Crupi, M.L.; Cotroneo, A.; Mondello, L. Comprehensive two-dimensional liquid chromatography combined with mass spectrometric detection in the analyses of triacylglycerols in natural lipidic matrixes. J. Chromatogr. A 2006c, 1112, 269-275.

Dugo, P.; Cacciola, F.; Kumm, T.; Dugo, G.; Mondello, L. Comprehensive multidimensional liquid chromatography: theory and applications. J. Chromatogr. A 2008a, 1184, 353-368.

Dugo, P.; Kumm, T.; Cacciola, F.; Dugo, G.; Mondello, L. Multidimensional liquid chromatographic separations applied to the analysis of food samples. J. Liq. Chromatogr. Relat. Technol. 2008b, 31, 1758-1807.

Dugo, P.; Cacciola, F.; Herrero, M.; Donato, P.; Mondello, L. Use of partially porous column as second dimension in comprehensive two-dimensional system for analysis of polyphenolic antioxidants. J. Sep. Sci. 2008c, 31, 3297-3308.

Dugo, P.; Herrero, M.; Kumm, T.; Giuffrida, D.; Dugo, G.; Mondello, L. Comprehensive normal-phase x reversed-phase liquid chromatography coupled to photodiode array and mass spectrometry detectors for the analysis of free carotenoids and carotenoid esters from mandarin. J. Chromatogr. A 2008d, 1139, 196-206.

Dugo, P.; Herrero, M.; Giuffrida, D.; Kumm, T.; Dugo, G.; Mondello, L. Application of comprehensive two-dimensional liquid chromatography to elucidate the native carotenoid composition in red orange essential oil. J. Agric. Food Chem. 2008e, 56, 3478-3485.

Dugo, P.; Cacciola, F.; Donato, P.; Airado-Rodriguez, D.; Herrero, M.; Mondello, L. Comprehensive two-dimensional liquid chromatography to quantify polyphenols in red wines. J. Chromatogr. A 2009a, 1216, 7483-7487.

Dugo, P.; Cacciola, F.; Donato, P.; Assis, R.A.; Caramao, E.B.; Mondello, L. High efficiency liquid chromatography techniques coupled to mass spectrometry for the characterization of mate extracts. J. Chromatogr. A 2009b, 1216, 7213-7221.

Dugo, P.; Herrero, M.; Giuffrida, D.; Donato, P.; Mondello, L. Epoxycarotenoids esters analysis in intact orange juices using two-dimensional comprehensive liquid chromatography. J. Sep. Sci. 2009c, 32, 973-980.

Erni, F.; Frei, R.W. Two-dimensional column liquid chromatographic technique for resolution of complex mixtures. J. Chromatogr. 1978, 149, 561–569.

Evans, C.R.; Jorgenson, J.W. Multidimensional LC-LC and LC-CE for high-resolution separations of biological molecules. Anal. Bioanal. Chem. 2004, 378, 1952-1961.

Fairchild J.N.; Horvath, K.; Guiochon, G. Theoretical advantages and drawbacks of on-line, multidimensional liquid chromatography using multiple columns operated in parallel. J. Chromatogr. A 2009, 1216, 6210–6217.

François, I.; Sandra, K.; Sandra, P. Comprehensive liquid chromatography: fundamental aspects and practical considerations-a review. Anal. Chim. Acta 2009a, 641, 14-31.

François, I.; Cabooter, D.; Sandra, K.; Lynen, F.; Desmet, G.; Sandra, P. Tryptic digest analysis by comprehensive reversed phase×two reversed phase liquid chromatography (RP-LC×2RP-LC) at different pH's. J. Sep. Sci. 2009b, 32, 1137–1144.

Giddings, J. C. Unified Separation Science; Wiley Interscience: New York, 1991.

Giddings, J.C. Two-dimensional separations: Concept and promise. Anal. Chem. 1984, 56, 1258A-1260A, 1262A.

Giddings, J.C. Concepts and comparisons in multidimensional separation. J. High Resolut. Chromatogr. Chromatogr. Commun, 1987, 10, 319-323.

Gritti, F.; Cavazzini, A.; Marchetti, N.; Guiochon, G. Comparison between the efficiencies of columns packed with fully and partially porous C18-bonded silica materials. J. Chromatogr. A 2007a, 1157, 289-303.

Gritti, F.; Guiochon, G. Comparative study of the performance of columns packed with several new fine silica particles: Would the external roughness of the particles affect column properties? J. Chromatogr. A 2007b, 1166, 30-46.

Guiochon, G.; Marchetti, N.; Mriziq, K.; Shalliker, R.A. Implementations of two-dimensional liquid chromatography. J. Chromatogr. A 2008, 1189, 109-168.

Guttman, A.; Varoglu, M.; Khandurina, J. Multidimensional separations in the pharmaceutical arena. Drug Discov. Today 2004, 9, 136-144.

Hajek, T.; Skerikova, V.; Cesla, P.; Vynuchalova, K.; Jandera, P. Multidimensional LC×LC analysis of phenolic and flavone natural antioxidants with UV-electrochemical coulometric and MS detection. J. Sep. Sci. 2008, 31, 3309-3328.

Harynuk, J.J.; Kwong, A.H.; Marriott, P.J. Modulation-induced error in comprehensive two-dimensional gas chromatographic separations. J. Chromatogr A 2008, 1200, 17–27.

Huidobro, A.L.; Pruim, P.; Schoenmakers, P.; Barbas, C. Ultra rapid liquid chromatography as second dimension in a comprehensive two-dimensional method for the screening of pharmaceutical samples in stability and stress studies. J. Chromatogr. A 2008, 1190, 182–190.

Herrero, M.; Jaime, L.; Martin-Alvarez, P.J.; Cifuentes, A.; Ibanez, E. Optimization of the extraction of antioxidants from Dunaliella salina microalga by pressurized liquids. J. Agric. Food Chem. 2006, 54, 5597-5603.

Herrero, M.; Ibanez, E.; Cifuentes, A.; Bernal, J. Multidimensional chromatography in food analysis. J. Chromatogr. A 2009, 1216, 7110-7129.

Hertog, M.G.L.; Hollman, P.C.H.; Van de Putte, B; Content of potentially anticarcinogenic flavonoids of tea infusions, wines, and fruit juices. J. Agric. Food Chem. 1993, 41, 1242–1246.

Holčapek, M.; Velinska, H.; Lisa, M.; Česla, P. Orthogonality of silver-ion and non-aqueous reversed-phase HPLC/MS in the analysis of complex natural mixtures of triacylglycerols. J. Sep. Sci. 2009, 32, 3672-3680.

Horie, K.; Kimura, H.; Ikegami, T.; Iwatsuka, A.; Saad, N.; Fiehn, O.; Tanaka, N. Calculating optimal modulation periods to maximize the peak capacity in two- dimensional HPLC. Anal. Chem. 2007, 79, 3764–3770.

Ikegami, T.; Hara, T.; Kimura, H.; Kobayashi, H.; Hosoya, K.; Cabrera, K.; Tanaka, N. Two-dimensional reversed-phase liquid chromatography using two monolithic silica C18 columns and different mobile phase modifiers in the two dimensions. J. Chromatogr. A 2006, 1106, 112–117.

Karger, B.L.; Snyder, L.R.; Horvath, C. An introduction to separation science; Wiley & Sons: New York, 1973.

Kivilompolo, M.; Hyotylainen, T. Comprehensive two-dimensional liquid chromatography in analysis of Lamiaceae herbs: Characterisation and quantification of antioxidant phenolic acids. J. Chromatogr. A 2007, 1145, 155-167.

Kivilompolo, M.; Oburka, V.; Hyotylainen, T. Comprehensive two-dimensional liquid chromatography in the analysis of antioxidant phenolic compounds in wines and juices. Anal. Bioanal. Chem. 2008, 391, 373-380.

Kivilompolo, M.; Hyotylainen, T. Comparison of separation power of ultra performance liquid chromatography and comprehensive two-dimensional liquid chromatography in the separation of phenolic compounds in beverages. J. Sep. Sci. 2008, 31, 3466-3472.

Jandera, P. Predictive calculation methods for optimization of gradient elution using binary and ternary solvent gradients. J. Chromatogr. 1989, 485, 113-141.

Jandera, P. Column selectivity for two-dimensional liquid chromatography. J. Sep. Sci. 2006, 29, 1763-1783.

Jandera P.; Cesla, P.; Hajek, T.; Vohralik, G.; Vynuchalova, K.; Fischer, J. Optimization of separation in two-dimensional high-performance liquid chromatography by adjusting phase system selectivity and using programmed elution techniques. J. Chromatogr A 2008, 1189, 207–220.

Leach, G.; Olivera, R.; Morais, R. Production of a carotenoid-rich product by alginate entrapment and fluid-bed drying of Dunaliella Salina. J. Sci. Food Agric. 1998, 76, 298-302.

Li, X.; Stoll D.R.; Carr, P.W. Equation for peak capacity estimation in two- dimensional liquid chromatography. Anal. Chem. 2009, 81, 845-850.

Liu, Z.; Lee, M.L. Comprehensive Two-dimensional Separations Using Microcolumns. J. Microcolumn Sep. 2000, 12, 241-254.

Lubda, D.; Cabrera, K.; Kraas, W.; Shafer, C.; Cunningham, D. New developments in the application of monolithic HPLC columns. LC GC Eur. 2001, 14, 730-734.

Marchetti, N.; Cavazzini, A.; Gritti, F.; Guiochon, G. Gradient elution separation and peak capacity of columns packed with porous shell particles. J. Chromatogr. A 2007, 1163, 203-211.

Mendes-Pinto, M.M.; Silva Ferreira, A.C.; Caris-Veyrat, C.; Guedes de Pinto, P. Carotenoid, chlorophyll, and chlorophyll-derived compounds in grapes and Port wines. J. Agric. Food Chem. 2005, 53, 10034-10041.

Mondello, L.; Tranchida, P.Q.; Stanek, V.; Jandera, P.; Dugo, G.; Dugo, P. Silver-ion reversed-phase comprehensive two-dimensional liquid chromatography combined with mass spectrometric detection in lipidic food analysis. J. Chromatogr. A 2005, 1086, 91–98.

Mondello, L.; Herrero, M.; Kumm, T.; Dugo, P.; Cortes, H.; Dugo, G. Quantification in comprehensive two-dimensional liquid chromatography. Anal. Chem. 2008, 80, 5418-5424.

Mondello, L.; Donato, P.; Cacciola, F.; Fanali, C.; Dugo, P. RP-LC×RP-LC analysis of a tryptic digest using a combination of totally porous and partially porous stationary phases. J. Sep. Sci. 2010, 33, 1454–1461.

Murphy, R.E.; Schure, M.R.; Foley, J.P. Effect of sampling rate on resolution in comprehensive two-dimensional liquid chromatography. Anal. Chem. 1998, 70, 1585-1594.

Nishino, H.; Tokuda, H.; Satomi, Y.; Masuda, M.; Bu, P.; Onozuka, M.; Yamaguchi, S.; Okuda, Y.; Takayasu, J.; Tsuruta, J.; et al. Cancer prevention by carotenoids. Pure Appl. Chem. 1999, 71, 2273-2278.

Opiteck, G.J.; Jorgenson, J.W.; Anderegg, R.J. Two-dimensional SEC/RPLC coupled to mass spectrometry for the analysis of peptides. Anal. Chem. 1997, 69, 2283–2291.

Pierce, K.M.; Hoggard, J.C.; Mohler, R.E.; Synovec, R.E. Recent advancements in comprehensive two-dimensional separations with chemometrics. J. Chromatogr A 2008, 1184, 341-352.

Pol, J.; Hohnova, B.; Hyotylainen, T. Characterization of Stevia Rebaudiana by comprehensive two-dimensional liquid chromatography time-of-flight mass spectrometry. J. Chromatogr. A 2007, 1150, 85-92.

Pol, J.; Hyotylainen, T. Comprehensive Two-Dimensional Liquid Chromatography Coupled with Mass Spectrometry. Anal. Bioanal. Chem. 2008, 391, 21-31.

Potts, L.W.; Stoll, D.W., Li, X., Carr, P.W. The impact of sampling time on peak capacity and analysis speed in on-line comprehensive two-dimensional liquid chromatography. J. Chromatogr. A 2010, 1217, 5700–5709.

Rao, A.V.; Rao, L.G. Carotenoids and human health. Pharmacol. Res. 2007, 55, 207-216.

Reichenbach, S.E.; Kottapalli, V.; Ni, M.; Visvanathan, A. Computer language for identifying chemicals with comprehensive two-dimensional gas chromatography and mass spectrometry. J. Chromatogr. A 2005, 1071, 263–269.

Sandra, K., Moshir M.; D'hondt, F.; Tuytten, R.; Verleysen, K.; Kas, K.; François, I.; Sandra, P. Highly efficient peptide separations in proteomics. Part 2: bi- and multidimensional liquid-based separation techniques. J. Chromatogr. B 2009, 877, 1019-1039.

Schoenmakers, P.J. Optimization of Chromatographic Selectivity; Elsevier: Amsterdam, 1986, 200.

Schoenmakers, P.J.; Vivo-Truyols, G.; Decrop, W.M.C. A protocol for designing comprehensive two-dimensional liquid chromatography separation systems. J. Chromatogr. A 2006, 1120, 282–290.

Schure, M.R. Advances in HPLC, LCGC Supplement, 2010.

Seeley, J.V. Theoretical study of incomplete sampling of the first dimension in comprehensive two-dimensional chromatography. J. Chromatogr. A 2002, 962, 21-27.

Shalliker, R.A.; Gray, M.J. Concepts and practice of multidimensional high- performance liquid chromatography. Adv. Chromatogr. 2006, 44, 177-236.

Shellie, R.A., Haddad, P.R. Comprehensive two-dimensional liquid chromatography. Anal. Bioanal. Chem. 2006, 386, 405-415.

Snodderly, M.D. Evidence for protection against age-related macular degeneration by carotenoids and antioxidant vitamins. Am. J. Clin. Nutr. 1995, 62, S1448-S1461.

Stroink, T.; Ortiz, M.C.; Bult, A.; Lingeman, H.; de Jong, G.J.; Underberg, W.J.M. On- line multidimensional liquid chromatography and capillary electrophoresis systems for peptides and proteins. J. Chromatogr. B 2005, 817, 49-66.

Stoll, D.R.; Cohen, J.D.; Carr, P.W. Fast, comprehensive online two-dimensional high performance liquid chromatography through the use of high temperature ultra-fast gradient elution reversed-phase liquid chromatography. J. Chromatogr. A 2006, 1122, 123-137.

Stoll, D.R.; Li, X.; Wang, X.; Carr, P.W.; Porter, S.E.G.; Rutan, S.C. Fast Comprehensive Two-Dimensional Liquid Chromatography. J. Chromatogr. A 2007, 1168, 3-43.

Stoll, D.R.; Wang, X.; Carr, P.W. Comparison of the practical resolving power of one- and two-dimensional high-performance liquid chromatography analysis of metabolomic samples. Anal. Chem. 2008, 80, 268–278.

Tanaka, N.; Kimura, H.; Tokuda, D.; Hosoya, K.; Ikegami, T.; Ishizuka, N.; Minakuchi, H.; Nakanishi, K.; Shintani, Y.; Furuno, M.; Cabrera, K. Simple and comprehensive two-dimensional reversed-phase HPLC using monolithic silica columns. Anal. Chem. 2004, 76, 1273–1281.

Toll, H.; Oberacher, H.; Swart, R.; Huber, C.G. Separation, detection, and identification of peptides by ion-pair reversed-phase high-performance liquid chromatography-electrospray ionization mass spectrometry at high and low pH. J. Chromatogr. A 2005, 1079, 274-286.

Tsuchiya, H.; Sato, M.; Kato, H.; Okubo, T.; Juneja, L.R.; Kim, M. Simultaneous determination of catechins in human saliva by high performance liquid chromatography. J. Chromatogr. B 1997, 703, 253-258.

Van der Klift, E.J.C.; Vivo-Truyols, G.; Claassen, F.W.; Van Holthoon, F.L.; Van Beek, T.A. Comprehensive two-dimensional liquid chromatography with ultraviolet, evaporative light scattering and mass spectrometric detection of triacylglycerols in corn oil. J. Chromatogr. A 2008, 1178, 43–55.

Way, W. The Reporter USA, 2007a, 25, 6–7.

Way, W.K. The Reporter Europe, 2007b, 28, 3-4.

8

MULTIPLE PARALLEL MASS SPECTROMETRY TECHNIQUES FOR LIPID AND VITAMIN D ANALYSIS

WILLIAM CRAIG BYRDWELL
USDA, ARS, FOOD COMPOSITION AND METHODS DEVELOPMENT LABORATORY, BELTSVILLE, MD

Introduction

Lipid analysis is a complex, even daunting, undertaking due to the wide range of classes containing fatty chains, and the variety of chain lengths and degrees of unsaturation (number of double bonds) encountered. Lipid classes range from non-polar molecules such as triacylglycerols (TAG) and sterols to polar molecules such as saccharolipids (liposaccharides) to zwitterionic or charged molecules such as phospholipids (PLs). Table 8.1 lists the lipids that were defined by "A Comprehensive Classification Scheme for Lipids" (Fahy et al., 2005) and its update (Fahy et al., 2009). Given the wide range of polarities, a single liquid chromatographic approach is not suitable for all classes of lipids. Typically, non-polar molecules such as triacylglycerols are separated using reversed-phase (RP) high performance liquid chromatography (HPLC) [often non-aqueous reversed-phase (NARP) HPLC], while polar molecules are often separated by normal-phase (NP) chromatography. Of course, this is merely a generalization, since some classes may be separated by either approach. For instance, a separation of phospholipid classes is often done by NP-HPLC, while molecular species within a class may be separated by RP-HPLC. Similarly, the class of secosteroids known as "vitamin D" may be separated from other classes by NP-HPLC, while species within the class, i.e., vitamin D_2 and vitamin D_3, may be separated by RP-HPLC. In some cases, the best separation is a compromise between a separation of classes and a separation of molecular species, such as the separation of sphingomyelin from dihydrosphingomyelin using an amine column.

Just as the wide range of polarities of lipid classes means that no one type of chromatography is universally applicable, it is also true that no one type of mass spectrometry technique is universally applicable to all lipids analyzed by LC-MS. Some molecules, such as phospholipids, which are easily ionized or are already ionic, are very amenable to electrospray ionization (ESI) MS. While other classes, such as triacylglycerols and secosteroids do not ionize well by ESI-MS unless special measures are taken. On the other hand,

Table 8.1. Lipid Classes as Defined by "A Comprehensive Classification Scheme for Lipids"[a] and LIPID MAPS[b].

Fatty Acyls [FA]	LM_ID[c]	Glycerophospholipids [GP]	
Fatty Acids and Conjugates	FA01	Glycerophosphocholines	GP01
Octadecanoids	FA02	Glycerophosphoethanolamines	GP02
Eicosanoids	FA03	Glycerophosphoserines	GP03
Docosanoids	FA04	Glycerophosphoglycerols	GP04
Fatty alcohols	FA05	Glycerophosphoglycerophosphates	GP05
Fatty aldehydes	FA06	Glycerophosphoinositols	GP06
Fatty esters	FA07	Glycerophosphoinositol monophosphates	GP07
Fatty amides	FA08	Glycerophosphoinositol bisphosphates	GP08
Fatty nitriles	FA09	Glycerophosphoinositol trisphosphates	GP09
Fatty ethers	FA10	Glycerophosphates	GP10
Hydrocarbons	FA11	Glyceropyrophosphates	GP11
Oxygenated hydrocarbons	FA12	Glyceroph osphoglycerophospho-glycerols	GP12
Fatty acyl glycosides	FA13	CDP-Glycerols	GP13
Other Fatty Acyls	FA00	Glycosylglycerophospholipids	GP14
Glycerolipids [GL]		Glycerophosphoinositolglycans	GP15
Monoradylglycerols	GL01	Glycerophosphonocholines	GP16
Diradylglycerols	GL02	Glycerophosphonoethanolamines	GP17
Triradylglycerols	GL03	Di-glycerol tetraether phospholipids	GP18
Glycosylmonoradylglycerols	GL04	Glycerol-nonitol tetraether phospholipids	GP19
Glycosyldiradylglycerols	GL05	Oxidized glycerophospholipids	GP20
Other Glycerolipids	GL00	Other Glycerophospholipids	GP00

(continued on next page)

these classes are readily ionized by atmospheric pressure chemical ionization (APCI) MS and produce valuable fragments in the ionization source. The two techniques of ESI-MS and APCI-MS have become the most popular ionization methods for coupling liquid chromatography to mass spectrometry. There are some classes of molecules, i.e. phospholipids, which produce excellent, but complementary, data by both techniques. A third atmospheric pressure ionization technique is gaining popularity, which is atmospheric pressure photoionization (APPI). This technique typically requires a dopant that is

Table 8.1. *(continued)*

Sphingolipids [SL]	LM_ID	**Saccharolipids [SL]**	
Sphingoid bases	SL01	Acylaminosugars	SL01
Ceramides	SL02	Acylaminosugar glycans	SL02
Phosphosphingolipids	SL03	Acyltrehaloses	SL03
Phosphonosphingolipids	SL04	Acyltrehalose glycans	SL04
Neutral glycosphingolipids	SL05	Other acyl sugars	SL05
Acidic glycosphingolipids	SL06	Other Saccharolipids	SL00
Basic glycosphingolipids	SL07	**Polyketides [PK]**	
Amphoteric glycosphingolipids	SL08	Linear polyketides	PK01
Arsenosphingolipids	SL09	Halogenated acetogenins	PK02
Other Sphingolipids	SL00	Annonaceae acetogenins	PK03
Sterol Lipids [ST]		Macrolides and lactone polyketides	PK04
Sterols	ST01	Ansamycins and related polyketides	PK05
Steroids	ST02	Polyenes	PK06
Secosteroids	ST03	Linear tetracyclines	PK07
Bile acids and derivatives	ST04	Angucyclines	PK08
Steroid conjugates	ST05	Polyether antibiotics	PK09
Other Sterol lipids	ST00	Aflatoxins and related substances	PK10
Prenol Lipids [Pr]		Cytochalasins	PK11
Isoprenoids	PR01	Flavonoids	PK12
Quinones and hydroquinones	PR02	Aromatic polyketides	PK13
Polyprenols	PR03	Non-ribosomal peptide/polyketide hybrids	PK14
Hopanoids	PR04	Other Polyketides	PK00
Other Prenol lipids	PR00		

[a]Fahy et al., 2005; Fahy et al., 2009; [b]LIPID MAPS, www.lipidmaps.org; [c]LIPID MAPS identifier

itself ionized by APPI, which then ionizes the analyte. In a few special cases this can produce better sensitivity toward some classes than APCI. However, in many cases the sensitivity is the same or less than APCI, thereby offering no clear advantage. Given that the use of a dopant makes for a more complicated experimental arrangement, the potential additional benefits often do not outweigh the inherent additional requirements. Furthermore, the APPI lamp is easily damaged, e.g., by stray cold solvent used to clean the ionization source. Thus, for the classes of molecules analyzed in our laboratory, APPI has not shown sufficient benefit to overcome the disadvantages, leading us to uninstall the APPI source from our combination APPI/APCI source.

Since no one MS technique is by itself ideal, and ESI-MS and APCI-MS produce complementary data, we have found it very valuable to obtain both types of data for every type of analyte that we can. Normally, this would mean that separate chromatographic runs would be conducted, with changing the ionization source between runs (as we did in early published results). There are some 'combination sources' available that allow switching between ESI and APCI modes, but these have several drawbacks. First is the cost of the combination source, which is certainly higher than either individual source. However, since a substantial commitment of resources is already being made for any LC-MS method, the cost differential of the ion source should not be considered a limiting factor. Instead, there are two other considerations that hinder widespread adoption of combination sources. The first is that the source conditions for each type of ionization are usually optimized to produce maximum signal and sensitivity for that source type, and the optimized conditions between the sources are different (i.e., APCI operates at elevated temperature, while ESI often optimizes at a lower temperature). Thus, compromise conditions must be used, such that neither source operates under optimal conditions. This means that maximum sensitivity is not obtained, and mass spectra may contain undesirable adducts. But more importantly, higher temperatures often lead to more fragmentation, meaning that ESI-MS may produce more fragments than is typical for this technique at lower temperatures. Spectra under compromise conditions are often not the same as either technique individually, which affects how results under these conditions compare to results in the literature. The second important consideration is the limited duty cycle of the mass spectrometer. As mass spectrometers become faster and faster, this is becoming less of a problem, but it seems that there is never enough time across a chromatographic peak to perform all of the scan functions desired [i.e., full-scan, selected reaction monitoring (SRM), selected ion monitoring (SIM), MS/MS, MS^3, MS^4, positive mode, negative mode, etc.]. It is simply a fundamental principle that, since the signal-to-noise ratio (S/N) increases with the square root of the number of scans (ideally), if the number of scans is less, the S/N is lower. Thus, if the duty cycle is divided between two different ionization types, the number of scans for each type is less and the S/N is less. If two ionization types are used with equivalent scan types, half as many scans can be obtained for each ionization mode, compared to each mode used individually. Therefore, the ideal situation would be to obtain data for each ionization type separately, resulting in the best S/N for each instrument.

Based on all of these considerations, we have developed techniques that employ multiple mass spectrometers acquiring data simultaneously in parallel using a variety of ionization modes and scan types to provide complementary data, while still maintaining optimal conditions for each, and also allowing the duty cycle of each mass spectrometer to be used for maximum effect, to provide the maximum amount of desired information about the analytes. Thus, ours are not examples of extreme chromatographic conditions (high temperature, UHPLC, etc.) as are others in this volume, but are examples of extreme detection methods (e.g., six detectors, including three mass spectrometers in parallel, single-wavelength and full scan UV, and simultaneous detection by an ELSD and a CAD). Our version of "extreme chromatography" is done by unconventional combinations of conventional instruments rather than techniques that require special instrumentation or components. Among the many benefits of the techniques demonstrated herein is the ability to use older and legacy instruments to provide valuable supplementary data, long after what might be considered their normal operational lifetimes.

Literature reports of liquid chromatography with detection by dual parallel mass spectrometry are scarce (search of parallel mass spectrometry AND ESI AND APCI). Our first application seems to be the first to combine APCI-MS and ESI-MS for simultaneous analysis of lipids. Most applications have used sequential ESI-MS and APCI-MS, or APCI and ESI in a mode-switching process to alternate between APCI and ESI using the combination sources mentioned above, which have the same shortcomings. A multimode ESI and APCI source by Agilent was discussed in the trade literature (Duncan & Perkins, 2005), but this source uses ESI for molecules that respond to that ionization and only directs neutral, unresponsive molecules to the APCI region of the source. In contrast, our approach is a truly parallel technique in which all molecules are directed to both ESI and APCI sources simultaneously.

In addition to the MS considerations, multiple options regarding the liquid chromatographic separation of lipids are shown. For some classes of lipids, a single separation is used that is coupled to multiple mass spectrometers, in what we define as LC1/MS2 or LC1/MS3 configurations, depending on whether two or three mass spectrometers are used in parallel. When disparate classes are present together, as in a total lipid extract such as that given by a chloroform/methanol extract (Bligh & Dyer, 1959; Folch et al., 1957), both NP-HPLC and RP-HPLC can be used in a column-switching arrangement, with one or more mass spectrometers acting as a detector(s) for each HPLC system. This is defined as an LC2/MS2 or LC2/MS*n* experimental arrangement. Several examples of LC*x*/MS*y* experiments are given here, including LC1/MS2, LC1/MS3, and LC2/MS2. But before the full value of these data can be appreciated, it is necessary to discuss the duty cycle of the mass spectrometer as it applies to lipidomics.

Holistic Lipidomics

As discussed by Spener et al. (2003), the full range of lipids in a system may be called the 'lipidome'. And when lipids are considered along with processes in which they are involved, including transport, enzymatic activity, co-factors, protein interactions, and other associated mechanisms, the whole area of the analysis of lipids and related anabolic and catabolic constituents comprises the field of 'lipidomics'. Within the area of lipidomics are the studies of sets of lipids within each class, such as the 'triacylglycerol lipidome' (Byrdwell, 2005b), called 'triacylglycerolomics' (Cvacka et al., 2006), the 'phospholipidome', called 'phospholipidomics' (Fonteh et al., 2006; Willmann et al., 2008), the 'sphingolipidome', called 'sphingolipidomics' (Maceyka et al., 2005; Merrill et al., 2005), and others. Multiple parallel mass spectrometry techniques discussed herein have been applied in each of these areas.

Lipidomics may be divided into two categories: survey lipidomics and targeted lipidomics. Survey lipidomics involves the use of full-scan MS, without the prior selection of masses for use in MS/MS. Conversely, targeted lipidomics may be used to indicate any type of analysis in which specific masses are pre-programmed into the mass spectrometer as distinct targets for analysis. Targeted analysis requires prior knowledge of all masses of interest, and can be carried out using selected ion monitoring (SIM), selected reaction monitoring (SRM), or any of the normal MS/MS modes that employ a precursor or product ion that is specified in advance, i.e. product ion scanning, precursor ion scanning, or neutral loss scanning. Targeted lipidomic approaches obviously require prior knowledge of the masses associated with all analytes of interest, and they provide a higher degree of specificity than survey approaches. Targeted lipidomics also almost always produces greater sensitivity, since the duty cycle of the mass spectrometer is used for a specific group of masses, allowing more time or more scans at each selected mass, leading to an improved S/N. Of course, the greatly reduced background noise of techniques such as SRM also contributes to an improved S/N. A good example of an extremely effective targeted lipidomics approach is the 'shotgun lipidomics' technique first demonstrated by Han and Gross (2005a; 2005b).

While extremely effective for known target analytes, the targeted approach has some limitations. First and most important, the technique does not identify species for which masses are not pre-programmed into the instrument. Examples are cited below that demonstrate the value of using a more holistic approach to take advantage of the sensitivity offered by a targeted technique while still providing survey data that allow detection and characterization of unexpected or unidentified species. Second, some targeted approaches require the use of newer, faster, more sensitive instruments, whereas the advantages of using older instruments as auxiliary detectors in a multiple parallel arrangement is also demonstrated. Third, targeted analysis often uses only a subset of the fragments produced either in the source or the collision cell, meaning that valuable structural information may go unexploited, or some of the signal simply goes to waste. Fourth, targeted data do not act as a repository of unexamined species that can be accessed at a future date. If a problem or question arises, full-scan data can be re-examined to identify masses corresponding to species that were initially not recognized or viewed as important.

This chapter demonstrates the advantages of obtaining targeted lipidomics data, to take advantage of the concomitant sensitivity and specificity, while at the same time conducting survey lipidomic analysis, to take advantage of the power of that technique for qualitative detection and/or identification of unexpected species. In short, without survey data in combination with targeted analyses, one 'never knows what one is missing'. The approach of combining targeted lipidomics with survey lipidomics, either on a single instrument or on multiple instruments in parallel, gives a more holistic data set that offers the advantages of both techniques. We refer to this combination of targeted and survey lipidomics as 'holistic lipidomics', since it provides 'the whole picture' of the lipidome. Of course lipidomics is not solely limited to mass spectrometry. Other combinations of targeted and survey lipidomic techniques, such as those that include NMR, would fall into the category of holistic lipidomics as we use the term.

It should be mentioned that others (Navas-Iglesias et al., 2009) have published a division of lipidomics into untargeted lipidomics (equivalent to survey lipidomics), targeted lipidomics, focused lipidomics and shotgun lipidomics. The reasons we do not use those divisions are i) selected ion monitoring (SIM) is not included in any category, ii) SRM and MRM are categorized differently from product ion, precursor ion, and neutral loss scanning, even though these all require similar prior knowledge and are targeted at specific molecules (or classes for neutral loss scanning), iii) shotgun lipidomics is treated as a separate category, although it typically employs neutral loss scanning and other conventional MS/MS modes, iv) shotgun lipidomics is shown as occurring prior to the ionization source (at the same level as LC/MS), leading to either untargeted, focused or targeted lipidomics. For these and other reasons, the divisions proposed by those authors have not been adopted in our lab. For the reasons discussed throughout this section, the terms survey lipidomics and targeted lipidomics are used as described in the preceding paragraphs.

Experimental

Numerous combinations of mass spectrometers are described that have been used for multiple parallel LC-MS experiments. The instruments used include an SSQ 710C single quadrupole mass spectrometer [Thermo Scientific (formerly Finnigan MAT), San Jose, CA] for APCI-MS; a TSQ 700 tandem sector quadrupole mass spectrometer (Thermo Scientific, San Jose, CA) for ESI-MS, APCI-MS, and MS/MS; an LCQ Deca ion trap mass spectrometer (Thermo Scientific, San Jose, CA) for ESI-MS^n (n= up to 4); a TSQ 7000 tandem sector quadrupole mass spectrometer (Thermo Scientific, San Jose, CA) for APCI-MS and selected ion monitoring (SIM); an LCQ Deca XP (Thermo Scientific, San Jose, CA) ion trap mass spectrometer for ESI-MS^n (n= up to 4); and a QTrap 5500 (AB SCIEX, Foster City, CA) for APCI-MS, SIM and selected reaction monitoring (SRM). Specific liquid chromatography and mass spectrometry conditions that have been previously reported are not given here. Only components not previously described in detail are given here.

The LCx/MSy Switching Panel

Early experiments in our lab used an ad-hoc arrangement, in which the contact closures from the LC systems, which provide the start signal for all other instruments, were 'daisy-chained' from one instrument to another. When we had an opportunity to set up a new laboratory, it was arranged with multiple parallel mass spectrometry experiments in mind. The instruments were situated with their ionization sources as close as practical to each other, with several available liquid chromatographs situated in close proximity. Thus, the whole lab was setup to facilitate LC*x*/MS*y* experiments. We have made progressive improvements in the arrangement, so now three liquid chromatographs (two for HPLC and one for UHPLC) provide contact closure inputs to a series of switches, shown in Fig. 8.1. These contact closures can be distributed to any of four mass spectrometers, as well as to an SS420X external analog-to-digital converter (ADC) [Grace (formerly Alltech, Inc.), Deerfield, IL], to the corona charged aerosol detector (CAD) (ESA, Inc., Chelmsford, MA) to provide an autozero signal, and to two spare switches used for other devices such as the AB140B syringe pump (Applied Biosystems, Foster City, CA). The liquid chromatographs are interconnected to allow the LC instruments to synchronize in column-switching experiments up to LC3/MS4.

The key to our multiple parallel mass spectrometry arrangement is the switch panel that distributes contact closure signals from the autosamplers. The switches are all center-off double-pole double-throw switches. Note that the terminals have the opposite orientation from the switch handle (the up position activates the lower terminal and vice versa). Twenty-gauge speaker wire was used for all connections. The first two switches have the contact closures from the relay contacts outputs of the autosamplers attached as inputs to the upper and lower terminals, and the center terminals act as the outputs to the other devices. The upper terminal of the first switch (SW1) has inputs coming from the Agilent 1200 HPLC autosampler (Agilent, Inc., Santa Clara, CA) relay contact board, and the lower terminal is the input from the Agilent 1290 UHPLC autosampler relay contact board. The second switch (SW2) selects either the relay contact from the Thermo Separation Products (TSP) AS3000 autosampler (Thermo Fisher Scientific, San Jose, CA) or a fixed closure (for testing and for immediate contact closure). The third switch (SW3) has the center output from SW1 attached as an input to its central terminal, with the upper terminal then attached to the remote input of the Agilent 1200 HPLC (pin 1 = ground, pin 3 = start), and the lower terminal attached to the remote input of the Agilent 1290 (pins 1 and 3). In this way, either the 1200 or the 1290 can start its counterpart, the 1290 or 1200.

The central terminal (output) of SW1 is then connected to the lower ter-

Fig. 8.1. Switching panel that joins incoming contact closures from LC autosamplers to the relay start inputs of various instruments, analog-to-digital converter (ADC), and other detectors. Switch 1 (SW1) selects either the Agilent 1200 HPLC or 1290 UHPLC, SW2 selects the TSP HPLC or a contact test. SW3 allows the 1200 to start the 1290 and vice versa. The start signal from SW1 is directed to any instrument by selecting the UP positions of SW4 to SW12. The start signal from SW2 is directed to any instrument by selecting the DOWN positions of SW4 to SW12.

minals (inputs) of the remaining nine switches (SW4-SW12) (soldered daisy-chained). The central terminal (output) of SW2 is connected to the upper terminals (inputs) of the remaining nine switches (SW4-SW12). The central terminals from switches SW4 to SW12 act as outputs to the QTrap 5500 mass spectrometer (SW4), TSQ 7000 mass spectrometer (SW5), LCQ Deca XP mass spectrometer (SW6), TSQ 700 mass spectrometer (SW7), TSP P4000 pump input (SW8), SS420X ADC start (SW9), and corona CAD autozero (SW10), with two spare switches (SW11, SW12). Using this panel, any LC*x*/MS*y* experiment from LC1/MS1 to LC3/MS4 can be selected.

Analog-to-Digital Converters

As mentioned above, our instruments include an SS420X ADC which is connected to the TSQ 7000 mass spectrometer. 1V analog signal outputs from an evaporative light scattering detector (ELSD 800, Grace, Deerfield, IL) and from the corona CAD are attached to two of the four available inputs to the 24-bit ADC. Data are acquired and processed using the XCalibur™ data acquisition system. This software package is the most user-friendly and feature-rich of the acquisition packages available in our lab.

For maximum flexibility, the signals from these two auxiliary detectors are also input to an Agilent 35900E, which is a 24-bit two-channel ADC. Data acquisition from the 35900E ADC on the Agilent 1200 is done using Chemstation software, but the signal from the corona CAD is also acquired (and the detector is controlled) by a USB connection to the Chemstation acquisition computer. The software treats these two signals from the same detector differently, and the 35900E ADC signal is cut off when it becomes negative.

The signals from the two auxiliary detectors are also input to the QTrap 5500 acquisition computer. Unfortunately, the only ADC compatible with that system is an older single-channel 16-bit ADC (PCI-DAQ16, National Instruments, Austin, TX). Therefore, a center-off double-pole double-throw switch (not shown) was connected to the screw terminal connector block to select either the ELSD or the corona CAD signal (to the outer terminals of the switch) for input to the DAQ16 (from the central terminal). Due to the limited resolution of this ADC and some software shortcomings, Analyst® 1.5.1 is used for acquisition of the ELSD or corona CAD only if neither of the other ADCs is participating in an experiment.

Because the auxiliary detectors are connected to three ADCs simultaneously, we retain maximum flexibility to mix-and-match liquid chromatographs and mass spectrometers in numerous combinations, while still having the option to acquire auxiliary data on the highest resolution ADC and best acquisition software package participating in any experiment.

Other Experimental Considerations

When numerous instruments are involved in an experiment, each with its own data acquisition system, extra considerations come into play. The first factor is that the system clocks should be synchronized as closely as possible to give run start times that are as close as possible across all computers. The start times should not be deleted from the headers of figures used for data presentation. In our early experiments that used a combination of Microsoft Windows acquisition/control systems with legacy Ultrix-based systems (a Unix clone), runs on different instruments showed up to a ten minute differential in start times since these systems were not synchronized. Chemstation causes a 0.2 minute delay from the time the Agilent 1200 system starts and the time that all other instruments connected to it start.

The TSQ 7000 operates using Xcalibur™ 1.2, which is the highest version compatible with the instrument. The LCQ Deca XP uses Xcalibur™ 1.3, so data, methods, and sequences acquired or developed on that instrument cannot be read on the TSQ 7000 computer. Therefore, to avoid re-entering information in sequence tables and processing methods, these should be developed on the lowest software revision and transferred to the computer using the higher version, which is backward compatible. All computers are networked to allow easy method and data transfer.

Results and Discussion

Dual Parallel Mass Spectrometry, LC1/MS2, for Phospholipid Analysis

We first applied dual parallel mass spectrometry to standards and to lipid extracts of human plasma and porcine eye lenses in 1998 (Byrdwell, 1998). We had previously reported the composition of human eye lenses by both APCI-MS and ESI-MS using separate chromatographic runs, and both techniques had provided valuable and complementary data (Byrdwell & Borchman, 1997). Therefore, the value of combining both techniques into a single run seemed obvious.

The early reports of the dual parallel mass spectrometry (LC1/MS2) used older instruments that required Instrument Control Language (ICL) procedures written in our lab to accomplish automatic detection of the most abundant ions and subsequent MS/MS of those ions. This can be seen in the very coarse alternating saw-tooth appearance of chromatograms as the instrument cycled between MS and MS/MS modes. Such functions are now routinely incorporated into instrument manufacturers' software, allowing such experiments to be done faster and more efficiently, requiring much less user expertise than early LC1/MS2 experiments. Fig. 8.2.A can be contrasted with the very rapid switching between modes seen in Fig. 8.12, which was obtained on a newer, faster instrument.

The data in this section were acquired using the SSQ710C for APCI-MS in parallel with the TSQ 700 for ESI-MS and MS/MS, both using Ultrix-based data acquisition systems. Fig. 8.2 shows the analysis of a model mixture of PL in proportions similar to those in the human eye lens [~70% sphingomyelin (SM), ~25% phosphatidylethanolamine (PE), ~5% phosphatidylcholine (PC)]. Because of the high background evident in Fig. 8.2.A, it was useful to extract out the full scans in the range of *m/z* 700-850, Fig. 8.2.B, to more easily visualize the PC and SM peaks. The full-scan mass spectrum in Fig. 8.2.D shows the protonated and natriated molecules, $[M+H]^+$ and $[M+Na]^+$, of long-chain dihydrosphingomyelin (DSM) species, over the range 20.11 to 21.11 min. The ESI-MS/MS spectrum of 24:0-DSM at *m/z* 817.8 shown in Fig. 8.2.E. Identities and masses of these and other DSM and SM species in these spectra can be seen at www.Sphingomyelin.com, so are not given here. Medium- and short-chain DSM species eluted next, overlapped with long-chain SM species. The spectrum of these is shown in Fig. 8.2.F, over the range 22.09 to 22.70 min. The mass at *m/z* 733.7 represents 18:0 DSM, while the larger masses represent long-chain SM, such as *m/z* 787.7, *m/z* 813.7 and *m/z* 815.7 which correspond with 22:0, 24:1 and 24:0 SM, respectively. The full-scan spectrum in Fig. 8.2.G, over the range 22.69-23.33 min, shows the medium- and short-chain SM species, such as 18:0 SM at *m/z* 731.7.

DSM made up ~11.5% of bovine brain SM used for the lens model. This is in contrast to human eye lens PLs, which contained ~77% DSM (still the only cellular system known for which DSM is the most abundant PL)

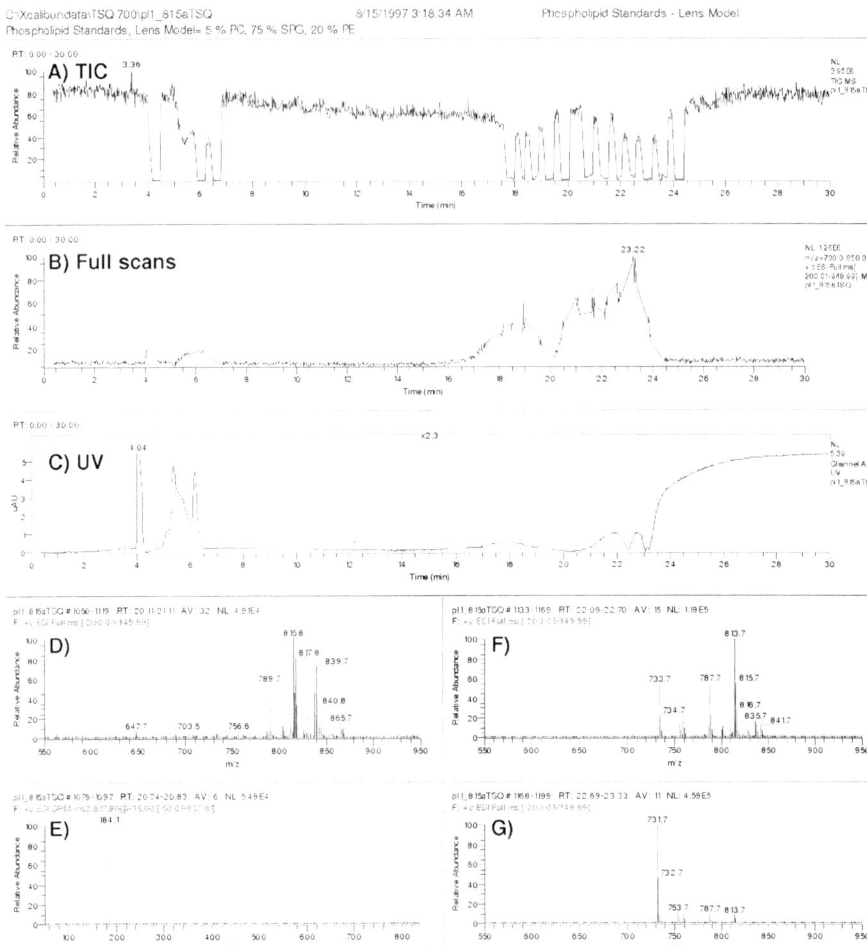

Fig. 8.2. First dual parallel LC-MS experiment. ESI-MS and MS/MS of commercial phospholipids. A) Total ion current chromatogram (TIC); B) Filtered chromatogram showing only full scans; C) Photodiode array detector (PDA) UV chromatogram; D) Average mass spectrum from 20.11 to 21.11 min, showing long-chain dihydrosphingomyelin (DSM) species; E) MS/MS spectrum of 24:0 DSM $[M+H]^+$ at *m/z* 817.8. F) Average spectrum from 22.09 to 22.70 min, showing medium- and short-chain DSM overlapped with long-chain SM species; G) Average spectrum from 22.69 to 23.33 min, showing medium- and short-chain SM species. Parallel APCI-MS run is given in Fig. 8.3.

(Byrdwell & Borchman, 1997). Saturated DSM species have unique masses that are 2 amu greater than their SM counterparts, making them readily identifiable (e.g. *m/z* 789.7 & 811.7 that represent the $[M+H]^+$ and $[M+Na]^+$, respectively, for 22:0-DSM, and *m/z* 817.7 & 839.7 for 24:0-DSM). Monounsaturated DSM species, however, have masses that are isobaric with saturated SM species, which is why so many literature reports have not identified unsaturated DSM species. The chromatographic resolution of DSM from SM is crucial to proper identification of unsaturated DSM species, especially since SM and DSM fragment differently (Byrdwell & Perry, 2006), so MS/MS is not reliable for distinguishing the two. Typical protonated and natriated ions from long-chain mono-unsaturated DSM species can be seen in Fig. 8.2.D at *m/z* 815.8 & 837.7, respectively, for 24:1-DSM and at *m/z* 843.8 & 865.7 for 26:1-DSM. Long-chain DSM molecular species, as well as other PL classes (PE and PC) have been shown in greater detail and discussed in depth elsewhere (Byrdwell, 1998).

The MS/MS spectrum of the *m/z* 817.7 precursor in Fig. 8.2.E is typical for phosphocholine-containing PL (PC, SM and DSM) and demonstrates that ESI-MS/MS often produces only the ion associated with the phosphocholine head group at *m/z* 184 and provides no direct structural information on the sphingolipid backbone structure. Herein lies one of the great benefits of the dual parallel mass spectrometer approach. Fig. 8.3 shows the APCI-MS data obtained in parallel from the same run. In Fig. 8.3.C, the ions in the range *m/z* 550–700 represent the $[M+H\text{-phosphocholine}]^+$ backbone of the DSM species. The fragments produced in the APCI source represent losses of both the phosphocholine and the natriated phosphocholine headgroup. Thus, *m/z* 606.6 came from both 22:0-DSM $[M+H]^+$ and $[M+Na]^+$, *m/z* 634.7 came from both 24:0-DSM $[M+H]^+$ and $[M+Na]^+$, and *m/z* 632.7 came from both 24:1-DSM $[M+H]^+$ and $[M+Na]^+$, etc. Thus, APCI-MS gave direct information about the sphingolipid backbone, but gave a low abundance of the intact protonated or natriated molecules, while ESI-MS gave almost exclusively the intact protonated and natriated molecules. These two types of complementary data were valuable for characterizing different structural subunits of the analytes and allowed the compositions of the molecular species of PE, PC, SM and DSM to be identified.

The benefit of using additional, and also complementary, auxiliary detectors can be seen by comparing the ELSD chromatogram in Fig. 8.3.B to the UV chromatogram in Fig. 8.2.C. The gradient solvent program caused a rising baseline with high noise level by UV detection, while the ELSD gave a good baseline and distinct chromatographic peaks. These experiments demonstrated the benefit of using dual parallel mass spectrometers with different ionization modes, as well as the value of having ELSD data in addition to conventional UV data. There were differences in the response of ESI-MS versus APCI-MS versus the ELSD for quantification of classes, which was discussed in detail in the initial report (Byrdwell, 1998).

More recently, we improved the separation of bovine brain SM and DSM to get better, though still partial, resolution between classes (DSM versus SM), and performed a thorough quantitative analysis of the molecular species in commercial bovine brain 'SM' present even at low levels, using numerous

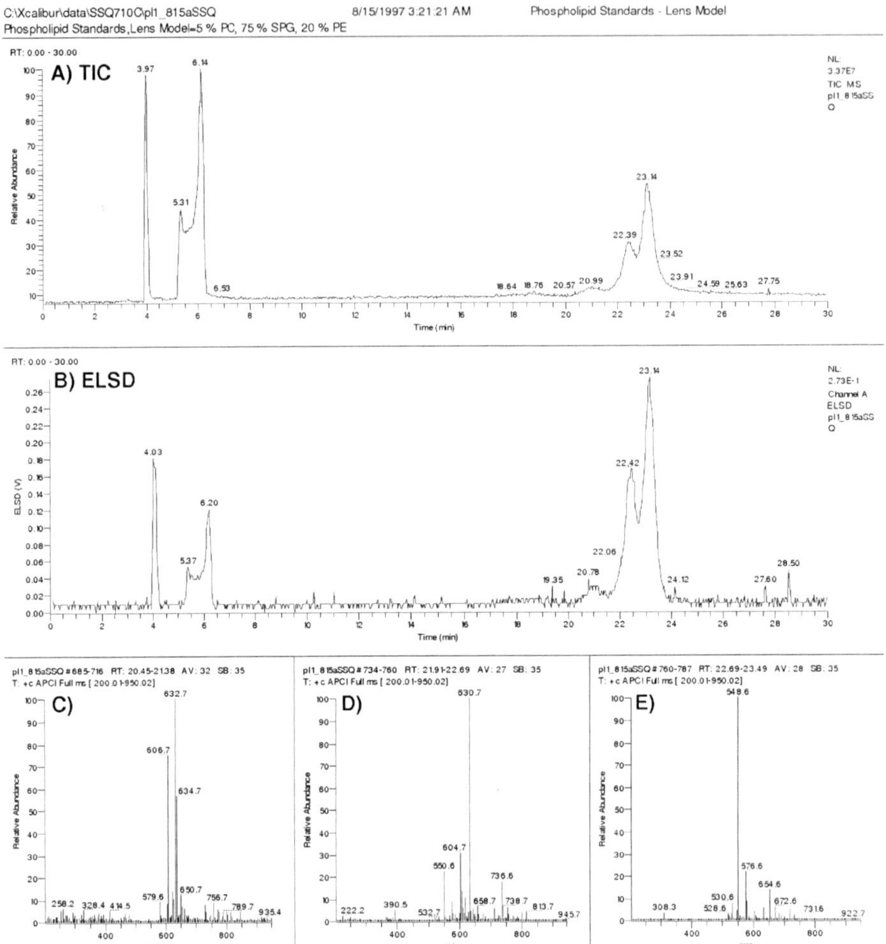

Fig. 8.3. First dual parallel LC-MS experiment. Parallel ESI-MS data are given in Fig. 8.2. A) APCI-MS TIC; B) Evaporative light scattering detector (ELSD) chromatogram; C) Average mass spectrum from 20.45 to 21.38 min, corresponding to Fig. 8.2.D; D) Average mass spectrum from 21.91 to 22.69 min, corresponding to Fig. 8.2.F; E) Average spectrum from 22.69 to 23.49, corresponding to Fig. 8.2.G.

combinations of APCI-MS and ESI-MS (both up to MS^3) both in parallel and not in parallel (Byrdwell & Perry, 2006). Runs were done in which the automated peak picking function was set to sequentially focus on smaller and smaller peaks, so that MS^n spectra were obtained for even the minor peaks (not shown). Fig. 8.4 shows the ESI-MS data (on a TSQ 700) for a dual parallel run, and Fig. 8.5 shows the corresponding APCI-MS run (on an LCQ Deca). In Fig. 8.4.C, the spectrum is magnified to bring the long-chain DSM species

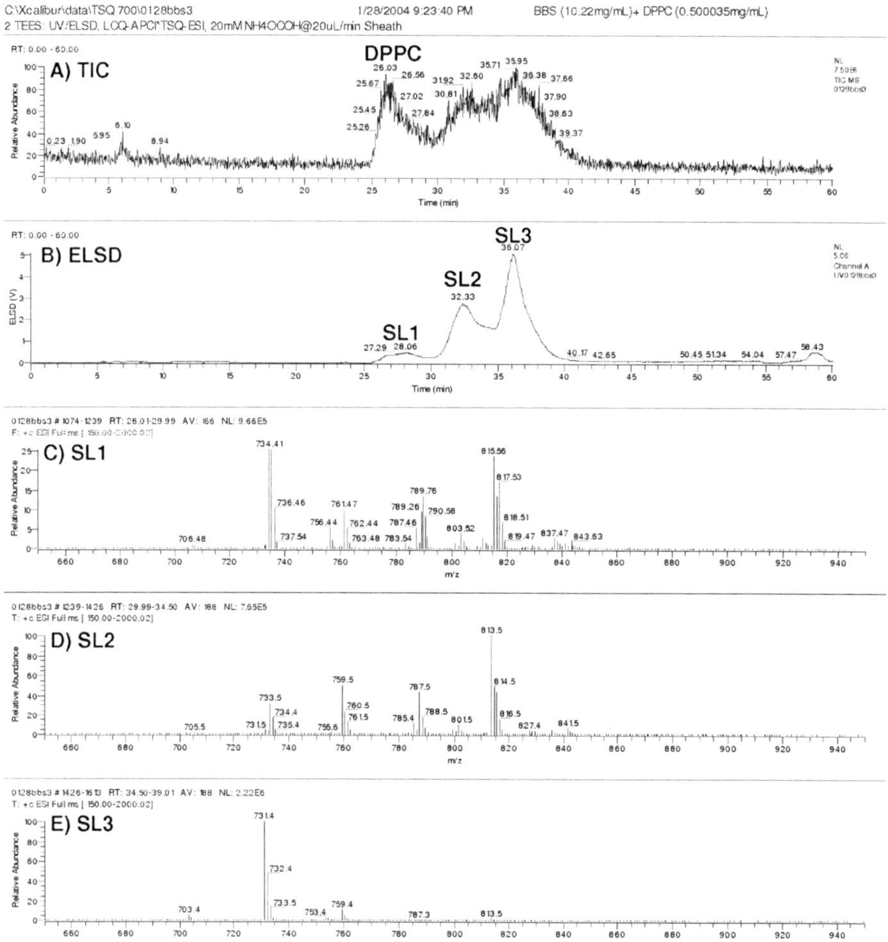

Fig. 8.4. ESI-MS of commercial bovine brain sphingomyelin. A) TIC; B) ELSD chromatogram; C) Average mass spectrum across sphingolipid peak 1 (SL1); D) Average spectrum across SL2; E) Average spectrum across SL3. Parallel APCI-MS data are shown in Fig. 8.5. DPPC is dipalmitoylphosphatidylcholine used as an internal standard.

to full scale, and the dipalmitoylphosphatidylcholine (DPPC) that was used as an internal standard was allowed to go off-scale. The differences in response between PC and the sphingolipids (SLs) by different detection techniques have been discussed in detail, along with quantification of classes by ^{31}P NMR (Byrdwell & Perry, 2006). The complementary nature of the intact protonated molecules seen by ESI-MS (Fig. 8.4) compared to the sphingolipid backbone formed by loss of the phosphocholine headgroup in APCI-MS (Fig. 8.5)

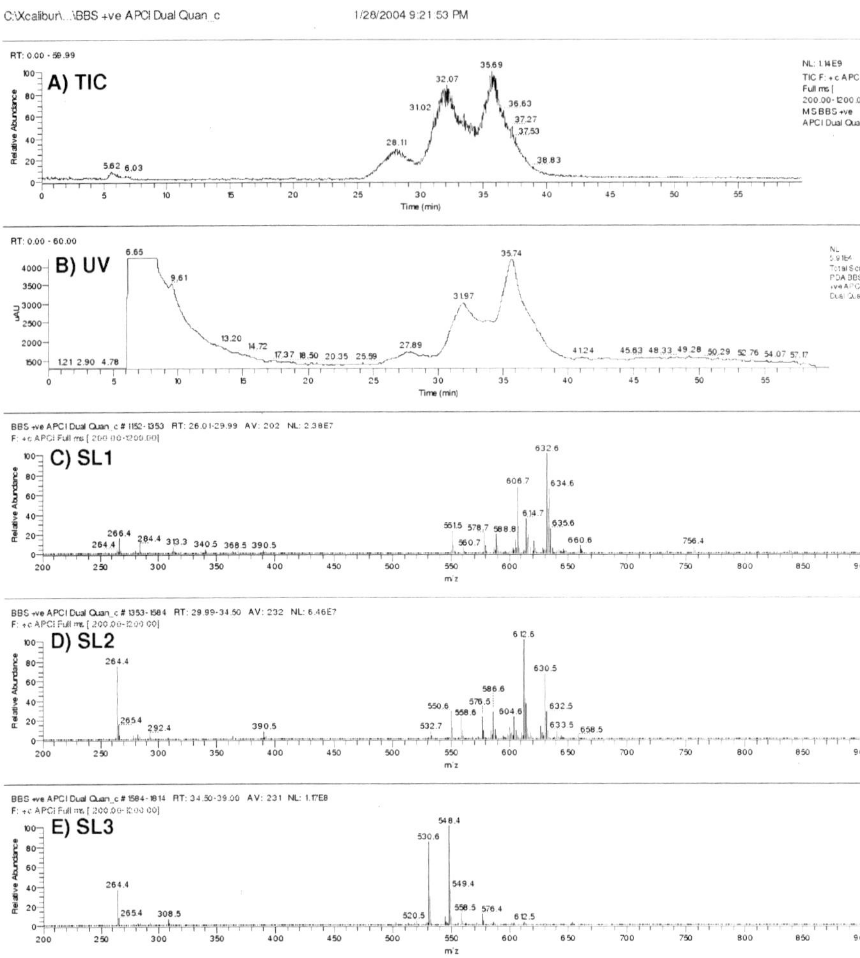

Fig. 8.5. APCI-MS of commercial bovine brain sphingomyelin. A) TIC; B) UV PDA; C) SL1; D) SL2; E) SL3. Parallel ESI-MS data are shown in Fig. 8.4.

allowed us to conclusively identify species present in small amounts with mutually confirmatory data by different techniques.

The same method was then applied to bovine milk SM and DSM to demonstrate that milk contains a substantial quantity of DSM, making it a dietary sphingolipid (Byrdwell & Perry, 2007). Fig. 8.6 shows ESI-MS data in both positive- and negative-ion modes (on the LCQ Deca) from a dual parallel run. As we reported, negative-ion mode analysis of SM and DSM produced almost exclusively $[M+45]^-$ formate adducts, by virtue of the ammonium formate electrolyte (Fig. 8.6.D–F). This adduct allowed the quaternary amine-contain-

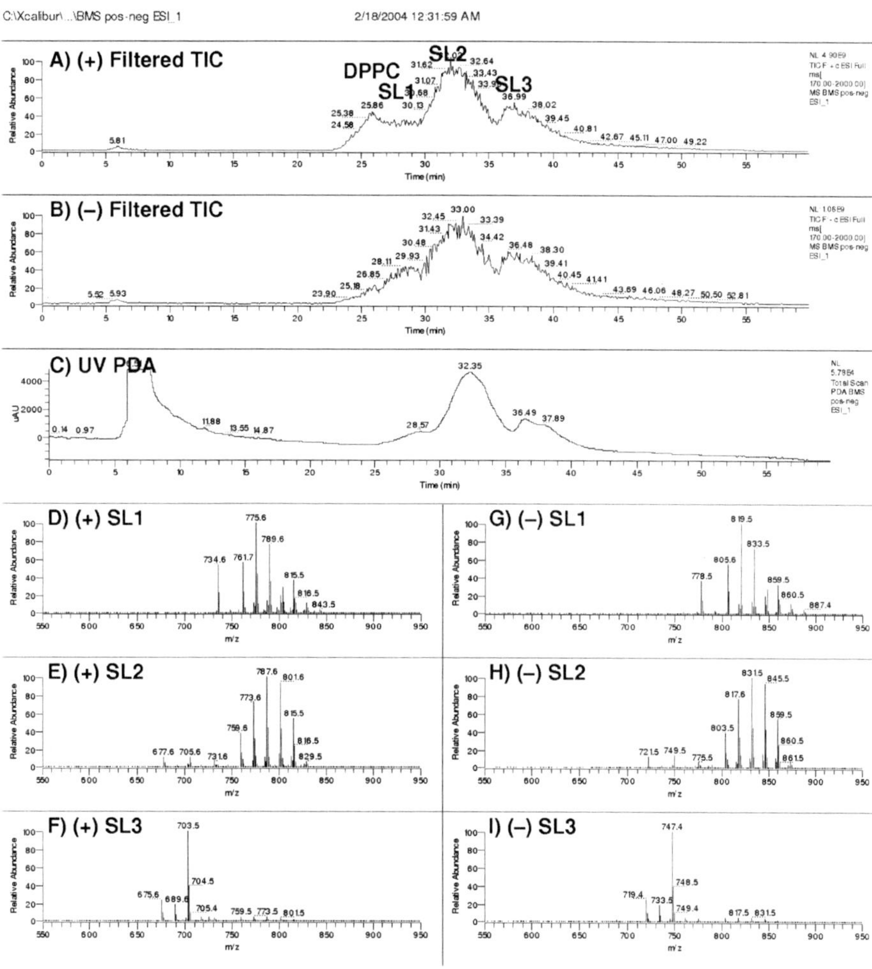

Fig. 8.6. ESI-MS of commercial bovine milk sphingomyelin. A) Positive-ion filtered TIC; B) Negative-ion filtered TIC; C) Chromatogram of PDA UV full scans; D) Average (+) mass spectrum across SL1; E) Avg. (+) spectrum of SL2; F) Avg. (+) spectrum of SL3; G) Avg. (–) mass spectrum across SL1; H) Avg. (–) spectrum of SL2; I) Avg. (–) spectrum of SL3. Parallel APCI-MS data in Fig. 8.7.

ing SM and DSM (as well as DPPC) to produce abundant ions in negative-ion mode that they would not otherwise produce, which provided additional confirmation of the molecular masses of the species. Therefore, there is a 44 amu difference between the $[M+H]^+$ in positive-ion spectra (Fig. 8.6D–F) compared to the $[M+OCOH]^-$ in negative-ion spectra (Fig. 8.6G–I). Of course, additional structural information was also provided by the APCI-MS data acquired in parallel (Fig. 8.7). As described above, these spectra gave base peaks from the

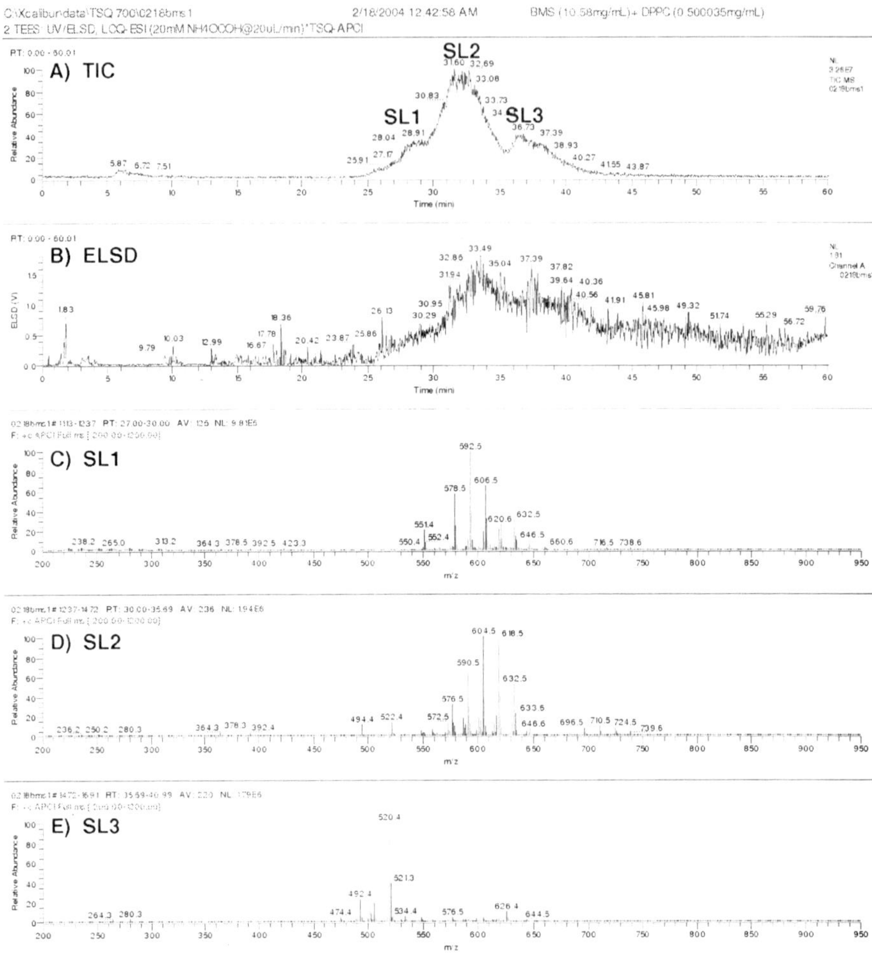

Fig. 8.7. APCI-MS of commercial bovine milk sphingomyelin. A) TIC; B) ELSD; C) Average APCI-MS mass spectrum across SL1; D) Avg. spectrum of SL2; E) Avg. spectrum of SL3.

sphingolipid backbones after loss of the phosphocholine moiety. These spectra show the larger number of odd-number carbon chain molecular species in bovine milk compared to other SM sources. Once again, these complementary data provided valuable mutually confirmatory masses and fragments that allowed a high degree of confidence in the qualitative and quantitative analysis of these sphingolipids. As with our other dual parallel MS reports, we also acquired data on two other auxiliary detectors, a photodiode array detector (PDA) and the ELSD.

Dual Parallel Mass Spectrometry, LC1/MS2, for Triacylglycerol Analysis

Triacylglycerols

Since triacylglycerols are not normally amenable to ESI-MS, because they do not contain a readily charged functional group, an electrolyte must normally be added to produce fruitful ionization. As first noted by Duffin and Henion (1991), a simple acid solution is not sufficient to produce a strong signal by ESI-MS, based on their infusion experiments (not LC-MS). In such cases, ubiquitous sodium caused $[M+Na]^+$ adducts that were often more abundant than the protonated molecules. In the early experiments using low-energy collision-induced dissociation (CID), inadequate fragmentation of $[M+Na]^+$ led to the choice of ammonium acetate as their preferred electrolyte. Other experiments using high energy (2 kV) collisionally activated dissociation (CAD) on a double-focusing magnetic sector instrument yielded valuable fragmentation patterns from sodium adducts and other alkali metals from TAG by infusion (Cheng et al., 1998). Sodium adducts were used to provide information on double bond positions on the fatty acyl chains. Kuksis and coworkers (Sjovall et al., 1997) performed LC-ESI-MS (not infusion) of TAG and TAG oxidation products (TAGOX) by adding a solution of ammonium acetate to the post-column flow. They discussed the formation of abundant $[M+NH_4]^+$ ions, as well as $[M+Na]^+$ ions from ubiquitous sodium. Hsu and Turk (1999) demonstrated the use of lithium adducts for TAG analysis by ESI-MS. Fragments allowed identification of TAG regioisomers. The MS/MS spectra were fairly complex, exhibiting a large number of both lithiated and non-lithiated fragments, with diacylglylcerol-like fragments, $[DAG]^+$, and $[DAG+Li]^+$ being dominant. The advantage of lithium adducts was that MS/MS of fragments produced by up-front fragmentation (produced in the API interface region) can provide fragments that allow localization of double bonds. Other early applications of ESI-MS of TAG have previously been described in detail (Byrdwell, 2005a), so that discussion will not be repeated here. Some general observations are these: sodium adducts do not fragment well under low-energy CAD conditions; lithium adducts fragment well and provide valuable information about the positions of FA chains on the glycerol backbone and may provide information on the positions of double bonds, but the mass spectra are more complex; ammonium adducts fragment well and produce simple ESI-MS/MS spectra that are similar to APCI-MS spectra, which also allow identification of regioisomers.

Our goal has always been to obtain the maximum amount of information in the fewest number of adducts and fragments possible. When numerous adducts are observed, this means that the population of molecules is split into several different pools of adducts, instead of being focused into a single adduct ion. This has two direct implications for quantification: i) the population of ions is divided, so the abundance of each ion is less than it would be if only one adduct ion is observed (two adducts present in equal amounts would be half the absolute abundance of one single adduct, all else being equal), and ii) when performing quantification, either the peaks from all adduct EICs must be integrated and added together (resulting in two, three, four or more times more time and work for integration) or else some adducts are ignored and do not contribute to quantification (leading to the skewing of quantification if all adducts are not produced in equal proportions from all TAGs, and also essentially wasting some of the population of ions). Similar considerations come into play when extensive fragmentation occurs in the ionization source or in the interface region. One alternative is to use the total ion current chromatogram (TIC), instead of the sum of extracted ion chromatograms, but chromatographic peaks for TAG are often not single TAGs with no overlap; plus, the TIC includes all noise peaks.

To accomplish our goal, we have selected ammonium formate as our electrolyte of choice. This electrolyte produces one single abundant ammonium adduct, [M+NH$_4$]; that actively suppresses sodium adduct formation, it does not contain an alkali metal that can accumulate and clog the interface heated capillary orifice, and it is easy to prepare and not particularly toxic. Most importantly, it allows a very sensitive analysis of both TAG and TAGOX present at low levels, and it provides an abundant precursor for MS/MS, which produces structurally informative diacylglycerol fragments that allow isobaric species to be differentiated and regioisomers to be identified. ESI-MS spectra of TAG also usually exhibit small [DAG]$^+$ peaks. The only drawback of ammonium formate (or ammonium acetate, etc.) is that ammonium adducts, like most other adducts, do not produce fragments that provide information regarding the localization of double bonds under low-energy CAD conditions. However, the many benefits of this electrolyte far outweigh this disadvantage, since the positions of double bonds in common fatty acids incorporated into most TAGs are known, and most other electrolytes generally suffer from this same disadvantage.

Fig. 8.8 demonstrates the use of ammonium formate solution (1:4 in acetonitrile) added as an electrolyte directly into the 'sheath liquid' port of an electrospray ionization source (sheath liquid was available on older ThermoElectron sources, but not in newer source designs) for analysis of canola oil triacylglycerols (Byrdwell & Neff, 2002). These data were obtained on an ion trap mass spectrometer, with MS^3 (not shown here), in parallel with APCI-MS/MS on a tandem sector quadrupole instrument, shown in Fig. 8.9. The first column of spectra in Fig. 8.8 represents full-scan spectra, which exhibit primarily [M+NH$_4$]$^+$ ions, with small [DAG]$^+$ fragments, the mass range of which is expanded in the insets. The second column is MS/MS spectra from the [M+NH$_4$]$^+$ precursor ions, with insets showing the expanded [DAG]$^+$ fragments. The ESI-MS/MS spectra can be compared to the APCI-MS spectra in the first column of Fig. 8.9. The same

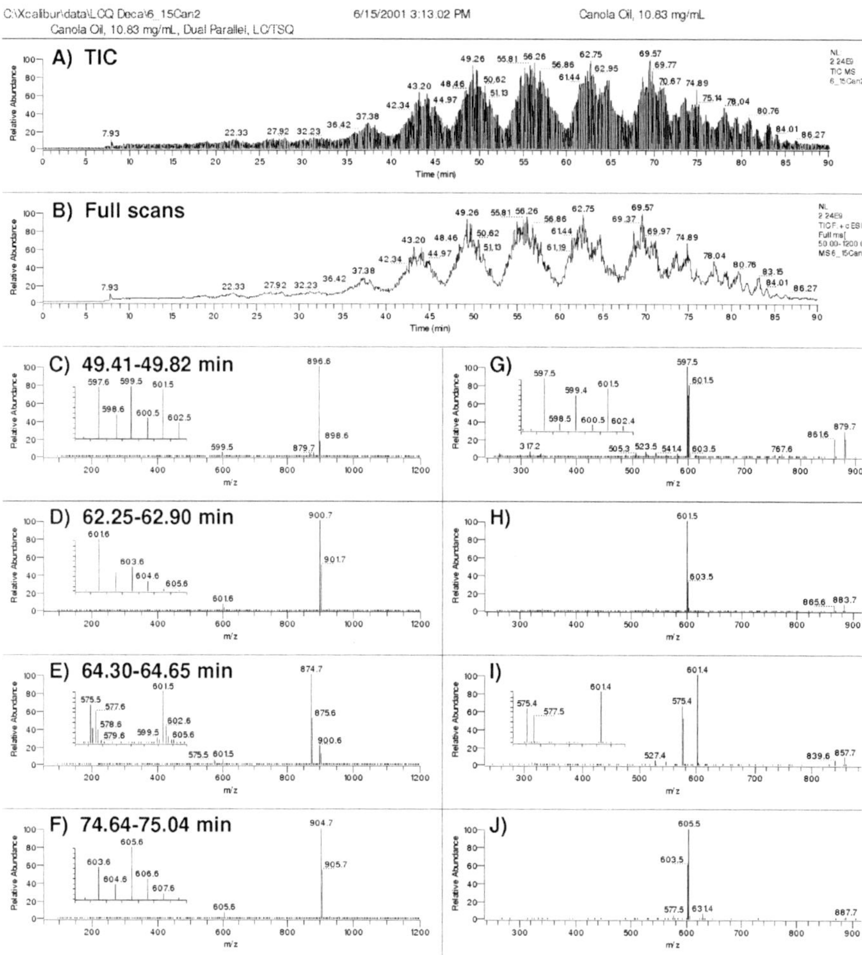

Fig. 8.8. ESI-MS$_N$ analysis of common commercial canola oil. A) TIC; B) Full-scan filtered TIC; C) Average mass spectrum across top of peak at 49.41-49.82 min; D) Avg. spectrum 62.25-62.90 min; E) Avg. spectrum 64.30-64.65 min; F) Avg. spectrum 74.64-75.04 min; G) MS/MS spectrum of *m/z* 896.6 peak in C), = OLLn; H) MS/MS of *m/z* 900.7 in D), = OOL; I) MS/MS *m/z* 874.6 in E), = POL; J) MS/MS of *m/z* 904.7 in F), = OOS. Ln = 18:3, L = 18:2, O = 18:1, S = 18:0, P = 16:0. Parallel APCI-MS run given in Fig. 8.9.

ions appear in both spectra, though the ESI-MS/MS spectra exhibit a larger $[M+H-H_2O]^+$ fragment.

The benefit of the dual parallel mass spectrometer approach is most apparent for TAG species having few sites of unsaturation (e.g., OOS in Fig. 8.8.F versus Fig. 8.9.G). The ESI-MS data provide both the abundant $[M+NH_4]^+$ ion, as well as small abundances of $[DAG]^+$ ions, whereas APCI-MS data exhibit little

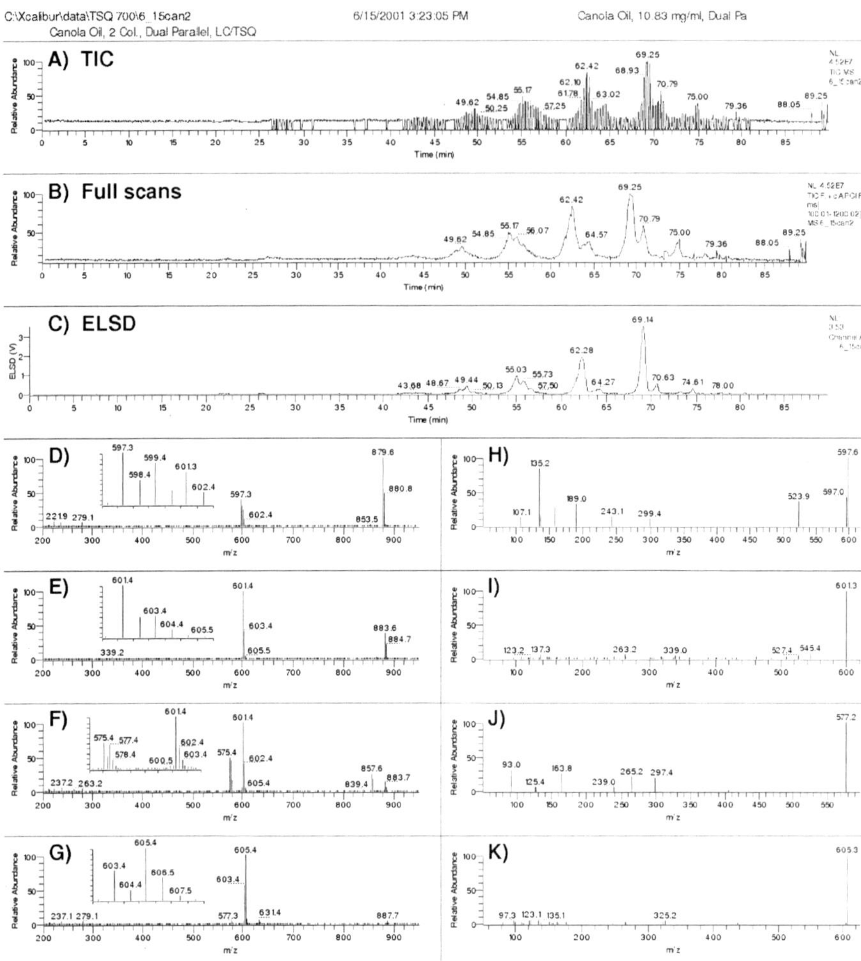

Fig. 8.9. APCI-MS/MS data for common commercially available canola oil. A) TIC; B) Full-scan filtered TIC; C) ELSD chromatogram; D) Average mass spectrum across top of peak at 49.32-49.62 min, = OLLn; E) Avg. spectrum 62.05-62.69 min, = OOL; F) Avg. spectrum 64.22-64.27 min, = POL; G) Avg. spectrum 74.38-74.74 min, = OOS; H) MS/MS spectrum of *m/z* 597.3, = $[LLn]^+$; I) MS/MS spectrum of *m/z* 601.4, = $[OL]^+$; J) MS/MS spectrum of *m/z* 577.4, = $[PO]^+$; K) MS/MS spectrum of *m/z* 605.4, = $[OS]^+$. Parallel ESI-MS^n data in Fig. 8.8.

or no $[M+H]^+$ from species with few sites of unsaturation. On the other hand, APCI-MS data provide abundant $[DAG]^+$ fragments that provide direct evidence for the FAs attached to the glycerol backbone. Furthermore, the ratios of the $[DAG]^+$ fragments by APCI-MS (or by ESI-MS/MS) can allow identification of specific regioisomers (Hsu & Turk, 1999; Hvattum, 2001; Laakso & Voutilainen, 1996; Mottram et al., 1997; Mottram et al., 2001; Mottram & Evershed, 1996).

The ESI-MS/MS/MS spectra and the APCI-MS/MS spectra of $[DAG]^+$ indicate a high degree of charge-remote fragmentation (e.g., Fig. 8.9.H and J), but some fragments, such as the acylium ion, $[RCO]^+$ and the $[RCO+74]^+$ ion ($=[RCOO+58]^+$) (Cheng et al., 1998), do provide confirmation of the individual FA chains. Overall, though, these spectra are less consistent and reproducible than ESI-MS, ESI-MS/MS, or APCI-MS spectra. Based on the spectra shown in Figs. 8.8 and 8.9, one could conclude that it would be possible to get by with only ESI-MS^n, since ESI-MS/MS spectra appear so similar to APCI-MS spectra. However, the TIC obtained by ESI-MS under these conditions exhibited a substantially higher response to polyunsaturated TAG (Fig. 8.8) than did APCI-MS, resulting in much larger early-eluting peaks than those seen in Fig. 8.9. Therefore, both techniques are, again, valuable and complementary.

Triacylglycerol Oxidation Products

APCI-MS spectra of TAGOX exhibit very useful, structurally informative fragments by APCI-MS, which allow a range of oxidation products to be identified, including epoxides, intact aldehydes, hydroperoxides, core aldehydes, and other breakdown products (Byrdwell & Neff, 1999). However, some classes, such as intact aldehydes and hydroperoxides produced only very low abundances of intact protonated molecules. This led to the need for ESI-MS of these species, while still retaining the valuable structural information provided by APCI-MS.

Fig. 8.10 shows the ESI-MS TICs and spectra (MS and MS/MS) for several triolein oxidation products formed after 6 hr of heating at frying temperature (190°C) [MS^3 shown in original report (Byrdwell & Neff, 2002)]. As with normal TAG, ESI-MS of TAGOX produced intact ammonium adducts that allowed facile identification of the molecular masses of the TAGOX, which immediately indicated the number of oxygen functional groups added to the normal TAG during oxidation. Fig. 8.10 shows spectra of OOS-epoxide (oxidation at a site of unsaturation), OOO-epoxide (oxidation not across a site of unsaturation), OOO-hydroperoxide, and an oxidized dimer (two cross-linked oxidized monomers), as well as MS/MS spectra of each of these. The value of ESI-MS data is especially evident in the mass spectrum of OOO-hydroperoxide, which produced only a very small intact protonated molecule by APCI-MS, but an abundant ammonium adduct base peak by ESI-MS. Another substantial advantage of ESI-MS for analysis of TAGOX is the sensitivity to species present even at low levels. While un-oxidized triolein is evident at 46.76 min in Fig. 8.10, the chromatographic peaks from the oxidation products are, in many cases, larger than the un-oxidized starting material, indicating better response to those species, even though they were present in much lower amounts than the starting material (as seen from the ELSD chromatogram in Fig. 8.11.C). APCI-MS data, seen in Fig. 8.11, showed less discrimination in response between classes and gave a TIC similar to the ELSD trace.

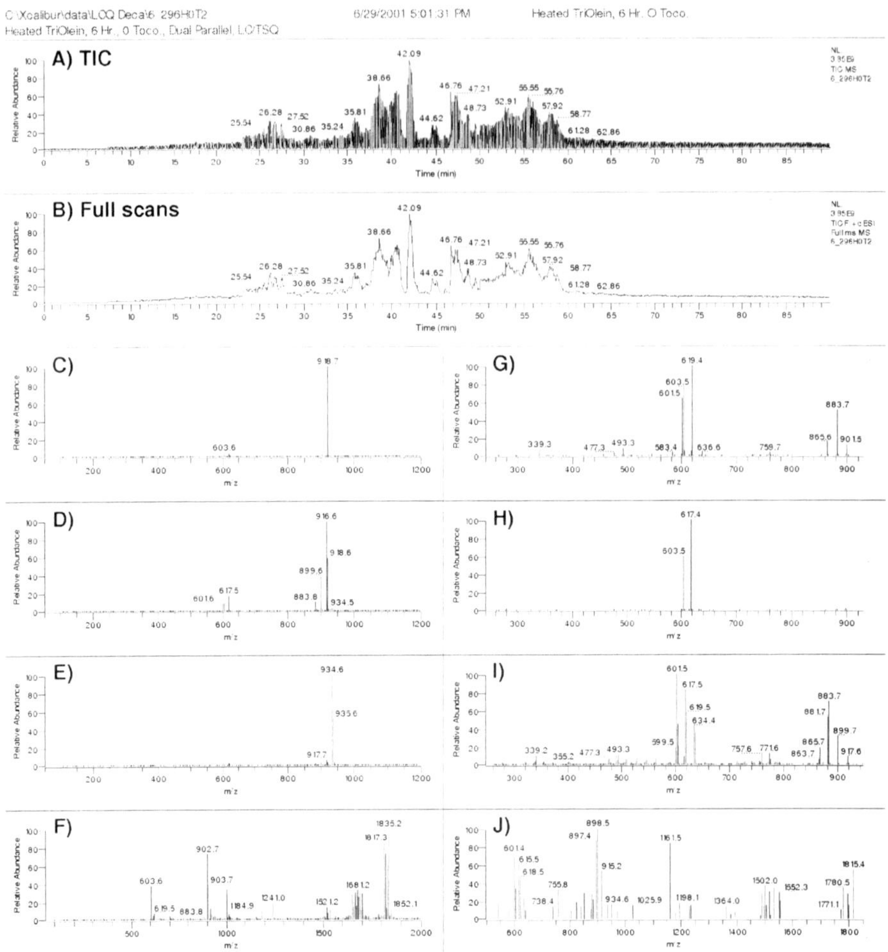

Fig. 8.10. ESI-MS and MS/MS of heated triolein oxidation products. A) TIC; B) Filtered full-scan TIC; C) Average mass spectrum of OOS-epoxide, 41.93-42.24 min; D) Avg. spectrum of OOO-epoxide, 38.61-39.44 min; E) Avg. spectrum of OOO-hydroperoxide, 35.76-35.81 min; F) Avg. spectrum of oxidized triolein dimer containing 3 additional oxygens, 52.91-53.95 min. Parallel APCI-MS data shown in Fig. 8.11.

Although ESI-MS/MS spectra were extremely helpful for structural elucidation, the limited number of MS/MS (and MS^3, not shown) spectra across each peak, and the less-than-effective fragmentation of some species in the ion trap (e.g. Fig. 8.10.H), meant that ESI-MS/MS was not, by itself, completely conclusive for identification of the many oxidation products formed. Once again, the APCI-MS data obtained in parallel proved to be a valuable complement to the ESI-MS^n data. While the ESI-MS spectra indicated how

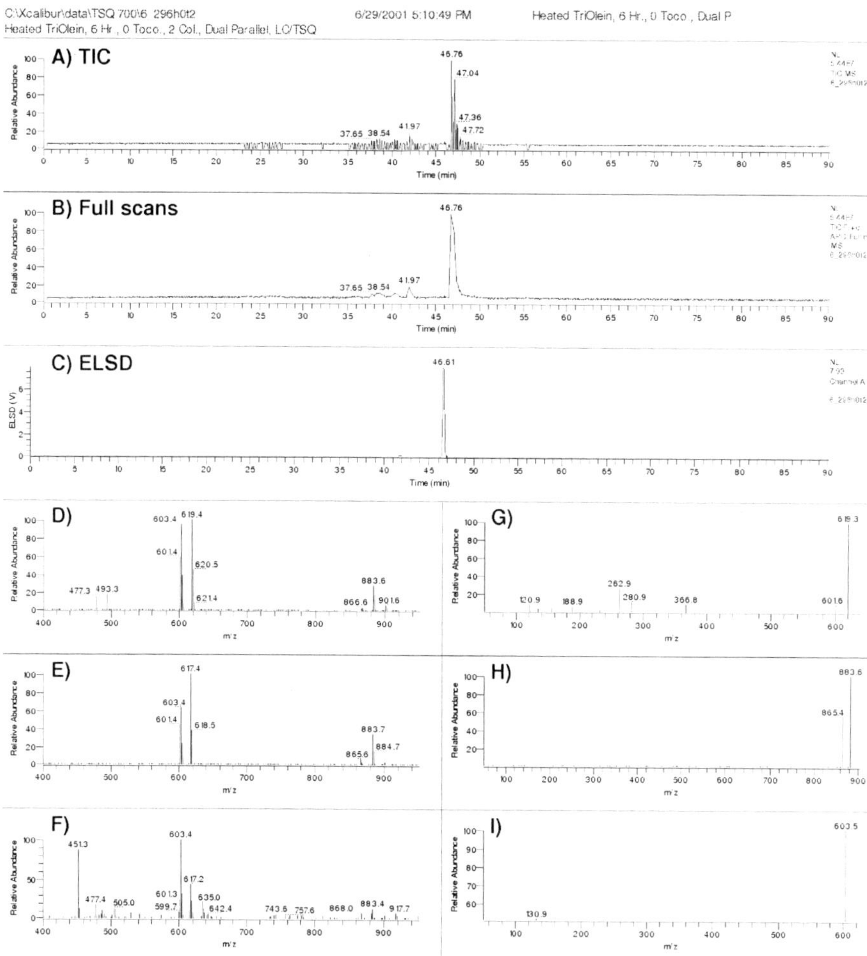

Fig. 8.11. APCI-MS and MS/MS of heated triolein oxidation products. A) TIC; B) Full-scan filtered TIC; C) ELSD; D) Average mass spectrum across OOS-epoxide peak, 41.92-42.27 min; E) Avg. spectrum of OOO-epoxide, 38.58-39.47 min; F) Avg. spectrum of OOO-hydroperoxide, 35.76-35.81 min; G) MS/MS spectrum of *m/z* 619.4; H) MS/MS spectrum of *m/z* 883.7; I) MS/MS spectrum of *m/z* 603.6. Parallel ESI-MS data given in Fig. 8.10.

many oxygen atoms were added to the overall molecule, the APCI-MS spectra helped to determine where on the diacylglycerol sub-units the oxygens occurred (on the same FA or different FA, etc.). The discussion of each of the peak assignments has been presented in detail previously (Byrdwell & Neff, 2002), in comparison to the earlier APCI-MS data for TAGOX (Byrdwell & Neff, 1999), and so is not discussed in detail here. The data here are presented simply to demonstrate the complementary nature of the two types of data obtained using a dual parallel mass spectrometer approach, and the valuable information contributed by each to allow identification of species not previously identified by online LC-MS.

Dual Parallel Mass Spectrometry, LC1/MS2, for Vitamin D Analysis

The examples of LC1/MS2 experiments above have focused on the benefit of obtaining data from two complementary ionization types, ESI and APCI. Another good reason to employ a dual parallel mass spectrometer approach is to maximize the duty cycle of one mass spectrometer for quantitative analysis, while using a second instrument for qualitative analysis. This was demonstrated in our analysis of vitamin D in orange juice (Byrdwell et al., 2011). Unfortunately, Vitamin D is one of the few lipids that does not form an ammonium adduct well, rendering ESI-MS generally ineffective for this class of molecules. It is, however, efficiently ionized by APCI (as are most classes of molecules), so this has become the technique of choice for LC-MS of vitamin D (vitamin D_2 and D_3 being the two most relevant to the diet). Therefore, we routinely use LC-APCI-MS for analysis of vitamin D in food extracts. APCI-MS is also effective for analyzing the metabolites most commonly used as biomarkers for clinical analysis, 25-OH vitamin D_2 and 25-OH vitamin D_3 (Tsugawa et al., 2005).

Selected ion monitoring is ~100 times more sensitive than selected reaction monitoring, since the former does not require fragmentation in a collision cell, which not an entirely efficient process. Our TSQ 7000 instrument is old enough that we can readily detect the low levels of vitamin D in food extracts by SIM, but not by SRM, so we routinely use SIM. However, the disadvantage of SIM, compared to SRM, is that it is less selective. Ions are formed in the source, and any molecules that give ions at *m/z* 385.3 ($[M+H]^+$) or *m/z* 367.3 ($[M+H-H_2O]^+$) for vitamin D_3 or at *m/z* 397.3 ($[M+H]^+$) or *m/z* 379.3 ($[M+H-H_2O]^+$) for vitamin D_2 give a signal by SIM, including isotopic peaks (e.g. the 2x ^{13}C peak from *m/z* 383.3 is common). Thus, in real extracts, peaks can appear in SIM chromatograms that do not arise from authentic analyte. Therefore, it is desirable to obtain full-scan mass spectra in addition to SIM data, to allow identification of chromatographically unresolved species. On extremely fast, extremely sensitive new instruments, both can be accomplished on the same instrument, but on older instruments, it is desirable (even necessary) to acquire the maximum number of scans across each peak to obtain the best S/N for quantification (as discussed in Introduction). Therefore, we have employed a dual parallel mass spectrometer technique in which SIM APCI-MS data are acquired on a TSQ 7000 instrument, while simultaneously recording full-scan APCI-MS data on an LCQ Deca XP ion trap instrument.

Ultraviolet detection is also very sensitive for vitamin D molecules, and is still widely used as a valuable adjunct to MS data. UV detection is even more sensitive than SIM on older instruments, so UV data is preferred, if it can be proven that the analyte peaks are pure and free from interfering species.

Fig. 8.12 shows the SIM APCI-MS and UV data for an orange juice analysis conducted on a Vydac 201TP54 C18 column. The UV at 265 nm chromatogram, Fig. 8.12.B, showed a slight hump on the side of the vitamin D_3 peak, which was the only chromatographic indication that there might be

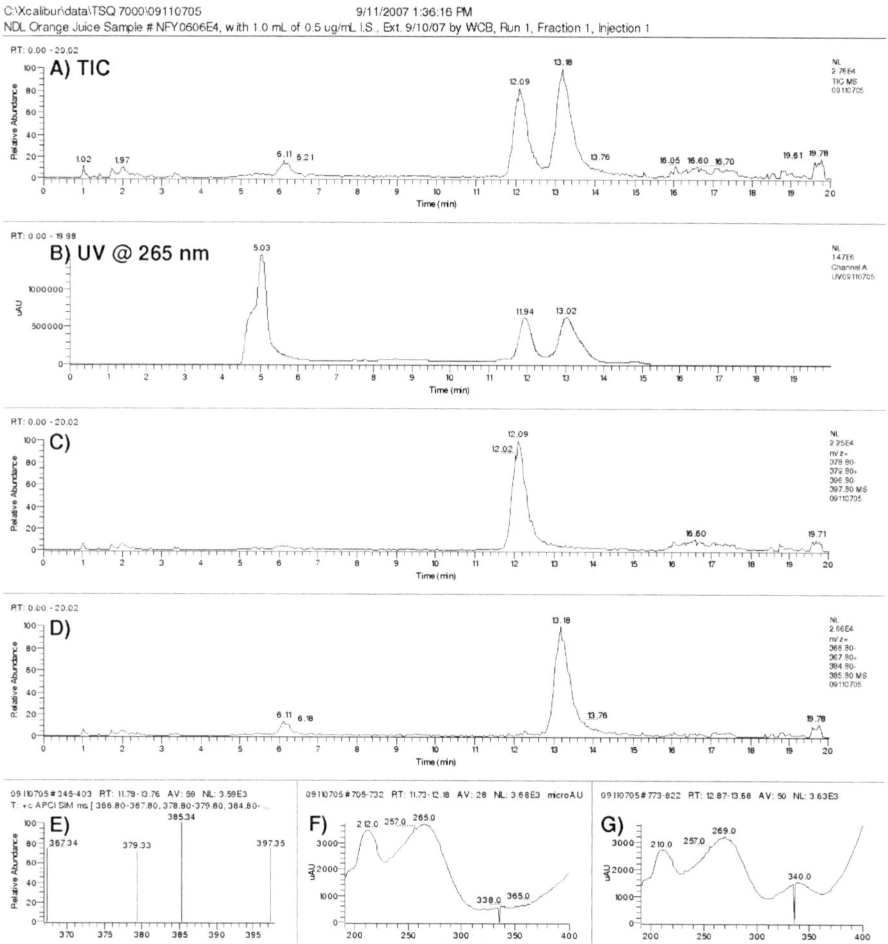

Fig. 8.12. Selected ion monitoring (SIM) APCI-MS and UV analysis of vitamin D in orange juice extract. A) SIM TIC; B) UV at 265 nm chromatogram; C) EIC of *m/z* 379.3 and 397.3 from vitamin D_2; D) EIC of *m/z* 367.3 and 385.3 from vitamin D_3; E) SIM spectrum; F) UV spectrum averaged 11.73-12.18 min, across vitamin D_2 peak; G) UV spectrum averaged 12.87-13.68 min, across vitamin D_3 peak. Parallel full-scan APCI-MS data shown in Fig. 8.13.

an overlapping species at that time. The peak for vitamin D_2 in Fig. 8.12.B was symmetric and appeared clean, with no underlying interferents. The full-scan UV spectrum, Fig. 8.12.G, did not match the spectrum of authentic vitamin D_3, since it exhibited a broad peak at 310–360 nm, and the normal peak maximum was shifted away from 265 nm. The UV spectrum across the vitamin D_2 peak, Fig. 8.12.F, had its maximum at 265 nm, but showed some absorbance at longer wavelength. Unfortunately, the SIM data obtained by APCI-MS gave no indication of potential overlap, as would be expected unless the interferent gave an ion at one of the four SIM masses monitored ($[M+H]^+$ and $[M+H-H_2O]^+$ for vitamin D_2 and D_3).

Fig. 8.13 shows the full-scan APCI-MS data from the LCQ Deca XP ion trap mass spectrometer obtained in parallel with the TSQ 7000 instrument. Although vitamin D species were not present in sufficient quantity to give noticeable peaks in the TIC in Fig. 8.13.A, they gave good peaks in the extracted ion chromatograms (EICs) associated with the two vitamin D molecules, shown in Figs. 8.13.B and 8.13.C. More importantly, the full-scan spectra averaged across the vitamin D_2 and vitamin D_3 peaks in Figs. 8.13.F and 8.13.G clearly showed peaks at higher masses than those given above for the vitamin D_3 analyte and D_2 internal standard. When the potential interferent masses of *m/z* 551.2 and *m/z* 553.2 were extracted out of the total ion current chromatogram, the EICs indicated the presence of species that overlapped vitamin D_2 and D_3. Thus, the use of the holistic approach of combining full-scan spectra with SIM data, on two different instruments in parallel, provided conclusive evidence that an overlapping interferent was present. Of course, this meant that the UV data were unreliable and should not be used, since the molar absorptivities at 265 nm of the interfering species were unknown.

Based on the full-scan data, we modified the chromatographic method to utilize an Inertsil ODS-2 C18 column, which had a much higher carbon load, and different selectivity. This column required a higher percentage of methanol to elute the analyte and internal standard, which had beneficial effects on the quality of the APCI-MS spectra, compared to those obtained using the former gradient, which was higher in acetonitrile. Fig. 8.14 shows the SIM APCI-MS and UV data from the improved chromatographic separation. The chromatographic peaks are sharper and more symmetric, and the UV spectra (Figs. 8.14.F and 8.14.G) across the vitamin D_2 and D_3 peaks are virtually identical to those from the calibration standards run in the same sequence of runs. The higher percentage of methanol also changed the ratio of $[M+H]^+$ and $[M+H-H_2O]^+$ ions, giving a larger abundance of the protonated molecule. While these data were encouraging, only the addition of full-scan APCI-MS data provided proof that the vitamin D peaks were pure.

Fig. 8.15 shows the full-scan APCI-MS data acquired in parallel with the SIM APCI-MS and UV data. The noise in the TIC in Fig. 8.15.A is lower, due to the additional percentage of methanol in the mobile phase, but the

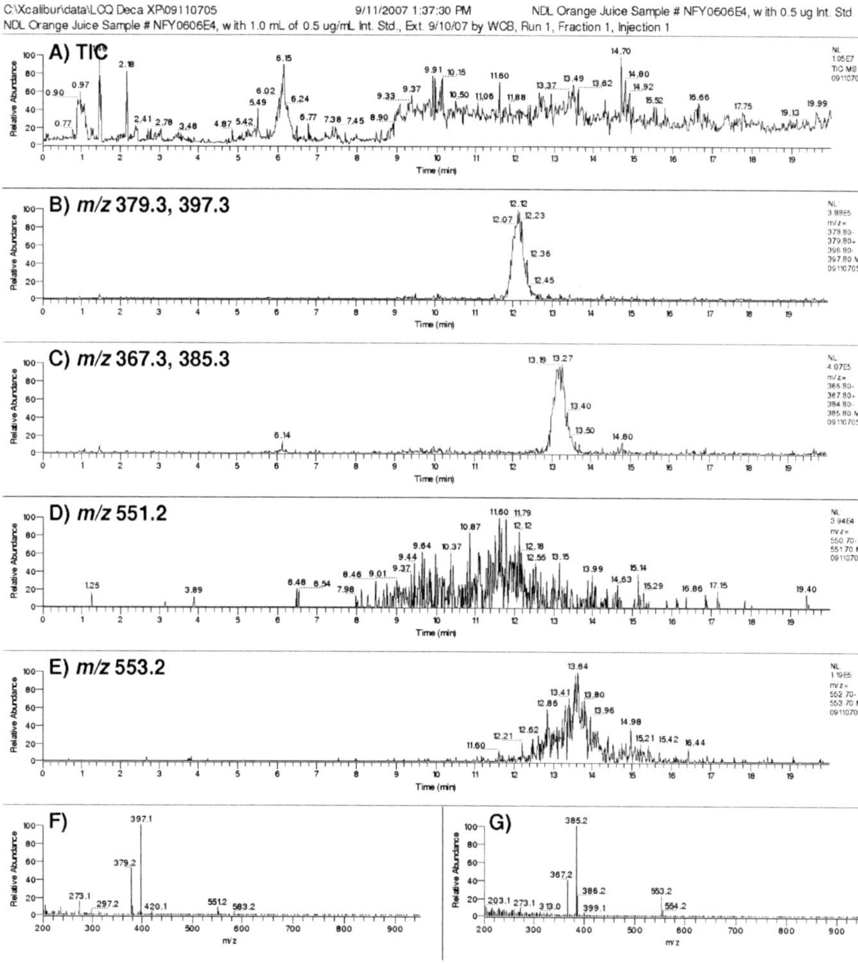

Fig. 8.13. Full-scan qualitative APCI-MS analysis of vitamin D in orange juice extract. A) TIC; B) EIC of *m/z* 379.3 and 397.3 from vitamin D_2; C) EIC of *m/z* 367.3 and 385.3 from vitamin D_3; D) EIC of *m/z* 551.2 seen in F); E) EIC of *m/z* 553.2 seen in G); F) Average mass spectrum 11.91-12.42 min; G) Avg. spectrum 12.97-13.47 min. Parallel SIM APCI-MS data given in Fig. 8.12.

vitamin D species were still not evident in the TIC. The EICs in Fig. 8.15.B and 8.15.C clearly showed the peaks arising from the internal standard and analyte. Most importantly, the full-scan mass spectra across these peaks showed that the peaks were free from interfering species (except low levels of ubiquitous column bleed). The masses at *m/z* 551.2 and *m/z* 553.2 can be seen in Fig. 8.15.D, proving that they no longer overlap the vitamin D analyte and internal standard. These data provided confidence in the UV data,

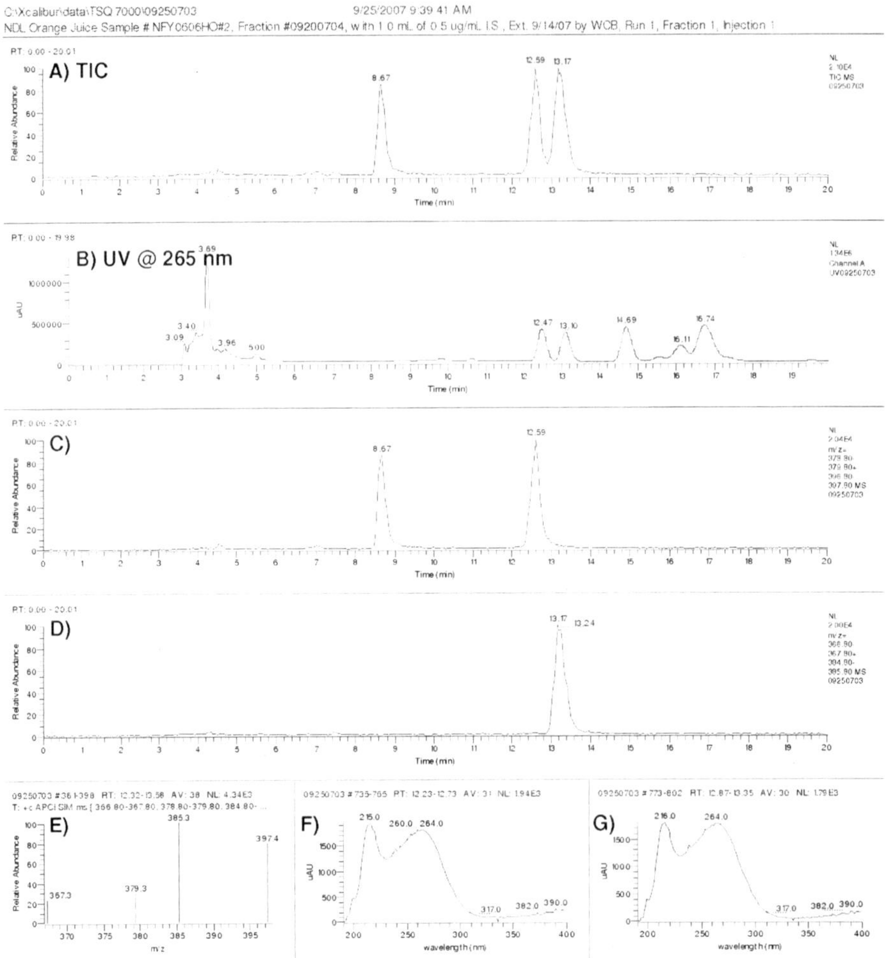

Fig. 8.14. SIM APCI-MS and UV analysis of vitamin D in orange juice extract. A) SIM TIC; B) UV at 265 nm chromatogram; C) EIC of *m/z* 379.3 and 397.3 from vitamin D_2; D) EIC of *m/z* 367.3 and 385.3 from vitamin D_3; E) SIM spectrum; F) UV spectrum averaged 12.23-12.73 min, across vitamin D_2 peak; G) UV spectrum averaged 12.87-13.35 min, across vitamin D_3 peak. Parallel full-scan APCI-MS data shown in Fig. 8.15.

and allowed it to be used for sensitive quantification. Additional details have been provided in the original report (Byrdwell et al., 2011). The data herein demonstrate that the use of a holistic approach combining targeted modes of analysis (i.e. SIM APCI-MS) with survey analysis (i.e. full-scan APCI-MS) allows reliable quantification and confirmation of the quality of the chromatographic separation.

Triple Parallel Mass Spectrometry, LC1/MS3, for Vitamin D and Triacylglycerol Analysis

As discussed in several preceding sections, we have always had limitations imposed by insufficient instrument speed and sensitivity. These limitations made dual parallel mass spectrometry not only beneficial, but necessary for many of our analyses. We recently acquired a QTrap 5500 mass spectrometer that theoretically offered the speed and sensitivity to obviate the need for multiple instruments in parallel. However, the data presented here demonstrate the value of continuing to use older legacy instruments in combination with the latest generation of instruments for a comprehensive holistic analysis.

A method has been developed that allows the complete elimination of all saponification and extraction associated with conventional vitamin D analysis (Byrdwell, 2011). Saponification is a common pre-treatment for samples containing fats or oils along with vitamin D, since conventional wisdom dictates that the former cause chromatographic overlap with the latter. Of course, during saponification all intact triacylglycerols are broken down, thereby allowing unimpeded analysis of vitamin D, but eliminating the possibility for a complete analysis of the acylglycerol components. The new method uses a QTrap 5500 in APCI-MS mode for quantitative analysis of vitamin D, an LCQ Deca XP for qualitative analysis (and/or quantitative analysis) of TAG and TAGOX by ESI-MS (with MS^4), a TSQ 7000 for quantitative analysis of TAG by APCI-MS, as well as UV detection for vitamin D, and an ELSD and Corona CAD, giving six detection methods simultaneously. The separation was conducted on an Agilent 1200 HPLC with two RP columns in series, similar to what we have done with TAG previously (Byrdwell & Neff, 2002). The chromatograms from the Agilent HPLC are not shown, since the focus here is on multiple parallel mass spectrometric detection. Flow was split after the UV detector by tees with different lengths and diameters of fused silica capillary tubing to direct lower flows to the mass spectrometers and higher flows to the 2-D auxiliary detectors, similar to what has been described in our earlier reports. The LC1/MS3 (triple parallel MS) experiments were easily configured by selecting the appropriate detectors on the switching panel pictured in Figure 8.1.

Fig. 8.16 shows the APCI-MS data from the new QTrap 5500 mass spectrometer, acquired during the analysis of oil-filled dietary supplements containing vitamin D_3. The method included full-scan spectra (Fig. 8.16.B), SIM (Fig. 8.16.D), and SRM of two transitions (Figs. 8.16.E and 8.16.F) each for vitamin D_2 and D_3, to allow a comparison of all of these, in the time range 0–28 min (in which vitamin D elutes), followed by full-scan mode from 28–130 min (Fig. 8.16.A, spectra in Figs. 8.16.G–8.16.J). This instrument was very sensitive for SIM and SRM analysis and allowed facile quantification of vitamin D with good S/N ratios. The greater sensitivity of SIM compared to SRM is seen in the maximum counts per second (cps) in the upper right corners of Figs. 8.16.D–8.16.F.

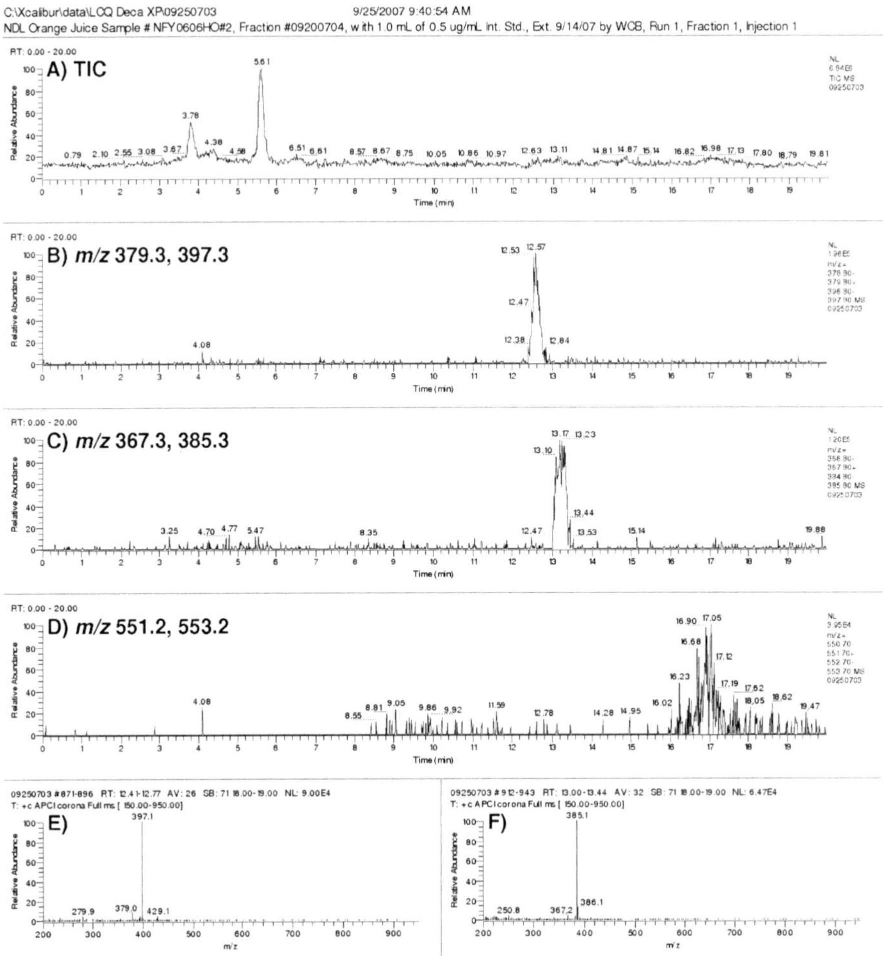

Fig. 8.15. Full-scan qualitative APCI-MS analysis of vitamin D in orange juice extract. A) TIC; B) EIC of *m/z* 379.3 and 397.3 from vitamin D_2; C) EIC of *m/z* 367.3 and 385.3 from vitamin D_3; D) EIC of *m/z* 551.2 and *m/z* 553.2; E) Average mass spectrum 12.41-12.77 min; F) Avg. spectrum 13.00-13.44 min. Parallel SIM APCI-MS data given in Fig. 8.14.

However, the instrument has an upper mass limit of *m/z* 1000 in enhanced MS (EMS) mode, which was required to avoid over-populating the linear trap, otherwise mass shifts and un-reproducible, unusable spectra resulted. Obviously, this mass range is not acceptable for analysis of many of the TAGOX seen in figures discussed in previous sections. Also, as will be shown elsewhere (manuscript in preparation), this mass range is also not useful for identification of some normal TAG containing very long-chain FA (VLCFA). Furthermore, the source design on that instrument led to two

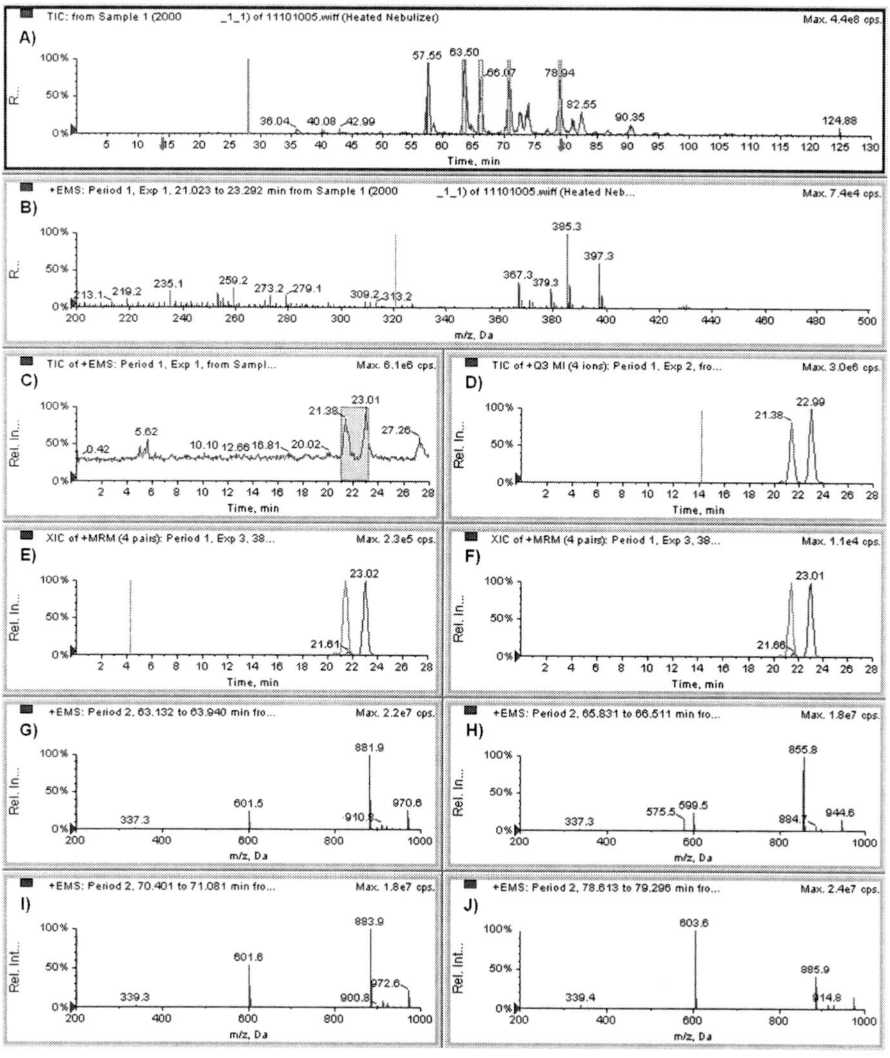

Fig. 8.16. QTrap 5500 APCI-MS analysis of dietary supplement gelcap. A) TIC; B) Average mass spectrum 21.02-23.29 min; C) TIC 0-28 min; D) SIM chromatogram 0-28 min; E) SRM transitions $[M+H]^+$ —> $[M+H-H_2O]^+$ for vitamin D_2 and D_3; F) Second SRM transition; G) Average mass spectrum 63.13-63.94 min, = LLO; H) Avg. spectrum 65.83-66.51 min, LLP; I) Avg. spectrum 70.40-71.08 min, OOL; J) Avg. spectrum 78.61-79.27 min, OOO. Parallel APCI-MS and ESI-MS^n data given in Fig.s 8.17 and 8.18.

substantial problems for TAG analysis: 1) abundant $[M+89]^+$ adducts were formed in the source from a combination of acetonitrile and dichloromethane in the eluent, and 2) the source was much more prone to 'sputtering' caused by accumulation of residue on the corona needle tip. The former problem became even more pronounced at the temperature of 400°C that was optimal for TAG, which is normally used on our other instruments for APCI-MS. Therefore, runs were conducted at 250°C, which was optimal for vitamin D but not for TAG. Thus, TAG could not be run under optimal conditions on that instrument. The latter problem caused the instrument to not be able to conduct a full sequence of 25 to 50 runs that the other instruments could run with no problems. Sequences of runs had to be paused periodically to clean the corona needle. The limited mass range, source design issues, lack of compatible multi-channel high resolution ADC, and several software issues (including the inability to easily produce digital graphics for publication) keeps this instrument from fulfilling its potential as the one instrument that can accomplish all of the types of scan functions and detection modes desired in our lab. There is still a need for multiple parallel mass spectrometers to obtain all of the types of data we desire.

Fig. 8.17 shows the ESI-MS^n data from the LCQ Deca XP ion trap mass spectrometer, obtained in parallel with the QTrap. The combination of an orthogonal source design and improved electrolyte conditions (to be reported elsewhere) produced a TIC, Fig. 8.17.A, that was much improved over that obtained on the LCQ Deca shown in Fig. 8.8. In fact, the ESI-MS TIC was very similar to the corona CAD chromatogram, Fig. 8.17.C, indicating the substantial response differential between unsaturated and saturated TAGs, seen in Fig. 8.10, was greatly reduced. The spectra in Figs. 8.17.E and 8.17.F, which are not background-subtracted, demonstrate the extremely clean, simple spectra provided by the ESI-MS of $[M+NH_4]^+$ of TAG on this ion trap instrument. The MS/MS spectra of the same TAG, Figs. 8.17.H and 8.17.I, are also simple and look very similar to the spectra produced by APCI-MS full-scan spectra. Due to space limitations, MS^3 and MS^4 spectra are not shown, although they were acquired.

The ESI-MS and MS/MS spectra in Fig. 8.17 provide valuable complementary data to that shown in Fig. 8.16, again demonstrating the value of the multiple parallel mass spectrometer technique. Furthermore, Fig. 8.17.G clearly demonstrates the need for the ESI-MS data on the LCQ Deca XP, on which data were obtained in the *m/z* 150–2000 range for dietary supplement sample runs, and in the range *m/z* 150–4000 for the column cleanup run at the end of sequences. Spectrum 8.17G arose from TAGOX dimers, which have masses from $[TAG1+TAG2+nO+NH_4]^+$ (n=0,1,2) ions in the *m/z* 1700–1800 range. Clearly these species cannot be observed on the QTrap instrument, which has an upper mass limit of 1000 amu in EMS mode and 1200 amu in Q3 mode. Of course, there is no possibility to observe TAGOX trimers (~*m/z* 2600-2700) and tetramers (~*m/z* 3500-3600) on the QTrap, or by APCI-MS. We reported the identification of high MW TAGOX species previously by ESI-MS^n (Byrdwell & Neff, 2004), and these were seen in the column cleanup runs at the end of each sequence of triple parallel mass spectrometry runs of the dietary supplements, to be reported elsewhere.

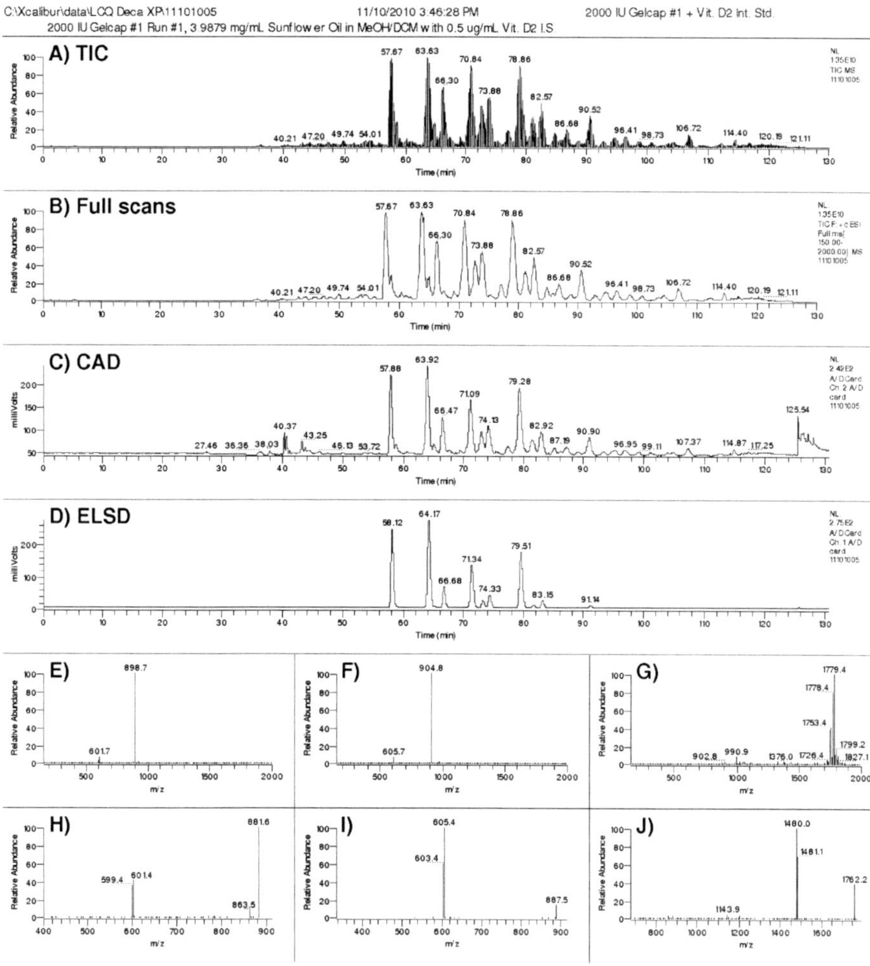

Fig. 8.17. LCQ Deca XP ESI-MS^n analysis of dietary supplement gelcap. A) TIC; B) Full-scan filtered TIC; C) Corona charged aerosol detector (CAD) chromagram; D) ELSD chromatogram; E) Average mass spectrum 63.08-64.18 min; F) Avg. spectrum 90.12-90.86 min; G) Avg. spectrum 119.0-121.0 min; H) MS/MS spectrum of *m/z* 898.7; I) MS/MS spectrum of *m/z* 904.7; J) MS/MS spectrum of *m/z* 1779.3. Parallel runs shown in Figs. 8.16 and 8.18.

Fig. 8.18 shows the data obtained from the TSQ 7000 mass spectrometer in APCI-MS mode, in parallel with the QTrap and LCQ Deca XP. The TIC in Fig. 8.18.A exhibits sharper, less distorted peaks that the APCI-MS TIC in Fig. 8.16.A (this is more apparent when the TICs are viewed full scale), since the QTrap was not operating at the optimal temperature for TAGs. The EIC of the $[M+H]^+$ and $[M+H-H_2O]^+$ ions for vitamins D_2 and D_3 in Fig. 8.18.B shows a

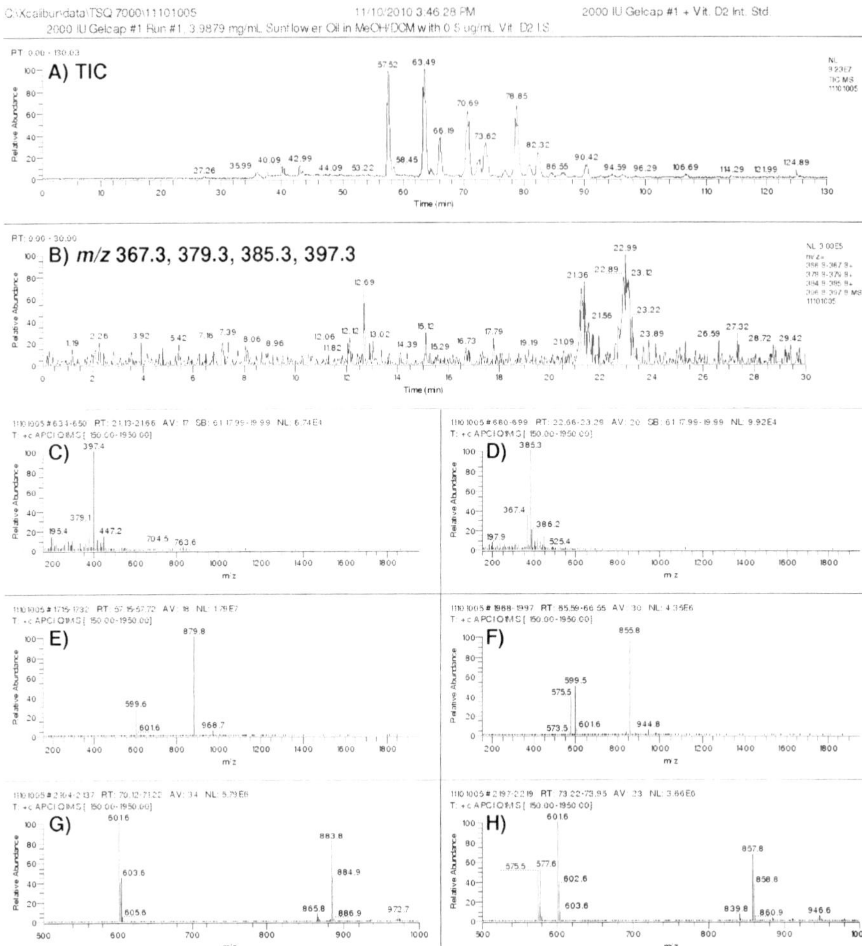

Fig. 8.18. TSQ 7000 APCI-MS data from dietary supplement gelcap. A) TIC; B) EIC of *m/z* 367.3, 379.3, 385.3, 397.3; C) Average mass spectrum 21.13-21.66, = vitamin D_2; D) Avg. spectrum 22.66-23.29, = vitamin D_3; E) Avg. spectrum 57.15-57.72, = LLL; F) Avg. spectrum 65.59-66.55, = LLP; G) Avg. spectrum 70.12-71.22, = OOL; H) Avg. spectrum 73.22-73.95, = POL. Parallel runs shown in Figs. 8.16 and 8.17.

much poorer S/N than the TIC in Fig. 8.16.C, highlighting the sensitivity of the QTrap instrument operating in SIM and SRM modes for vitamin D analysis. However, the APCI-MS spectra appear similar on both instruments (Figs. 8.18.B and 8.18.C compared to Fig. 8.16.B). The APCI-MS spectra of TAGs on the TSQ7000 exhibit smaller $[M+89]^+$ adducts and show mass spectra the same as in the voluminous published literature on LC-APCI-MS of TAGs.

Thus, the APCI-MS data on the TSQ 7000 gave better TICs and cleaner spectra for TAGs, but insufficient sensitivity for vitamin D.

The triple parallel MS experiments, therefore, provided numerous benefits over analysis on a single instrument. The speed and sensitivity of the QTrap allowed SIM and SRM to be compared and contrasted and was extremely effective for vitamin D analysis. However, since it was not operating at optimal conditions for TAGs, and TAGs were present at higher levels that caused problems on the QTrap because of its high sensitivity, the QTrap was not ideal for analysis of these molecules. On the other hand, the TSQ 7000 was operated at optimal conditions for TAGs, and gave cleaner, simpler spectra. This made the TSQ 7000 excellent for TAG analysis. To complement both of these, the LCQ Deca XP gave very clean ESI-MS spectra with high sensitivity and a very broad mass range that allowed VLCFA-containing TAG and TAGOX to be identified with ease. Furthermore, the 24-bit 4-channel ADC data acquired on the LCQ Deca XP instrument provided additional auxiliary detection methods. Thus, these three mass spectrometers operating in different scan modes with two different types of ionization allowed a truly holistic and comprehensive analysis of both vitamin D and TAGs in a single run.

Dual Liquid Chromatography with Dual Parallel Mass Spectrometry, LC2/MS2, for Total Lipid Analysis

Total lipid extracts, such as those by Folch et al. (1957) and Bligh and Dyer (1959) give mixtures of both polar lipids (e.g., phospholipids) and non-polar lipids (e.g., triacylglycerols). These typically require different chromatographic methods for analysis, as described above. We demonstrated a column-switching technique that allowed all lipids, both polar and non-polar, to be analyzed by both NP- and RP-HPLC with ESI-MS^n and ESI-MS/MS, respectively, from a single injection (Byrdwell, 2003). The data cited in the sections above demonstrated the benefit of ESI-MS for analysis of TAG ammonium adducts, especially for saturated species and those with few sites of unsaturation. Therefore, ESI-MS with MS/MS on a TSQ 700 was used for detection of TAGs in a commercial bovine brain total lipid extract separated by RP-HPLC, while at the same time (+) ESI-MS, MS/MS, and MS^3, as well as (-) ESI-MS, scans were acquired on an LCQ Deca ion trap mass spectrometer from polar lipids separated by NP-HPLC. The LC2/MS2 column-switching arrangement that incorporates both NP-HPLC and RP-HPLC should not be confused with what has come to be known as 'comprehensive two-dimensional liquid chromatography'. The latter technique is characterized by the fact that every molecular species that elutes from one column is directed to a second column that represents a second dimension of separation, and all species have retention times on two columns. This topic is expertly covered in detail by Paola Dugo in Chapter 7 of this volume.

Fig. 8.19.A shows the TIC of the phospholipids of a bovine brain extract, reflecting the alternating (+), (+) MS/MS, (+) MS^3 and (-) spectra. Fig. 8.19.B shows the EIC of (–) full scans. The mass spectra in Figs. 8.19.C through 8.19.N show alternating (+) and (–) full scan spectra. The different ionization proclivities of several classes of polar lipids are evident. Figs. 8.19.G and

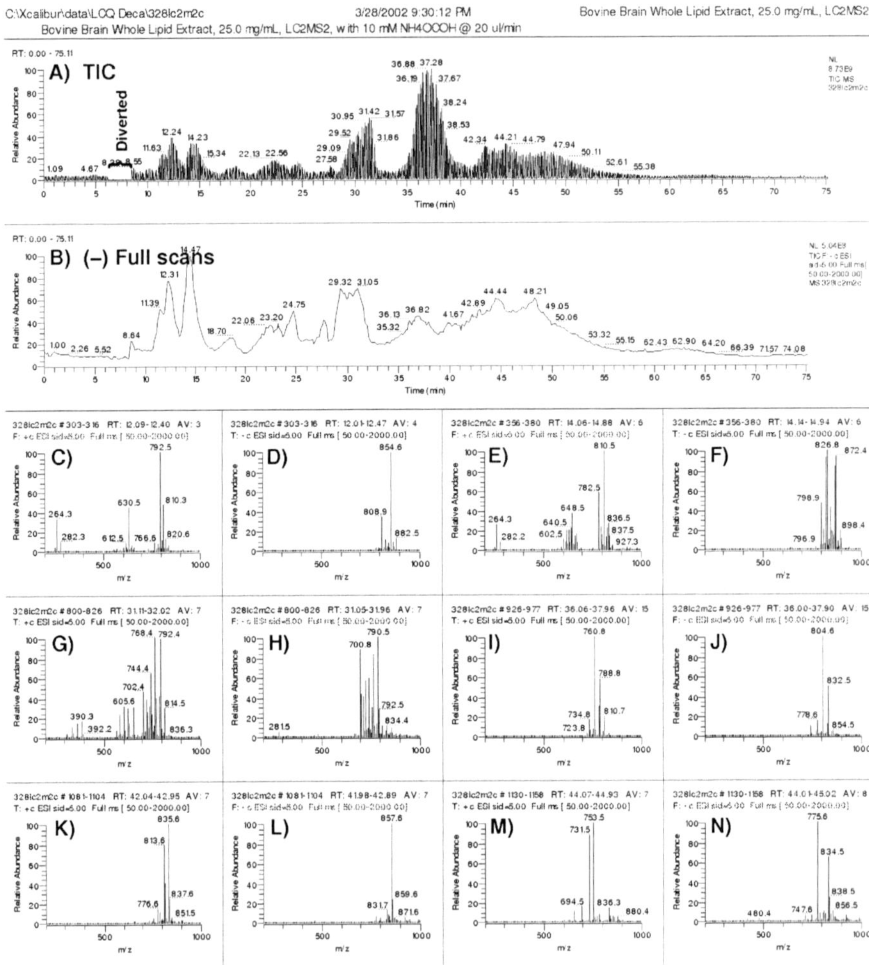

Fig. 8.19. LCQ Deca (+) ESI-MS^n and (–) ESI-MS data from polar lipids of commercially available bovine brain total lipid extract, separated by NP-HPLC. A) TIC; B) (-) Filtered TIC; C) Average (+) mass spectrum 12.09-12.40 min, = glycosphingolipids; D) Avg. (–) spectrum 12.01-12.47 min; E) Avg. (+) spectrum 14.06-14.88 min, = glycosphingolipids; F) Avg. (–) spectrum 14.14-14.94 min; G) Avg. (+) spectrum 31.11-32.02 min, = PE and PE plasmalogen; H) Avg. (–) spectrum 31.05-31.96 min; I) Avg. (+) spectrum 36.06-37.96 min, = PC; J) Avg. (–) spectrum 36.00-37.90 min; K) Avg. (+) spectrum 42.04-42.95 min, = long-chain SM; L) Avg. (–) spectrum 41.95-42.89 min; M) Avg. (+) spectrum 44.07-44.93 min, = medium-chain SM + PS; N) Avg. (–) spectrum 44.01-45.02 min. Orthogonal run of diverted neutral lipids shown in Fig. 8.20.

8.19.H show the tendency of phosphatidylethanolamine (PE) and PE plasmalogen species to form $[M+H]^+$ and $[M-H]^-$ ionic molecules. In contrast, the phosphocholine-containing PLs showed $[M+H]^+$ and $[M+Na]^+$ ions in (+) mode, and formate adducts, $[M+OCOH]^-$, $=[M+45]^-$, in negative mode, as discussed in Triacylglycerols above. The SM species shown in Figs. 8.19.K and 8.19.M showed a more abundant $[M+Na]^+$ ion than $[M+H]^+$ (24:1 SM has $[M+H]^+ = m/z$ 813.6 in Fig. 8.19.K, and 18:0 SM has $[M+H]^+ = m/z$ 731.5 in Fig. 8.19M). On the other hand, phosphatidylcholine (PC) species in Fig. 8.19.I gave a larger $[M+H]^+$ than $[M+Na]^+$.

By using a survey approach instead of a targeted approach, we were able to identify several galactosphingolipids and sulfatogalactosphingolipids, two species of which are shown in Figs. 8.19.C through 8.19.F. Homologues were identified that differed only by the presence of a hydroxyl group. Since hydroxyl groups are readily lost in positive-ion mode, even with soft ionization, the spectra were so similar as to make differentiating the species challenging. The negative-ion spectra provided critical information, in the form of both intact $[M-H]^-$ and $[M+45]^-$ ions, that allowed the molecular weights to be identified. These ions, plus the fragments at m/z 264.3 and 282.3 that are characteristic of sphingolipids in (+) mode (Figs. 8.19.C and 8.19.E), in combination with the ESI-MS^n, allowed the structures to be elucidated. Detailed discussion of the spectra and the corresponding proposed structures were given previously (Byrdwell, 2003). These data showed that sphingolipids formed ammonium adducts by coordination of the formate anion with a charged backbone amine, as well as the typical coordination of formate with the phosphocholine quaternary amine typical of PC (Fig. 8.19.J).

The survey method also allowed overlapping classes to be differentiated. For example, Fig. 8.19.N shows a peak at m/z 834.5 that had an entirely different chromatographic profile than the EIC (not shown) of m/z 775.6, which is the base peak in Fig. 8.19.N. The m/z 834.5 ion was identified as the 40:6-phosphatidylseriene (PS) molecular species, which was confirmed by the protonated molecule at m/z 836.3 in Fig. 8.19.M. Detailed treatment of the range of PS molecular species was given in the original report (Byrdwell, 2003). With such complex samples, it would certainly have been beneficial to have acquired APCI-MS data to complement the ESI-MS^n data. This would have required a third mass spectrometer in an LC2/MS3 configuration, since our second mass spectrometer was already being used to obtain ESI-MS data for non-polar lipids on the RP-HPLC system that was coupled to the NP-HPLC system in this LC2/MS2 experiment.

The neutral lipid bolus that was virtually un-retained on the NP-HPLC column came off the column in the range 6 to 8.25 min, and was diverted to the head of the RP-HPLC column. The alternating ESI-MS and MS/MS TIC is shown in Fig. 8.20.A, and the full-scan EIC is in Fig. 8.20.B. The largest peak represents a ceramide (Cer), which is the same as a sphingolipid backbone, but lacks a polar head group and so was not substantially retained on the

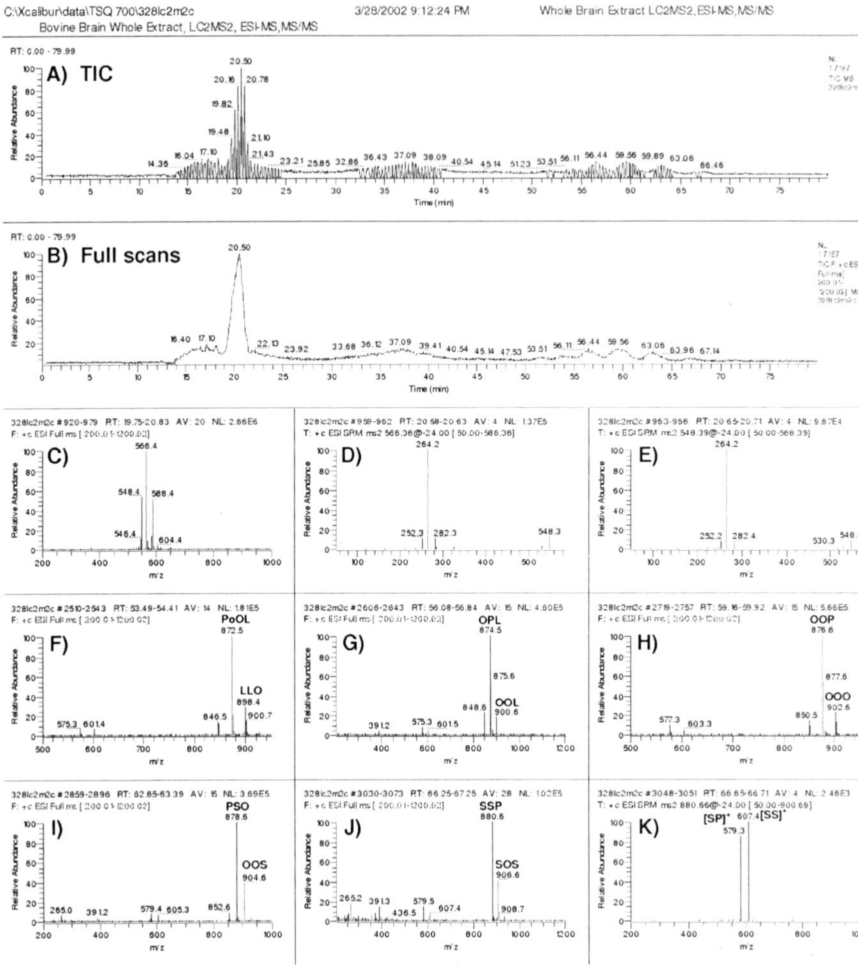

Fig. 8.20. TSQ 700 APCI-MS data from neutral lipids of commercially available bovine brain lipid extract diverted in column-switching experiment, separated by RP-HPLC. A) TIC; B) Full-scan filtered TIC; C) Average mass spectrum across peak at 20.50 min (19.75-20.83 min), = 18:0 Ceramide; D) MS/MS spectrum of *m/z* 566.4 in C); E) MS/MS spectrum of *m/z* 548.4 in C); F) Avg. spectrum 53.49-54.41 min, = PoOL; G) Avg. spectrum 56.08-56.84 min, = OPL; H) Avg. spectrum 59.16-59.92, = OOP (+PSL); I) Avg. spectrum 62.65-63.39, = PSO; J) Avg. spectrum 66.25-67.25 min, = SSP; K) MS/MS spectrum of *m/z* 880.7. Po = palmitoleic acyl chain, 16:1. Orthogonal run of polar lipids given in Fig. 8.19.

NP-HPLC system. The dominant species (largest peak in Fig. 8.20.B) was 18:0 Cer, the ESI-MS mass spectrum of which is shown in Fig. 8.20.C. It exhibited a protonated molecule at *m/z* 566.4, a dehydration product, $[M+H-H_2O]^+$ at *m/z* 548.4, and a natriated molecule, $[M+Na]^+$ at *m/z* 588.4. The loss of H_2O was so facile that it occurred even under the very mild ESI source conditions, and it caused the ESI-MS/MS spectrum of *m/z* 566.4, Fig. 8.20.D, to appear virtually the same as the MS/MS spectrum of *m/z* 548.4, Fig. 8.20.E. Ceramide has a second hydroxyl group (from its serine precursor), in addition to that next to the 4,5 *trans* site of unsaturation (formed by *3-ketosphinganine reductase* during synthesis), so it gave a fragment at *m/z* 530.3 in both spectra, from a second loss of water. Of course, the ions at *m/z* 264.3 and 282.3 were characteristic of molecules containing the ceramide backbone, similar to the sphingolipids seen in Fig. 8.19.C and 8.19.E.

Since the mobile phase contained ammonium formate to facilitate adduct formation, $[M+NH_4]^+$ of TAGs were observed. The spectra in Figs. 8.20.F through 8.20.K represent average spectra across the tops of the major TAG peaks eluted in the range of 50 to 70 min. These spectra indicate (and EICs of $[DAG]^+$ confirm) that the most abundant FA was 16:0. The spectra in Figs. 8.20.J and 8.20.K again demonstrate the benefit of $[M+NH_4]^+$ of TAGs. The saturated TAG SSP does not produce a useful abundance of $[M+H]^+$ by APCI-MS, but by ESI-MS, the $[M+NH_4]^+$ was the base peak. This ion provided the molecular weight and acted as an excellent precursor ion, giving the ESI-MS/MS spectrum in Fig. 8.20.K, which showed the $[DAG]^+$ fragments that conclusively identify the $[PS]^+$ and $[SS]^+$ structural sub-units of the TAG (P=palmitoyl moiety, 16:0; S= stearoyl moiety, 18:0).

The LC1/MS2 experiment served as a demonstration of the possibility to identify both polar and non-polar lipids in a total lipid extract in one injection. When only one MS was used for detection of TAGs and other neutral lipids (compared to an LC1/MS2 experiment), ESI was preferred because of the definitive $[M+NH_4]^+$ adducts and informative MS/MS spectra. This experiment employed an older TSQ 700 that required that the acquisition program be written manually. Modern instruments make these types of experiments much easier, since this type of automated peak-picking and automatic MS/MS is now a standard function. Easier-to-use software, electronic switching valves built into most modern instruments, and a switching system such as that pictured in Fig. 8.1 make experiments combining multiple liquid chromatographs with different combinations of mass spectrometers into LC*x*/MS*y* within reach of any determined chromatographer. The benefits of the wealth of information provided far outweigh the challenges in the initial set-up. Plus, once the switching system is constructed, it allows a wide variety of combination of LC with MS using different ionization modes. If our current research provided the opportunity to continue work on total lipid extracts, it would be very straightforward to add APCI-MS to both the NP-HPLC analysis of PLs and the RP-HPLC of TAGs and neutral lipids in an LC2/MS4 configuration.

Conclusions

The data above have clearly shown the benefits of obtaining multiple types of data from different ionization modes to thoroughly characterize molecules of interest. No single technique is completely satisfactory—all have some shortcomings. For instance, while ESI-MS is very effective for many classes of molecules, as long as the appropriate electrolyte is used, some classes, such as vitamin D simply do not provide good ESI-MS spectra. For such classes, APCI-MS is still an indispensable tool. Sometimes the shortcoming is instrument-specific, such as a limited mass range or other instrument short-comings. Even with fast and sensitive instruments, there will always be more combinations of scan types (SIM, SRM, MS/MS, MS^3, MS^4), polarity (positive- and negative-ion) and ionization type (ESI and APCI, as well as APPI) than can be performed on a single instrument. It seems that some compromise must always be made. However, when two or more instruments are available, each can be run under its own optimal conditions (i.e., source temperature) to allow the best acquisitions conditions for each instrument. Thus, optimal data can be obtained without compromise.

One large hurdle to using multiple parallel mass spectrometry for LC detection seems to be a mental hurdle. Many, if not most, people may be intimidated by the prospect of wiring the contact closures between all of the instruments and wiring the auxiliary detectors to the necessary ADCs. But fortunately, modern instruments are making the process easier. Virtually all instruments can be synchronously started via contact closure.

The second apparent hurdle may be simply having multiple mass spectrometers at one's disposal. This hurdle can be due to limited resources that do not allow the purchase of multiple instruments, or it can be due to a shared instrument environment (core facility) that does not allow a researcher exclusive access to instruments. The first problem (instrument cost) is a problem that tends to solve itself over time, as a laboratory inevitably upgrades to newer instruments. Certainly, instrument manufacturers intentionally try to hasten the upgrade process by rendering older instruments officially obsolete and stop providing service. In such cases, those researchers who perform their own service and are intimately familiar with the internal workings of their instruments have a clear advantage. Those with the necessary aptitude can keep instruments running long after they have been made officially obsolete, since third-party suppliers of parts continue to make components available.

Obviously, in a production environment multiple parallel MS experiments may not be practical. However, in a research environment these types of LC*x*/MS*y* experiments serve as terrific training opportunities for eager graduate students. Here again, mechanical and electrical aptitude dictate whether such experiments are feasible.

Another hurdle is the fact that the more instruments that are involved in an experiment (especially older instruments), the harder it is to have them all operating without error or glitch at the same time. Nowadays, our routine experiments require four computers operating Microsoft Windows-based acquisition systems. Just getting the software to operate flawlessly for run after run in a sequence is challenging. It was certainly a daunting task to get six detectors running in the LC1/MS3 experiments described in Triple Parallel

Mass Spectrometry, LC1/MS3, for Vitamin D and Triacylglycerol Analysis. For method validation, some sequences were run that used five 130 min runs in quintuplicate (5×5=25 sample runs) plus bracketed sets of ~30 min calibration runs (three calibration levels) before, between, and after sample runs (3×6=18 runs) plus a 70 min column cleanup run. Such sequences had to run for multiple days around the clock without any single component failing. As might be expected, we have many gigabytes of data from partial sequences that never made it to completion. We now routinely use three samples in triplicate (=3×3 sample runs), plus three calibration standards bracketed before, between and after sample runs (=3×4 calibration runs), plus a cleanup run, to somewhat simplify the analyses. Admittedly, such complex runs are not for the fainthearted. Nevertheless, the switching system we constructed makes numerous combinations of experiments as easy as they can be, given the other complexities.

In short, for those few researchers who have the time, instruments, aptitude and patience to perform LC*x*/MS*y* experiments, the wealth of data they provide makes all of the effort worthwhile. The sense of accomplishment and satisfaction that results when a room full of instruments springs to life in a unified symphony of data acquisition is its own irreplaceable reward.

References

Bligh, E.G.; Dyer, W.J. A rapid method of total lipid extraction and purification. Can. J. Biochem. Physiol. 1959, 37, 911-917.

Byrdwell, C.; Borchman, D. Liquid chromatography/mass-spectrometric characterization of sphingomyelin and dihydrosphingomyelin of human lens membranes. Ophthalmic Res. 1997, 29, 191-206.

Byrdwell, W.C. Dual Parallel Liquid Chromatography/Dual Mass Spectrometry (LC2/MS2) of Bovine Brain Total Lipid Extract. J. Liq. Chromatogr. Relat. Technol. 2003, 26, 3147-3181.

Byrdwell, W.C. Qualitative and Quantitative Analysis of Triacylglycerols by Atmospheric Pressure Ionization (APCI and ESI) Mass Spectrometry Techniques. Modern Methods for Lipid Analysis by Liquid Chromatography/Mass Spectrometry and Related Techniques, 1; W.C. Byrdwell, Ed.; AOCS Press: Champaign, IL, 2005a; pp. 298-412.

Byrdwell, W.C. The bottom-up solution to the triacylglycerol lipidome using atmospheric pressure chemical ionization mass spectrometry. Lipids 2005b, 40, 383-417.

Byrdwell, W.C. 'Dilute-and-Shoot' Method for Vitamin D Analysis in Dietary Supplement Capsules. FASEB J. 2011, 25, In Press.

Byrdwell, W.C.; Exler, J.; Gebhardt, S.E.; Harnly, J.M.; Holden, J.M.; Horst, R.L.; Patterson, K.Y.; Phillips, K.M.; Wolf, W.R. Liquid chromatography with ultraviolet and dual parallel mass spectrometric detection for analysis of vitamin D in retail fortified orange juice. J. Food Comp. Anal. 2011, 24, In Press. DOI:10.1016/j.jfca.2010.09.020.

Byrdwell, W.C.; Neff, W.E. Non-volatile products of triolein produced at frying temperatures characterized using liquid chromatography with online mass spectrometric detection. J. Chromatogr. A 1999, 852, 417-432.

Byrdwell, W.C.; Neff, W.E. Electrospray Ionization MS of High M.W. TAG Oligomers. J. Am. Oil Chem. Soc. 2004, 81, 13-26.

Byrdwell, W.C.; Perry, R.H. Liquid chromatography with dual parallel mass spectrometry and 31P nuclear magnetic resonance spectroscopy for analysis of sphingomyelin and dihydrosphingomyelin. I. Bovine brain and chicken egg yolk. J. Chromatogr. A 2006, 1133, 149-171.

Byrdwell, W. Dual parallel mass spectrometers for analysis of sphingolipid, glycerophospholipid and plasmalogen molecular species. Rapid Commun. Mass Spectrom. 1998, 12, 256-272.

Byrdwell, W.; Neff, W.E. Dual parallel electrospray ionization and atmospheric pressure chemical ionization mass spectrometry (MS), MS/MS and MS/MS/MS for the analysis of triacylglycerols and triacylglycerol oxidation products. Rapid Commun. Mass Spectrom. 2002, 16, 300-319.

Byrdwell, W.; Perry, R.H. Liquid chromatography with dual parallel mass spectrometry and 31P nuclear magnetic resonance spectroscopy for analysis of sphingomyelin and dihydrosphingomyelin. II. Bovine milk sphingolipids. J. Chromatogr. A 2007, 1146, 164-185.

Cheng, C.; Gross, M.L.; Pittenauer, E. Complete Structural Elucidation of Triacylglycerols by Tandem Sector Mass Spectrometry. Anal. Chem. 1998, 70, 4417-4426.

Cvacka, J.; Krafkova, E.; Jiros, P.; Valterova, I. Computer-assisted interpretation of atmospheric pressure chemical ionization mass spectra of triacylglycerols. Rapid Commun. Mass Spectrom. 2006, 20, 3586-3594.

Duffin, K.L.; Henion, J.D.; Shieh, J.J. Electrospray and tandem mass spectrometric characterization of acylglycerol mixtures that are dissolved in nonpolar solvents. Anal. Chem. 1991, 63, 1781-1788.

Duncan, W.P.; Perkins, P.D. LC-MS with simultaneous electrospray and atmospheric pressure chemical ionization. Am. Lab. 2005, 37, 28-33.

Folch, J.; Lees, M.; Sloane-Stanley, G.H. A simple method for the isolation and purification of total lipides from animal tissue. J. Biol. Chem. 1957, 226, 497-509.

Fonteh, A.N.; Harrington, R.J.; Huhmer, A.F.; Biringer, R.G.; Riggins, J.N.; Harrington, M.G. Identification of disease markers in human cerebrospinal fluid using lipidomic and proteomic methods. Dis. Markers 2006, 22, 39-64.

Han, X.; Gross, R.W. Shotgun lipidomics: Electrospray ionization mass spectrometric analysis and quantitation of cellular lipidomes directly from crude extracts of biological samples. Mass. Spectrom. Rev. 2005a, 24, 367-412.

Han, X.; Gross, R.W. Shotgun lipidomics: Multidimensional MS analysis of cellular lipidomes. Expert Rev. Proteomics 2005b, 2, 253-264.

Hsu, F.F.; Turk, J. Structural characterization of triacylglycerols as lithiated adducts by electrospray ionization mass spectrometry using low-energy collisionally activated dissociation on a triple stage quadrupole instrument. J. Am. Soc. Mass Spectrom. 1999, 10, 587-599.

Hvattum, E. Analysis of triacylglycerols with non-aqueous reversed-phase liquid chromatography and positive ion electrospray tandem mass spectrometry. Rapid Commun. Mass Spectrom. 2001, 15, 187-190.

Laakso, P.; Voutilainen, P. Analysis of triacylglycerols by silver-ion high-performance liquid chromatography-atmospheric pressure chemical ionization mass spectrometry. Lipids 1996, 31, 1311-1322.

Maceyka, M.; Milstien, S.; Spiegel, S. Sphingosine kinases, sphingosine-1-phosphate and sphingolipidomics. Prostag. Other Lipid Med. 2005, 77, 15-22.

Merrill, J.; Sullards, M.C.; Allegood, J.C.; Kelly, S.; Wang, E. Sphingolipidomics: High-throughput, structure-specific, and quantitative analysis of sphingolipids by liquid chromatography tandem mass spectrometry. Methods 2005, 36, 207-224.

Mottram, H.R.; Crossman, Z.M.; Evershed, R.P. Regiospecific characterisation of the triacylglycerols in animal fats using high performance liquid chromatography-atmospheric pressure chemical ionisation mass spectrometry. Analyst 2001, 126, 1018-1024.

Mottram, H.R.; Evershed, R.P. Structure analysis of triacylglycerol positional isomers using atmospheric pressure chemical ionisation mass spectrometry. Tetrahedron Lett. 1996, 37, 8593-8596.

Mottram, H.R.; Woodbury, S.E.; Evershed, R.P. Identification of triacylglycerol positional isomers present in vegetable oils by high performance liquid chromatography/atmospheric pressure chemical ionization mass spectrometry. Rapid Commun. Mass Spectrom. 1997, 11, 1240-1252.

Navas-Iglesias, N.; Carrasco-Pancorbo, A.; Cuadros-RodrÃ-guez, L. From lipids analysis towards lipidomics, a new challenge for the analytical chemistry of the 21st century. Part II: Analytical lipidomics. Trend. Anal. Chem. 2009, 28, 393-403.

Sjovall, O.; Kuksis, A.; Marai, L.; Myher, J.J. Elution factors of synthetic oxotriacylglycerols as an aid in identification of peroxidized natural triacylglycerols by reverse-phase high- performance liquid chromatography with electrospray mass spectrometry. Lipids 1997, 32, 1211-1218.

Spener, F.; Lagarde, M.; Géloën; A; Record, M. What is lipidomics? Eur. J. Lipid Sci. Technol. 2003, 105, 481-482.

Tsugawa, N.; Suhara, Y.; Kamao, M.; Okano, T. Determination of 25-hydroxyvitamin D in human plasma using high-performance liquid chromatography-tandem mass spectrometry. Anal. Chem. 2005, 77, 3001-3007.

Willmann, J.; Leibfritz, D.; Thiele, H. Hyphenated tools for phospholipidomics. J. Biomol. Tech. 2008, 19, 211-216.

9

Nano-liquid Chromatographic Separations

Chiara Fanali[1], María Asensio-Ramos[2], Javier Hernández-Borges[2], Anna Rocco[3], and Salvatore Fanali[3]

[1]University Campus-Biomedico, Via Alvaro del Portillo, 21, 00128 Rome, Italy; [2]Departamento de Química Analítica, Nutrición y Bromatología, Facultad de Química, Universidad de La Laguna (ULL), Avenida Astrofísico Francisco Sánchez s/n, 38206 La Laguna (Tenerife), España; and [3]Institute of Chemical Methodologies, Italian National Council of Research, Area della Ricerca di Roma I, Via Salaria Km 29,300, 00015 Monterotondo Scalo, Rome, Italy.

Introduction

Nano-liquid chromatography (nano-LC), first introduced by Karlsson and Novotny in 1988 (Karlsson & Novotny, 1988), is a miniaturized liquid chromatographic technique where the separation of analytes takes place on a capillary column containing selected stationary phases (SPs). Even though a standard definition of nano-LC has not been finalized, usually this name is assigned to those techniques employing capillaries of I.D. <100 μm and flow rates in the range 50–800 nL/min. In contrast, capillary-LC (CLC) makes use of 100–320 μm I.D. columns and flow rates in the range 1–100 μL/min. Since the boundary between nano-LC and CLC overlaps, both of them will be considered in this chapter. Furthermore, other techniques which employ capillaries, such as capillary zone electrophoresis (CZE) and Capillary Electrochromatography (CEC), will be discussed and compared due to the similarity of their set up.

In nano-LC, the relatively low flow-rate represents an interesting advantage over conventional techniques allowing the perfect coupling with mass spectrometry (MS). In addition: costs are reduced due to the limited use of mobile phases and consequently low waste makes this technique eco-friendly. In addition peak dilution during the chromatographic run is reduced with an increase of mass sensitivity. Finally high efficiency, good resolution and short analysis times are currently obtained (Karlsson & Novotny, 2988; Szumski & Buszewski, 2002; Hernández-Borges et al., 2007a).

Because of its features and advantages, nano-LC has been applied mainly in analytical chemistry for the analysis of a large number of compounds currently investigated in various fields such as proteomics, pharmaceutical, chiral, food and beverage, environmental samples, etc.

The aim of this publication is to illustrate the state of the art of nano-LC by reviewing the basic principles, the instrumentation currently used and its advantages over conventional high performance liquid chromatography

(HPLC). Specific applications of nano-LC are also reported and described in this chapter. They cover not only the chromatographic separation of several compounds but also the validation of the method and quantitative aspects both applied to the analysis of real samples of different origin. In this chapter, nano-LC performed in micro-chip will only be mentioned because, due to its importance for the implementation of miniaturized techniques, requires a separate chapter for a thorough discussion.

Miniaturization in Liquid Chromatography

Theoretical Considerations

Improved Efficiency by Decreasing Column Internal Diameter (I.D.)

In their study about the separation efficiency of packed columns with small inner diameter, Karlsson and Novotny (1988) evaluated packed columns in the I.D. range 44–265 μm. In this study the system variance (σ^2_{sys}) can be described by the following equation:

$$\sigma^2_{sys} = \sigma^2_{ext} + \sigma^2_{col} \tag{1}$$

Where σ^2_{ext} and σ^2_{col} are the external and column variance, respectively.

The external dispersion is accounted for the broadening caused by the different parts of the LC system with the exclusion of the column. In order to minimize σ^2_{ext} the column was directly connected with the sample valve injection. In addition the authors recorded several peaks in the same LC run moving the detector along the column. This approach was used to show the increase of peak width with column length having a constant external band broadening.

It has been reported that the extra column variance should influence the decrease of column plate numbers not more than 10% (Yeung, 1985; Gluckman & Novotny, 1985). Therefore there is a maximum extra column variance acceptable as reported in the following equation:

$$\sigma_{e^2(acc)} \leq \sigma^2_{col} * 0.10 \leq \frac{\pi^2 L^2 r^4 \varepsilon^2 (1 + k)^2}{N} \tag{2}$$

where N is the number of theoretical plates, L the column length, k the retention factor, ε the column permeability and r the column radius.

From this equation it is evident that the most important parameter, influencing the column variance, is the radius of the column. Consequently when using columns with very low I.D. there is a requirement of a strong reduction of extra-column effect. This effect can be controlled by selecting an appropriate design of the nano-LC system such as I.D. and length of connecting tubes, injection valve (type and volumes), detector (cell volume and electronic time constant), etc.

In nano-LC, cell volumes are in the order of a few nL, while injected sample volumes are usually in the range of 10–60 nL. However, when appropriate experimental conditions are selected, larger volumes can be injected utilizing the focusing effect in order to increase the sensitivity.

For plate-height evaluation Kalsson and Novotny used the following equation:

$$u_s^2 M2_{sys} = M2_{ext} + HL \tag{2}$$

where $M2$ is the second central moment and u_s the migrating zone speed. Reporting $u_s^2 M2$ vs. column length (mm) at different mobile phase flow rates, straight lines are obtained. The slopes of the lines give the plate-heights, while the intercepts offer data about external band broadening.

To study the influence of the column I.D., Karlsson and Novotny (1988) applied the equation studied by Kennedy and Knox (1972) based on reduced parameters h (minimum plate height) and v (linear velocity):

$$h = Av^{0.33} + B/v + Cv \tag{3}$$

B term (longitudinal diffusion) slightly increased by decreasing the I.D. column, while A (flow dispersion) and C (resistance to mass transfer) decreased by decreasing the column I.D. The lowest h values were obtained using the capillary column of 44 μm I.D. This work was the starting of capillary/nano-LC that later on was further studied not only from the theoretical point of view but also developing technology and methodology.

Chromatographic Dilution

Compounds analyzed by chromatography are transported by the mobile phase towards the detector. It is well-known that even if the sample is injected as a narrow zone, due to chromatographic dilution, a broadening effect is introduced. Consequently, this effect must be minimized to obtain the highest efficiency.

The following equation about chromatographic dilution has been proposed (Vissers et al., 1997):

$$D = \frac{c_o}{c_{max}} = \frac{\sigma \pi r^2 (1 + k)\sqrt{2\pi L H}}{V_{inj}} \tag{4}$$

where c_o and c_{max} are the concentration of the sample at the injection and at the peak maximum, respectively; r and L are the column radius and length, respectively; H is the plate-height, ε the column porosity and k the retention factor.

As can be observed in this equation, to have a low dilution, it is necessary to maximize the sample concentration at the peak maximum. This can be obtained either by injecting a high volume of sample and/or using a shorter column and/or a system with lower plate-height, lower porosity. Finally, the use of columns with low radius, due to the square root dependence, will give certainly the lowest chromatographic dilution. Therefore, injecting the same sample volume and using columns of different radius with SPs of the same characteristics, passing from an I.D. of 4.6 mm to ones of 300 µm or 75 µm it can be calculated an increase in peak height of 235- and 3800-fold, respectively.

Capillary Columns

The choice of the appropriate chromatographic column, considering the length, the I.D. and the SP type is fundamental for the success of the undertaken analytical separation. For long, in liquid chromatography (LC) the geometry of the column was under investigation especially to get a better performance in terms of efficiency and to speed up analyses (Chen & Horvath, 1995; Poppe, 1997). The use of chromatographic columns with smaller and smaller I.D. aroused lively interest in the scientific community for several accessible advantages such as easy coupling with MS, high efficiency, low consumption of reagents and resultant low environmental pollution (Abian et al., 1999; Saito et al., 2004; Szunski & Beszewski, 2002). This attention has been even increased since the recent studies on several *omics* branches of biology have been done on large level, industrially as well (Svec, 2002; Dong, 2006; Mechref & Novotny, 2010; Mikulovic & Mechtler, 2006). As a result, miniaturized separation techniques, employing chromatographic columns with reduced I.D. (so called capillary columns), i.e., CLC/nano-LC, CEC, have been recognized worldwide as a valid alternative to classical analytical techniques, specifically HPLC and GC (Gübitz & Schmid, 2008; Asensio-Ramos et al., 2009; Fu et al., 2003).

CEC, generally described as a hybrid technique exploiting advantages of both HPLC (i.e., selectivity) and CZE (i.e., high efficiency) (Fu et al., 2003), will be considered as well because it makes use of the same capillary columns employed in CLC/nano-LC. The only exception is the requirement of particular properties of the SP, that in CEC has to assure an electroosmotic flow (EOF) (Svec, 2002). Theoretical considerations about this technique is out of the aim of this chapter, but they can be found in several reviews (Robson et al., 1997; Rathore, 2002). However, procedures developed to obtain capillary columns in CEC can be adopted for CLC/nano-LC and vice versa, and for this reason it is possible to consider parallel the evolution of these miniaturized techniques, at least considering the separation medium.

The I.D. of capillary columns is in the range of 100–500 μm for CLC, 10–100 μm for nano-LC (Vissers et al., 1997; Saito et al., 2004), and 10–350 μm for CEC (Eeltink et al., 2003; Robson et al., 1997; Colón et al., 1997). Usually columns consist of a fused silica capillary or a polyether ether ketone (PEEK) tube (Szunski & Buszewski, 2002; Cappiello et al., 1991). Fused silica capillaries are preferred owing to several benefits, including UV transparency, higher mechanical strength than polymeric materials (PEEK) and flexibility due to the polyimide coating (Szumski & Buszewski, 2002; Saito et al, 2004). In any case, some precautions have to be taken also when working with these columns in order to preserve their performance and not to destroy them. For example, the presence of zones where the polyimide coating has been removed (i.e., at detection window or frits) makes the capillary very fragile (Saito et al., 2004).

In CLC/nano-LC, as well as in CEC, three kinds of capillary columns can be employed: packed, monolithic and open tubular (OT) ones. Each type will be discussed separately.

All mechanisms capable of selectively delaying the analytes and described elsewhere for LC (for instance, partition, adsorption, ion exchange, affinity chromatography) are exploited in CLC, nano-LC and CEC, too. Furthermore, independently on the type of capillary column selected, the need of very small amount of stationary and mobile phase has made accessible the use of expensive materials (i.e., chiral SP (CSP) or additives) commercially available, to improve the selectivity of the chromatographic column (Healy et al., 2001; Rocco & Fanali, 2009b). Nevertheless, the great majority of separations performed are based in reversed phase mode (Hernández-Borges et al., 2007a; Pursch & Sander, 2000).

Finally, the reduced dimensions of capillary columns make them particularly suitable for temperature studies (Saito et al., 2004).

Packed Capillary Columns

Packed capillary columns are widely used due to the considerable variety of particulate SPs existing for HPLC. These SPs allow the separation of analytes by partition, size exclusion, adsorption, affinity and/or ion exchange mechanism and, arising from a exceedingly established technique such as HPLC, have well defined packing parameters (regular particle and pore size, improved bonding chemistry) (Hernández-Borges et al., 2007a; Svec, 2002). The selection of the packing material is done considering the nature of analytes (basic, acidic or neutral compounds, chiral or achiral, with high or low molecular weight). SPs with a 3–5 μm particle diameter are mostly employed. More recently, particles of smaller size (1.5–1.8 μm), developed to obtain higher efficiency and selectivity, and specific for ultra performance liquid chromatography (UPLC), have also been utilized in nano-LC and CEC (Robson et al., 1997). In this last technique, the use of an electroosmotic flow

(EOF) as driving force avoids the occurrence of high back pressure, even when very small particles are employed (Svec, 2002).

Dwelling upon CEC, it is worth noting that SPs have to guarantee specific interactions for the analytes and, at the same time, a high EOF, indispensable as driving force (Eeltink & Svec, 2007). Usually, SPs are constituted by bonded silica material, and residual silanol groups (Si-OH), ionisable at pH higher than 3, are good promoters for a high EOF (Pursch & Sander, 2000; Colón et al., 1997). However, the same silanol groups are also responsible for peak tailing due to their strong adsorptive properties (Fanali et al., 2007). Unfortunately, most SPs used in CEC are tailored for chromatography and therefore, residual silanol groups are further derivatized (end-capped) to improve the chromatographic performance (Svec, 2002). As a consequence, the resulting SPs provide a low EOF and are less suitable for CEC applications (Mistry et al., 2002).

The cost and the still low availability of commercial capillary columns force most of researchers to prepare packed columns by themselves (Mistry et al., 2002). However, packing and retaining a SP in narrow bore capillaries can have difficult results and requires dedicated apparatuses as well as a certain experience in order to warrant a good stability and reproducibility between columns (Boughtflower et al., 1995; Fermier & Colón, 1998; Saito et al., 2004; Eeltink & Svec, 2007).

Different packing procedures are reported in literature: the slurry packing is the most widespread, but also dry packing, packing with supercritical CO_2, by gravity, by centripetal force and by electromigration of charged silica beads were proposed as alternative to pumping the slurry suspension (Szumski & Buszewski, 2002; Colón et al., 2000). In any case, special attention has to be paid to avoid undesirable empty space volume and to obtain a stable packing and a good performance of the column (i.e., selecting the appropriate slurry packing solvent and impact velocity, using ultrasonic bath to prevent particle aggregation, etc.) (Saito et al., 2004; Fermier & Colón, 1998). In this section the principles of each approach will be briefly illustrated, while detailed practical operations can be found in literature. Initially, also "drawn-packed columns" were proposed, whose preparation includes selection of columns with large bore, packing of underivatized material, pulling of the column at high temperature with a glass drawing machine up to the final diameter, and derivatization of packing material (Knox & Grant, 1991). However, this procedure has not gained much interest, likely because long columns are employed and consequently the analyses are time-consuming.

Slurry packing method is highly diffused because it has been introduced to prepare conventional HPLC columns, and therefore it is well-known and largely described in literature (Svec, 2002). The term "slurry" refers to a suspension of the particulate SP in an appropriate solvent, which is placed in a reservoir and then pushed into the capillary column by a solvent delivery

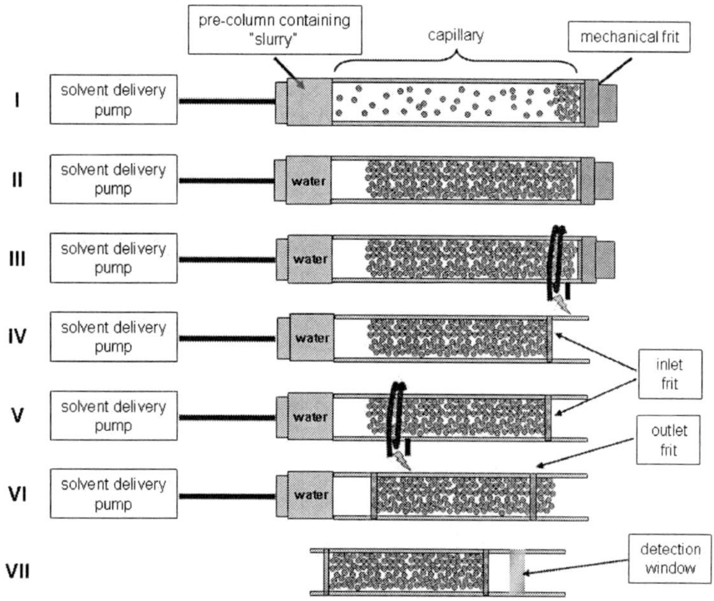

Fig. 9.1. Schematic representation of the preparation of a capillary column by the "slurry" packing method. I: packing of the SP; II: washing with water; III: inlet frit preparation with heated wire; IV: removing of mechanical frit and flushing out of SP particles between inlet frit and mechanical frit; V: outlet frit preparation; VI: removal of excess SP connecting the capillary column with the opposite side; VII: removal of polyimide layer for the detection window.

pump (see Fig. 9.1). During the filling, a temporary frit is positioned at the end of the capillary to retain the adsorbent. Usually a slurry solvent with low viscosity (i.e., acetone) is preferred for a faster packing (Svec, 2002). Anyway, a solvent suitable for one SP may not be compatible with another one. Due to the fact that the slurry is a suspension, ultrasonic bath or magnetic stirring can help to avoid agglomeration of particles into the reservoir, that can obstruct the packing (Colón et al., 2000).

Dry packing is simply performed by pumping the particulate medium using gas pressure (nitrogen) (Crescentini et al., 1988). In order to reduce static charge and aggregation, the packing material is conditioned with ethanol (EtOH) vapors for several hours. Even if this method requires a long time for packing and gives good efficiency, it is not much frequently employed.

The use of supercritical CO_2 as fluid to pump particulate SP is justified by its low viscosity that allows packing longer capillary columns than the ones obtained with the slurry method. During the packing, supercritical conditions have to be maintained so the column as well is immersed in an ultrasonic bath at a temperature above the critical temperature of CO_2 (Malik et al., 1993).

Packing by gravity makes use of gravitational force to introduce adsorbent particles into the capillary column. This is connected with one end to a syringe containing the slurry and with the other end to a temporary frit. The capillary is filled by sedimentation of particles in at least 10–12 h (Reynolds et al., 1998).

Electrokinetic packing, developed by Yan (1993; Yan et al., 1995) avoids the use of pressure because particles are driven into the column by means of an EOF. Furthermore charged particles move with their own electrophoretic mobility and in such way even small particles can be packed without generating high back pressure. The low cost of implementation and the possibility to simultaneously pack several capillaries are other advantages of this method. Finally, Colón et al. (2000) introduced the use of centripetal forces to pack capillary columns. Centripetal acceleration of particles through the column causes their sedimentation at the fritted end. The packing is very fast and also in this case it is possible to pack numerous capillaries at once (Fermier & Colón, 1998).

Whatever the method of packing employed, retaining frits are required to keep the SP restricted into the capillary column. Functional frits have to resist the high pressure applied and, at the same time, allow the crossing of mobile phase and analytes through the column (Colón et al., 2000). Consequently frits fabrication represents a big challenge for the final performance of the column. Inlet and outlet frits are usually prepared in situ by sintering the packed material with a heated wire. As already mentioned, this procedure causes the removal of polyimide coating at the frit zone and makes the column more fragile. Furthermore a bad sinterization can compromise the flow through the column and change the properties of the packing material where the heating has been applied, generating a non-homogeneous packing (Legido-Quigley et al., 2003). This occurrence is particularly detrimental in CEC because it produces irregularity in EOF, which can lead to bubble formation (Rapp & Bayer, 2000). Also the reproducibility of frits is a problematical task and special attention has to be paid at the time and temperature of sintering. Furthermore, frit fabrication depends on column I.D., particle size and type of packing material (Colón et al., 2000). For all these drawbacks, different solutions have been suggested to retain the SP avoiding the sintering of silica. Externally and internally tapered columns have been proposed as well as columns with inlet, outlet or both monolithic frits (Lord et al., 1997; Chourhary et al., 1998; Chen et al., 2000). Entrapment of particulate material by sintering and sol-gel technology (a hybrid between packed and monolithic column) can be also mentioned as an alternative to frit formation (Colón et al., 2000).

After all, troubles in frit fabrication and the above-mentioned practical disadvantages related to packing procedure, have led many researchers to consider monolithic media (Puy et al., 2007).

Monolithic Capillary Columns

Several aspects related to monolithic material make it attractive for the implementation of capillary columns complementary to packed ones. First of all, the generally simple *in-situ* synthesis of monolithic bed, offered by the capillary format of the columns, does not require frits (Mistry et al., 2002; Maruška & Kornyšova, 2004; Roux et al., 2007). The continuous porous structure of monoliths, due to the presence of macropores and mesopores of different sizes, allows the use of high flow rates due to the lower backpressure generated compared to packed columns (Tanaka et al., 2002). The wide availability of monomers yields a separation medium with unique chromatographic properties.

Monolithic columns can be obtained by polymerization of organic monomers or by sol-gel process in case of silica based ones. In the first case, after an appropriate derivatization of the inner capillary wall for the anchoring of the polymer, the capillary is filled with a polymeric mixture and the radical polymerization can be accomplished thermally or photochemically (Maruška & Kornyšova, 2004; Svec & Huber, 2006). The polymeric mixture is composed by functional monomers, a crosslinker, a porogenic agent and an inert solvent. Each of them can affect the porous structure of the monolithic material. Once the polymerization is complete, unreacted moieties are washed out flushing the capillary column with a proper solvent. Depending on the type of monomers, it is possible to distinguish water soluble co-monomers (acrylamide-based monoliths), where a salt is used as porogen, and co-monomers soluble in organic solvent (polystyrene- and methacrylate-based monoliths), where polymerization takes place in the presence of a porogen co-solvent (usually 2-PrOH) (Mistry et al., 2002; Svec, 2010). In fig. 9.2 it is reported the reaction mechanism for the methacrylate-based polymerization. For each mixture it is present the proper initiator and accelerator. When monolithic beds have to be employed in CEC, a charged monomer is included in the polymeric mixture to assure the EOF (Eeltink & Svec, 2007). Even if characterized by an easy preparation, such kind of continuous bed presents low efficiency and poor reproducibility (Myers & Bartle, 2002).

Silica monolithic SPs are obtained with a more complicated procedure, but their properties are very attractive. Especially, they possess a well defined pore size distribution of both macropores (few micrometers in size) and mesopores (few nanometers in size), which allow respectively high flow rates and high specific area. Moreover, the thin skeleton reduces the mass transfer resistance, allowing it to maintain high efficiencies at high flow rates (Tanaka et al., 2002). Their preparation (sol-gel process) is based on several steps and assures an independent control of the size of the silica skeleton and through-pores (Tanaka et al., 2002; Roux et al., 2007): hydrolythic polymerization of an alkoxy-silane (i.e., tetramethoxysilane, TMOS) in an acidic medium;

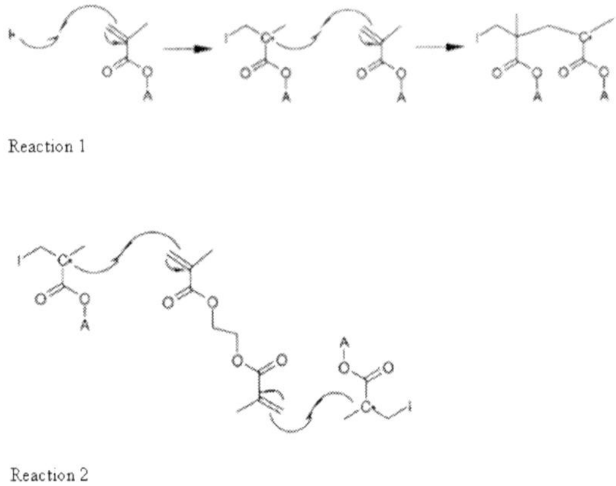

Fig. 9.2. Reaction mechanism for the polymerization of methacrylate. Reaction 1: radical initiation of monomer and growing polymer chain, A is a terminal group. Reaction 2. Monomer radical polymerization, cross-linking with ethylene dimethacrylate. Reprinted from (Legido-Quigley et al., 2003) with permission from John Wiley & Sons.

condensation of silica in presence of poly(ethylene glycol) (PEG) as porogen, and phase separation between silica/PEG system and water; separation of the resulting gel into solid- and liquid-rich domains that further react with each other to give the gel its final structure (syneresis); ageing and heat treatment, involved in tailoring mesopores; drying, to remove the liquid; chemical modification. This last step is needed to give the desired surface feature for chromatographic separation and the functional groups can be introduced directly by incorporation of functional monomers in the polymeric mixture (Wu et al., 2008).

Usually, silica-based monoliths offer higher efficiencies than those obtained by polymerization of organic polymers and are more resistant to organic solvents.

Polymeric mixtures are fit for the preparation of molecularly imprinted polymers (MIPs), which have been used for several applications in the analytical field, such as SPs for solid-phase extraction (SPE) and chromatography, sensors, etc. (Liu et al., 2005). The way to produce molecularly imprinted SPs consists of the addition of a template (imprint) molecule in the polymeric mixture. Before starting polymerization, functional groups on the template interact with those on the monomer(s), and when the reaction begins monomers result arranged around the template. During the polymerization, the template, depending on its structure, can react, forming covalent bonds, or give rise to secondary interactions with other monomers (Tan & Remcho, 1998). If the imprinting is successful, after the removal of the imprint molecule, the polymer network will posses cavities having shape, size and functional groups complementary to those of the template, showing a remarkable affinity towards it (Simon et al., 2007). To increase the surface area, MIPs can be grounded. However these particles tend to have an extremely broad size and shape distribution (Tan & Remcho, 1998). Several approaches have been used to obtain molecularly imprinted particles with a controlled size and shape distribution. For example, suspension polymerization has been utilized to obtain monodispersed MIP particles. Submicron scale MIP particles have also been prepared by precipitation and emulsion polymerization. An improved methodology is miniemulsion polymerisation, where the diffusion process in the continuous phase is suppressed and a stable emulsion of homogeneous size is generated (Nilsson & Nilsson, 2006; Nilsson et al., 2004).

When the aim of the research is the separation of a chiral compound, one of its enantiomers is used as template (Tan & Remcho, 1998; Ou et al., 2007). MIP technology is also used as an alternative to affinity chromatography, avoiding some disadvantages related to ligands, such as their toxicity or low stability (Nováková & Vlèková, 2009; Haginaka, 2009).

Monolithic material has also been employed in order to obtain restricted-access media (RAM). This type of support allows the on-column separation of analytes from complex matrices (i.e., human plasma), being characterized by a "bimodal surface topochemistry", where size exclusion chromatography (SEC) is combined with adsorption chromatography. SEC prevents the access of large biomolecules into the small pores of the RAM structure and consequently these undesired compounds elute from the column at the void volume. Molecules with a low molecular weight, instead, penetrate through active sites responsible of chromatographic retention (Boos & Fleischer, 2001; Jarmalavièienë et al., 2003).

Open Tubular Capillary Columns

Open tubular (OT) columns where the SP is directly bonded or adsorbed to the inner wall of the capillary usually have an I.D. between 10 and 60 μm (Ishii et al., 1988). The adsorption of the SP, or modifier, can occur dynamically (weak interactions, the inner modifier is added to the mobile phase) or physically (modifier strongly adsorbed). Modifiers utilized for the preparation of adsorbed SPs can be grouped in: (i) cationic surfactants, (ii) polymeric surfactants, and (iii) charged polymers (Kapnissi-Christodoulou et al., 2003). The layer of SP can be fixed also by covalent bonding and/or cross-linking. This approach offers a long capillary lifetime, but it usually requires a more complicated coating procedure (Nyholm & Markides, 1998).

OT columns are characterized by low convective dispersion, good efficiency and high permeability. For this reason, conditioning of the columns is fast and their length can be increased without the occurrence of long analysis time (Swart et al., 1997). They can be used in gas chromatography (GC), supercritical fluid chromatography (SFC) and in OT-LC, giving the possibility of a unified approach for separations, reached by employing a single column in all chromatographic modes (Wistuba & Schurig, 2000b).

Longer columns can supply the poor selectivity typical of OT columns. In fact they possess a very low phase ratio, and for best performance an I.D. ≤ 20 μm is recommended. However, these small I.D. aggravate the sensitivity when optical detection methods are employed (Bruin et al., 1990; Schurig & Mayer, 2001). Another drawback, related to the low phase ratio, is the low sample loading capacity of the column. As a result, column overloading can easily occur causing peak asymmetry and decreasing efficiency. To enlarge the wall surface available for the coating with the SP and increase the phase ratio, the chemical etching of the inner wall has been proposed (Kapnissi-Christodoulou et al., 2003). The etching process increases the surface area of the capillary by a factor of up to 1000. This reaction is carried out by using a solution of ammonium hydrogen difluoride and is dependent on the time and temperature of exposure. Other chemical treatments and increasing the film thickness (porous layer OT, immobilized organic polymer) have been proposed for enhancing phase ratio (Swart et al., 1997). However, reproducible creation of a thick SP coating on the inner walls of small-diameter columns is associated with considerable practical difficulties. OT columns with dynamic nanoparticles as pseudostationary phases and sol-gel surface coatings have also been suggested (Dong et al., 2008; Guihen & Glennon, 2004).

Nano Flow Generation

As previously mentioned, due to the reduced dimension of the capillary I.D. and to the use of small packed particles in nano-LC and CLC, the direct use of conventional pumping systems is not allowed due to the high back-pressure that is usually obtained. Therefore with these miniaturized techniques it is necessary to reduce the mobile phase flow at levels of nL/min. If monolithic SPs are used, the flow rate can be increased to μL/min levels because they have higher permeability than packed columns.

Nowadays, different commercial devices that automatically generate a repeatable and constant delivery of solvents (also under gradient elution modes) are available from a number of manufacturers. Originally, commercialized instrumentation was based on the use of split-flow systems (the most common), which is also greatly applied in laboratory-made instrumentation because of its simplicity (Hernández-Borges et al., 2007a). The device is assembled connecting a micro-pump to a T union with zero dead-volume; the other two ends are connected to the packed capillary and to an empty tube of suitable length and I.D. The split ratio can be changed by modifying the length and the I.D. of the empty capillary (waste). The method is highly appropriate for obtaining nano/micro-flows in the isocratic mode but under laboratory-made instrumentation it frequently fails (especially in nano-LC) to achieve a reproducible gradient elution due to the change of the backpressure caused by the different viscosity of the mixed solvents. In addition, the mobile phase mixture is transferred into the column with a small delay. Current commercialized systems based on flow splitting normally use electronic flow control devices to maintain a constant flow while viscosity of the mobile phase changes (http://www.agilent.com). One of its main disadvantages is that solvent saving is not that high because it requires flow rates higher than those commonly used.

Several works have also proposed the use of conventional pumps with the aid of suitable switching valves with or without split flows (Cappiello et al., 2003a; Takeuchi et al., 1987; Deguchi et al., 2004; Ito et al., 2004; Ito et al., 2005; Natsume et al., 2002). This is the case of the work by Cappiello et al. (2003a) in which an electronically-actuated, computer controlled, multiposition HPLC valve is used. The valve has 14 ports and six loops, each loop containing a selected mixture of solvents of different strength. The electric actuation allowed selecting which reservoir to be on-line with the column for a desired time, generating a specific mobile phase gradient. The lowest flow tested was 250 nL/min. This device was also used in a later work by the same group for the determination of nitro-polycyclic aromatic hydrocarbons (PAHs), pesticides and human hormones showing good precision, linearity and sensitivity (Cappiello et al., 2003b).

Dedicated splitless miniaturized pumps (syringe or reciprocating pumps), on the contrary, have the advantage of low solvent consumption, but they are less commonly available. An example is a commercialized nano-pump (http://www.dionex.com) in which the continuous direct flow is delivered by four pistons (two per solvent channel). While one piston is refilled the other provides suitable flow delivery. After the refilling cycle, the refilled piston is pressurized and takes over the flow delivery. At these very low flow rates, suitable mobile phase degassing is of even higher importance.

Injection Devices

When working in nano-LC where the I.D. and the total volume of the columns are relatively low, the injected sample volume is one of the most important parameters to be carefully considered in order to avoid problems with sensitivity as well as with overloading related to low and high sample injection, respectively. Clearly the injected volume should be as low as possible to avoid band broadening. However, a low sample volume will reduce the sensitivity. Therefore there must be a compromise allowing the highest efficiency with the best signal response.

The maximum volume that can be injected in a chromatographic system can be calculated utilizing the following equation proposed by Martin et al. (1975) in LC:

$$V_M = \theta \frac{\pi}{4} d_c \varepsilon (1 + k) \sqrt{LH} \qquad (5)$$

This equation shows that the maximum cell volume depends on the diameter of the capillary (d_c), on its length (L) and plate height (H). Obviously, decreasing the column diameter makes it necessary to lower the injection volumes. Therefore in nano-LC injection volumes in the range 10–100 nL are currently used unless a focusing approach is done in order to increase the sensitivity. Valve injectors with fixed loops of 40, 60 and 100 nL are commercially available. For volumes lower than 20 nL a split vent has been inserted between the valve and the column (Chervet et al., 1996). Alternatively the sampling procedure can be controlled by regulating the injection time (Claessens et al., 1990; Fanali et al., 2010). This can also be done automatically, and it has been successfully applied to the autosamplers available with almost all commercial nano-LC instrumentations where analytes can be injected with very high precision.

Detection Systems

As previously mentioned, although miniaturized techniques exhibit high mass sensitivity due to the reduction of the column I.D., concentration sensitivity is relatively low because the injected sample volume is limited. The detectors usually employed in electrodriven capillary and CLC/nano-LC techniques are an adaptation of those currently employed in conventional techniques. They include UV, fluorescence and conductivity detectors. The latter detector was mainly applied in capillary isotachophoresis (CITP) where the separated zones exhibit a different electrical resistance with good signal. Good sensitivity can be obtained by employing laser induced fluorescence detectors (LIF) where a laser source is used (García-Cañas et al., 2002). Microchip equipment normally employs LIF detectors due to the need of high sensitivity and UV absorption problems related to the material used for chip preparation. Microchip tools operated with conductivity detectors have been studied by Kaniansky's group (Masár et al., 2009).

To avoid dispersion problems during detection the volume of the flow cell must be adapted to the size of the peak volumes eluting from the column. This is done by employing the detector cell with the lowest possible volume. Best results are currently obtained with on-column detectors where the path-length is the capillary I.D. Finally, acquisition of data must be very fast in order to collect peaks zones with low width (1–2 s).

Absorption detectors, previously demonstrated in combination with electrodriven techniques, have been widely used in CLC/nano-LC achieving good results. Generally in these devices the UV light is focused on the capillary cell in order to have the highest sensitivity. Recently, commercial detectors, including UV and diode array (DAD), capable of offering very high sensitivity are available. The use of an extended-path cell, well-known in conventional HPLC, has also been proposed in miniaturized techniques. Applications employing bubble or zeta-cells (Desiderio et al., 1999; Hampel, 2000; Steiner & Steiner, 2000) have been done in electrodriven techniques, while in CLC/nano-LC detector cells with extended path length are usually employed especially with commercial instrumentation.

Improving Sensitity in Nano-LC

Even if nano-LC is a highly mass sensitive chromatographic technique, the concentration detection limit is very questionable when compared to conventional HPLC where higher sample volumes are currently injected (5–20 μL instead of 10–60 nL in nano-LC). In addition, in nano-LC the path length of the UV detector cell is very low and usually corresponds to the I.D. of the capillary used (≤ 100 μm). Therefore, in order to increase the sensitivity when using this miniaturized technique, some approaches have been proposed to overcome the mentioned drawbacks allowing analysts to apply nano-LC to real samples analysis. Among the different solutions proposed, i) sensitive and specific detectors, e.g., mass spectrometers, ii) UV-detector cells with extended path lengths, iii) on-column focusing (Mills et al., 1997; Cappiello et al., 2002b; Hernández-Borges et al., 2007a; He & Lee, 1998; Molander et al., 2002; D'Orazio & Fanali, 2008; Buonasera et al., 2009) and iv) pre- and/or 2D-column pre-concentration are employed (Vissers et al., 1997).

On-column focusing is a quite simple approach that does not need special equipment. In fact the sample is dissolved in a solvent mixture with poor elution capability towards analyzed compounds compared to the mobile phase. Consequently a sample plug on the top of the column is formed even if a relatively high volume is injected. To show the feasibility of the on-column focusing by using nano-LC, some selected applications are reported here as an example. Cappiello et al. 2002b utilized a capillary column of 75 μm I.D. coupled with MS for trace level determination of organophosphorus pesticides. Even though only a few nanoliters of sample are normally injected in such systems to avoid overload, the authors injected about 500 nL, decreasing the limits of detection (LODs) about 100-fold; the mobile phase was modified with water. In order to improve the sensitivity, on-column focusing was also studied for the chiral resolution of β-blockers such as atenolol, propranolol, oxprenolol and metoprolol. The enantiomers were separated by nano-LC in a capillary column packed with silica modified with vancomycin. Racemic drugs were dissolved in pure MeOH (1500 nL injected), while the mobile phase used contained ammonium acetate pH 4.5/water/MeOH (1:4:95, v/v/v). LODs and limits of quantification (LOQs) for each alprenolol enantiomer were 9 and 15.6 ng/mL, respectively. The method was validated and applied to the analysis of an extracted plasma sample (D'Orazio & Fanali, 2008). Fig. 9.3 shows the separation of racemic alprenolol in its enantiomers employing a silica column modified with vancomycin.

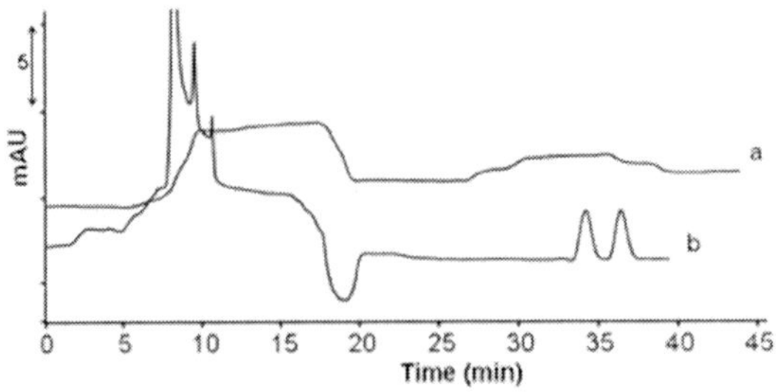

Fig. 9.3. Enantiomeric resolution of alprenolol by nano-LC. Mobile phase: 500 mM ammonium acetate pH 4.5/water/MeOH 1:4:95 by volume; flow rate, 150 nL/min. Injection of 1500 nL of racemic 500 ng/mL alprenolol dissolved in (a) mobile phase and (b) MeOH. Reprinted from (D'Orazio & Fanali, 2008) with permission from Elsevier.

Hyphenation of Nano-LC with Other Techniques

Hyphenation of LC with other techniques is a challenging topic for researchers, but it can give additional advantages on using the separation method alone. Hyphenation of miniaturized techniques with other ones offering, e.g., possibilities of i) characterization of analytes or ii) two dimensional separation, is very important especially when complex matrices have to be investigated. Among these techniques, MS and nuclear magnetic resonance (NMR) are the most advanced tools recently studied. The hyphenation of CLC-MS and nano-LC-MS is currently performed and documented by several practical applications also reported in this chapter, while CLC or nano-LC coupled with NMR include only a few reports.

CLC/nano-LC-NMR

Wu et al. (1994) first reported the hyphenation of CZE with NMR; later, Albert's group (1996) demonstrated the possibility to couple CLC and CEC with NMR (Pusecker et al., 1998; Albert et al., 1999). In the report by Albert et al. (1996), HNMR spectra were recorded in both stopped-flow and continuous-flow mode. A 180 μm I.D. capillary was positioned into a micro-probe on the NMR spectrometer and vitamin A was analyzed. Stopped-flow was instead used for 2D NMR spectra. The coupling of a commercial CLC instrument with NMR has been shown by Lacey et al. (2001) analyzing terpenoid compounds in a 320 μm I.D. capillary packed with 3 μm C_{18} particles. The detection was done in stopped-flow mode, achieving sensitivity sufficiently high for the detection of trace impurities (sub-nmol). Fig. 9.4 shows a scheme of CLC-NMR. In a recent review paper, Kuhnle et al. (2009) summarized the data available in the literature dealing with hyphenation of micro-scale LC with NMR.

CLC/nano-LC-MS

The coupling of LC with MS has been done utilizing atmospheric pressure ionization (API) sources where the sample eluted by the mobile phase at

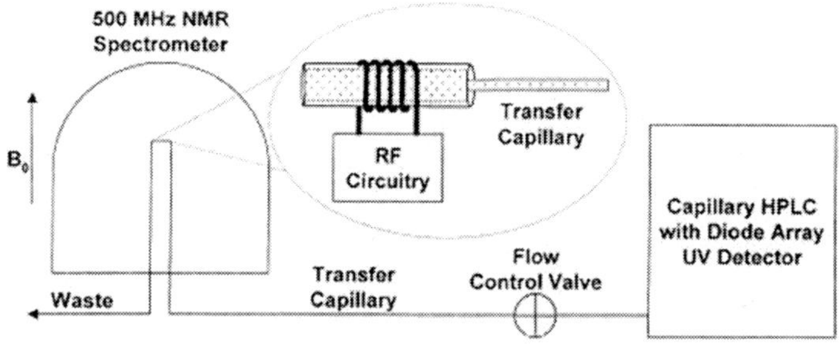

Fig. 9.4. Schematic representation of the cHPLC–NMR set-up in which the cHPLC instrument (C_{18} column, 15 cm × 0.32 mm) with a diode array UV detector is located ~4 m from the 11.7 T NMR magnet. The flow control valve is used to halt the flow so that peaks of interest can be positioned in the microcoil NMR detection cell (V_{obs} ~1.1 μl) for long data acquisitions. Reprinted from (Lacey et al., 2001) with permission form Elsevier.

atmospheric pressure is ionized and transferred into the MS. In LC, electrospray and atmospheric pressure chemical ionization (ESI and APCI, respectively) are used. The ESI source is the most applied with miniaturized techniques such as chromatographic and electrodriven (CZE and CEC) allowing the analysis of compounds with charged or chargeable groups. This type of source can be easily used for compounds with high molecular weight such as proteins, peptides and other polymers because multiply charged ions are formed and detected. In addition small ions are also analyzed.

The APCI source is very useful especially for the analysis of uncharged compounds and can be advantageously utilized in LC-MS. However it presents some drawbacks in CLC or nano-LC mainly due to the reduced flow rate. The use of this source is rare with these techniques although some studies document its applicability (Barefoot & Reiser, 1987; Hutton & Games, 1997; Östman et al., 2006). On the other hand the use of an electron ionization source (EI) for nano-LC hyphenation with MS was reported by Cappiello et al. (2003b), analyzing three different classes of compounds, namely nitro-PAHs, pesticides and human hormones. In this work a conventional HPLC pump was used supplemented by a split system to obtain nL/min flows.

Because of the relatively low flow rates used in CLC/nano-LC, nano-ESI interfaces have to be used: they are commercially available and allow one to obtain high sensitivity. The interface is quite simple and it can also be assembled in the laboratory.

Fig. 9.5 shows a scheme of the nano-LC hyphenation with MS. The pump system with a split for reducing the flow rate to nL/min is connected to an injection valve. The packed capillary column, joined directly to the valve is connected with the nano-ESI interface where a capillary tip is positioned. Usually, the interface is moved carefully using an x, y, z manipulator and can be monitored with a camera to achieve the best ionization conditions. The

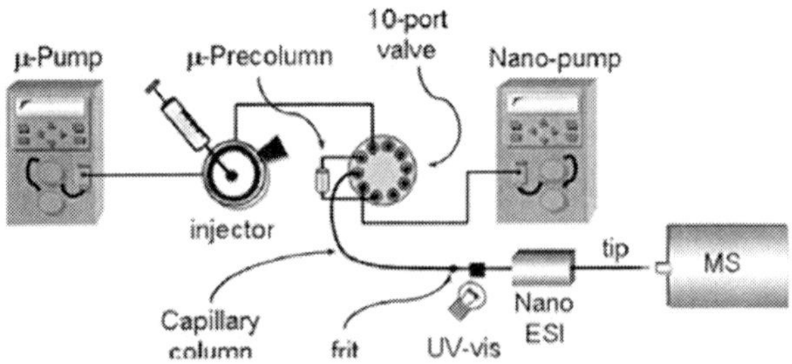

Fig. 9.5. Schematic representation of the column-switching system coupled with mass spectrometry. Reprinted from (D'Orazio et al., 2008).

distance from the MS orifice and the applied voltage as well as the composition of the mobile phase must be carefully controlled.

The capillary column and the tip are set into a union that can be of different materials, either stainless steel or PEEK polymer. With the metal union the voltage is applied directly while with the polymer a wire is necessary for connection with the MS power supply. The commercial interfaces can also employ a "guest metal" or metal-coated tips. When using the latter, although good results can be obtained, some drawbacks can be expected mainly related to the instabilty of the coating under the influence of the relatively high voltage applied (1.5–2.5 kV). Capillary tips can be prepared in the laboratory, using fused silica capillaries of 25 or 50 μm I.D.: one end is worked with fine sandpaper to obtain the most appropriate shape (Fanali et al., 2007; D'Orazio et al., 2008).

The coupling of CEC with MS has also been demonstrated utilizing different ESI interfaces, e.g., sheathless (Warriner et al., 1998; Zheng & Shamsi, 2003), coaxial sheath flow (Desiderio & Fanali, 2000; Bragg et al., 2008), and liquid-junction (L.-J.) (Lee et al., 1989; D'Orazio & Fanali, 2010). Among the used interfaces, the last one seems to be very promising because it minimizes the drawbacks derived by the use of the high electric field applied, bubble formation, ion signal stability and sensitivity. With the use of L.-J. interface, LODs of some studied pesticides were obtained at ng/mL levels (D'Orazio & Fanali, 2010).

Applications

The application of an analytical technique over time clearly demonstrates its working potential and capacity. Therefore, suitable review of the literature is of high importance. In this section, different applications of both techniques in proteome research, environmental, pharmaceutical and food analysis, among others, are reviewed and commented upon, paying particular attention to the most recent works.

Protein/Peptide Analysis

Proteomic/peptidomic research, which involves protein/peptide analysis in cells, tissues or body fluids, is a very important and challenging research field in which miniaturized techniques like nano-LC and CLC have found an important application area. This is probably due to the low amount of sample available for these analyses. Peptide mapping, protein sequencing or post-translational protein modification analysis are the main objectives to accomplish, which also require the use of MS.

Table 9.A shows some recent examples of the use of nano-LC or CLC for proteomic/peptidomic analysis. In most of these cases, the techniques have been used only as a specific part of the work, since a good number of sample pre-treatment steps (i.e., protein extraction and digestion, cell fractionation, etc.) are frequently necessary at this level.

As part of many analyses, proteins are frequently transformed to peptides by digestion with a specific enzyme (proteases like trypsin—the most common) since MS on peptides is easier and more sensitive. When digestion is performed early in the sample preparation step and then peptides are introduced into the mass spectrometer, the strategy is called *bottom-up proteomics*, which is the most commonly used method. Obviously, proteins can be firstly separated by gel electrophoresis or chromatography and then digested. Peptide mass fingerprinting and tandem MS (MS/MS) are the two approaches for protein identification using this strategy. On the contrary, when intact protein protonated molecules are introduced in the mass analyzer (by ESI), the strategy is called *top-down proteomics* (Wehr, 2006). In this last case, which is also more recent, charge state manipulation through gas phase ion–ion interactions as well as the use of instruments with high mass measurement accuracy are necessary for the identification of product ion masses from multiply charged precursor ions. In the *bottom-up* approach, only a small fraction of the protein sequence is identified, while the *top-down* method provides potential access to the complete protein sequence as well as the ability to locate and characterize post-translational modifications. The fact that *top-down proteomics* uses direct MS analysis of intact proteins, has generally required protein fractionation and purification to be accomplished off-line (generally by CE).

Separation technology for *bottom-up proteomics* has been previously reviewed (Wehr, 2001; Wehr, 2003; Wysocki et al., 2005; Bogdanov & Smith, 2005). Depending on the sample complexity the set-up will comprise a single column or multiple columns for multidimensional separations (if peak capacity of a single column is exceeded), in which nano-LC or CLC play an important role (Tao et al., 2010). Numerous and interesting applications for *bottom-up* 2D-LC using nano-LC of peptide mixtures have been published in the last few years. Regarding the employed approaches, on the one hand, off-line coupling can take place, in which the eluent of the first column (first dimension)

Table 9.A. Nano-LC and Capillary-LC Applications on Proteomics

Analyte	Matrix	Sample pretreatment	Column	Detector	Mobile phase	Observations	References
Peptides	Lysis, reduction, alkylation, tryptic digest (trypsin), desolation, SPE (C_{18} disk)	Mouse brain tissue	1D: PS-DVB (15 cm × -; 3 μm) 2D: C_{18} (15 cm × 100 μm; 3 μm)	MS/MS (IT)	1D: ACN/water/ ammonium hydroxide (Gradient) 2D: ACN/water/ acetic acid (Gradient)	Use of off-line 2D-LC. Comparison with SCX and SAX for the first dimension. SPE (C_{18} disk) also before nano-LC for desalting/ concentration	(Nakamura et al., 2008)
Peptides	*Escherichia coli*	Digestion	1D: SCX (15 cm × 1 mm; 5 μm) 2D: C_{18} (15 or 50 cm × 75 μm; 3 μm)	UV (214 nm) MS/MS (IT)	1D: ACN/water/ phosphate buffer/NaCl (Gradient) 2D: ACN/water/ TFA (Gradient)	Use of off-line 2D-LC. Comparison with 1D-LC	(Eeltink et al., 2009)
Phosphopeptides	Mouse liver tissue	Lysis, tryptic digestion (trypsin)	1D: C_{18} (25 cm × 2.1 mm; 5 μm) 2D: C_{18} (12 cm × - ; 5 μm)	DAD MS/MS (IT)	1D: ACN/water/ ammonium formate (Gradient) 2D: ACN/water/ formic acid (Gradient)	Use of off-line 2D-LC	(Song et al., 2010)
Proteins	Human parotid saliva	Isolation of parotid exosomes, tryptic digestion (trypsin)	1D: SCX (3 cm × 100 μm; 5 μm) 2D: C_{18} (10 cm × 100 μm; 3 μm)	MS/MS (IT)	ACN/water/ formic acid/ ammonium acetate (Gradient)	Use of on-line 2D-LC with one column	(Gonzalez-Begne et al., 2009)

Analyte	Matrix	Sample pretreatment	Column	Detector	Mobile phase	Observations	References
Peptides	Yeast	Protein extraction, reduction, alkylation, tryptic digestion (trypsin)	1D: monolithic SCX (7 cm × 150 µm) 2D: C_{18} (12 cm × 75 µm; 5 µm)	MS/MS (IT)	1D: water/ ammonium acetate (Gradient) 2D: ACN/water/ formic acid/ ammonium acetate (Gradient)	Use of on-line 2D-LC using coupled columns	(Wang et al., 2007)
Peptides	*Corynebacterium glutamicum*	Protein extraction, tryptic digestion (trypsin)	1D: C_{18} (15 cm × 2 mm; 3 µm) 2D: PS-DVB monolith (6 cm × 100 µm)	UV (280 nm) MS (IT)	1D: ACN/TEA/ acetic acid (Gradient) 2D: ACN/water/ TFA (Gradient)	Use of on-line 2D-LC using a preconcentration column (PS-DVB, 1 cm × 200 µm). Comparison with 2D-LC using a SCX column as first dimension	(Delmotte et al., 2007)
Peptides	Mouse liver tissue	Differential centrifugation separation of nuclei fractions, purification (Nycodenz density gradient centrifugation), section cutting (transmission electron microscopy), lysis, SDS-PAGE, reduction, alkylation	1D: SCX (15 cm × 300 µm) 2D: C_{18} (15 cm × 75 µm; -)	MS/MS (Q-TOF)	1D: water/ ammonium acetate (Gradient) 2D: ACN/water/ formic acid (Gradient)	Use of on-line 2D-LC with column switching. Use of trap column (C_{18}, 5 mm × 300 µm) for desalting	(Zhang et al., 2006)

(continued on next page)

Table 9.A. *(continued)*

Analyte	Matrix	Sample pretreatment	Column	Detector	Mobile phase	Observations	References
Peptides	Mixture of known proteins and nuclear matrix protein sample isolated from human Jurkat cells	Tryptic digestion (trypsin)	1D: SCX (15 mm × 500 μm) 2D: C_{18} (15 cm × 75 μm; 3 μm)	MS/MS (IT)	1D: water/ ammonium acetate/formic acid (Gradient) 2D: ACN/water/ formic acid (Gradient)	Use of on-line 2D-LC with column switching. Use of trap column (C_{18}, 5 mm × 300 μm)	(Mituloviæ et al., 2004)
Peptides	Prolactinomas	Immuno-laser capture microdissection, cell lysis, SDS-PAGE gel protein separation, digestion (trypsin)	1D: SCX (2.4 cm × 180 μm; 5 μm) 1D: C_{18} (20 cm × 75 μm; 1.7 μm)	MS/MS (IT)	1D: ACN/ ammonium formate (Gradient) 2D: ACN/water/ formic acid (Gradient)	Use of on line 2D-LC with a trap column (C_{18}, 2 cm × 180 μm; 5 μm) between the SCX and the C_{18} columns	(Liu et al., 2010)
Peptides	Lake and chlorinated tap water	Filtration	C_{18} (15 cm × 320 μm; 3.5 μm)	UV (210 nm) MS (TOF)	ACN/water/TFA (Gradient)	Use of a pre-column (C_{18}, 3 cm × 320 μm; 10 μm) and a large volume injection column switching system. LOD: 1 ng/mL	(Bjellaas et al., 2004)
Peptides	Myoglobin	Digestion (trypsin immobilized in carbonyldiimidazol agarose)	Monolithic C_{18} (15 cm × 100 μm) Particulate C_{18} (16 cm × 100 μm; 5 μm)	MS/MS (IT)	ACN/water/TFA (Gradient)	Comparison between monolithic and particulate C_{18} capillary columns	(Rozenbrand et al., 2006)

Analyte	Matrix	Sample pretreatment	Column	Detector	Mobile phase	Observations	References
Peptides	In vitro endothelial cells protein secretion due to exposure to smoking-machine-generated cigarette smoke	Exposure of endothelial cells to cigarette smoke total particulate matter, collection of supernatant, protein concentration, desalination, denaturation, digestion (trypsin)	C_{18} (5 cm)	MS/MS (IT/FTICR)	ACN/water/formic acid (Gradient)	Use of a trap column (C8, 5.1. cm × 300 μm)	(Flora et al., 2008)
Peptides	Equine heart myoglobin, bovine blood hemoglobin, albumin from chicken egg white	Reduction, alkylation (hemoglobin and albumin), dissolution, in-line digestion (pepsin)	Polymeric monolith (20 cm × 100 μm)	MS (TOF)	ACN/water/formic acid (Gradient)	Use of an in-line system containing porous polymer monoliths for protein digestion, peptide concentration and nano-LC	(Geiser et al., 2008)
Whole-cell lysates from *Aspergillus flavus*	—	—	C_{18} (15 cm × 75 μm; 5 μm)	MS (FTICR)	ACN/water (Gradient)	Top-down analysis. Use of a trap cartridge (C_{18})	(Collier et al., 2008)
Glycoproteins and glycans	—	Denaturation, hydrolysis, derivatization (2-aminobenzamide)	Surface bound lectin monolithic silica (20/27 cm × 100 μm) Surface bound polar cyano monolithic silica (25/33 cm × 100 μm)	UV (-)	Lectin: water/Bis-Tris/NaCl/N-acetyl-D-glucosamine (Isocratic) Cyano: ACN/water/ammonium acetate/sulfated β-CD	Evaluation of two lectin monolithic columns. Comparison of CEC with nanoLC for polar cyano monolithic column	(Zhong & El Rassi, 2009).

(continued on next page)

Table 9.A. *(continued)*

Analyte	Matrix	Sample pretreatment	Column	Detector	Mobile phase	Observations	References
D-Asp-containing proteins	Mouse brain tissue	Protein extraction, D-aspartil endopeptidase treatment or control, 2D gel electrophoresis, in-gel digestion	C_{18} (150 cm × 75 μm; 3 μm)	MS/MS (IT)	ACN/water/ formic acid (Gradient)	Use of an enrichment column (C_{18}, 5 cm × 300 μm; 5 μm). Nano-LC used for identification purposes	(Sakai-Kato et al., 2009)
Vasoactive peptide hormones	Rat plasma (before and after acute angiotensin-converting enzyme inhibition)	Purification of plasma peptides (MeOH precipitation or off-line HPLC fractionation)	C_{18} (660 nL × 75 μm)	MS (IT)	MeOH/formate/ TFA (Gradient)	Use of an enrichment column (40 nL; 5 μm) and nano-LC-chip	(Lortie et al., 2009)
Phosphorylated peptides	—	—	C_{18} (15 cm × 300 μm; 3.5 μm)	ICP-MS MS/MS (Q-IT)	ACN/water/TFA (Gradient)	Use of a pre-column (C_{18}, 5 mm × 300 μm; 5 μm). LODs for phosphorus: 6 μg/L	(Pröfrock & Prange, 2009)
Phosphoproteins	Human cerebral spinal fluid	Protein spin concentration, tryptic digestion	CLC: C_{18} (15 cm × 500 μm; 5 μm) nano-LC-chip: C_{18} (43 mm × 75 μm; 5 μm)	CLC: ICPMS nano-LC-chip: MS (IT)	ACN/water/ formic acid (Gradient)	Use of CLC and nano-LC-chip to identify phosphoproteins. Use of an enrichment column in nano-LC-chip (C_{18}, 4 mm × 75 μm; 5 μm)	(Kroening et al., 2010)

Analyte	Matrix	Sample pretreatment	Column	Detector	Mobile phase	Observations	References
Caseinophosphopeptides	Sodium caseinate	Tryptic digestion	C_{18} (15 cm × 75 µm; 3 µm)	MS/MS (Q-TOF)	ACN/water/ formic acid (Gradient)	Use of a pre-column (C_{18})	(Zhu & Fitzgerald, 2010)
Intact proteins (ribonuclease A from bovine pancreas, carbonic anhydrase from bovine erythrocytes, lysozyme from hen egg white, myoglobyn from horse heart and trypsinogen from bovine pancreas), *E. Coli* protein sample	—	—	PS-DVB monolith (5 cm × 200 µm; small pore size) PS-DVB monolith (25 cm × 200 µm; small pore size)	UV (214 nm)	ACN/water/TFA (Gradient)	Top-down analysis. Study of the influence of the column properties (length, pore size) and LC conditions in the separation performance. Comparison with packed column (C4, 5 cm × 200 µm; 5 µm). Use of a z-shaped flow cell	(Detobel et al., 2010)
Glycopeptides	Human and bovine α1-acid glycoproteins	Tryptic digestion (trypsin), reduction, alkylation, on-column enrichment with cellulose (SPE)	C_{18} (2, 5, 10 cm × 75 µm; -) Graphitized carbon (5 cm × 180 µm; -)	MS (IT-orbitrap)	ACN/water/ formic acid (Gradient)	Validation with more complex standard protein digest mixtures	(Snovida et al., 2010)

is taken and reinjected in the second column (second dimension) (Eeltink et al., 2009; Nakamura et al., 2008; Song et al., 2010). This approach permits more facility in terms of elution conditions. On the other hand, an on-line approach, which is also called *multidimensional protein identification technology* (MudPIT), can take place with the two packing materials inside a single binary column (Gonzalez-Begne et al., 2009). However, the multidimensional analysis performed in a single column has several disadvantages such as low inter-column reproducibility or limited choice of mobile phases. On-line approach can also occur with coupled columns (Wang et al., 2007) or with two different separation columns by using trap columns or valves between them, that is, column switching (Delmotte et al., 2007; Zhang et al., 2006; Mitulović et al., 2004; Liu et al., 2010). This is also a very common technique for large volume injections (focusing) of peptides. In fact most peptides are readily dissolved in aqueous solutions and they are eluted from the reversed-phase column employing an ACN-water gradient. Normally, in 2D-LC with the exception of on-line binary colum approaches, the first dimension is carried out in a conventional HPLC column, while the second is carried out in a micro or a nano-column (Tao et al., 2010).

Regarding the SPs, as several authors have described (Tao et al., 2010; Zhang et al., 2010), among the different 2D-LC modes available, 2D strong cation exchanger (SCX)-RPLC is highly preferred for *bottom-up* analysis. This is because of the high loading capacity, resolution, as well as peak capacity [at low pH (tryptic) peptides bear a positive charge assuring binding to the SCX column] (Eeltink et al., 2009; Gonzalez-Begne et al., 2009; Wang et al., 2007; Zhang et al., 2006; Mitulović et al., 2004; Liu et al., 2010). An example of this approach was recently developed by Eeltink et al. (2009) who examined the potential of an off-line 2D-LC (SCX × RPnano-LC) coupled with MS/MS via an ESI source for the determination of a tryptic digest of a mixture of proteins. Cytochrome *c*, transferrin, bovine serum albumin, β-galactosidase, alcohol dehydrogenase, lysozyme, β-casein, bovine serum albumin and monoclonal antibody, and an *Escherichia coli* digest were analyzed. The first dimension was carried out in a conventional 15 cm × 1 mm SCX column with a linear salt gradient of 5 mM dihydrogenphosphate, 1 M NaCl solution at pH 3, while the second dimension took place on a 15 or 50 cm × 75 μm C_{18} packed capillary column with a gradient of ACN, water and trifluoroacetic acid (TFA). After suitable collection, preconcentration/desalting, and injection of the fractions obtained in the previous ion-exchange chromatography, a comparison in terms of peak capacity was done between 1D-RPnano-LC and 2D-SCX-RPnano-LC. Fig. 9.6 shows the off-line SCX × RPnano-LC separation of an *E. Coli* digest, resulting in the tryptic fragments randomly distributed within the elution window, a fact that demonstrates the high degree of orthogonality of the approach.

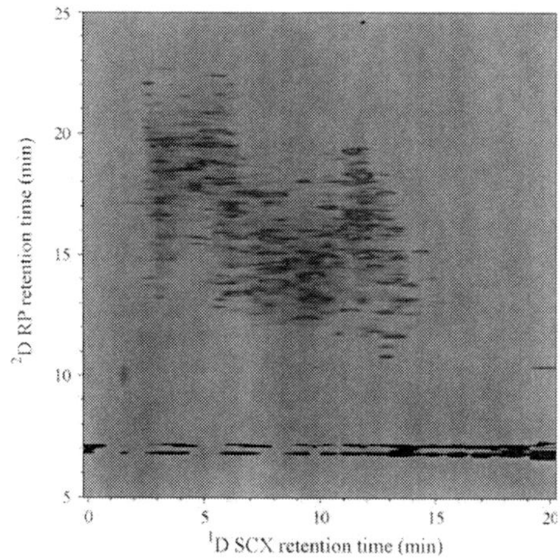

Fig. 9.6. Retention plot of the off-line SCX × RPLC separation of a tryptic digest of *E. Coli*. ^{1}D SCX conditions: injection volume = 2 µL, flow rate = 50 µL/min; mobile phase A: 95:5% H_2O:ACN, 5 mM NaH_2PO_4 buffer pH = 3; mobile phase B: mobile phase A + 1 M NaCl; gradient = 4-55% B; column temperature = 25°C; UV detection at 214 nm. Collecting 0.5 min ^{1}D fractions during the 20 min ^{1}D analysis. Desalting for 5 min at 40 µL/min. ^{2}D RP conditions: injection volume = 20 µL, flow rate = 750 nL/min; mobile phase A: 0.05% aqueous TFA, mobile phase B: 80:20% ACN:H2O containing 0.04% TFA; gradient = 2-55% B; gradient time = 20 min; wash step at 90% B for 2.5 min; equilibration time = 12.5 min; column temperature = 25°C; UV detection at 214 nm, MS/MS detection via ESI interfacing. Reprinted from (Eeltink et al., 2009) with permission from Elsevier.

Despite the fact that SCX columns are normally preferred for the first dimension separation, the truth is that they have some disadvantages. Among them the need of desalting the fractions, the limitation in separation efficiency, the relatively narrow elution windows of peptides with the same charge, and the use of buffered and non-volatile salt solutions have to be considered. As an alternative and to improve the orthogonality of the analysis, several authors have used different approaches, such as a RP or ion-pair RP column with alkaline elution, combined with a RP or ion-pair RP column with acidic elution. In this sense, Delmotte et al. (2007) employed a conventional C_{18}-RP column using high pH elution conditions coupled off-line with an ion-pair PS-DVB-RP capillary column using low pH eluents for the analysis of the proteome of *Corynebacterium glutamicum* and compared the method with traditional SCX-RPLC. Using the first approach, a general improvement concerning a decrease in total analysis time and an increase in the number of unique peptides and proteins identified was obtained. Nakamura et al. (2008) off-line coupled a conventional narrow-bore polymer-based RP (PolyRP) column using alkaline elution with a C_{18} nano-LC column using acidic conditions for shotgun proteomics of mouse brain tissue after suitable sample

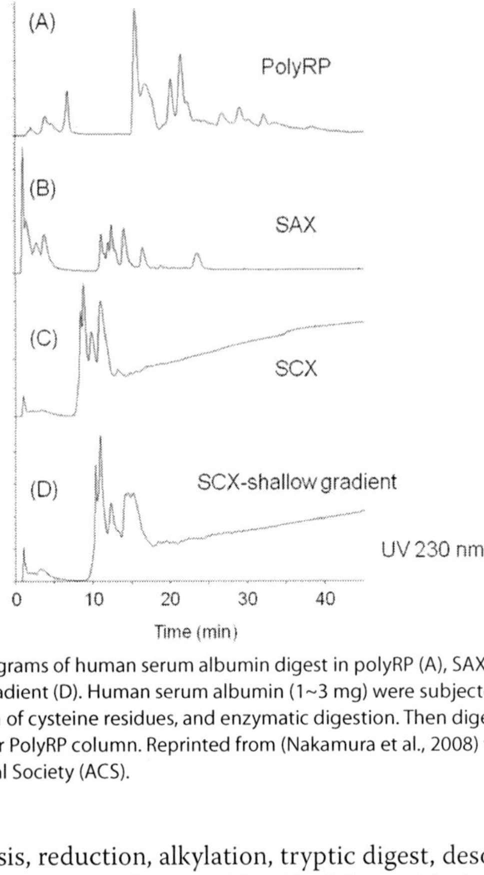

Fig. 9.7. UV chromatograms of human serum albumin digest in polyRP (A), SAX (B), SCX (C), and SCX by shallow gradient (D). Human serum albumin (1~3 mg) were subjected to reduction, carbamidomethylation of cysteine residues, and enzymatic digestion. Then digested samples were injected to SCX, SAX, or PolyRP column. Reprinted from (Nakamura et al., 2008) with permission from The American Chemical Society (ACS).

pre-treatment (lysis, reduction, alkylation, tryptic digest, desolvation and SPE). In this case, even more peptides were identified than with the conventional SCX-RPLC technique. Authors compared peptide separation profiles among three different columns in the first dimension by using tryptic albumin (real biological samples are too complicated to observe peak shapes). As can be seen in Fig. 9.7, more peaks were found in the PolyRP and strong anion exchanger (SAX) separation and even SAX identified more peptides. However, PolyRP showed better peak shapes and has the advantages of not using salts.

Besides regarding the use of one dimensional CLC or nano-LC for the determination of peptides, in the last few years, a good number of works have been developed in this field. Table 9.A shows a selection of them. Different SPs have been assayed and, despite the fact that C_{18} columns have been the most commonly used (Bjellaas et al., 2004; Rozenbrand et al., 2006; Flora et al., 2008; Collier et al., 2008; Sakai-Kato et al., 2009; Lortie et al., 2009; Pröfrock & Prange, 2009; Kroening et al., 2010; Zhu & Fitzgerald, 2010; Snovida et al., 2010), monolithic materials have also attracted great interest (Rozenbrand et al., 2006; Geiser et al., 2008; Detobel et al., 2010).

An interesting approach that has been recently developed is the one reported by Geiser et al. (2008) who prepared two monolithic columns coupled in-line: a monolith used as a bioreactor by *in-situ* copolymerization of 2-vinyl-4,4-dimethylazlactone and ethylene dimethacrylate as support for covalent immobilization of pepsin for protein digestion and, then, a second poly(lauryl methacrylate-*co*-ethylene dimethacrylate) monolith with a different porous structure for peptide concentration and separation by nano-LC. To demonstrate the viability of the method, three proteins greatly differing in molecular sizes were used. Myoglobin (which does not include cysteine residues) was directly dissolved in 10 mM aqueous HCl at pH 2 and injected in the system, while the other two proteins, hemoglobin and albumin (both containing cysteine residues) were reduced, alkylated and then dissolved in the same manner as myoglobin. As a clear example, Fig. 9.8 shows a chromatogram of peptides obtained after the in-line digestion and separation of a solution of 1 μM myoglobin, where 11 different peaks can be observed. A total of 15 peptides were identified with the help of MS detection with a mass deviation of 2 Da or less. Results when using this enzymatic reactor for the digestion of more complex proteins, e.g., albumin and hemoglobin, were more difficult to interpret. However the authors demonstrated that the immobilized pepsin reactor also permitted their digestion.

Rozenbrand et al. (2006) made a comparison between a 15 cm × 100 μm monolithic and a 16 cm × 100 μm particulate RP silica (5 μm) capillary column for the separation of a myoglobin digest. High flow rates, up to 2.8 μL/min, could be applied when using the monolithic column, while with the

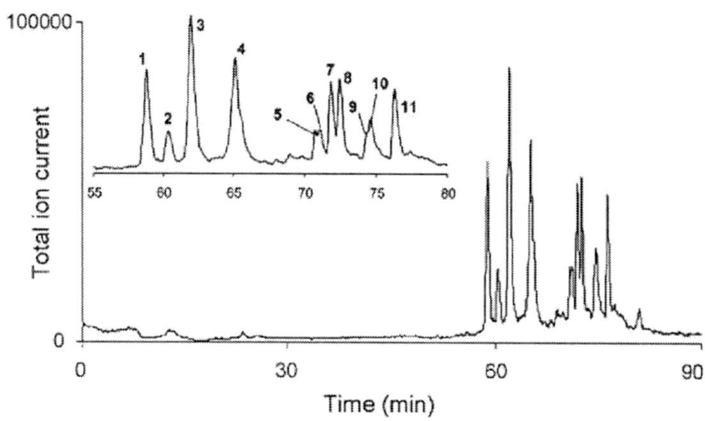

Fig. 9.8. Separation of peptides resulting from the in-line digestion of 2 pmol of myoglobin in 15 cm × 100 μm I.D. immobilized pepsin reactor using a 20 cm × 100 μm I.D. monolithic column. Myogloblin concentration in solution 1 μmol/L, injection volume 2 μL; mobile phase: 2% formic acid in water for 25 min pumped through the complete system followed by switch of the valve and use of a 0.80% linear gradient of ACN gradient in 2% aqueous formic acid in 40 min. Flow rate of 500 nL/min. Reprinted from (Geiser et al., 2008) with permission from Elsevier.

particulate C_{18} column this was impossible due to the high back-pressure. Compared to other nano-LC methods to separate myoglobin digests, a drastic diminution of the analysis time (4 min *vs* 30 min) was obtained using this monolithic column.

Top-down proteomics is not so much employed as *bottom-up analysis*, because it is generally less effective for protein identification than peptide level measurement. Proteins are selected for analysis by MS without the need of proteolysis, obtaining at the end useful information about protein masses, structures and modifications (Bogdanov & Smith, 2005). In this sense, recently Detobel et al. (2010) employed a poly(styrene-*co*-divinylbenzene)-based monolithic column to separate intact proteins and studied the effects of different column parameters (length, morphology) and gradient conditions (flow rate, gradient duration, temperature) on peak capacity and analysis time. They observed that, as expected, the peak capacity increased with the square root of the column length, while decreasing the macropore size of the monolith resulted in an increase of the peak capacity. However, depending on the separation, a trade-off between peak capacity and analysis time has to be made. For example, the authors showed that for the separation of a mixture of ribonuclease A, carbonic anhydrase, lysozyme, myoglobin and trypsinogen, a 5 cm × 200 μm monolithic column with small pore size provided the maximum peak capacity. For more complex separations such as an *E. Coli* protein sample, a 5 cm × 200 μm monolith provided the highest peak capacity in the shortest time.

Environmental Analysis

From a detailed review of the current literature it is clear that proteomics is the field in which nano-LC and CLC have found their biggest application area. Even so, important steps are being developed to extend such applications to other fields of importance like environmental analysis in which most works have been focused in the pesticide field (see Table 9.B). Most of these works have been developed to only demonstrate the capability of the technique for pesticide separation, not for real sample applications. An example is the work of Rosales-Conrado et al. (2004) in which the enantiomeric separation of three phenoxy acid herbicides, namely mecoprop, dichlorprop and fenoprop was studied using a 75 μm I.D. capillary column packed with vancomycin-modified silica particles of 5 μm. Enantioseparation is of particular interest since phenoxy acid herbicide (*R*) isomers show much higher herbicide activity and different metabolism than their (*S*)-forms (Garrison et al., 1996). An aqueous mobile phase containing 85% MeOH, 5% of 500 mM ammonium acetate pH 4.5 buffer and 10% water was found appropriate to fully enantioresolve the three studied herbicides with a high repeatability. Fanali et al. (2006) also studied the enantiomeric separation of eight acidic herbicides (mecoprop, dichlorprop, fenoprop, fluazifop, flamprop, diclofop, fenoxaprop and haloxyfop) and some pharmaceutical acidic compounds (naproxen, suprofen, ibuprofen, carprofen, cicloprofen and ketoprofen) by using a 100 μm I.D. column packed with tert-butylbenzoylated tartardiamide CSP (CHI-TBB). The use of 89/10/1 (v/v/v) *n*-hexane/2-PrOH/acetic acid as mobile phase (normal phase mode) provided a good resolution for mecoprop, dichlorprop, diclofop and fenoxaprop. However, when 8/4/87/1 (v/v/v/v) 2-PrOH/CH_2Cl_2/*n*-hexane/acetic acid was used, the resolution of some of the enantiomers improved. Capillary temperature influence was also tested between 10 and 40°C for mecoprop and haloxyfop with the first mobile phase. Their chromatographic retention was shown to be an exothermic process and, therefore, rising temperature provided a decrease in enantiomers resolution.

Regarding the use of these techniques for the analysis of pesticides in real samples of environmental interest, very few applications have been developed (Cappiello et al., 2002b; Famiglini et al., 2005; Cháfer-Pericás et al., 2006; Cháfer-Pericás, 2007; Famiglini et al., 2008; Mohamed, 2009; Armenta & Lendl, 2010; León-González et al., 2001; Rosales-Conrado et al., 2002), most of them dealing with the analysis of water samples. This is the case of the work by Cappiello et al. (2002b) in which ten organophosphorus pesticides, commonly distributed in local sugar beet cultivations, were analyzed using a new direct EI interface. In this case, separation was performed on a C_{18} column (15 cm × 75 μm; 3 μm particles) after pre-concentration of 2 L of water in a graphitized carbon black SPE cartridge; recovery values close to 100% were obtained. Direct-EI was found to provide necessary linearity, ruggedness, reproducibility of response, and readily interpretable spectra from commercially available sources, for the positive identification of the target pesticides. Another interesting application for the analysis of pesticides in real samples was developed by Rosales-Conrado et al. (2002) for the determination of two chlorophenoxy acid herbicides and their esters in soils using CLC-UV with large volume injection and temperature gradient. After ultrasound assisted extraction of 1 g of the samples with ACN and 5 mM H_2SO_4 solutions, 20 μL of the collected extracts that had been diluted to 100 mL were injected on a C_{18} packed capillary column of 15 cm × 300 μm I.D. Isocratic elution with a 56:44 v/v mixture of MeOH and 0.8% H_3PO_4 aqueous solution and a temperature gradient up to 65°C was used. Temperature was found to be an important factor due to the small diameter of the column. Recoveries obtained were between 72 and 97% at two spiked levels with LODs between 0.3 and 0.5 mg/kg.

In some cases, the determination of pesticides has been carried out together with other analytes of environmental importance (Famiglini et al., 2005; Campíns-Falcó et al., 2010). This is the case of the analysis of endocrine disrupting compounds like phenols (i.e., bisphenol A and alkylphenols), PAHs (i.e., naphthalene, pyrene, anthracene, phenanthrene and fluoranthene), hormones (testosterone, estradiols, diethylstilbestrol and mestranol) as well as an

Table 9.B. NanoLC and CLC Applications on Pesticide Analysis

Pesticides	Matrix	Sample pretreatment	Column	Detector	Mobile phase	Observations	References
Carbaryl, carbendazim, chorpropham, propanil (among other compounds)	—	—	C_{18} (15 cm × 75 or 50 μm; 3 μm)	MS (IT)	ACN/water/ ammonium acetate/formic acid (Gradient)	Nano-LC coupled to an IT originally designed for GC-MS	(Moneti et al., 2001)
Triazines (hydroxyatrazine, cyanazine, simazine, atrazine, methoprotryn, propazine, terbutylazine, terbutryn)	Ground water	SPE (anion exchange column)	C_{18} (15 cm × 300 μm; 3 μm)	UV (238 nm)	ACN/water/ ammonium acetate (Isocratic)	Use of on-column focusing. LOD: 0.009-0.71 μg/L	(León-González et al., 2001)
Carbaryl, propanyl (among other compounds)	—	—	C_{18} (15 cm × 75 μm; 3 μm)	MS (Q)	ACN/ water (Gradient)	LC-MS interface evaluation. LODs: 0.05-1.5 ng	(Cappiello et al., 2002a)
Chlorophenoxy acid herbicides (2,4-D, 2,4-DB) and their esters (2,4-D-1-butyl ester, 2,4-D-1-methyl ester)	Soil	USE (ACN), USE (diluted sulfuric acid)	C_{18} (15 cm × 300 μm; 3 μm)	UV (232 nm)	MeOH/ phosphoric acid (Isocratic)	Use of large volume injection and temperature gradient. LODs: 0.3-0.5 μg/g	(Rosales-Conrado et al., 2002)

Pesticides	Matrix	Sample pretreatment	Column	Detector	Mobile phase	Observations	References
^{14}C-labeled azoxystrobin	—	—	C_{18} (15 cm × 322 μm; 3 μm)	Radio-chemical	ACN/water/acetic acid (Isocratic)	Use of a post column flow-splitter with coaxially transported make-up flow that increases the split flow rate to a flow compatible with commercially available radiochemical flow cell	(Onisko, 2002)
Dimethoate, paraoxon, azinphos-methyl, parathion-methyl, malathion, azinphos-ethyl, parathion-ethyl, diazinon, phorate, phoxim	Waters	SPE (non-porous graphitized carbon black)	C_{18} (15 cm × 75 μm; 3 μm)	MS (Q)	ACN/water (Isocratic)	LODs: 3 ng/L	(Cappiello et al., 2002b)

(continued on next page)

Table 9.B. *(continued)*

Pesticides	Matrix	Sample pretreatment	Column	Detector	Mobile phase	Observations	References
Monuron, chlortoluron, fluometuron, isoproturon, propham, linuron, chlorbromuron (among other compounds)	—	—	C_{18} (15 cm × 75 μm; 3 μm)	MS (Q)	ACN/ water (Gradient)	LODs: 0.012-4.5 ng	(Cappiello et al., 2003b)
(±) Mecoprop, (±) dichlorprop, (±) fenoprop	—	—	Vancomycin-modified chiral silica stationary phase (21 cm × 100 μm; 5 μm)	UV (195 nm)	MeOH/water/ ammonium acetate (Isocratic)	Enantiomeric separation. LODs: 1.4-4.2 μg/mL	(Rosales-Conrado et al., 2004)
Methomyl, aldicarb, cyanazine, atrazine, carbaryl, propazine, terbutryn, lindane (among other compounds)	Marine water	SPE (Oasis HLB)	C_{18} (15 cm × 75 μm; 3 μm)	MS (Q)	ACN/water (Gradient)	LODs: 0.4-118.7 ng/L	(Famiglini et al., 2005)

Pesticides	Matrix	Sample pretreatment	Column	Detector	Mobile phase	Observations	References
(±) Mecocrop, (±) dichlorprop, (±) fenoprop, (±) fluazifop, (±) haloxyfop, (±) diclofop, (±) flamprop, (±) fenoxaprop (among other compounds)	—	—	*Tert*-butyl-benzoylated tartardiamide-modified silica stationary phase (23 cm × 100 μm)	UV (195 nm)	n-hexane/2-PrOH/acetic acid (Isocratic)	—	(Fanali et al., 2006)
Simazine, atrazine, propazine, ametryn, terbutryn	Waters	In-tube SPME (comparison between a C_{18} packed column and an open column coated with PDMS)	C_{18} (15 cm × 500 μm; 5 μm)	UV (230 nm)	ACN/water (Isocratic)	Comparison with on-fibre SPME coupled to conventional HPLC. LODs (in-tube: 0.025-0.5 μg/L	(Cháfer-Pericás et al., 2006)
Fensulfothion, fenamiphos, parathion, malathion, fenitrothion, chlorfenvinphos, fenthion, fonofos, chlorpyrifos, trifluraline	Waters	In-tube SPME (open columns coated with PDMS/ diphenylpolysiloxane (95/5), 30 cm × 250 μm, 0.25 μm thickness and 30 cm × 100 μm, 0.1 μm thickness)	C_{18} (15 cm × 500 μm; 5 μm)	UV (230 nm) FD (program)	ACN/water (Gradient)	Evaluation of triazines, PAHs, nonylphenol, OCPs and PBDEs as possible interferents. LODs: 0.1-10 μg/L (in-tube SPME) vs. 50-1000 μg/L (direct injection)	(Cháfer-Pericás et al., 2007)

(continued on next page)

Table 9.B. *(continued)*

Pesticides	Matrix	Sample pretreatment	Column	Detector	Mobile phase	Observations	References
Lindane, methoxychlor, β-endosulfan, α-endosulfan, *o,p'*-DDD, *p,p'*-DDD, heptachlor, *o,p'*-DDT, *p,p'*-DDT, DDE, HCB, aldrin	Waters	SPE (C_{18})	C_{18} (15 cm × 75 μm; 3.5 μm)	MS (Q)	ACN/water (Gradient)	First alternative LC-MS method to typical GC-MS for OCPs analysis. LODs: 0.044-0.33 μg/L	(Famiglini et al., 2008)
Methomyl	Irrigation water	SPE (Oasis HLB)	C_{18} (25 cm × 4.6 mm; 5 μm)	MS (IT)	ACN/water/ formic acid (Isocratic)	Study of the pesticide degradation by the novel bacterial strain *Stenotrophomonas maltophilia* M1. LOQ: 0.02 μg/L	(Mohamed, 2009)
Thiourea, fenuron, metoxuron, monuron, fluometuron, monolinuron, chlorotoluron, diuron, metobromuron, chlorobromuron, linuron (among other compounds)	—	—	C_8 hybrid silica monolith (27.2 cm × 150 μm)	UV (254 nm)	ACN/water (Isocratic)	*In-situ* synthesized C_8 hybrid silica monolith. Focus on reduction of the sources of radial heterogeneity	(Roux et al., 2009)
Chlortoluron, atrazine, diuron, terbuthylazine	River water	SPE (C_{18})	C_{18} (15 cm × 300 μm; 3 μm)	UV (224, 248 nm), vibrational spectro-scopic detection (mid-IR and Raman)	MeOH/water (Gradient)	Comparison with another described method in the literature. LODs: 2 ng	(Armenta & Lendl, 2010)

alkaloid (caffeine) and dimethylphthalate together with a group of multiclass pesticides in marine water samples (Famiglini et al., 2005). In this work, 200 mg Oasis HLB SPE cartridges were used as a pre-concentration step with 1 L of sample prior to nano-LC-EI-MS. LODs achieved were in the 0.4–118.7 ng/L range and recovery values were higher than 79%. Campíns-Falcó et al. (2010) also determined phosphorus, chlorine and nitrogen containing pesticides together with some phenols, phthalates and PAHs in waters from the Mediterranean coast. On-line coupling of in-tube solid phase microextraction (SPME) in the injection valve to CLC-DAD was applied. In this work, authors used a conventional GC capillary column, 95% methyl–5% phenyl substituted backbone, 80 cm × 320 µm I.D., 3 µm film thickness. The in-tube SPME procedure was compared with the performance of a column-switching device by means of a 35 mm × 0.5 mm I.D., 5 µm particle size C_{18} column. The best results were achieved by employing in-tube SPME with LODs in the 0.008–0.2 µg/L range and 70–116% recovery values. No matrix effect was found. In another recent work by the same group, two faecal sterols, namely cholesterol and coprostanol were determined in wastewater samples as tracers of human sewage contamination using a titania-based capillary polymeric column (Moliner-Martínez et al., 2010). The titanium dioxide (titania)-based column allowed the separation of the analytes in a shorter time (10 min) and with better chromatographic profiles, which in turn resulted in better detectability than using a conventional C_{18} capillary column. Analyte enrichment was also carried out by in-tube SPME (TRB-5 GC capillary column) which allowed the determination of both compounds down to 10 and 1.2 µg/L for coprostanol and cholesterol, respectively.

Other environmental applications involve the determination of bisphenol A in air samples (Sabatini et al., 2005), drugs like β-lactam antibiotics (i.e., ampicillin, amoxicillin, penicillin V, penicillin G, cloxacillin, oxacillin, dicloxacillin, nafcillin, piperacillin and clavulanic acid) (Bailón-Pérez et al., 2009), diazepam (van der Ven et al., 2004) or sulphamethoxazole, bezafibrate, metoprolol, carbamazepine and bisoprolol (Pitarch et al., 2004). In addition, nitrophenols (Zhu et al., 2001), priority EPA phenols (Chung & Lee, 2001) and PAHs (Zheng et al., 2007) in water samples were also determined. The chemical warfare agent mustard has also been analyzed by CLC in soil (D'Agostino et al., 2004), as well as sarin which was determined in contaminated snow (D'Agostino et al., 2002). Most of these different applications have been developed by CLC instead of nano-LC. In the case of the work by Zhu et al, (2001), CLC-UV was used to determine the nitrophenols 2-nitrophenol, 2,4-dinitrophenol, 3-nitrophenol, 3,4-dinitrophenol and 4-nitrophenol in seawater samples after pre-concentration with a liquid-liquid-liquid microextraction (LLLME) procedure. Nitrophenols were initially extracted from water using a 2 cm × 600 µm I.D. polypropylene hollow fiber membrane of 200 µm wall thickness, using 0.1 M HCl as the donor phase and 0.1 M NaOH as the acceptor phase under a stirrer speed of 1000 rpm and an extraction time of 50 min. In this case, the separation after the extraction was carried out in a 3 µm particle C_{18} column (15 cm × 300 µm) using an isocratic mobile phase of 40:60 (v/v) ACN/water with 75 mM sodium acetate-glacial acetic acid. Due to the necessity of having the compounds in their neutral form to be determined in CLC, this last buffer had to be added to the mobile phase in order to neutralize the NaOH that was previously used as the acceptor solution in LLLME.

On the other hand, Zheng et al. (2007) used in-tube SPME to preconcentrate six PAHs (toluene, naphthalene, biphenyl, fluorene, phenanthrene and fluoranthene). An octyl-functionalized silica monolithic column synthesized by an acid/base-catalyzed hydrolysis/co-condensation of tetraethoxysilane and *n*-octyltriethoxysilane, was coupled to CLC. The analytical column was a 25 cm × 100 µm capillary filled with 3 µm particle size-C_{18} and elution was carried out in isocratic mode using a MeOH/water (85:15, v/v) mixture. LODs in the range 2.4–8.1 ng/mL were recorded for all analytes. These values were more than 300-fold lower than those obtained without SPME pre-concentration.

Pharmaceutical Analysis

When talking about pharmaceutical analysis, HPLC is still the technique preferred for drug separation, quantitation and identification. That is why this type of analyses has also found a good field of application in nano-LC and CLC. Excellent results were achieved mainly for chiral applications that will be later discussed. Table 9.C shows some research articles that employ these two nanoscale techniques in chiral and non-chiral pharmaceutical analysis, especially focusing on real sample applications during the last decade.

Regarding non-chiral analysis (Lanckmans et al., 2006; Fanali et al, 2007; Bailón-Pérez et al., 2009; Marchesini et al., 2008; Hsieh et al., 2009) developed a methodology to determine the anti-epileptic drug oxcarbazepine and its active metabolite 10,11-dihydro-10-hydroxycarbamazepine in dialysate samples; 2-methyl-5H-dibenz(*b,f*)azepine-5-carboxamide was used as an internal standard. Microbore column of 1 mm I.D., 300 and 75 µm I.D. capillary columns packed with C_{18} particles were compared using ACN/water/formic acid gradient elution and ESI-MS/MS detection. Due to the sensitivity of ESI-MS detection to matrix effects, it was studied comparing the standards dissolved in water, Ringer's solution (salt solution) and a blank dialysate matrix. Using column switching for desalting and pre-concentration of the dialysates and not exceeding 1 µL as injection volume, no differences were found between Ringer's solution and blank dialysate matrix in CLC and nano-LC. Detection sensitivity was 5- and 50-fold better for CLC-MS/MS and nano-LC-MS/MS, respectively, with respect to microbore dimensions. However, accuracy and precision were found better for microbore LC-MS/MS. Another example of non-chiral analysis of pharmaceutical compounds is the work by Fanali et al.

Table 9.C. NanoLC and CLC Applications on Pharmaceutical Analysis

Pharmaceutical	Matrix	Sample pretreatment	Column	Detector	Mobile phase	Observations	References
Oxcarbazepine, 10,11-dihydro-10-hydroxycarbamazepine, 2-methyl-5H-diben(b,f) azepine-5-carboxamine	Dialysate	Desalting	C_{18} (15 cm × 300 μm; 5 μm) C_{18} (15 cm × 75 μm; 5 μm)	MS (QqQ)	ACN/water/ formic acid (Gradient)	Use of column switching for desalting and preconcentration. Comparison between nano-LC, CLC and microbore. LODs: 20-50 pg/mL (CLC), 3-9 pg/mL (nano-LC)	(Lanckmans et al., 2006)
β-blockers (nadolol, oxprenolol, alprenolol, propranolol)	Pharmaceutical formulations (analysis of oxprenolol)	Sonication (MeOH/water), centrifugation, dilution	C_{18} bidentate (23 cm × 75 μm; 5 μm)	UV (205 nm) MS (IT)	MeOH/water/ ammonium acetate (Isocratic)	Comparison between sheathless nano-spray and sheath-flow interfaces. LODs were lower for nano-spray	(Fanali et al., 2007)
(±) Naproxen	—	—	C_{18} (22 cm × 75 μm; 5 μm)	UV (232 nm)	ACN/water/ sodium acetate/ methyl-β-cyclodextrine (Isocratic)	Development of the methodology as conventional HPLC	(Healy et al., 2001)
(±) Mephobarbital, (±) hexobarbital, (±) barbituric acids, (±) benzoin, (±) carprofen	—	—	Permethylated-β-cyclodextrin-modified chiral monolith (17-25 cm × 100 μm)	UV (230 nm)	MeOH/water/2-(*N*-morpholino) ethanesulfonic acid (Isocratic)	Comparison with CEC-UV. Higher efficiency for CEC	(Wistuba & Schurig, 2000a)

(continued on next page)

Table 9.C. *(continued)*

Pharmaceutical	Matrix	Sample pretreatment	Column	Detector	Mobile phase	Observations	References
(±) Ambucetamide, (±) aminoglutethimide, (±) etozolin, (±) nifurtimox, (±) norgestrel, (±) omeprazole, (±) thalidomide analog (EM-12), (±) Tröger's base	—	—	Amylose tris(3,5-dimethylphenyl-carbamate)-coated on silica gel (24 cm × 100 µm; 5 µm)	UV (214 nm)	ACN/water/ammonium acetate (Isocratic)	Comparison with CEC-UV	(Chank-vetadze et al., 2002)
(±) Norgestrel	Pharmaceutical formulation	Dissolution in MeOH	Cellulose tris(3,5-dichlorophenyl-carbamate)-coated on silica gel (24 cm × 100 µm; 5 µm)	UV (254 nm)	MeOH/ammonium acetate (Isocratic)	Comparison with CEC-UV	(Chank-vetadze et al., 2003)
(±) Loxiglumide	Pharmaceutical formulation	—	Hepta-Tyr-modified silica stationary phase/amino silica stationary phase (3:1 w/w) (7 cm × 75 µm; 5 µm)	UV (214 nm)	ACN/water/sodium phosphate (Isocratic)	Comparison with CEC-UV using the same column. LODs: 0.3-0.5 µg/mL	(Fanali et al., 2004b)

Pharmaceutical	Matrix	Sample pretreatment	Column	Detector	Mobile phase	Observations	References
Basic compounds: (±) alprenolol, (±) atenolol, (±) metoprolol, (±) oxprenolol, (±) pindolol, (±) propranolol; acidic compounds: (±) 2-[(5′-benzoyl-2′-hydroxyl)phenyl] propionic acid, (±) 2-[(4′-benzoyloxy-2′-hydroxy)phenyl] propionic acid, (±) ketoprofen, (±) indoprofen, (±) suprofen	Pharmaceutical formulation (analysis of metoprolol)	Dilution in MeOH, sonication, filtration	Vancomycin-modified silica chiral stationary phase (22.8 cm × 75 μm; 5 μm)	UV (195 nm)	ACN/MeOH/water/ammonium acetate (Isocratic)	Different mobile phase composition for basic or acidic compounds. LOD: 1.5 μg/mL (metoprolol)	(D'Orazio et al., 2005)
(±) Atenolol	Human urine	Dilution, alkalinization, LLE (chloroform/2-propanol)	Teicoplanin-modified silica chiral stationary phase (22.8 cm × 75 μm; 5 μm)	MS (IT)	MeOH/ACN/water/ammonium acetate (Isocratic)	Comparison with nano-LC-UV (205 nm) CE-UV. LOD: 50 ng/mL	(D'Orazio & Fanali, 2006)
Non-steroidal anti-inflammatory drugs ((±) naproxen, (±) suprofen, (±) ibuprofen, (±) carprofen, (±) cicloprofen, (±) ketoprofen) (among other compounds)	—	—	*Tert*-butyl-benzoylated tartardiamide-modified silica stationary phase (23 cm × 100 μm)	UV (195 nm)	*n*-hexane/acetone/acetic acid, *n*-hexane/i-PrOH/acetic acid, *n*-hexane/2-PrOH/acetic acid (Isocratic)	-	(Fanali et al., 2006)

(continued on next page)

Table 9.C. *(continued)*

Pharmaceutical	Matrix	Sample pretreatment	Column	Detector	Mobile phase	Observations	References
(±) Alprenolol, (±) propranolol, (±) oxprenolol, (±) metoprolol	Human plasma (alprenolol)	Addition of 1 M NaOH and dichloromethane	Vancomycin-modified silica particles (25 cm × 75 μm; 5 μm)	UV (205 nm)	MeOH/water/ammonium acetate (Isocratic)	Use of on-line focusing. LODs: 9-15.6 ng/mL (alprenolol)	(D'Orazio & Fanali, 2008)
(±) Naproxen, (±) indoprofen, (±) ketoprofen, (±) flurbiprofen, (±) carprofen, (±) cicloprofen, (±) flunoxaprofen, (±) suprofen	—	—	C_{18} (10 cm × 100 μm; 3 μm)	UV (205 nm)	ACN/water/sodium acetate/heptakis (2,3,6-tri-*O*-methyl)-β-cyclodextrin (Isocratic)	Use of smaller amounts of mobile phase chiral selector than in classical HPLC methods	(Rocco & Fanali, 2009b)
(±) Flavanone, (±) etozoline, (±) lopirazepam, (±) temazepam	—	—	Amylose tris(5-chloro-2-methylphenylcarbamate) coated on silica gel (25 cm × 100 μm; 5 μm)	UV (214 nm)	MeOH/ACN/water/ammonium acetate (Isocratic)	Study of the coating on aminopropylsilanized silica. Comparison with CEC-UV	(Fanali et al., 2010)

(2007) in which the separation of a group of β-blocker drugs (nadolol, oxprenolol, alprenolol and propranolol), using terbutaline as internal standard, was studied. The separation of basic compounds in RP-C_{18}-based HPLC is not easy due to the strong adsorption between analytes and free silanols. In this work, two different 75 μm I.D. columns packed with 23 cm of C_{18} with free silanol and bidentate C_{18} without free silanol groups were compared employing the same mobile phase composition in isocratic elution: 500 mM ammonium acetate at pH 4.5/water/MeOH (1:8:91 v/v/v).

Fig. 9.9 shows that, as expected, the bidentate C_{18} column provided better results concerning peak tailing, efficiency and also repeatability and sensitivity. Furthermore, different detection systems were assayed in this research, namely, UV-vis and MS with different interfaces. ESI sheath-liquid and nano-ESI interfaces offered a better performance than the UV-vis detector. However, best results were achieved using nano-ESI: LODs of 0.05 and 1.8 μg/mL for nadolol were observed by employing nano-ESI and ESI, respectively. After validation the developed nano-LC-nano-ESI-MS methodology was applied to the determination of oxprenolol in a pharmaceutical preparation.

Although, as it has been reported, several works have dealt with the determination of non-chiral pharmaceuticals in the last years, the truth is that the vast majority of the research that takes place in this field is focused on chiral analysis (Hernández-Borges et al., 2007a), as the following section demonstrates.

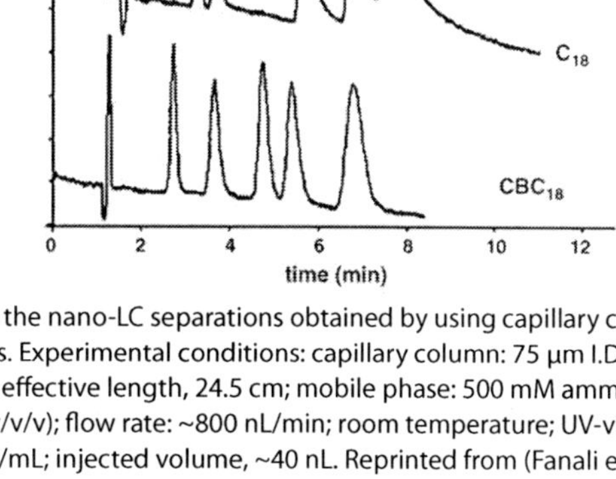

Fig. 9.9. Comparison of the nano-LC separations obtained by using capillary columns packed with C_{18} and C_{18} bidentate SPs. Experimental conditions: capillary column: 75 μm I.D. × 375 μm O.D.; packed length, 23.0 cm, effective length, 24.5 cm; mobile phase: 500 mM ammonium acetate pH 4.5-H_2O-MeOH (1:8:91, v/v/v); flow rate: ~800 nL/min; room temperature; UV-vis: 205 nm; samples concentration, 20-90 μg/mL; injected volume, ~40 nL. Reprinted from (Fanali et al., 2007) with permission from Elsevier.

Chiral Analysis

Enantiomeric separations are of crucial importance in numerous fields, especially in pharmaceutical analysis. It has been reported that very often, one of the enantiomers of a given substance can exhibit different therapeutic effect than the other and in some cases can be even dangerous for health. Miniaturized techniques have become an interesting working path in chiral analysis due to the smaller waste of expensive CSPs and mobile phases with valuable chiral additives. Table 9.C shows some applications of these techniques in pharmaceutical analysis (Chankvetadze et al., 2002; Chankvetadze et al., 2003; Fanali et al., 2004b; Fanali et al., 2005b; D'Orazio et al., 2005; D'Orazio & Fanali, 2006; Fanali et al., 2006; D'Orazio & Fanali, 2008; Rocco & Fanali, 2009b; Fanali et al., 2010).

Polysaccharide derivatives coated into silica gel are one of the most used CSPs in HPLC. In fact, they have been also applied in the last few years in nano-LC. In this sense, some authors have employed derivatives of cellulose or amylose, such as cellulose tris(3,5-dichlorophenylcarbamate) (Chankvetadze et al., 2003), amylose (3,5-dimethylphenylcarbamate) (Chankvetadze et al., 2002), amylose tris(5-chloro-2-methylcarbamate) (Fanali et al., 2010), etc. In one of these works, Chankvetadze et al. (2003) determined the enantiomeric purity of the contraceptive drug levonorgestrel in a pharmaceutical formulation using a capillary column of 100 µm packed with 24 cm of cellulose tris(3,5-dichlorophenylcarbamate) coated onto 5 µm silica gel particles. Using a mobile phase consisting of 5 mM ammonium acetate in MeOH in an isocratic mode, the percentage of polysaccharide coated into silica gel was studied between 5–25% (w/w) (see Fig. 9.10) in nano-LC-UV and CEC-UV modes. Fig. 9.10 shows that separation selectivity increased with increasing loading of the chiral selector onto silica gel, getting a baseline enantioseparation for nano-LC only between 15–25%.

Also, silica particles modified with glycopeptide antibiotics such as hepta-Tyr antibiotic (Fanali et al., 2004b), vancomycin (D'Orazio et al., 2005; D'Orazio & Fanali 2008) or teicoplanin (D'Orazio & Fanali, 2006) have been used to pack the capillary columns (Fanali et al., 2004b); (D'Orazio et al., 2005; D'Orazio & Fanali, 2008; D'Orazio & Fanali, 2006). For instance, Fanali et al. (2004b) used a 75 µm I.D. column packed with 7 cm of 3:1 w/w hepta-Tyr antibiotic modified silica SP/amino-silica particles for the resolution of the enantiomers of loxiglumide, a drug proposed for the treatment of gastrointestinal diseases. The same column was employed both for nano-LC-UV and CEC-UV, using the same electrophoresis instrumentation. Aminosilica particles were mixed with the CSP with the aim of increasing the reversed EOF in CEC. Good enantiomeric resolution was achieved in less than 5 min with LODs between 0.3 and 0.5 µg/mL by nano-LC using a mobile phase composed of a mixture of 1:1 v/v 5 mM sodium phosphate pH 6/ACN. Finally, the methodology was applied for the determination of D-loxiglumide in a phar-

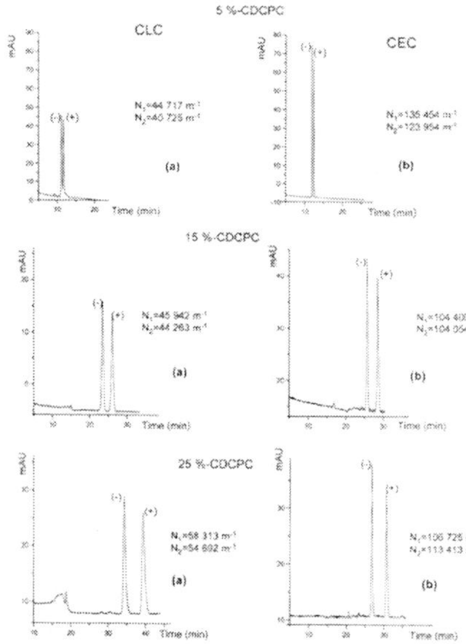

Fig. 9.10. Enantioseparation of norgestrel on a capillary column containing different amounts of cellulose tris(3,5-dichlorophenylcarbamate) (5, 15 and 25%, w/w) in nano-LC (a) and CEC (b) mode. Separation capillary was 100 μm I.D. with 24 cm packed and 31 cm total length. The mobile phase was 5 mM ammonium acetate in MeOH. The applied pressure was 12 bar in nano-LC and the applied voltage was 12 kV in CEC mode. Reprinted from (Chankvetadze at al., 2003) with permission from Elsevier.

maceutical formulation. D'Orazio et al. (2006) employed a teicoplanin-modified silica SP to enantioseparate atenolol and determine it in human urine samples of a patient under therapy. A suitable sample dilution with water, alkalinisation with 8 M sodium hydroxide and liquid-liquid extraction (LLE) with chloroform/2-PrOH (4:1 v/v) was applied. (-)-Ψ-nor-ephedrine was the internal standard and a MeOH/ACN/500 mM ammonium acetate pH 4.5 (60:39:1 v/v/v) was the eluent used in isocratic elution mode. Comparing the results obtained with UV detection (zeta cell) and MS detection with respect to the method sensitivity, lower LODs (50 ng/mL for each enantiomer) were obtained with MS detection. Fig. 9.11 shows the chromatogram of the analysis of a urine sample, as well as the confirmative MS spectra of the substances. The authors also developed a study of the urinary excretion of atenolol enantiomers *vs.* time, that is, the kinetics of its excretion.

Other CSP have been used, e.g., *tert*-butylbenzoylated tartadiamide-modified silica (Fanali et al., 2006) or even a permethylated-β-cyclodextrin (CD)-modified chiral monolith (Wistuba & Schurig, 2000a). In this last case, the authors developed a work in which a group of chiral compounds, i.e., methobarbital, hexobarbital, some barbituric acids, benzoin and caprofen, among others, were enantioseparated using a chiral rod monolithic SP that

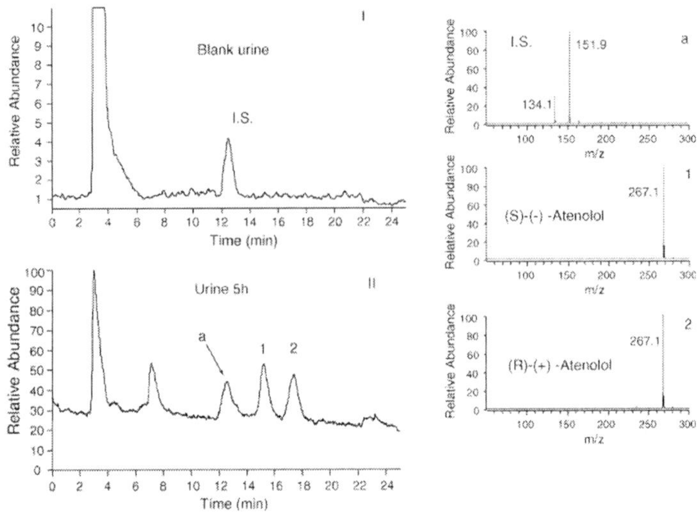

Fig. 9.11. Nano-LC-ESI-MS chromatogram of the analysis of human urine extract containing atenolol enantiomers and the internal standard (-)-Ψ-nor-ephedrine. (I) Blank urine with I.S. 50 μg/mL; (II) urine under atenol therapy with I.S. 50 μg/mL. (a) and (1) and (2) MS spectra of I.S., *S*-(-)- and *R*-(+)-atenolol, respectively. Mass range 50-300 *m/z*. Packed capillary with teicoplanin SP (75 μm I.D.; 23 cm packed); mobile phase: MeOH-ACN-500 mM ammonium acetate pH 4.5 (60:39:1 v/v/v); UV detection at 205 nm; flow rate: 900 nL/min; injection 60 nL of extracted urine sample; concentrations of internal standard and racemic atenolol were 20 and 40 μg/mL, respectively. Reprinted from (D'Orazio & Fanali, 2006) with permission from Elsevier.

was prepared by polymer-coating a silica monolithic packed column with a permethylated-β-CD. The columns (17–25 cm × 100 μm) were mainly assayed in rod-CEC-UV and the most favourable composition of the mobile phase was found to be 20 mM 2-(*N*-morpholino) ethanesulfonic acid buffer at pH 6.0/MeOH 7:3 (v/v). Employing the same conditions, the column was used in nano-LC and both methods were compared. Efficiency was 2–2.5 times higher in rod-CEC than in rod-nano-LC. Despite the fact that similar chiral separation factors were obtained, the resolution factors were higher in the rod-CEC mode. In general, the number of applications regarding the use of monolithic materials for chiral separations in nano-LC is relatively low (Hernández-Borges et al., 2007a) compared to CLC (Wistuba, 2010).

CSPs have been, without any doubt, the preferred approach in enantiomers analysis. However, when talking about conventional HPLC, the addition of a chiral selector in the mobile phase can also be considered. In nano-LC, this approach has not been fully explored, although it has been tested in a few works (Healy et al., 2001; Rocco & Fanali, 2009b), for example the one developed by Rocco & Fanali (2009b), in which a group of non-steroidal anti-inflammatory drugs (naproxen, indoprofen, ketoprofen, flurbiprofen, carprofen, cicloprofen, flunoxaprofen and suprofen) were enantioseparated by nano-LC-UV using a 10 cm × 100 μm C_{18} packed column adding to the mobile phase a methylated-β-CD [heptakis (2,3,6-tri-*O*-methyl)-β-CD]. After selecting the adequate SP and the packing length, several factors regarding the mobile phase that affected the retention factor and the resolution were considered. The pH or the percentage of organic modifier (ACN) was evaluated. The content of an organic modifier is really important because it may reduce the inclusion-complexation between enantiomers and CDs due to the affinity of the solvent for the chiral selector. Finally, the most favourable mobile phase was found to be composed of 30% v/v ACN, 30 mM of chiral selector and pH 3.0. In this work, association constants between drugs and CDs were also determined. Although similar methods exist using HPLC, this nano-LC methodology has the clear advantage of using small amounts of chiral additive resulting in a very cheap approach.

Food Analysis

Food analysis is the application field in which latest works have been proposed using nano-LC or CLC. Table 9.D compiles the different features of these works, focusing basically on food allergens (Weber et al., 2006; Chassaigne et al., 2007; Monaci et al., 2010) or food components such as fatty acids (Yoo & Mcguffin, 1992), phenolic compounds (Blanco Gomis et al., 2001), biogenic amines (Hernández-Borges et al., 2007b; Sont et al., 2004), proteins (Kwon, 2004; Coda et al., 2008; Bianco et al., 2009; Sawaskipuksa et al., 2009; Monaci et al., 2010), peptides (Gil-Agusti et al., 2005; Rizzelo et al., 2007; Coda et al., 2008; Escudero et al., 2010), as well as on food contaminants or residues like pesticides (Lucas-Delfa et al., 2000; Halvorsen et al., 2000; Onisko et al., 2002; Rosales-Conrado et al., 2005; Buonasera et al., 2009) or pharmaceuticals (Marchesini et al., 2008; Bailón-Pérez et al., 2009; Hsieh et al., 2009). From a detailed review of the literature it is clear that a good number of works have been reported.

Sterols are an important class of organic molecules present in either plants (*phytosterols*) or animals (*zoosterols*, the best representative of which is cholesterol). In this sense, Rocco and Fanali (2009a) developed the first application of nano-LC for the analysis of five main sterols (phytosterols) like brassicasterol, stigmasterol, campesterol, cholesterol and β-sitosterol in extra-virgin olive oil (EVOO) employing a capillary column of 15 cm and 100 μm I.D. packed with silica hydride-based RP-C_{18} sub-2 μm particles. Separation was achieved in about 20 min using 100% MeOH as mobile phase. Although these compounds represent a minor part of lipids in vegetable oils, their quantification can be useful to establish oil origin and to reveal intentional adulterations. Extraction of phytosterols was assayed by SPE with C_{18} cartridges, after suitable saponification and LLE of the unsaponified fraction with diethyl ether. Recovery values ranged between 90% and 103%, with RSDs within 5%

Table 9.D. Nano-LC and CLC Applications on Food Analysis

Analytes	Matrix	Sample pretreatment	Column	Detector	Mobile phase composition	Observations	References
Milk allergens	Cookies, orange sherbet, lacto-free ice cream, Enfalac regular, oatmeal cereal, cookies arrowroot	Extraction (Tris, Tween), ion exchange chromatography (DEAE Sepharose Fast Flow resin), desalting. Use of trap column in the LC system	C_{18} (10 cm × 75 µm; 5 µm)	MS (Q-TOF)	ACN/water/ formic acid (Gradient)	Use of three pumps: two for binary gradient and the other for sample focusing and precolumn desalting. Comparison with ELISA. LOD: 1.25 mg/L	(Weber et al., 2006)
Allergens	Processed peanuts	Extraction (Tris, NaCl, EtOH/water), Smith copper/ bicinchoninic assay, trypsin digestion	C_{18} (15 cm × 150 µm; 3 µm)	MS/MS (Q-TOF)	ACN/water/ formic acid (Gradient)	Use of a precolumn (C_{18}, 150 mm × 300 µm; 5 µm). On-column focusing. LODs: 10-40 ng	(Chassaigne et al., 2007)
Allergens	Processed peanuts	Extraction (Tris, NaCl, EtOH/water), Smith copper/ bicinchoninic assay, trypsin digestion	C_{18} (15 cm × 150 µm; 3 µm)	MS/MS (Q-TOF)	ACN/water/ formic acid (Gradient)	Use of a precolumn (C_{18}, 150 mm × 300 µm; 5 µm). On-column focusing. LODs: 10-40 ng	(Chassaigne et al., 2007)
Satured and unsatured fatty acids (lauric, tridecanoic, myristic, pentadecanoic, palmitic, myristoleic, palmitoleic, oleic, linoleic, gammalinolenic, homogammalinolenic, arachidonic, 5,8,11,14,17-eicosapentaenoic and 4,7,10,13,16,19-docosa-hexaenoic)	Fish oil dietary supplements	Saponification, solvent extraction (chloroform), derivatization (4-bromomethyl-7-methoxycoumarin)	C_{18} (96, 152 or 178 cm × 200 µm; 3 µm)	LIF (ex: 325 nm; em: 420 nm)	MeOH/ ACN/water (Isocratic)	LODs: fM level	(Yoo & Mcguffin, 1992)

Analytes	Matrix	Sample pretreatment	Column	Detector	Mobile phase composition	Observations	References
Phenolic compounds ((+)-catechin, (-)-epicatechin, phloridzin, chlorogenic acid, glycosides of quercetin (isoquercetin, hyperin, quercitrin, avicularin, rutin), procyanidins, phloretin xyloglucoside)	Apple juice	Centrifugation, SPE (C_{18})	C_{18} (25 cm × 300 μm; 5 μm)	UV (280 nm)	MeOH/water (Gradient)	A C_{18} column (25 cm × 0.4 cm, 3 μm) used as a pressure restrictor. LOD: 9-97 pg	(Blanco Gomis et al., 2001)
Biogenic amines (ethanolamine, methylamine, tryptamine, 2-phenylethylamine, putrescine, cadaverine, histamine, tyramine, spermidine, spermine)	Red, white wine	Treatment with PVP, derivatization (dansyl-chloride), SPE (C_{18})	C_{18} bidentate (23 cm × 100 μm; 5 μm)	UV (250 nm)	ACN/ water/TEA (Gradient)	Use of a precolumn (C_{18}, 5 mm × 300 μm; 100 Å, 5 μm). LODs: 18.3-48.3 ng/mL	(Hernández-Borges et al., 2007b)
Histamine	Water, alcoholic beverages (whisky, red wine, apple cider), rat brain tissue	Pre-column derivatization (4-fluoro-7-nitro-2,1,3-benzoxadiazole)	C_{18} (5 cm × 75 μm; 5 μm)	MS/MS (IT)	ACN/water/ formic acid (Isocratic)	Use of nanoelectrospray emitters packed with (C_{18}). Preconditioning of the column with water/1-heptanesulfonic acid sodium. LOD: 0.1 ng/mL	(Song et al., 2004)
Proteins	White wine	Extraction of protein, protein separation, digestion	C_{18} (5 cm × 75 μm; 5 μm)	MS (IT)	ACN/water/ acetic acid (Gradient)	Soluble proteins profiling	(Kwon, 2004)

(continued on next page)

Table 9.D. *(continued)*

Analytes	Matrix	Sample pretreatment	Column	Detector	Mobile phase composition	Observations	References
Antifungal proteins and peptides	Bread	Digestion in gel (trypsin) of the protein bands cut from gels	C_{18} (15 cm × 75 μm)	MS/MS (IT)	ACN/water/ formic acid (Gradient)	Use of nano-LC for identification. Trapping and desalting of the peptides were carried out on an enrichment column (C_{18}, 0.5 cm × 0.3 μm)	(Coda et al., 2008)
Proteins	Strawberry	Protein extraction, precipitation	C_{18} (4.3 cm × 75 μm; 5 μm)	MS/MS (IT)	ACN/water/ formic acid (11 salt steps) (Gradient)	Use of 2-D gel electrophoresis, followed by 2-D nano-LC-ESI-IT-MS/MS or 1-D SDS-PAGE coupled with nano-LC-ESI-IT-MS/MS. Also use of a SCX capillary column (3.5 cm × 300 μm; 3.5 μm)	(Bianco et al., 2009)
Proteins	*Pithecellobium dulce* seeds	Protein extraction, 2-D gel electrophoresis, gel image analysis, in-gel digestion	C_{18} (15 cm × 75 μm; 100 Å)	Q-TOF	ACN/water/ formic acid (Gradient)	Use of nano-LC to analyse the most abundant protein spots. Use of a precolumn (1 mm × 300 μm; 100 Å) for desalting and concentration	(Sawasdi-puksa et al., 2009)
Allergenic milk protein markers	Fine white wine	Protein precipitation, filtration, reduction, digestion	C_{18} (15 cm × 320 μm; 5 μm)	MS/MS (Q-TOF)	ACN/water/ formic acid (Gradient)	Detection and identification of markers of potentially allergenic milk proteins used as wine fining agents. LOD: 50 μg/mL (casein)	(Monaci et al., 2010)

Analytes	Matrix	Sample pretreatment	Column	Detector	Mobile phase composition	Observations	References
Peptides	Green coffee bean	Protein extraction (with acetone/TCA/ β-mercapto-ethanol; or with lysis solution; or with acetic acid/urea/CTAB solution)	C_{18} (10 cm × 75 μm; 3.5 μm)	MS/MS (IT)	ACN/water/ formic acid (Gradient)	Use of a precolumn (C_{18} 0.5 cm × 300 μm; 5 μm). Classification of coffee bean species with 2-D maps	(Gil-Agusti et al., 2005)
Gliadin peptides	Fermented wheat flour	Extraction (Tris, HCl, EtOH), enzymatic digestion, off-line SCX-LC	C_{18} (15 cm × 320 μm; 5 μm)	MS (Q-TOF)	ACN/water/ formic acid (Gradient)	LODs	(Rizzelo et al., 2007)
Peptides	Pork meat	*In vitro* digestion of pork meat using pepsin and pancreatin	C_{18} (15 cm × 75 μm; -)	MS (Q-TOF)	ACN/ water/TFA (Gradient)	Use of nano-LC for separation and identification of peptides contained in the deproteinized protein hydrolysate. Use of a precolumn (C_{18}, 0.5 cm × 300 μm)	(Escudero et al., 2010)
Chlorophenoxy acid herbicides (2,4-D, MCPA, 2,4-DP, 2,4-DB, MCPB, 2,4,5-TP), dicamba (3,6-dichloro-2-methoxybenzoic acid)	Green bean	Hydrolysis and SPE (SAX)	C_{18} (15 cm × 300 μm; 5 μm)	UV (232 nm)	MeOH/ water/ phosphoric acid (Gradient and isocratic)	2,4-DB and MCPB partially resolved, but lower analysis times with gradient elution. LODs: 5-20 ng (isocratic) and 10-20 ng (gradient)	(Lucas-Delfa et al., 2000)

(continued on next page)

Table 9.D. *(continued)*

Analytes	Matrix	Sample pretreatment	Column	Detector	Mobile phase composition	Observations	References
Fenpyroximate	Apple	SFE (CO_2)	C_{18} (25 or 35 cm × 320 μm; 5 μm)	UV (260 nm)	ACN/water/ ammonium acetate (Isocratic)	On-column focusing. Columns prepared using supercritical CO_2 as slurry medium. LOD: 0.03 mg/kg	(Halvorsen et al., 2000)
Fonofos metabolites	Lettuce, beet, wheat	Solvent extraction (ACN/ water), extraction of residual aqueous solution (ethyl acetate)	C_{18} (15 cm × 322 μm; 3 μm)	UV (210 nm), MS (QqQ)	ACN/water/ TFA/glycerol (Isocratic)	On-column focusing	(Onisko et al., 2002)
Chlorophenoxy acid herbicides (2,4-D, 2,4,5-TP, MCPA, 2,4-DP, 2,4-DB, MCPP, MCPB) and their esters (2,4-D-1-methyl ester, 2,4-D-1-butyl ester)	Apple juice	SPE (MCX)	C_{18} (15 cm × 300 μm; 5 μm)	UV (232 nm)	MeOH/ water/ phosphoric acid (Gradient)	On-column focusing. LODs: 0.018-0.005 mg/kg	(Rosales-Conrado et al., 2005)
Fensulfothion, fenamiphos, profenofos, fonofos, isofenphos, dialifos, sulprofos, prothiofos	Baby foods	LLE (ethyl acetate), SPE (C_{18})	Phenyl (25 cm × 100 μm; 3 μm)	UV (205 nm)	ACN/water (Isocratic)	Studies phenyl, C_{18} and CN silica particles as stationary phases. LODs: 4.4-37.5 ng/mL	(Buonasera et al., 2009)
Enrofloxacin, ciprofloxacin	Chicken muscle	Homogenization (water), filtration, dilution	C_{18} (20 cm × 50 μm; 3 μm) (nano-LC) C_{18} (43 mm, 75 × 50 μm cross section; 3.5 μm) (chip-based nano-LC)	MS (TOF)	ACN/water/ formic acid (Gradient)	Use of trapping columns. Use of surface plasmon resonance-inhibition biosensor immunoassay coupled to nano-LC-MS interfaced with a reusable recovery chip. Feasibility of combining two biosensor chips with chip-based nano-LC chip TOF-MS system	(Marchesini et al., 2008)

Analytes	Matrix	Sample pretreatment	Column	Detector	Mobile phase composition	Observations	References
Antibiotics (penicillin G, ampicillin, nafcillin, dicloxacillin, amoxicillin, cloxacillin, oxacillin, penicillin V, piperacillin, clavulanic acid)	Water, milk, chicken muscle	Water: SPE (Oasis HLB) Milk and meat: solvent extraction (ACN), SPE (Oasis HLB+Alumina)	C_{18} (15 cm × 500 μm; 5 μm)	DAD (220 nm)	ACN/ water/TFA (Gradient)	LODs: 0.04-1.40 μg/L	(Bailón-Pérez et al., 2009)
Antibiotics (amoxicillin, ampicillin, penicillin G, penicillin V, oxacillin, cloxacillin, nafcillin, dicloxacillin)	Pharmaceuticals, milk, porcine tissues	Pharmaceuticals: sonication (water), filtration Milk: defatting (*n*-hexane), SPE (C_{18}) Porcine tissues: solvent extraction (MeOH)	C_{18} (10 cm × 100 μm; 3 μm)	UV (200 nm) MS (IT)	ACN/water/ formic acid (Gradient)	Comparison with C_8 particles as packing material. Use of a z-shaped flow cell. Employment of on-column polymeric frits. LODs: 2.27-4.06 μg/L (UV), 0.01-0.51 μg/L (MS)	(Hsieh et al., 2009)
Phytosterols (brassicasterol, cholesterol, stigmasterol, campesterol, β-sitosterol)	Extra-virgin olive oil	Saponification, LLE (diethyl ether), dissolution, SPE (C_{18})	C_{18} (15 cm × 100 μm; sub-2 μm)	UV (195 nm)	MeOH (Isocratic)	Evaluation of different stationary phases (C8, phenyl, cholesteryl, cyano). LODs: 0.78 μg/mL	(Rocco & Fanali, 2009a)
Tocopherols (α-T, α-T-acetate, γ-T, δ-T), butylated hydroxytoluene	Pharmaceutical preparation	Solvent extraction (hexane)	C_{18} (23 cm × 100 μm; 3 μm)	UV (205 nm)	ACN/MeOH/ water (Isocratic)	Also analysis of human serum. Comparison with 5 μm C_{18} particles as packing material. LODs: 4-6 μg/mL	(Fanali et al., 2004a)
Tocopherols (α-T, β-T, γ-T, δ-T), tocotrienols (α-T_3, β-T_3, γ-T_3, δ-T_3)	Vegetable oils	Solvent extraction (MeOH 0.1% BHT, MeOH/2-propanol)	C_{18} monolithic (15 cm × 100 μm)	UV (295 nm)	ACN/MeOH/ water/ acetic acid (Isocratic)	LODs (Ts): 0.07-0.16 μg/ mL	(Cerretani et al., 2010)

(continued on next page)

Table 9.D. *(continued)*

Analytes	Matrix	Sample pretreatment	Column	Detector	Mobile phase composition	Observations	References
Organic acids (quinic, malic, shikimic, citric)	Apple juice	Ceramic crossflow microfiltration or ultrafiltration by zirconium oxide membrane	C_{18} (30 cm × 320 μm; 5 μm)	UV (206 nm)	Phosphate buffer (Isocratic)	LODs: 0.04-2.9 ng	(Blanco et al., 1996)
Fumonisin toxins (FB_1, FB_2, FB_3, two isomers of partially hydrolyzed FB_1, hydrolyzed FB_1, *N*-acetyl FB_2)	Corn	Solvent extraction (ACN/ water), SPE (C_{18})	C_{18} (20 cm × 320 μm; 3 μm)	MS (Qq)	ACN/water/ TFA/glycerol (Gradient)	Preinjector splitter (C_{18}, 1.2 mm i.d. stainless steel tubing). LODs: 0.1 μg/g	(Xie et al., 1997)
Sudan azo-dyes (I, II, II, IV)	Hot chilli tomato sauce	Solvent extraction (acetone)	C_{18} (15 cm × 320 μm; 5 μm)	MS (Q-TOF)	ACN/ acetone/ water/ formic acid (Gradient)	Use of a precolumn (C_{18}, 1 cm × 320 μm; 5 μm). Comparison with LC-TQ-MS. LODs: 0.4-1.1 μg/g	(Calbiani et al., 2004)
Glycyrrhizin, glycyrrhetic acid	Licorice roots, candies	Solvent extraction (ethanol/water)	C_{18} (23 cm × 75 μm; 5 μm)	UV (254 nm)	Water/ MeOH/ACN/ acetic acid (Gradient)	LODs: 1.25 and 0.16 μg/ mL	(Fanali et al., 2005a)
Ochratoxin A	Grape (must, acinus)	Extraction (chloroform), lyophilization, immunoaffinity column	C_{18} butyl silica (15 cm × 180 μm; 5 μm)	MS (IT)	ACN/water/ formic acid (Gradient)	Comparison with HPLC-FD. LOD: 1 pg/g	(Timperio et al., 2006)

Analytes	Matrix	Sample pretreatment	Column	Detector	Mobile phase composition	Observations	References
Heterocyclic aromatic amines (MeIQx, norharman, harman, harmine)	Ready-to-eat cooked ham (irradiated and non-irradiated by en electron-beam radiation source)	Adition of NaOH, SPE (diatomaceous earths), SPE (MCX)	C_8 (15 cm × 300 μm; 3 μm)	DAD (250, 265 nm)	ACN/water/ ammonium acetate (Gradient)	Comparison with C_{18} capillary column. On-column focusing. LODs: 0.1-0.3 μg/L	(Rosales-Conrado et al., 2008)
Flavonoids (catechin, epicatechin, eriodictyol, naringenin, isorhamnetin, kaempferol, quercetin, quercetin-3-*O*-glucoside, quercetin-3-*O*-rutinoside, quercetin-3-*O*-galactoside, isorhamnetin-3-*O*-glucoside, isorhamnetin-3-*O*-rutinoside, isorhamnetin-3-*O*-galactoside, kaempferol-3-*O*-glucoside, kaempferol-3-*O*-rutinoside, kaempferol-3-*O*-galactoside, naringenin-7-*O*-glucoside, dihydroxykaempferol)	Almond skin	Defatting (hexane), filtration, USE (MeOH/water)	C_{18} (15 cm × 300 μm; 3 μm)	UV-MS (285 nm; IT)	MeOH/ ACN/water/ formic acid (Gradient)	Comparison with HPLC-DAD getting a reduction of the analysis time with capillary-LC. LODs: 0.02-0.13 μg/L	(Hughey et al., 2008)
D,L-amino acids (arginine, proline, alanine, leucine, serine, phenylalanine, asparagine, glutamic acid, aspartic acid)	Orange juice	Centrifugation, adjustment of pH of the supernatant, derivatization (FITC)	Vancomycin-modified silica-diol (25 cm × 75 μm; 5 μm)	UV (205 nm) MS (Q-IT)	MeOH/ water/ ammonium formate (Isocratic)	Use of a precolumn (C_{18}, 5 mm × 300 μm; 100 Å, 5 μm). Use of a "z" cell. LODs: 3-8 ng/mL	(D'Orazio et al., 2008)

(continued on next page)

Table 9.D. *(continued)*

Analytes	Matrix	Sample pretreatment	Column	Detector	Mobile phase composition	Observations	References
Chlorogenic acid	Honeysuckle	MAE (70% ethanol)	C_{18} (15 cm × 75μm; 3.5 μm)	MS (IT-TOF)	ACN/water/ formic acid (Gradient)	LOD: 0.5 ng/mL	(Hu et al., 2009)
Histone H3 and H4	Soybean	Nuclei extraction, histone protein in-gel digestion	C_{18} (15 cm × 75 μm; 3 μm)	MS/MS (MALDI-TOF)	ACN/ water/TFA (Gradient)	Use of nano-LC to identify variants of the histones	(Wu et al., 2009)
Alkylbenzenes (eugenol, isoeugenol, eugenol methyl ether, myristicin, anethole, estragole), related flavour compounds (coumarin, pulegone)	Basil, nutmeg, anis, clove, cassia	USE and maceration (MeOH)	C_{18} (25 cm × 500 μm; 5 μm)	DAD (215 and 260 nm)	ACN/water (Isocratic)	On-column preconcentration. LODs: 97-148 ng/mL (alkylbenzenes) and 9.5-14.2 ng/mL (flavour compounds)	(Avila et al., 2009)
Glutathionylated, cysteinylated precursors of 3-mercaptohexan-1-ol and 4-mercapto-4-methylpentan-2-one (G3MH, G4MMP, Cys3MH, Cys4MMP)	White grape juice	Micro-column cation exchange chromatography (DOWEX 50WX4-100), SPE (C_{18})	C_{18} (10 cm × 75 μm; 5 μm)	MS/MS	ACN/water/ formic acid (Gradient)	Use of stable isotope dilution assay. LODs: 0.02-1.44 nmol/L	(Roland et al., 2010)

and LOQs of 1.56 mg/L. Accurate confirmation of the analytes was performed by coupling the system with a mass spectrometer via a nano-ESI interface. Ammonium acetate was added to the mobile phase to improve analyte ionization in the MS.

Fanali et al. (2004a) also studied the potential of the technique for the separation of several isomers of the lipid soluble vitamin E (α, γ, δ-tocopherol, T, α-T-acetate) and an antioxidant compound, butylated hydroxytoluene (BHT) used to prevent tocopherol's autoxidation, using fused silica capillaries of 100 μm packed with RP18 silica particles of 3–5 μm and ACN-MeOH (70:30. v/v) as mobile phase. Efficiency and resolution were strongly influenced by the particle diameter, which were higher using 3 μm particles. The sensitivity of the method (4–6 mg/L) was sufficient to apply the optimized method to the analysis of α, γ, δ-tocopherol in a biological sample (human serum) and a pharmaceutical sample due to the relatively high content of Vitamin E. Later on, Cerretani et al. (2010) also applied the technique with UV detection for the determination of the eight naturally occurring forms of vitamin E, α-, β-, γ- and δ-tocopherol and α-, β-, γ- and δ-tocotrienols (T_3), in sixteen vegetable oils of different botanical origins. A 150 mm × 100 μm I.D. silica monolithic column was used. Oil samples were extracted twice with 10 mL of MeOH containing 0.1% BHT (w/v) as an antioxidant and once again with 10 mL of a MeOH/2-PrOH mixture 80:20 (v/v). After evaporation and reconstitution in MeOH, 10 nL were injected into the nano-LC system. Because of reversed-phase characteristics of the monolithic column, it was unable to resolve the β-tocopherol and γ-tocopherol isomers present in the oil samples, and consequently, the sum of the concentrations of both isomers was considered for quantification. LODs achieved ranged between 0.07 and 0.16 mg/L. Important differences between the content of tocopherols were found for some of the analyzed samples.

The analysis of biogenic amines by nano-LC has been proposed by Hernández-Borges et al. (2007b). It is well known that biogenic amines are present in protein-rich foods or in fermented foods and beverages, representing a toxicological risk as well as serving as indicators of food quality and how well the food has been preserved. Hernández-Borges et al. (2007b) developed a method for the determination of ten biogenic amines (ethanolamine, methylamine, tryptamine, 2-phenylethylamine, putrescine, cadaverine, histamine, tyramine, spermidine and spermine) in red and white wine. In this case, samples were treated with polyvinylpyrrolidone and derivatized with dansylchloride (Dns-Cl). After extraction of the samples by a C_{18}-SPE cartridge they were injected in a bidentated C_{18} 23 cm × 100 I.D. column. An on-line clean-up step using a C_{18} trapping column for the successful elimination of the excess of Dns-Cl was also performed. UV detection at 250 nm was used.

Fig. 9.12 shows a nano-LC chromatogram of a standard mixture of the ten biogenic amines (Fig. 9.12.A) as well as a nano-LC chromatogram of a

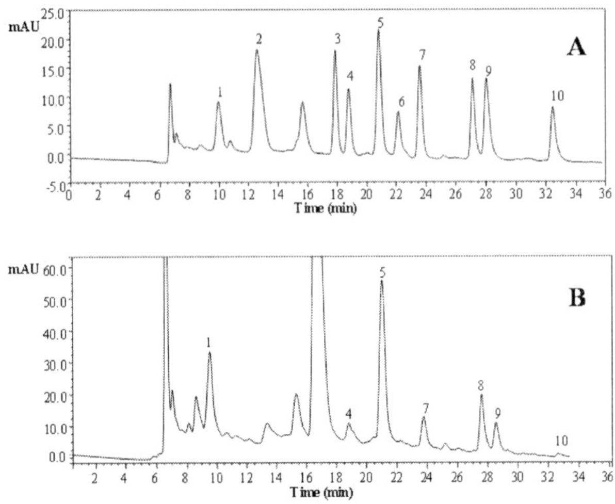

Fig. 9.12. A) Nano-LC-UV chromatogram of the separation of a standard mixture of 10 amines. Peak identification: (1) Ethanolamine; (2) Methylamine; (3) Tryptamine; (4) 2-phenylethylamine; (5) Putrescine; (6) Cadaverine; (7) Histamine; (8) Tyramine; (9) Spermidine; (10) Spermine. Concentration: 2 μg/mL of each amine except for peak 4 (1.1 μg/mL), peak 6 (0.8 μg/mL), peak 7 (2.8 μg/mL), peak 8 (1.2 μg/mL) and peak 9 (3.8 μg/mL). UV detection at 250 nm. Column: Cogent® bidentate C_{18} type C silica (23 cm packed, 100 μm I.D. × 375 μm o.d., 5 μm). Trapping column: C_{18} PepMap100 μ-pre-column cartridge (5 μm, 100 Å, 300 μm I.D. × 5 mm). Mobile phase A: 10/90 (v/v) ACN/0.23% (w/v) TEA pH 5.0; mobile phase B: 90/10 (v/v) ACN/0.23% (w/v) TEA pH 5.0. Gradient conditions: See Table 9.B. Injection: 1 μL. B) Nano-LC-UV chromatogram resulting from the analysis of a red wine sample after PVP treatment, Dns-Cl derivatization and off-line SPE extraction. Redrawn from (Hernández-Borges et al., 2007) with permission from Elsevier.

red wine sample (Fig. 9.12.B) after the whole extraction procedure. LODs achieved ranged between 18.3 μg/L for putrescine and 48.3 μg/L for spermidine, while calibration curves showed good linearity ($R>0.996$). The amine content of several wine samples was also determined. The work represents the first separation and determination of these biogenic amines in food samples. Only in an early work by Song et al. (2004) one biogenic amine (histamine) was analyzed in water and alcoholic beverages (whisky, red wine and apple cider) as well as in rat brain tissue but, in this case, derivatization with 7-fluoro-4-nitrobenzoxadiazole was carried out.

The analysis of pesticides in foods is also a topic of special concern, especially in order to verify their absence or presence below or above maximum residue limits (MRLs) established by current legislation. Buonasera et al. (2009) analyzed eight organophosphorus pesticides by nano-LC-UV in five different baby foods (apple juice, three homogenized fruits, i.e., apple, pear and plum, and homogenized zucchini). A LLE and SPE clean-up was used to treat the samples. In this case, different silica-based SPs, namely CN, C_{18} and phenyl silica were initially investigated. Fig. 9.13 shows a compari-

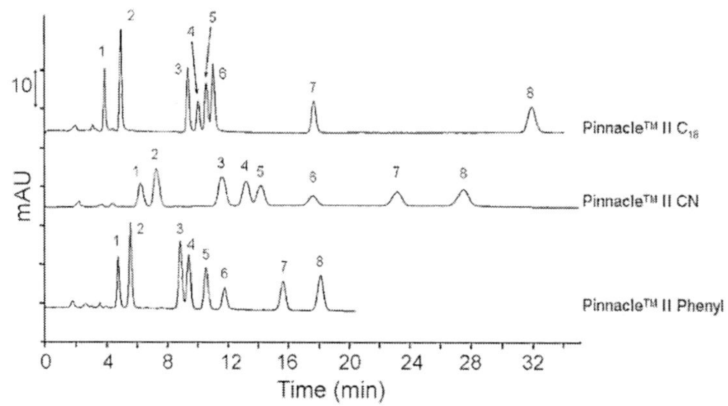

Fig. 9.13. Separation of the eight studied organophosphorus pesticides (OPPs), on three different SPs. Peak identification: (1) fensulfothion, (2) fenamiphos, (3) profenofos, (4) fonofos, (5) isofenphos, (6) dialifos, (7) sulprofos, and (8) prothiofos. Capillary column: 100 µm ID, 25.0 cm packed, 25.5 cm effective length. SPs: Pinnacle™ II CN, Pinnacle™ II C18, Pinnacle™ II Phenyl. Mobile phase: ACN/water, 70/30, 45/55 and 60/40 (v/v), respectively. Flow rate, 360 nL/min; room temperature. Concentration of standard OPPs: 50 µg/mL in water/ACN 50/50 (v/v). Injection volume: 60 nL. Detection wavelength: 205 nm. Reprinted from (Buonasera et al., 2009) with permission from Elsevier.

son of the separation capacity of the three columns tested. As can be seen, although Pinnacle™ II C_{18} provided a sufficient separation (32 min) of peaks 4-6, corresponding to fonofos, isofenphos and dialifos, they were not baseline resolved, and it was not possible to improve their resolution. On Pinnacle™ II CN and Pinnacle™ II Phenyl, due to the lower carbon load values, a better separation was achieved by diminishing the percentage of organic solvent in the mobile phase. However, only the latter SP was selected due to its better performance in terms of efficiency and analysis time. This was mainly by π-π interactions between the column and the analytes. On-column focusing of a large injection volume was employed to increase the sensitivity, which was between 4.4 and 37.5 µg/L. Extraction of pesticides from spiked baby food samples was found to be highly satisfactory for homogenized fruits and the apple juice with good recovery values except for sulprofos and prothiofos, which ranged between 37–54% and 47–67%, respectively. In the case of the homogenized zucchini, which contained 2.5% of olive oil, low recovery values were obtained and further investigations had to be pursued to extend the application of the method to fatty foods.

Other examples of the analysis of pesticides in fruits or vegetables have been proposed by Lucas-Delfa et al. (2000) for the analysis of chlorophenoxy acid herbicides and dicamba by SPE-CLC-UV in green beans. Halvorsen et al. (2000) analyzed a single pesticide, fenpyroximate, in apples by SFE-CLC-UV, while Onisko et al. (2002) identified metabolites of the thiophosphonate insecticide fonofos in lettuce, beet and wheat by solvent extraction-CLC-MS

(U-^{14}C-^{13}C^{6}-labeled fonofos and a radioactivity monitor were used for metabolism studies). Another interesting application is the work by Rosales-Conrado et al. (2005) in which chlorophenoxy acid herbicides and some of their esters were analyzed in apple juice samples by SPE-CLC-UV. Clean-up and preconcentration of the samples were carried out in an Oasis MCX polymer (polydivinylbenzene-N-vinylpyrrolidone) cartridge. LODs, between 0.005 and 0.018 mg/kg, allowed the determination of these compounds in apple juice samples around the levels permitted by the EU regulations, with recoveries in the range 84–99% (RSD<4%).

Regarding the analysis of pharmaceuticals in food, only a few works have dealt with this matter (Marchesini et al., 2008; Bailón-Pérez et al., 2009; Hsieh et al., 2009). For example, Hsieh et al. (2009) studied the nano-LC-UV or nano-LC-MS simultaneous determination of a group of eight penicillin antibiotics. Amoxicillin, ampicillin, penicillin G, penicillin V, oxacillin, cloxacillin, nafcillin and dicloxacillin, in diverse matrices such as pharmaceuticals, milk and porcine tissues, after suitable extraction and preconcentration steps, were investigated. A thorough study was carried out for the comparison of the packing SP materials (3 μm C_{18} and C_8 particles) and also for the material types of the on-column polymeric outlet frit (polystyrene-based and polymethacrylate-based monoliths). In this last case, using C_{18} particles as SP, the chromatograms showed that no differences were obtained when either the polystyrene- or polymethacrylate-based monolith were used as outlet frits (see Fig. 9.14). Pore size and pore size distribution was determined by mercury intrusion, finally selecting the polystyrene-based monolithic frit due to its smaller pore size and narrower pore size distribution. An additional study about the separation ability of the frit itself was carried out (Fig. 9.14), resulting in a poor resolution that demonstrated that its unique function was retention of the packing material inside the column. Other nano-LC-MS parameters, such as mobile phase, flow rate, capillary voltage and temperature of dry gas were studied and an in-line nylon filter was used to connect the column with UV and MS detectors.

Miscellaneous

CLC and nano-LC have been applied for the determination of other compounds apart from the fields of proteomic, environmental, pharmaceutical and food analysis. Although for these other applications SPs such as C_8 have been used (De Andrés et al., 2010), C_{18} has still been the most commonly used for nano or capillary separations in the last few years of, for example, ecdysteroids (Blais et al., 2010), minor DNA-mephalan adducts (Hoes et al., 2000), insulin (Ho et al., 2008), phosphatidylcholine and phosphatidylethanolamine (Ahn et al., 2008), yessotoxin (Ruppén Cañás et al., 2006), 9-(alkylthio) acridines (Srbek et al., 2004), etc. Ruppén Cañas et al. (Ruppén Cañás et al., 2006) studied the determination of a toxin (yessotoxin) in marine phytoplank-

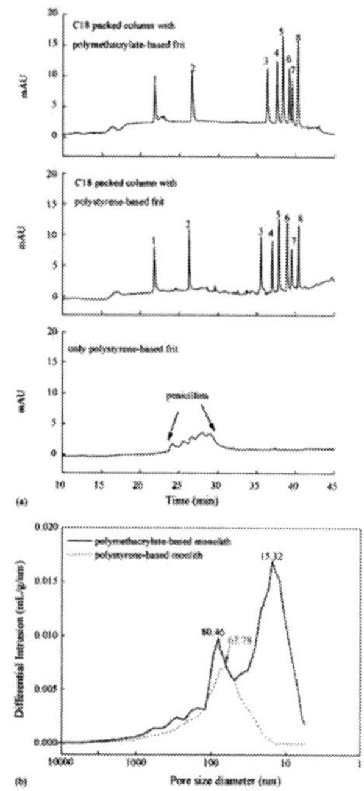

Fig. 9.14. Effect of different on-column frits on penicillin separation. (a) The chromatograpms of nano-LC-UV were obtained by a 10 cm C_{18} packed column with polystyrene-, polymethacrylate-based monolithic frit, or by only polystyrene-based frit without C_{18} particles. The distribution of the polymeric frit pore size measured by mercury intrusion porosimeter is shown in (b). An initial 0% ACN (i.e., 0.1% formic acid aqueous solution) was used, then it was linearly increased to a final of 90% ACN in 20 min with a flow rate of 0.2 μL/min, and this was employed as the gradient elution program in the nano-LC system. The column temperature was kept at 25°C; 200 nm wavelength was used for penicillin detection; and the amount of sample injection was maintained at 0.05 μL. amoxicillin (1), ampicillin (2), penicillin G (3), penicillin V (4), oxacillin (5), cloxacillin (6), nafcillin (7) and dicloxacillin (8). Reprinted from (Hsieh et al., 2009) with permission from Elsevier.

ton by nano-LC-MS using a Qq-time-of-flight (TOF) mass spectrometer with a nano-ESI. The technique was applied to the analysis of yessotoxin in *Protoceratium reticulatum* using a C_{18} column of 75 μm I.D. and 3 μm particles. Phytoplankton cells were picked and extracted with a MeOH/water mixture prior to their injection into the LC system. Yessotoxin contents in the analyzed samples were found to be between 0.011 and 0.020 ng/cell (LOD of the method was found to be 0.5 pg). Very recently, Flender et al. (2010) developed a method based on a direct-EI interface for the analysis of boronic acids, a series of compounds that are used as coupling reagents to produce drugs, agrochemicals or herbicides, and are insufficiently amenable to usual MS methods. Separations were carried out in a monolithic C_{18} nanocolumn of

60 cm × 50 μm. The eluent of the column was directly introduced into the heated electron impact ion source of the MS, getting LODs up to 200 pg and very robust results.

Other SPs have also been employed, such as monolithic silica coated with poly(octadecyl methacrylate) (Núñez et al., 2007) or macroporous monolithic poly(butyl methacrylate-*co*-ethylene dimethacrylate) (Nischang & Brüggemann, 2010). In this last case, small molecules such as benzene, toluene, ethylbenzene, propylbenzene, butylbenzene and pentylbenzene were separated in a 100 μm monolithic poly(butyl methacrylate-*co*-ethylene dimethacrylate) column. The monolith was synthesized via a free radical copolymerization of mono and divinyl monomeric precursors in the presence of porogenic diluents; elution was done with an isocratic mobile phase composition of ACN/water 70:30 (v/v). The authors demonstrated that the performance of this column was strongly affected by the partition and adsorption factors of the analytes under study, due to gel porosity of the SP.

Chiral analysis of other compounds different than pharmaceutical products has been carried out using, for instance, a L-prolinamide-modified monolithic silica column (Chen & Hobo, 2001), a physically adsorbed CSP of avidin on a monolithic silica column (Liu et al., 2002), a β-CD containing organic polymer monolithic column (Guerrouache et al., 2009), a phenylpropyl-β-CD SP silica packed column (Si-Ahmed et al., 2010a; Si-Ahmed et al., 2010b), etc. Guerrouache et al. (2009) prepared a macroporous cross-linked organic polymer on *N*-acryloxysuccinimide (NAS) and ethylene dimethacrylate (EDMA) inside a 31 cm × 75 μm I.D. fused silica capillary. The succinimide groups on the monolith surface provided reactive sites to functionalize with propargylamine (NAS-*co*-EDMA). Then, β-CD was inserted in this azido-reactive polymer and the polymer was used as a CSP for the separation of flavonone enantiomers. A scheme of this reaction can be seen in Fig. 9.15, as well as the capillary electrochromatogram and nano-LC chromatogram of the separation of flavonone enantiomers using a MeOH/borate buffer (5 mM, pH 8.2) 40:60 (v/v) as mobile phase and UV detection at 214 nm.

Glycomic investigations, aimed at a total structural analysis of the glycans released from glycoproteins, cannot be conceived without the use of MS. That is why both miniaturized techniques, nano-LC and CLC, have become good options for analyte separation during the last decade (Mechref & Novotny, 2010) using either capillary amide-based columns (Wuhrer et al., 2004) or graphitized carbon columns (Ruhaak et al., 2009; Karlsson et al., 2004; Snovida et al., 2010). Also very recently, carbon-clad zirconia columns (Karlsson & Thomsson, 2010) were found appropriate for analyte separation and especially to increase sensitivity compared to conventional Hypercarb-HPLC columns, the most commonly used. Snovida et al. (2010) developed an on-column method for glycopeptide enrichment with cellulose SPE,

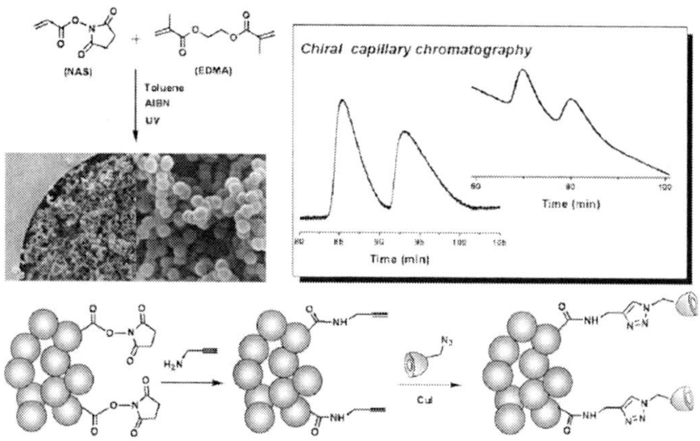

Fig. 9.15. Scheme of the reaction that takes place in the synthesis of the macroporous cross-linked organic polymer based on succinimide ester reactivity. Capillary electrochromatogram (lower) and nano-LC chromatogram (upper) of the separation of flavonone enantiomers. CEC and nano-LC with MeOH-borate buffer (5 mM, pH 8.2) 40:60 (v/v) and detection at 214 nm. CEC conditions: columns length, 21 cm to detector, 31.2 cm overall, injections 10 kV for 5 s, running voltage 20 kV at 25°C. Nano-LC conditions: columns length: 31 cm, flow rate 60 nL/min, injections 4 nL. Reprinted from (Guerrouache et al., 2009) with permission from Elsevier.

verifying this enrichment by nano-LC-MS/MS combining electron-transfer dissociation and collision-activated dissociation fragmentation. Regarding the nano-LC conditions, after digestion and preconcentration, samples were injected in a C_{18} or a graphitized carbon column and analytes eluted with a mixture of water/ACN/formic acid gradient. Another interesting work dealing with analysis of glycoproteins is the one developed by Luo et al. (2009), who prepared a high efficiency 2.5 m × 10 μm I.D. amine-bonded poly(vinylbenzyl chloride-divinylbenzene) hydrophilic interation liquid chromatography-porous layer open tubular (HILIC-PLOT) column for the determination of *N*-linked glycans released from glycoproteins. The polymer layer was attached to the walls of the capillary followed by modification with ethylenediamine to create a hydrophilic interaction SP. Furthermore, the OT configuration allowed high flow rates and low back pressure, with a consequent improvement in ESI/MS sensitivity.

A silica-based monoliths' surface can be modified with the aim of attaching a chromatographic ligand for providing the proper retention mechanism for a certain group of analytes. This was reported by Zhong and El Rassi (2009) who developed a modification to carry out affinity chromatography or polar chromatography. At the beginning, a sol-gel technique was used to build a silica-based monolith and then, consecutive γ-glycidoxypropylsilyl sublayers were anchored to the surface with the aim of reacting with another top interactive layer, in this case, lectins and a cyano polar ligand. The first column showed good affinity separation of monosaccharides and glycoproteins, while the other cyano silica-based monolith showed good separation of *N*-glycans by both CEC and nano-LC.

Conclusions and Future Trends

Miniaturization in analytical chemistry has been developed in the last decade in order to offer reliable separation methods capable of analyzing the different compounds present in various complex matrices. Theoretical principles, technological solutions, method validation and several applications have been developed for the nano-scale separation techniques including both electro-driven and chromatographic ones. Both tools offer several advantages over the conventional ones. Among them, high efficiency separation, high resolution, short analysis time, easy coupling with MS, use of minute volumes of both samples and mobile phases are the most attractive benefits for those working in the analytical field. In this chapter the current state of the art of CLC/nano-LC has been reviewed focusing on i) the general theoretical principles documenting the advantages in using these technique, ii) problems related to the instrumentation to be used and iii) selected applications in different fields such as proteomic, environmental and agrochemical.

Although the literature offers a valuable number of applications in various fields and commercial instrumentation is available, it cannot be certainly concluded that nano-LC is mature enough to be used routinely. In fact some drawbacks mainly related to the sensitivity must be carefully considered. This is mainly due to the use of low volumes of samples injected (nL). However sample pre-concentration, on-column focusing or two dimensional separation are the currently proposed approaches for increasing sensitivity.

Future developments in CLC/nano-LC include the realization of new SPs or the employment of those proposed for ultra high pressure liquid chromatography (UPLC) with particles diameter < 2 μm, and new interfaces for the coupling with other tools such as MS or NMR. The development of microdevices is another important research field for separation science, arising from the "lab-on-a-chip" concept, and valuable results have already been obtained. Microdevices offer the possibility to analyze even more minute amounts of samples, to further reduce analysis time, to integrate both sample pretreatment and analytical separation in the same instrument (μ-total analysis system, μ-TAS), to achieve low limits of detection (LODs). Moreover, high throughput multichannel analysis devices for parallel experiments can be implemented and successfully applied for peptide mapping, DNA hybridization analysis, and PCR analysis (Hernández-Borges et al., 2007a; Vázquez & Paull, 2010; Faure, 2010; Ohno et al., 2008).

Many icrodevices have been developed for electrophoretic techniques, while downsizing the chromatographic system has faced several technical challenges, such as the integration of on-chip injectors, mechanical valves and the introduction of the SP into narrow channels (Hernández-Borges et al., 2007a; Faure, 2010). Concerning the last drawback, the use of monolithic material, characterized by the advantageous *in situ* polymerization and high permeability, has given a new chance for an easier implementation of chromatographic microdevices (Vázquez & Paull, 2010). Further advances in micro-fabrication techniques may lead to a wider number of commercialized microfluidic-LC chips and offer new separative solutions for application fields where extremely small quantities of sample are available and fast analyses are required.

References

Abian, J.; Oosterkamp, A.J.; Gelpi, E. Comparison of conventional, narrow-bore and capillary liquid chromatography/mass spectrometry for electrospray ionization mass spectrometry: practical considerations. J. Mass Spectrom. 1999, 34, 244-254.

Ahn, E.J.; Kim, H.; Chung, B.C.; Kong, G.; Moon, M.H. Quantitative profiling of phosphatidylcholine and phosphatidylethanolamine in a steatosis/fibrosis model of rat liver by nanoflow liquid chromatography/tandem mass spectrometry. J. Chromatogr. A 2008, 1194, 96-102.

Albert, K.; Dachtler, M.; Glaser, T.; Handel, H.; Lacker, T.; Schlotterbeck, G.; Strohschein, S.; Tseng, L.H.; Braumann, U. On-line coupling of separation techniques to NMR. J. High Resolut. Chromatogr. 1999, 22, 135-143.

Albert, K.; Schlotterbeck, G.; Tseng, L.-H.; Braumann, U. Application of on-line capillary high-performance liquid chromatography-nuclear magnetic resonance spectrometry coupling for the analysis of Vitamin A derivatives. J. Chromatogr. A 1996, 750, 303-309.

Armenta, S.; Lendl, B. Capillary liquid chromatography with off-line mid-IR and raman micro-spectroscopic detection: analysis of chlorinated pesticides at ppb Levels. Anal. Bioanal. Chem. 2010, 397, 297-308.

Asensio-Ramos, M.; Hernández-Borges, J.; Rocco, A.; Fanali, S. Food Analysis: A continuous challenge for miniaturized separation techniques. J. Sep. Sci. 2009, 32, 3764-3800.

Ávila, M.; Zougagh, M.; Escarpa, A.; Ríos, A. Determination of alkylbenzenes and related flavour compounds in food samples by on-column preconcentration-capillary liquid chromatography. J. Chromatogr. A 2009, 1216, 7179-7185.

Bailón-Pérez, M.L.; García-Campaña, A.M.; del Olmo-Iruela, M.; Gamiz-Gracia, L.; Cruces-Blanco, C. Trace determination of 10 beta-lactam antibiotics in environmental and food samples by capillary liquid chromatography. J. Chromatogr. A 2009, 2216, 8355-8361.

Barefoot, A.C.; Reiser, R.W. Packed capillary liquid chromatography-mass spectrometry using both direct-coupling and moving-belt interface. J. Chromatogr. A 1987, 398, 221-226.

Bianco, L.; Lopez, L.; Scalone, A.G.; Di Carli, M.; Desiderio, A.; Benvenuto, E.; Perrotta, G. Strawberry proteome characterization and its regulation during fruit ripening and in different genotypes. J. Proteomics 2009, 72, 586-607.

Bjellaas, T.; Holm, A.; Molander, P.; Tørnes, J.A.; Greibrokk, T.; Lundanes, E. Trace determination of peptides in water samples using packed capillary liquid chromatography with UV and MS detection and characterization of peptide oxidation roducts by MS. Anal. Bioanal. Chem. 2004, 378, 1021-1030.
Blais, C.; Blasco, T.; Maria, A.; Dauphin-Villemant, C.; Lafont, R. Characterization of ecdysteroids in drosophila melanogaster by enzyme immunoassay and nano-liquid chromatography-tandem mass spectrometry. J. Chromatogr. B 2010, 878, 925-932.
Blanco Gomis, D.; Fraga Palomino, N.; Mangas Alonso, J.J. Capillary liquid chromatographic determination of neutral compounds in apple juices. Anal. Chim. Acta 2001, 426, 111-117.
Blanco, D.; Qintanilla, M.E.; Mangas, J.J.; Gutierrez, M.D. Determination of organic acids in apple juice by capillary liquid chromatography. J. Liq. Chromatogr. Relat. Technol. 1996, 19, 2615-2621.
Bogdanov, B.; Smith, R.D. Proteomics by FTIR Mass Spectrometry: Top Down and Bottom Up. Mass Spectrom. Rev. 2005, 24, 168-200.
Boos, K.S.; Fleischer, C.T. Multidimensional on-line solid-phase extraction (SPE) using restricted access materials (RAM) in combination with molecular imprinted polymers (MIP). Fresenius J. Anal. Chem. 2001, 371, 16-20.
Boughtflower, R.J.; Underwood, D.; Paterson, C.J. Capillary electrochromatography - some important considerations in the preparation of packed capillaries and the choice of mobile phase buffers. Chromatographia 1995, 40, 329-335.
Bragg, W.; Norton, D.; Shamsi, S.A. Optimized separation of beta-blockers with multiple chiral centers using capillary electrochromatography-mass spectrometry. J. Chromatogr. B 2008, 875, 304-316.
Bruin, G.J.M.; Tock, P.P.H.; Kraak, J.C.; Poppe, H. Electrically driven open-tubular liquid chromatography. J. Chromatogr. 1990, 517, 557-572.
Buonasera, K.; D'Orazio, G.; Fanali, S.; Dugo, P.; Mondello, L. Separation of organophosphorous pesticides by using nano-liquid chromatography. J. Chromatogr. A 2009, 1216, 3970-3976.
Calbiani, F.; Careri, M.; Elviri, L.; Mangia, A.; Zagnoni, I. Accurate Mass Measurements for the confirmation of sudan zo-Dyes in hot chilli products by capillary liquid chromatography-electrospray tandem uadrupole orthogonal-acceleration time of flight mass spectrometry. J. Chromatogr. A 2004, 1058, 127-135.
Campíns-Falcó, P.; Verdú-Andrés, J.; Sevillano-Cabeza, A.; Herráez-Hernández, R.; Molins-Legua, C.; Moliner-Martinez, Y. In-tube solid-phase microextraction coupled by in valve mode to capillary LC-DAD: Improving detectability to multiresidue organic pollutants analysis in several whole waters. J. Chromatogr. A 2010, 1217, 2695-2702.
Cappiello, A.; Famiglini, G.; Fiorucci, C.; Mangani, F. Variable-gradient generator for micro- and nano-HPLC. Anal. Chem. 2003a, 75, 1173-1179.
Cappiello, A.; Famiglini, G.; Mangani, F.; Palma, P. A Simple approach for coupling liquid chromatography and electron ionization mass spectrometry. J. Am. Soc. Mass Spectrom. 2002a, 13, 265-273.
Cappiello, A.; Famiglini, G.; Mangani, F.; Palma, P.; Siviero, A. Nano-high-performance liquid chromatography-electron ionization mass spectrometry approach for environmental analysis. Anal. Chim. Acta 2003b, 493, 125-136.

Cappiello, A.; Famiglini, G.; Palma, P.; Mangani, F. Trace level determination of organophosphorus pesticides in water with the new direct-electron ionization LC/MS interface. Anal. Chem. 2002b, 74, 3547-3554.

Cappiello, A.; Palma, P.; Mangani, F. New materials and packing techniques for micro-HPLC packed capillary columns. Chromatographia 1991, 32, 389-391.

Cerretani, L.; Lerma-García, M.J.; Herrero-Martínez, J.M.; Gallina-Toschi, T.; Simó-Alfonso, E.F. Determination of tocopherols and tocotrienols in vegetable oils by nanoliquid chromatography with ultraviolet-visible detection using a silica monolithic column. J. Agric. Food Chem. 2010, 58, 757-761.

Chankvetadze, B.; Kartozia, I.; Yamamoto, C.; Okamoto, Y.; Balschke, G. Comparative study on the application of capillary liquid chromatography and capillary electrochromatography for investigation of enantiomeric purity of the contraceptive drug levonorgestrel. J. Pharm. Biomed. Anal. 2003, 30, 1897-1906.

Chankvetadze, L.; Kartozia, I.; Yamamoto, C.; Chankvetadze, B.; Blaschke, G.; Okamoto, Y. Enantioseparation in capillary liquid chromatography and capillary electrochromatography using amylose tris(3,5-dimethylphenylcarbamate) in combination with aqueous organic mobile phase. J. Sep. Sci. 2002, 25, 653-660.

Chassaigne, H.; Nørgaard, J.V.; van Hengel, A.J. Proteomics-based approach to detect and identify major allergens in processed peanuts by capillary LC-Q-TOF (MS/MS). J. Agric. Food Chem. 2007, 55, 4461-4473.

Cháfer-Pericás, C.; Herráez-Hernández, R.; Campíns-Falcó, P. On-fibre solid-phase microextraction coupled to conventional liquid chromatography versus in-tube solid-phase microextraction coupled to capillary liquid chromatography for the screening analysis of triazines in water samples. J. Chromatogr. A 2006, 1125, 159-171.

Cháfer-Pericás, C.; Herráez-Hernández, R.; Campíns-Falcó, P. In-tube solid-phase microextraction-capillary liquid chromatography as a solution for the screening analysis of organophosphorus pesticides in untreated environmental water samples. J. Chromatogr. A 2007, 1141, 10-21.

Chen, H.; Horvath, C. High-speed high performance liquid chromatography of peptides and proteins. J. Chromatogr. A 1995, 705, 3-20.

Chen, J.R.; Dulay, M.T.; Zare, R.N.; Svec, F.; Peters, E. Macroporous photopolymer frits for capillary electrochromatography. Anal. Chem. 2000, 72, 1224-1227.

Chen, Z.; Hobo, T. Chemically L-prolinamide-modified monolithic silica column for enantiomeric separation of dansyl amino hcids and hydroxyl ccids by capillary electrochromatography and μ-high performance liquid chromatography. Electrophoresis 2001, 22, 3339-3346.

Chervet, J.P.; Ursem, M.; Salzmann, J.P. Instrumental requirements for nanoscale liquid chromatography. Anal. Chem. 1996, 68, 1507-1512.

Choudhary, G.; Horvath, C.; Banks, J.F. Capillary electrochromatography of biomolecules with on-line electrospray ionization and time-of-flight mass spectrometry. J. Chromatogr. A 1998, 828, 469-480.

Chung, Y.; Lee, K. Separation and determination of eleven environmental protection agency priority phenols by reversed-phase capillary high-performance liquid chromatography with indirect fluorophotometric detection and solid phase extraction. Microchem. J. 2001, 69, 143-152.

Claessens, H.A.; Burcinova, A.; Cramers, C.A.; Mussche, Ph.; van Tilburg, C.C.E. evaluation of injection systems for open tubular liquid chromatography. J. Microcol. Sep. 1990, 2, 132-137.

Coda, R.; Rizzello, C.G.; Nigro, F.; De Angelis, M.; Arnault, P.; Gobbetti, M. Long-term fungal inhibitory activity of water-soluble extracts of phaseolus vulgaris cv. pinto and sourdough lactic acid bacteria during bread storage. Appl. Environ. Microbiol. 2008, 74, 7391-7398.

Collier, T.S.; Hawkridge, A.M.; Georgianna, D.R.; Payne, G.A.; Muddiman, D.C. Top-down identification and quantification of stable isotope labeled proteins from aspergillus flavus using online nano-flow reversed-pase liquid chromatography coupled to a LTQ-FTICR mass spectrometer. Anal. Chem. 2008, 80, 4994-5001.

Colón, L.A.; Maloney, T.D.; Fermier, A.M. Packing columns for capillary electrochromatography. J. Chromatogr. A 2000, 887, 43-53.

Colón, L.A.; Reynolds, K.J.; Alicea-Maldonado, R.; Fermier, A.M. Advances in capillary electrochromatography. Electrophoresis 1997, 18, 2162-2174.

Crescentini, G.; Bruner, F.; Mangani, F.; Yafeng, G. Preparation and evaluation of dry-packed capillary columns for high-performance liquid chromatography. Anal. Chem. 1988, 60, 1659-1662.

D'Agostino, P.A.; Chenier, C.L.; Hancock, J.R. Packed capillary liquid chromatography-electrospray mass spectrometry of snow contaminated with sarin. J. Chromatogr. A 2002, 950, 149-156.

D'Agostino, P.A.; Hancock, J.R.; Chenier, C.L. Packed capillary liquid chromatography-electrospray ionization (tandem) mass spectrometry of mustard hydrolysis products in soil. J. Chromatogr. A 2004, 1058, 97-105.

D'Orazio, G.; Aturki, Z.; Cristalli, M.; Quaglia, M.G.; Fanali, S. Use of vancomycin chiral stationary phase for the enantiomeric resolution of basic and acidic compounds by nano-liquid chromatography. J. Chromatogr. A 2005, 1081, 105-113.

D'Orazio, G.; Cifuentes, A.; Fanali, S. Chiral nano-liquid chromatography-mass spectrometry applied to amino acids analysis for orange juice profiling. Food Chem. 2008, 108, 1114-1121.

D'Orazio, G.; Fanali, S. Use of teicoplanin stationary phase for the enantiomeric resolution of atenolol in human urine by nano-liquid chromatography-mass spectrometry. J. Pharm. Biomed. Anal. 2006, 40, 539-544.

D'Orazio, G.; Fanali, S. Enantiomeric separation by using nano-liquid chromatography with on-column focusing. J. Sep. Sci. 2008, 31, 2567-2571.

D'Orazio, G.; Fanali, S. Coupling capillary electrochromatography with mass spectrometry by using a liquid-junction nano-spray interface. J. Chromatogr. A 2010, 1217, 4079-4086.

De Andrés, F.; Zougagh, M.; Castañeda, G.; Ríos, A. Determination of heterocyclic amines in urine samples by capillary liquid chromatography with evaporated light-scattering detection. Anal. Bioanal. Chem. 2010, 397, 223-231.

Deguchi, K.; Ito, S.; Yoshioka, S.; Ogada, I.; Takeda, A. Nanoflow gradient generator for capillary high-performance liquid chromatography. Anal. Chem. 2004, 76, 1524-1528.

Delmotte, N.; Lasaosa, M.; Tholey, A.; Heinzle, E.; Huber, C.G. Two-dimensional reversed-phase × ion-pair reversed-phase HPLC: an alternative approach to high-resolution peptide separation for shotgun proteome analysis. J. Proteom. Res. 2007, 6, 4363-4373.

Desiderio, C.; Fanali, S. Capillary electrochromatography and capillary electrochromatography-electrospray mass spectrometry for the separation of non-steroidal anti-inflammatory drugs. J. Chromatogr. A 2000, 895, 123-132.

Desiderio, C.; Rudaz, S.; Raggi, M.A.; Fanali, S. Enantiomeric separation of fluoxetine and norfluoxetine in plasma and serum samples with high detection sensitivity capillary electrophoresis. Electrophoresis 1999, 20, 3432-3438.

Detobel, F.; Broeckhoven, K.; Wellens, J.; Wouters, B.; Swart, R.; Ursem, M.; Desmet, G. Parameters affecting the separation of intact proteins in gradient-elution reversed-phase chromatography using poly(stirene-co-divinylbenzene) monolithic capillary columns. J. Chromatogr. A 2010, 1217, 3085-3090.

Dong, M.W. HPLC Columns and Trends: Modern HPLC for Practicing Scientist. John Wiley & Sons: Hoboken, NJ, 2006, pp. 47-76.

Dong, X.; Wu, R.; Dong, J.; Wu, M.; Zhu, Y.; Zou, H. A Mesoporous silica nanoparticles immobilized open tubular capillary column with a coating of cellulose tris(3,5-dimethylphenyl-carbamate) for enantioseparation in CEC. Electrophoresis 2008, 29, 3933-3940.

Eeltink, S.; Dolman, S.; Swart, R.; Ursem, M.; Schoenmakers, P.J. Optimizing the peak capacity per unit time in one-dimensional and off-line two-dimensional liquid chromatography for the separation of complex peptide samples. J. Chromatogr. A 2009, 1216, 7368-7374.

Eeltink, S.; Rozing, G.P.; Kok, W.T. Recent applications in capillary electrochromatography. Electrophoresis 2003, 24, 3935-3961.

Eeltink, S.; Svec, F. Recent advances in the control of morphology and surface chemistry of porous polymer-based monolithic stationary phases and their application in CEC. Electrophoresis 2007, 28, 137-147.

Escudero, E.; Sentandreu, M.A.; Toldrá, F. Characterization of peptides released by in vitro digestion of pork meat. J. Agric. Food Chem. 2010, 58, 5160-5165.

Famiglini, G.; Palma, P.; Pierini, E.; Trufelli, H.; Cappiello, A. Organochlorine pesticides by LC-MS. Anal. Chem. 2008, 80, 3445-3449.

Famiglini, G.; Palma, P.; Siviero, A.; Rezai, M.A.; Cappiello, A. Determination of endocrine disrupting compounds in marine water by nanoliquid chromatography/direct-electron ionization mass spectrometry. Anal. Chem. 2005, 77, 7654-7661.

Fanali, S.; Aturki, Z.; D'Orazio, G.; Raggi, M.A.; Quaglia, M.G.; Sabbioni, C.; Rocco, A. Use of nano-liquid chromatography for the analysis of glycyrrhizin and glycyrrhetic acid in licorice roots and dandies. J. Sep. Sci. 2005a, 28, 982-986.

Fanali, S.; Aturki, Z.; D'Orazio, G.; Rocco, A. Separation of basic compounds of pharmaceutical interest by using nano-liquid chromatography coupled with mass spectrometry. J. Chromatogr. A 2007, 1150, 252-258.

Fanali, S.; Aturki, Z.; Kasicka, V.; Raggi, M.A.; D'Orazio, G. Enantiomeric separation of mirtazapine and its metabolites by nano-liquid chromatography with UV-absorption and mass spectrometric detection. J. Sep. Sci. 2005b, 28, 1719-1728.

Fanali, S.; Camera, E.; Chankvetadze, B.; D'Orazio, G.; Quaglia, M.G. Separation of tocopherols by nano-liquid dhromatography. J. Pharm. Biomed. Anal. 2004a, 35, 331-337.

Fanali, S.; D'Orazio, G.; Lomsadze, K.; Samakashvili, S.; Chankvetadze, B. Enantioseparations on amylose tris(5-chloro-2-methylphenylcarbamate) in nano liquid chromatography and capillary electrochromatography. J. Chromatogr. A 2010, 1217, 1166-1174.

Fanali, S.; D'Orazio, G.; Quaglia, M.G.; Rocco, A. Use of a hepta-tyr antibiotic modified silica stationary phase for the enantiomeric resolution of D,L-loxiglumide by electrochromatography and nano-liquid chromatography. J. Chromatogr. A 2004b, 1051, 247-252.

Fanali, S.; D'Orazio, G.; Rocco, A. Use of tert-butylbenzoylated tartardiamide chiral stationary phase for the enantiomeric resolution of acidic compounds by nano-liquid chromatography. J. Sep. Sci. 2006, 29, 1423-1431.

Faure, K. Liquid chromatography on chip. Electrophoresis 2010, 31, 2499-2511.

Fermier, A.M.; Colón, L.A. Capillary electrochromatography in columns packed by centripetal forces. J. Microcol. Sep. 1998, 10, 439-447.

Flender, C.; Leonhard, P.; Wolf, C.; Fritzsche, M.; Karas, M. Analysis of boronic acids by nano liquid chromatography-direct electron ionization mass spectrometry. Anal. Chem. 2010, 82, 4194-4200.

Flora, J.W.; Edmiston, J.; Secrist, R.; LI, G.; Rana, G.S.J.B.; Langston, T.B.; McKinney, W. Identification of in vitro differential cell secretions due to cigarette smoke condensate exposure using nanoflow capillary liquid chromatography and high-resolution mass spectrometry. Anal. Bioanal. Chem. 2008, 391, 2845-2856.

Fu, H.; Huang, X.; Jin, W.; Zou, H. The separation of biomolecules using capillary electrochromatography. Curr. Opin. Biotechnol. 2003, 14, 96-100.

García-Cañas, V.; Gonzalez, R.; Cifuentes, A. Detection of cinetically modified maize by the polymerase chain reaction and capillary gel electrophoresis with UV detection and laser-induced fluorescence. J. Agric. Food Chem. 2002, 50, 1016-1021.

Garrison, A.W.; Schmitt, P.; Martens, D.; Kettrup, A. Enantiomeric selectivity in the environmental degradation of dichlorprop as determined by high-performance capillary electrophoresis. Environm. Sci. Technol. 1996, 30, 2449-2455.

Geiser, L.; Eeltink, S.; Svec, F.; Fréchet, J.M.J. In-line system containing porous polymer monoliths for protein digestion with immobilized pepsin, peptide preconcentration and nano-liquid chromatography separation coupled to electrospray ionization mass spectroscopy. J. Chromatogr. A 2008, 1188, 88-96.

Gil-Agusti, M.T.; Campostrini, N.; Zolla, L.; Ciambella, C.; Invernizzi, C.; Righetti, P.G. Two-dimensional mapping as a tool for classification of green coffee bean species. Proteomics 2005, 5, 710-718.

Gluckman, J.C.; Novotny, M.V. Sources of extracolumn band-broadening in microcolumn liquid chromatography. microcolumn separations. M.V. Novotny, D. Ishi, Eds.; Journal of Chromatography Library; Elsevier: Amsterdam, 1985, 30, pp. 57-72.

Gonzalez-Begne, M.; Lu, B.; Han, X.; Hagen, F.K.; Hand, A.R.; Melvin, J.E.; Jates, J.R.I. Proteomic analysis of human parotid gland exosomes by multidimensional protein edentification technology (MudPIT). J. Proteom. Res. 2009, 8, 1304-1314.

Guerrouache, M.; Millot, M.-C.; Carbonnier, B. Functionalization of macroporous organic polymer monolith based on succinimide ester reactivity for chiral capillary chromatography: a cyclodextrin click approach. Macrom. Rapid Commun. 2009, 30, 109-113.

Guihen, E.; Glennon, J.D. Recent highlights in stationary phase design for open-tubular capillary electrochromatography. J. Chromatogr. A 2004, 1044, 67-81.

Gübitz, G.; Schmid, M.G. Chiral separation by capillary electromigration techniques. J. Chromatogr. A 2008, 1204, 140-156.

Haginaka, J. Molecularly imprinted polymers as affinity-based separation media for sample preparation. J. Sep. Sci. 2009, 32, 1548-1565.

Halvorsen, B.L.; Thomsen, C.; Greibrokk, T.; Lundanes, E. Determination of fenpyroximate in apples by supercritical fluid extraction and packed capillary liquid chromatography with UV detection. J. Chromatogr. A 2000, 880, 121-128.

Hampel, G. Strategies to improve the sensitivity in capillary electrophoresis for the analysis of drugs in biological fluids. Electrophoresis 2000, 21, 691-698.

He, Y.; Lee, H.K. Trace analysis by combined use of off-line solid-phase extraction, on-column sample focusing and U-shape flow cell in capillary liquid chromatography. J. Chromatogr. A 1998, 808, 79-86.

Healy, L.O.; Murrihy, J.P.; Tan, A.; Cocker, D.; McEnery, M.; Glennon, J.D. Enantiomeric separation of R,S-naproxen by conventional and nano-liquid chromatography with methyl-beta-cyclodextrin as a mobile phase additive. J. Chromatogr. A 2001, 924, 459-464.

Hernández-Borges, J.; Aturki, Z.; Rocco, A.; Fanali, S. Recent applications in nano-liquid chromatography. J. Sep. Sci. 2007a, 30, 1589-1610.

Hernández-Borges, J.; D'Orazio, G.; Aturki, Z.; Fanali, S. Nano-liquid chromatography analysis of dansylated biogenic amines in wines. J. Chromatogr. A 2007b, 1147, 192-199.

Ho, E.N.M.; Wan, T.S.M.; Wong, A.S.Y.; Lam, K.K.H.; Stewart, B.D. Doping control analysis of insulin and its analogues in equine plasma by liquid chromatography-tandem mass spectrometry. J. Chromatogr. A 2008, 1201, 183-190.

Hoes, I.; Van Dongen, W.; Lemière, F.; Esmans, E.L.; Van Bockstaele, D.; Berneman, Z.N. Comparison between capillary and nano liquid chromatography-electrospray mass spectrometry for the analysis of minor DNA-melphalan adducts. J. Chromatogr. B 2000, 748, 197-212.

Hsieh, S.-H.; Huang, H.-Y.; Lee, S. Determination of eight penicillin antibiotics in pharmaceuticals, milk and porcine tissues by nano-liquid chromatography. J. Chromatogr. A 2009, 1216, 7186-7194.

Hu, F.; Deng, C.; Liu, Y.; Zhang, X. Quantitative determination of chlorogenic acid in honeysuckle using microwave-assisted extraction followed by nano-LC-ESI mass spectrometry. Talanta 2009, 77, 1299-1303.

Hughey, C.A.; Wilcox, B.; Minardi, C.S.; Takehara, C.W.; Sundararaman, M.; Were, L.M. Capillary liquid chromatography-mass spectrometry for the rapid identification and quantification of almond flavonoids. J. Chromatogr. A 2008, 1192, 259-265.

Hutton, K.A.; Games, D.E. The analysis of β-blockers by capillary liquid chromatography/atmospheric pressure chemical ionization mass spectrometry. Rapid Commun. Mass Spectr. 1997, 11, 735-744.

Ishii, D.; Takeuchi, T.; Wada, A. Introduction to microscale high-performance liquid chromatography. Microscale Columns, D. Ishii, Ed.; VCH Publisher: New York, 1988, pp. 33-69.

Ito, S.; Yoshioka, S.; Ogata, I.; Takeda, A.; Yamashita, E.; Deguchi, K. Nanoflow gradient generator for capillary high-performance liquid chromatography–nanoelectrospray mass spectrometry. J. Chromatogr. A 2004, 1051, 19-23.

Ito, S.; Yoshioka, S.; Ogata, I.; Yamashita, E.; Nagai, S.; Okumoto, T.; Ishiia, K.; Ito, M.; Kaji, H.; Takao, K., et al. Capillary high-performance liquid chromatography/electrospray ion trap time-of-flight mass spectrometry using a novel nanoflow gradient generator. J. Chromatogr. A 2005, 1090, 178-193.

Jarmalavièienë, R.; Kornyšova, O.; Westerlund, D.; Maruška, A. Non-particulate (continuous bed or monolithic) restricted-access reversed-phase media for sample clean-up and separation by capillary-format liquid chromatography. Anal. Bioanal. Chem. 2003, 377, 902-908.

Kapnissi-Christodoulou, C.P.; Zhu, X.; Warner, I.M. Analytical separations in open-tubular capillary electrochromatography. Electrophoresis 2003, 24, 3917-3934.

Karlsson, G.N.; Wilson, N.L.; Wirth, H.-J.; Dawes, P.; Joshi, H.; Packer, N.H. Negative ion graphitized carbon nano-liquid chromatography/mass spectrometry increases sensitivity for glycoprotein oligosaccharide analysis. Rapid Commun. Mass Spectrom. 2004, 18, 2282-2292.

Karlsson, K.E.; Novotny, M. Separation efficiency of slurry-packed liquid chromatography microcolumns with very small inner diameters. Anal. Chem. 1988, 60, 1662-1665.

Karlsson, N.G.; Thomsson, K.A. Salivary MUC7 Is a major carrier of blood group I type O-linked oligosaccharides serving as the scaffold for sialyl Lewis x. Glycobiology 2009, 19, 288-300.

Kennedy, G.J.; Knox, J.H. The performance of packings in high performance liquid chromatography (HPLC) I. Porous and surface layered supports. J. Chromatogr. Sci. 1972, 10, 549-556.

Knox, J.H.; Grant, I.H. Electrochromatography in packed tubes using 1.5 to 50 μm silica gels and ODS bonded silica gels. Chromatographia 1991, 32, 317-328.

Kroening, K.K.; Kuhlmann, J.; Easter, R.; Clark, J.F.; Pyne-Geithman, G.; Caruso, J.A. Protein phosphorylation studies of cerebral spinal fluid for potential biomarker development. Metallomics 2010, 2, 334-341.

Kuhnle, M.; Holtin, K.; Albert, K. Capillary NMR detection in separation science. J. Sep. Sci. 2009, 32, 719-726.

Kwon, S.W. Profiling of soluble proteins in wine by nano-high-performance liquid chromatography/tandem mass spectrometry. J. Agric. Food Chem. 2004, 52, 7258-7263.

Lacey, M.E.; Tan, Z.J.; Webb, A.G.; Sweedler, J.V. Union of capillary high-performance liquid chromatography and microcoil nuclear magnetic resonance spectroscopy applied to the separation and identification of terpenoids. J. Chromatogr. A 2001, 922, 139-149.

Lanckmans, K.; Van Eeckhaut, A.; Sarre, S.; Smolders, I.; Michotte, Y. Capillary and nano-liquid chromatography-tandem mass spectrometry for the quantification of small molecules in microdialysis samples: comparison with microbore dimensions. J. Chromatogr. A 2006, 1131, 166-175.

Lee, E.D.; Muck, W.; Henion, J.D.; Convey, T.R. Liquid junction coupling for capillary zone electrophoresis/ion spray mass spectrometry. Biomed. Envirom. Mass Spectrom. 1989, 18, 844-850.

Legido-Quigley, C.; Marlin, L.D.; Manz, A.; Smith, N.W. Advances in capillary electrochromatography and micro-high performance liquid chromatography monolithic columns for separation science. Electrophoresis 2003, 24, 917-944.

León-González, M.E.; Pérez-Arribas, L.V.; Polo Díez, L.M.; Panis, C.; San Andrés, M.P. Determination of triazine herbicides by capillary liquid chromatography with on-column focusing and temperature gradient. Anal. Chim. Acta 2001, 445, 29-34.

Liu, H.; Row, K.H.; Yang, G. Monolithic molecularly imprinted columns for chromatographic separation. Chromatographia 2005, 61, 429-432.

Liu, Y.; Wu, J.; Yan, G.; Hou, R.; Zhuang, D.; Chen, L.; Pang, Q.; Zhu, J. Proteomic analysis of prolactinoma cells by immuno-laser capture microdissection combined with online two-dimensional nano-scale liquid chromatography/mass spectrometry. Proteome Sci. 2010, 8, 2-11.

Liu, Z.; Otsuka, K.; Terabe, S.; Motokawa, M.; Tanaka, N. Physically adsorbed chiral stationary phase of avidin on monolithic silica column for capillary electrochromatography and capillary liquid chromatography. Electrophoresis 2002, 23, 2973-2981.

Lord, G.A.; Gordon, D.B.; Myers, P.; King, B.W. Tapers and restrictors for capillary electrochromatography and capillary electrochromatography mass spectrometry. J. Chromatogr. A 1997, 768, 9-16.

Lortie, M.; Bark, S.; Blantz, R.; Hook, V. Detecting low-abundance vasoactive peptides in plasma: progress toward absolute quantitation using nano liquid chromatography-mass spectrometry. Anal. Biochem. 2009, 394, 164-170.

Lucas-Delfa, M.A.; Pérez-Arribas, L.V.; Navarro-Villoslada, F.; León-González, M.E.; Polo-Díez, L.M. Determination of chlorophenoxy acid herbicide residues by capillary reversed-phase liquid chromatography. J. Liq. Chromatogr. Relat. Technol. 2000, 23, 755-767.

Luo, Q.; Rejtar, T.; Wu, S.-L.; Karger, B.L. Hydrophilic interaction 10 µm I.D. porous layer open tubular columns for ultratrace glycan analysis by liquid chromatography-mass spectrometry. J. Chromatogr. A 2009, 1216, 1223-1231.

Malik, A.; Li, W.; Lee, M.L. Preparation of long packed capillary columns using carbon dioxide slurries. J. Microcol. Sep. 1993, 5, 361-369.

Marchesini, G.R.; Buijs, J.; Haasnoot, W.; Hooijerink, D.; Jansson, O.; Nielen, M.W.F. Nanoscale affinity chip interface for coupling inhibition SPR immunosensor screening with nano-LC TOF MS. Anal. Chem. 2008, 80, 1159-1168.

Martin, M.; Eon, C.; Guiochon, G. Study of the pertinency of pressure in liquid chromatography II problems in equipment design. J. Chromatogr. 1975, 108, 229-241.

Maruška, A.; Kornyšova, O. Continuous beds (monoliths): Stationary shases for liquid chromatography formed using the hydrophobic interaction-based phase separation mechanism. J. Biochem. Biophys. Methods 2004, 59, 1-48.

Masár, M.; Sydes, D.; Luc, M.; Kaniansky, D.; Kuss, H.-M. Determination of ammonium, calcium, magnesium, potassium and sodium in drinking waters by capillary zone electrophoresis on a column coupling chip. J. Chromatogr. A 2009, 1216, 6252-6255.

Mechref, Y.; Novotny, M.V. Miniaturized separation techniques in glycomic investigations. J. Chromatogr. B 2010, 841, 67-78.

Mikulovic, G.; Mechtler, K. HPLC Techniques for proteomics analysis-a short overview of latest developments. Brief Funct. Genomic Proteomic 2006, 5, 249-260.

Mills, M.J.; Maltas, J.; Lough, W.J. Assessment of injection volume limits when using on-column focusing with microbore liquid chromatography. J. Chromatogr. A 1997, 759, 1-11.

Mistry, K.; Krull, I.; Grinberg, N. Capillary electrochromatography: an alternative to HPLC and CE. J. Sep. Sci. 2002, 25, 935-958.

Mituloviæ, G.; Stingl, C.; Smoluch, M.; Swart, R.; Chervet, J.-P.; Steinmacher, I.; Gerner, C.; Mechtler, K. Automated, on-line two-dimensional nano liquid chromatography tandem mass spectrometry for rapid analysis of complex protein digests. Proteomics 2004, 4, 2545-2557.

Mohamed, M.S. Degradation of methomyl by the novel bacterial strain stenotrophomonas maltophilia M1. Electron. J. Biotechnol. 2009, 12, 1-11.

Molander, P.; Thomassen, A.; Kristoffersen, L.; Greibrokk, T.; Lundanes, E. Simultaneous determination of citalopram, fluoxetine, paroxetine and their metabolites in plasma by temperature-programmedpacked capillary liquid chromatography with on-column focusing of large injection volumes. J. Chromatogr. B 2002, 766, 77-87.

Moliner-Martínez, Y.; Herráez-Hernández, R.; Molins-Legua, C.; Campíns-Falcó, P. Improving analysis of apolar organic compounds by the use of a capillary titania-based column: application to the direct determination of faecal sterols cholesterol and coprostanol in wastewater samples. J. Chromatogr. A 2010, 1217, 4682-4687.

Monaci, L.; Losito, I.; Palmisano, F.; Visconti, A. Identification of allergenic milk proteins markers in fined white wines by capillary liquid chromatography-electrospray ionization-tandem mass spectrometry. J. Chromatogr. A 2010, 1217, 4300-4305.

Moneti, G.; Pieraccini, G.; Bartolucci, G.; Fuida, G.; Seraglia, R.; Traldi, P. Direct coupling of a nano-high-performance liquid chromatography column to an ion trap designed for a gas chromatography/mass spectrometry system. Rapid Commun. Mass Spectrom. 2001, 15, 1609-1617.

Myers, P.; Bartle, K. Miniaturization in LC–MS. LCGC Europe 2002, 2, 2-5.

Nakamura, T.; Kuromitsu, J.; Oda, Y. Evaluation of comprenhensive multidimensional separations using reversed-phase, reversed-phase liquid chromatography/mass spectrometry for shotgun proteomics. J. Proteom. Res. 2008, 7, 1007-1011.

Natsume, T.; Yamauchi, Y.; Nakayama, H.; Shinkawa, T. A direct nanoflow liquid chromatography - tandem system for interaction proteomics. Anal. Chem. 2002, 74, 4725-4733.

Nilsson, C.; Nilsson, S. Nanoparticle-based Pseudostationary phases in capillary electrochromatography. Electrophoresis 2006, 27, 76-83.

Nilsson, J.; Spégel, P.; Nilsson, S. Molecularly imprinted polymer formats for capillary electrochromatography. J. Chromatogr. B 2004, 804, 3-12.

Nischang, I.; Brüggemann, O. On the Separation of small molecules by means of nano-liquid chromatography with methacrylate-based macroporous polymer monoliths. J. Chromatogr. A 2010, 1217, 5389-5397.

Nováková, L.; Vlèková, H. A review of current trends and advances in modern bioanalytical methods: chromatography and sample preparation. Anal. Chim. Acta 2009, 656, 8-35.

Núñez, O.; Ikegami, T.; Miyamoto, K.; Tanaka, N. Study of monolithic silica capillary column coated with poly(octadecyl methacrylate) for the reversed-phase liquid chromatographic separation of some polar and non-polar compounds. J. Chromatogr. A 2007, 1175, 7-15.

Nyholm, L.M.; Markides, K.E. Column preparation for reversed-phase high-temperature open tubular column liquid chromatography. J. Chromatogr. A 1998, 813, 11-20.

Ohno, K.; Tachikawa, K.; Manz, A. Microfluidics: applications for analytical purposes in chemistry and biochemistry. Electrophoresis 2008, 29, 4443-4453.

Onisko, B.C. Radiochemical detection for packed capillary liquid chromatography-mass spectrometry. J. Am. Soc. Mass Spectrom. 2002, 13, 82-84.

Onisko, B.C.; Tambling, D.R.; Gorder, G.W.; Diaz, D.G.; Ericson, J.L.; Prisbylla, M.P.; Spillner, C.J. Identification of fonofos metabolites in latuca sativa, beta vulgaris, and triticum aestivum by packed capillary flow fast atom bombardment tandem mass spectrometry. J. Agric. Food Chem. 2002, 50, 1922-1928.

Ou, J.; Li, X.; Feng, S.; Dong, J.; Dong, X.; Komg, L.; Ye, M.; Zou, H. Preparation and evaluation of a molecularly imprinted polymer derivatized silica monolithic column for capillary electrochromatography and capillary liquid chromatography. Anal. Chem. 2007, 79, 639-646.

Östman, P.; Jäntti, S.; Ketola, R.A.; Franssila, S.; Kotiaho, T.; Kostiainen, R. Capillary liquid chromatography–microchip atmospheric pressure chemical ionization–mass spectrometry. Lab Chip 2006, 6, 948-953.

Pitarch, E.; Hernandez, F.; ten Hove, J.; Meiring, H.; Niesing, W.; Dijkman, E.; Stolker, L.; Hogendoorn, E. Potential of capillary-column-switching liquid chromatography-tandem mass spectrometry for the quantitative trace analysis of small molecules—application to the on-line screening of drugs in water. J. Chromatogr. A 2004, 1031, 1-9.

Poppe, H. Some Reflections on speed and efficiency of modern chromatographic methods. J. Chromatogr. A 1997, 778, 3-21.

Pröfrock, D.; Prange, A. Compensation of gradient related effects when using capillary liquid chromatography and inductively coupled plasma mass spectrometry for the absolute quantification of phosphorylated peptides. J. Chromatogr. A 2009, 1216, 6706-6715.

Pursch, M.; Sander, L.C. Stationary phases for capillary electrochromatography. J. Chromatogr. A 2000, 887, 313-326.

Pusecker, K.; Schewitz, J.; Gfrorer, P.; Tseng, L.H.; Albert, K.; Bayer, E. On-line coupling of capillary electrochromatography, capillary electrophoresis, and capillary HPLC with nuclear magnetic resonance spectroscopy. Anal. Chem. 1998, 70, 3280-3285.

Puy, G.; Rouz, R.; Demesmay, C.; Rocca, J.-L.; Iapichella, J.; Galarneau, A.; Brunel, D. Infuence of the hydrothermal treatment on the chromatographic properties of monolithic silica capillaries for nano-liquid chromatography or capillary electrochromatography. J. Chromatogr. A 2007, 1160, 150-159.

Rapp, E.; Bayer, E. Improved column preparation and performance in capillary electrochromatography. J. Chromatogr. A 2000, 887, 367-378.

Rathore, A.S. Theory of electroosmotic flow, retention and separation efficiency in capillary electrochromatography. Electrophoresis 2002, 23, 3827-3845.

Reynolds, K.J.; Maloney, T.D.; Fermier, A.M.; Colon, L.A. Capillary electrochromatography in columns packed by gravity - preliminary study. Analyst 1998, 123, 1493-1495.

Rizzelo, C.G.; De Angelis, M.; Di Cagno, R.; Camarca, A.; Silano, M.; Losito, I.; De Vincenzi, M.; De Bari, M.D.; Palmisano, F.; Maurano, F.; et al. Highly efficient gluten degradation by lactobacilli and fungal proteases during food processing: new perspectives for celiac disease. Appl. Environ. Microbiol. 2007, 73, 4499-4507.

Robson, M.M.; Cikalo, M.G.; Myers, P.; Euerby, M.R.; Bartle, K.D. Capillary Electrochromatography - a Review. J. Microcol. Sep. 1997, 9, 357-372.

Rocco, A.; Fanali, S. Capillary electrochromatography without external pressure assistance. Use of packed columns with a monolithic inlet frit. J. Chromatogr. A 2008, 1191, 263-267.

Rocco, A.; Fanali, S. Analysis of phytosterols in extra-virgin olive oil by nano-liquid chromatography. J. Chromatogr. A 2009a, 1216, 7173-7178.

Rocco, A.; Fanali, S. Enantiomeric separation of acidic compounds by nano-liquid chromatography with methylated-beta-cyclodextrin as a mobile phase additive. J. Sep. Sci. 2009b, 32, 1696-1703.

Roland, A.; Vialaret, J.; Moniatte, M.; Rigou, P.; Razungles, A.; Schneider, R. Validation of a nanoliquid chromatography-tandem mass spectrometry method for the identification and the accurate quantification by isotopic dilution of glutathionylated and cysteinylated precursors of 3-mercaptohexan-1-ol and 4-mercapto-4-methylpentan-2-one in white grape juices. J. Chromatogr. A 2010, 1217, 1626-1635.

Rosales-Conrado, N.; León-Gonzáles, M.E.; Pérez-Arribas, L.V.; Polo-Díez, L.M. Determination of heterocyclic aromatic amines by capillary high-performance liquid chromatography with diode array detection in ready-to-eat cooked ham treated with electron-beam irradiation. Anal. Bioanal. Chem. 2008, 391, 1433-1442.

Rosales-Conrado, N.; León-González, M.E.; D'Orazio, G.; Fanali, S. Enantiomeric separation of chlorophenoxy acid Herbicides by nano liquid chromatography-UV detection on a vancomycin-based chiral stationary phase. J. Sep. Sci. 2004, 27, 1303-1308.

Rosales-Conrado, N.; León-González, M.E.; Pérez-Arribas, L.V.; Polo-Díez, L.-M. determination of chlorophenoxy acid herbicides and their esters in soil by capillary high performance liquid chromatography with ultraviolet detection, using large volume injection and temperature gradient. Anal. Chim. Acta 2002, 470, 147-154.

Rosales-Conrado, N.; León-González, M.E.; Pérez-Arribas, L.V.; Polo-Díez, L.M. Capillary liquid chromatography of chlorophenoxy acid herbicides and their esters in apple juice samples after preconcentration on a cation exchanger based on polydivinylbenzene-N-vinylpyrrolidone. J. Chromatogr. A 2005, 1076, 202-206.

Roux, R.; Jaoudé, M.A.; Demesmay, C. Improvement of chromatographic performance of in-situ synthesized hybrid C_8 silica monoliths by reduction of structural radial heterogeneities. J. Chromatogr. A 2009, 1216, 3857-3863.

Roux, R.; Puy, G.; Demesmay, C.; Rocca, J.-L. Synthesis of propyl-functionalized hybrid monolithic silica capillaries and evaluation of their performances in nano-LC and CEC. J. Sep. Sci. 2007, 30, 3035-3042.

Rozenbrand, J.; van Bennekom, W.P.; Unger, K.K.; de Jong, G.J. Fast LC Separation of a myoglobin digest: a case study using monolithic and particulate RP 18 silica capillary columns. Anal. Bioanal. Chem. 2006, 385, 1055-1061.

Ruhaak, R.L.; Deelder, A.R.; Wuhrer, M. Oligosaccharide analysis by graphitized carbon liquid chromatography–mass spectrometry. Anal. Bioanal. Chem. 2009, 394, 163-174.

Ruppén Cañás, I.; Hamilton, B.; Fernández Amandi, M.; Furey, A.; James, K.J. Nano liquid chromatography with hybrid quadrupole time-of-flight mass spectrometry for the determination of yessotoxin in marine phytoplankton. J. Chromatogr. A 2006, 1056, 253-256.

Sabatini, L.; Barbieri, A.; Violante, F.S. Development and validation of a capillary high-performance liquid chromatography/electrospray tandem mass spectrometric method for the quantification of bisphenol A in air samples. rapid commun. Mass Spectrom. 2005, 19, 3468-3472.

Saito, Y.; Jinno, K.; Greibrokk, T. Capillary columns in liquid chromatography: between conventional columns and microchips. J. Sep. Sci. 2004, 27, 1379-1390.

Sakai-Kato, K.; Kinouchi, T.; Fujii, N.; Imai, K.; Utsunomiya-Tate, N. Screening system for D-Asp-containing proteins using D-aspartyl endopeptidase and two-dimensional gel electrophoresis. Amino Acids 2009, 36, 125-129.

Sawasdipuksa, N.; Sumner, L.W.; Lei, Z.; Sangvanich, P. Proteome analysis of pithecellobium dulce seeds using two-dimensional gel electrophoresis and tandem mass spectrometry. J. Sci. Food Agric. 2009, 89, 1284-1291.

Schurig, V.; Mayer, S. Separation of enantiomers by open capillary electrochromatography on polysiloxane-bonded permethyl-beta-cyclodextrin. J. Biochem. Biophys. Methods 2001, 48, 117-141.

Si-Ahmed, K.; Tazerouti, F.; Badjah-Hadj-Ahmed, A.Y.; Aturki, Z.; D'Orazio, G.; Rocco, A.; Fanali, S. Analysis of hesperetin enantiomers in human urine after ingestion of blood orange juice by using nano-liquid chromatography. J. Phar. Biomed. Anal. 2010a, 51, 225-229.

Si-Ahmed, K.; Tazerouti, F.; Badjah-Hadj-Ahmed, A.Y.; Aturki, Z.; D'Orazio, G.; Rocco, A.; Fanali, S. Optical isomer separation of flavanones and flavanone glycosides by nano-liquid chromatography using a phenyl-carbamate-propyl-β-cyclodextrin chiral stationary phase. J. Chromatogr. A 2010b, 1217, 1175-1182.

Simon, R.; Collins, M.E.; Spivak, D.A. Shape selectivity versus functional group pre-organization in molecularly imprinted polymers. Anal. Chim. Acta 2007, 591, 7-16.

Snovida, S.I.; Bodnar, E.D.; Viner, R.; Saba, J.; Perreault, H. A simple cellulose column procedure for selective enrichment of glycopeptides and characterization by nano LC coupled with electron-transfer and high-energy collisional-dissociation tandem mass spectrometry. Carbohydr. Res. 2010, 345, 792-801.

Song, C.; Ye, M.; Han, G.; Jiang, X.; Wang, F.; Yu, Z.; Chen, R.; Zou, H. Reversed-phase- liquid chromatography approach with high orthogonality for multidimensional separation of phosphopeptides. Anal. Chem. 2010, 82, 53-56.

Song, Y.; Quan, Z.; Liu, Y.-M. Assay of histamine by nano-liquid chromatography/tandem mass spectrometry with a packed nanoelectrospray emitter. Rapid Commun. Mass Spectrom. 2004, 18, 2818-2822.

Srbek, J.; Bosáková, Z.; Tesarová, E.; Suchánková, J.; Coufal, P.; Nemcová, I. Method of determination of 9-(alkylthio) acridines in urine by capillary liquid chromatography. Chromatographia 2004, 60, S37-S41.

Steiner, F.; Scherer, B. Instrumentation for capillary electrochromatography [Review]. J. Chromatogr. A 2000, 887, 55-83.

Svec, F. Capillary electrochromatography: a rapidly emerging separation method. advances in biochemical engineering/biotechnology: modern advances in chromatography. T. Shepers, R. Freitag, Eds.; Springer-Verlag: Berlin Heidelberg, 2002, 76, pp. 1-47.

Svec, F. Porous Polymer Monoliths: Amazingly vide Variety of techniques enabling their preparation. J. Chromatogr. A 2010, 1217, 902-924.

Svec, F.; Huber, C.G. Monolithic materials: promises, challenges, achievements. Anal. Chem. 2006, 78, 2100-2107.

Swart, R.; Kraak, J.C.; Poppe, H. Recent progress in open tubular liquid chromatography. TRAC-Trends. in Anal. Chem. 1997, 16, 332-342.

Szumski, M.; Buszewski, B. State of the art in miniaturized separation techniques. Crit. Rev. Anal. Chem. 2002, 32, 1-46.

Takeuchi, T.; Niwa, T.; Ishii, D. Stepwise gradient elution using switching valves in micro high-performance liquid-chromatography. J. Chromatogr. 1987, 405, 117-124.

Tan, Z.J.; Remcho, V.T. Molecular imprint polymers as highly selective stationary phases for open tubular liquid chromatography and capillary electrochromatography. Electrophoresis 1998, 19, 2055-2060.

Tanaka, N.; Kobayashi, H.; Ishizuka, N.; Minakuchi, H.; Nakanishi, K.; Hosoya, K.; Ikegami, T. Monolithic silica columns for high-efficiency chromatographic separations. J. Chromatogr. A 2002, 965, 35-49.

Tao, D.; Zhang, L.; Shan, Y.; Liang, Y.; Zhang, Y. Recent advances in micro-scale and nano-scale high-performance liquid-phase chromatography for proteome research. Anal. Bioanal. Chem. 2010a.

Timperio, A.M.; Magro, P.; Chilosi, G.; Zolla, L. Assay of ochratoxyn A in grape by high-pressure liquid chromatography coupled on line with an ESI-mass spectrometry. J. Chromatogr. B 2006, 832, 127-133.

van der Ven, K.; Van Dongen, W.; Maes, B.U.W.; Esmans, E.L.; Blust, R.; De Coen, W.M. Determination of diazepam in aquatic samples by capillary liquid chromatography-electrospray tandem mass spectrometry. Chemosphere 2004, 57, 967-973.

Vázquez, M.; Paull, B. Review on recent and advanced applications of monoliths and related porous polymer gels in micro-fuidic devices. Anal. Chim. Acta 2010, 668, 100-113.

Vissers, J.P.C.; Claessens, H.A.; Cramers, C.A. Microcolumn liquid chromatography: Instrumentation, detection and applications. J. Chromatogr. A 1997, 779, 1-28.

Wang, F.; Dong, J.; Jiang, X.; Ye, M.; Zou, H. Capillary trap column with strong cation-exchange monolith for automated shotgun proteome analysis. Anal. Chem. 2007, 79, 6599-6606.

Warriner, R.N.; Craze, A.S.; Games, D.E.; Lane, S.J. Capillary electrochromatography mass spectrometry—a comparison of the sensitivity of nanospray and microspray ionization techniques. Rapid Commun. Mass Spectrom. 1998, 12, 1143-1149.

Weber, D.; Raymond, P.; Ben-Rejeb, S.; Lau, B. Development of a liquid chromatography-tandem mass spectrometry method using capillary liquid chromatography and nanoelectrospray ionization-quadrupole time of flight hybrid mass spectrometer for the detection of milk allergens. J. Agric. Food Chem. 2006, 54, 1604-1610.

Wehr, T. Separation technology in proteomics. LCGC North America 2001, 19, 702-711.

Wehr, T. Affinity selection techniques for proteomic studies. LCGC North America 2003, 21, 274-284.

Wehr, T. Top-Down versus bottom-up approaches in proteomics. LC-GC North America 2006, 24, 1004.

Wistuba, D. Chiral silica-based monoliths in chromatography and capillary electrochromatography. J. Chromatogr. A 2010, 1217, 941-952.

Wistuba, D.; Schurig, V. Enantiomer separation by capillary electrochromatography on a cyclodextrin-modified monolith. Electrophoresis 2000a, 21, 3152-3159.

Wistuba, D.; Schurig, V. Enantiomer separation of chiral pharmaceuticals by capillary electrochromatography. J. Chromatogr. A 2000b, 875, 255-276.

Wu, N.; Peck, T.L.; Webb, A.G.; Magin, R.L.; Sweedler, J.V. 1H-NMR Spectroscopy on the nanoliter scale for static and on-line measurements. Anal. Chem. 1994, 66, 3849-3857.

Wu, R.; Hu, L.; Wang, F.; Ye, M.; Zou, H. Recent development of monolithic stationary phases with emphasis on microscale chromatographic separation. J. Chromatogr. A 2008, 1184, 369-392.

Wu, T.; Yuan, T.; Tsai, S.-N.; Wang, C.; Sun, S.-M.; Lam, H.-M.; Ngai, S.-M. Mass spectrometry analysis of the variants of histone H3 and H4 of soybean and their post-translational modifications. MBC Plant Biol. 2009, 9.

Wuhrer, M.; Koeleman, C.A.M.; Da, M.; Kokke, C.H. Normal-phase nanoscale liquid chromatography-mass spectrometry of underivatized oligosaccharides at low-femtomole sensitivity. Anal. Chem. 2004, 76, 833-838.

Wysocki, V.H.; Resing, K.A.; Zhang, Q.; Cheng, G. Mass spectrometry of peptides and proteins. Methods 2005, 35, 211-222.

Xie, W.; Mirocha, C.J.; Chen, J. Detection of two naturally occurring structural isomers of partially hydrolyzed fumonisin B_1 in corn by on-line capillary liquid chromatography-fast atom bombardment mass spectrometry. J. Agric. Food Chem. 1997, 45, 1251-1255.

Yan, C. Electrokinetic packing of capillary columns. U.S. Patent 08/142917, October 29, 1993.

Yan, C.; Dadoo, R.; Zhao, H.; Zare, R.N.; Rakestraw, D.J. Capillary electrochromatography - analysis of polycyclic aromatic hydrocarbons. Anal. Chem. 1995, 67, 2026-2029.

Yeung, E.S. Optical detectors for microcolumn liquid chromatography. microcolumn separations. M.V. Novotny, D. Ishi, Eds.; Journal of Chromatography Library; Elsevier: Amsterdam, 1985, Vol. 30, pp. 135-158.

Yoo, J.S.; Mcguffin, V.L. Determination of fatty acids in fish oil dietary supplements by capillary liquid chromatography with laser-induced fluorescence detection. J. Chromatogr. A 1992, 627, 87-96.

Zhang, J.; Xu, X.; Shen, H.; Zhang, X. Analysis of nuclear proteome in C57 mouse liver tissue by a nano-flow 2-D-LC-ESI-MS/MS Approach. J. Sep. Sci. 2006, 29, 2635-2646.

Zhang, X.; Fang, A.; Riley, C.P.; Wang, M.; Regnier, F.E.; Buck, C. Multi-dimensional liquid chromatography in proteomics- A review. Anal. Chim. Acta 2010, 664, 101-113.

Zheng, J.; Shamsi, S. Combination of chiral capillary electrochromatography with electrospray ionization mass spectrometry: Method development and assay of warfarin enantiomers in human plasma. Anal. Chem. 2003, 75, 6295-6305.

Zheng, M.M.; Lin, B.; Feng, Y.Q. Hybrid Organic–inorganic octyl monolithic column for in-tube solid-phase microextraction coupled to capillary high-performance liquid chromatography. J. Chromatogr. A 2007, 1164, 48-55.

Zhong, H.; El Rassi, Z. Monolithic silica capillary columns having immobilized lectins and surface bound polar functionalities for lectin affinity and normal phase nano-LC and CEC of glycoconjugates, respectively. J. Sep. Sci. 2009, 32, 1642-1653.

Zhu, L.Y.; Zhu, L.; Lee, H.K. Liquid-liquid-liquid microextraction of nitrophenols with a hollow fiber membrane prior to capillary liquid chromatography. J. Chromatogr. A 2001, 924, 407-414.

Zhu, Y.-S.; Fitzgerald, R.J. Direct nanoHPLC-ESI-QTOF- MS/MS Analysis of tryptic caseinophosphopeptides. Food Chem. 2010, 123, 753-759.

10

Comprehensive 2D GC Methodologies for the Analysis of Lipids

Giorgia Purcaro, Peter Quinto Tranchida, and Luigi Mondello
Dipartimento Farmaco-Chimico, University of Messina, Italy

Introduction

One of the main challenges of separation scientists, in general, is to unravel the chemical profiles of complex, naturally-occurring samples. Among the analytical techniques, gas chromatography (GC) is one of the most widely employed for such a purpose. Gas chromatography was invented approximately 60 years ago and, since then, a considerable degree of progress has been made (James & Martin, 1952). The most important steps toward the increase of separation power in the GC field have been:

1) the replacement of the packed column with the more efficient capillary one in 1958 by Golay (1958), enabling an approximate ten-times increase in separation power;
2) the introduction of comprehensive two-dimensional GC (GC × GC) in 1991 by Liu and Philips (1991).

In GC × GC experiments, the separation process is carried out on two columns of different selectivity. An interface defined as modulator is located between the two capillaries and generates a separation peak capacity which is roughly equal to the product of the two peak capacities (peak capacity can be defined as "the number of peaks that can be potentially located in the one or two dimensional space, with a specific resolution value") in each dimension. Comprehensive GC can be considered an extension of conventional heart-cut multidimensional GC, which involves the separation of specific fractions on two different columns with similar lengths. A comprehensive GC separation is obtained when all the components from the first dimension (1D), usually a conventional column (e.g., 25–30 m × 0.25 mm ID × 0.25 μm), are re-analysed on the second dimension (2D), commonly a short micro-bore capillary segment (e.g., 1–2 m × 0.10 mm ID × 0.10 μm). Second-dimension analyses are carried out in a rapid manner, with duration equal to what is defined as the modulation period (typically in the 4–8 sec range).

The introduction of GC × GC has not only enabled a better understanding of complex matrices but has also opened a new insight into so-called "well-known" samples. The main advantages of comprehensive 2D GC, over conventional 1D GC, can be summed up in five points: i) speed: in terms of number of peaks resolved per unit of time; ii) selectivity: since two stationary phases are used; iii) separation power: increased resolving power, roughly equal to the product of the two column peak capacities; iv) sensitivity: thanks to the re-concentration step during modulation; v) structured chromatograms: formation of chemically-similar compound patterns when homologous series of compounds are present.

In the present chapter, GC × GC fundamentals will be discussed, along with the main applications in the lipid analysis field.

Instrumentation and Fundamental Operational Parameters

GC × GC systems can be constructed using the same equipment employed for conventional 1D GC. Samples are introduced by an injector onto the first conventional capillary column where the first separation occurs; the eluate is then fractionated and re-injected through the modulator onto a second capillary column coated with a different stationary phase for further separation. The two columns can be mounted in a single oven, or in two different ones, the latter option providing a higher degree of flexibility during method optimization. Since the separation on the second column is a very fast GC one, ideally it should last a modulation period (to abide the requirement for a comprehensive analysis). Dedicated detectors with high acquisition rates have negligible internal volumes, and rapid rise times are required to accurately re-construct the narrow chromatography bands generated.

Table 10.A. Most Employed Nomenclature in the GC × GC Field.

Term	Definition
Modulation frequency (f_M)	Number of modulations per unit of time.
Modulation number (n_M)	The number of modulations for a given first-dimension peak.
Single-stage modulation	Accumulation and focusing during a one-series process at one location in the modulator.
Dual-stage modulation	Accumulation and focusing during two successive series of processes at two locations in the modulator.
Wrap-around	The occurrence of second dimension peaks with retention times that exceed the modulation time.
Modulation phase (F_M)	The pattern of modulated peaks caused by the time relationship between peak distribution and the pulsing process of the modulator in a comprehensive two-dimensional system.
In-phase modulation	The modulation phase that produces a symmetrical sequence of peaks with a single maximum pulse peak.
Out-of-phase modulation	Any phase that produces a non-symmetrical peak-pulse distribution.

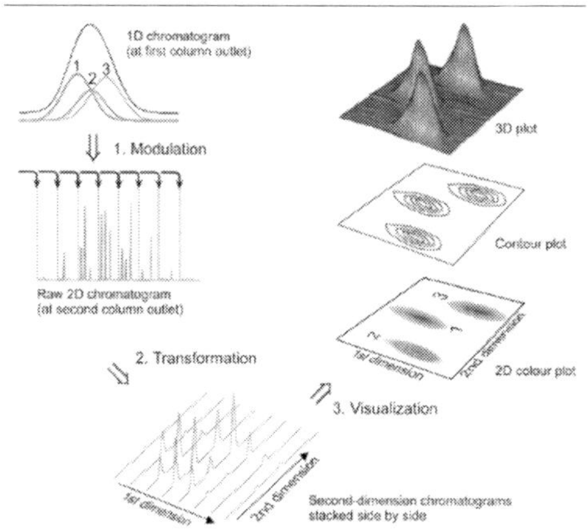

Fig. 10.1. Scheme representing GC × GC data manipulation and visualization. Reproduced with permission from (Adahchour et al., 2006).

Before entering into GC × GC details, a brief summary of the most employed terminologies in the field is given in Table 10.A, as reported by Schoenmakers and co-workers (2003).

The entire GC × GC process requires data elaboration and visualization (Fig. 10.1). Although basic information on the chromatographic performance can be deducted by expert operators from the modulated raw signal, a transformation process is necessary to visualize and elaborate the results. Dedicated software packages stack second-dimension chromatograms side by side and, considering the modulation time, derive the first and second dimension retention times for each peak. Peak areas are attained by summing the areas relative to each modulated peak while signal intensity is considered as the height of the tallest modulated peaks and can be visualized in a different way: I) by the means of colors (color plot), II) by contour lines (contour plot), III) by a 3D plot.

Modulators

The modulator, which is the heart of the GC × GC system, enables the isolation, re-concentration, and re-injection of chromatography bands from a primary to a secondary column in a continuous and sequential manner. Ideally, the separation on the second dimension must end before re-injection. The time required to complete a single cycle of events is called the modulation period (P_M). The latter parameter is very important since it must be sufficiently low to maintain the resolution achieved on the primary column but high enough to avoid a loss of sensitivity and to enable the complete elution of the compounds from the second column before the following injection, thus avoiding wrap-around. However, wrap-around is a problem only in specific analytical situations (*e.g.*, in the case of structured chromatograms, or if co-elutions arise with constituents of the later fraction). Murphy et al. (1998) studied the effects of the modulation period, concluding that first-dimension resolution can be preserved only if every eluting peak is sampled at least 3 or 4 times, in the case of in-phase and out-of-phase sampling, respectively. Therefore, in most GC × GC experiments the operational conditions enable the attainment of 15–20 sec 1D peak widths, and the application of 4–6-sec modulation periods.

The different types of modulators that have been developed can be divided in three main classes: I) heat-based, II) cryogenic and III) flow. The following discussion can be considered as a brief overview rather than an exhaustive review of all the modulators developed.

Heat-based Modulators

The very first modulator (Liu & Phillips, 1991) consisted of a 15-cm segment of thick-film capillary column divided equally into two-stages and painted with an electrically-conductive film and looped outside the oven under room temperature conditions (Fig. 10.2).

The primary column eluate was focused by the thick stationary-phase film in the first part of the capillary segment; then, the analyte band was remobilized by the application of a 20 msec electrical pulse directed to the first segment of the modulator. After the electrical pulse terminated, the segment cooled down to the initial trapping temperature, and the electrical pulse was applied to the second segment of the modulator with a 100 msec delay

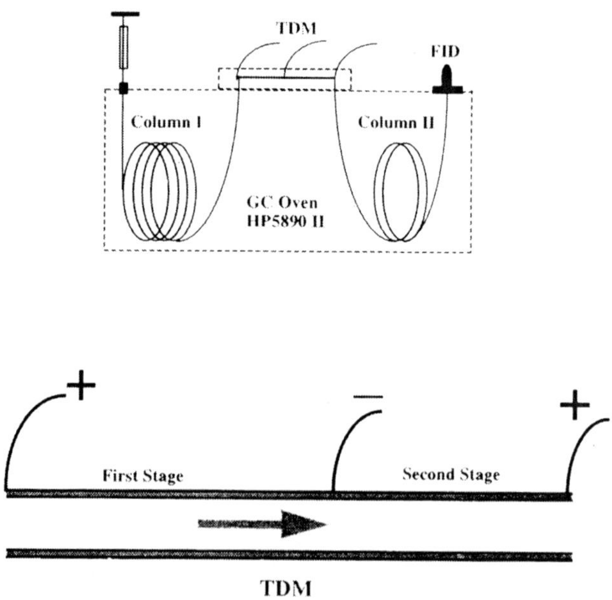

Fig. 10.2. The first modulator. Reproduced with permission from (Liu & Phillips, 1994).

between stages. The second pulse launched the entrapped analytes onto the second column. The described procedure prevented breakthrough of analytes in the modulator. Although the results obtained with the primordial modulator were outstanding (Liu & Phillips, 1994; Phillips & Xu, 1995; Venkatramani & Phillips, 1993), the device was also characterized by low run-to-run repeatability and by excessive fragility.

The first commercial modulator, also developed by Philips (Phillips & Ledford, 1996; Phillips et al., 1999), was the rotating thermal modulator, better known as a "sweeper" (Fig. 10.3). It consisted of a moving metallic slotted heater with a gap to allow the capillary column to pass through. The two columns were connected by using an accumulation section (A in Fig. 10.3), an intermediate thicker-film column which retarded the travel of solutes through this section, and a short section of an uncoated column (P in Fig. 10.3) (called a pigtail) which then delivered the analytes to the second column. The analytes were retained in the accumulation segment until the rotating sweeper (at about 100°C above the oven temperature) forced them forward in a focused band through rotation across columns A and P.

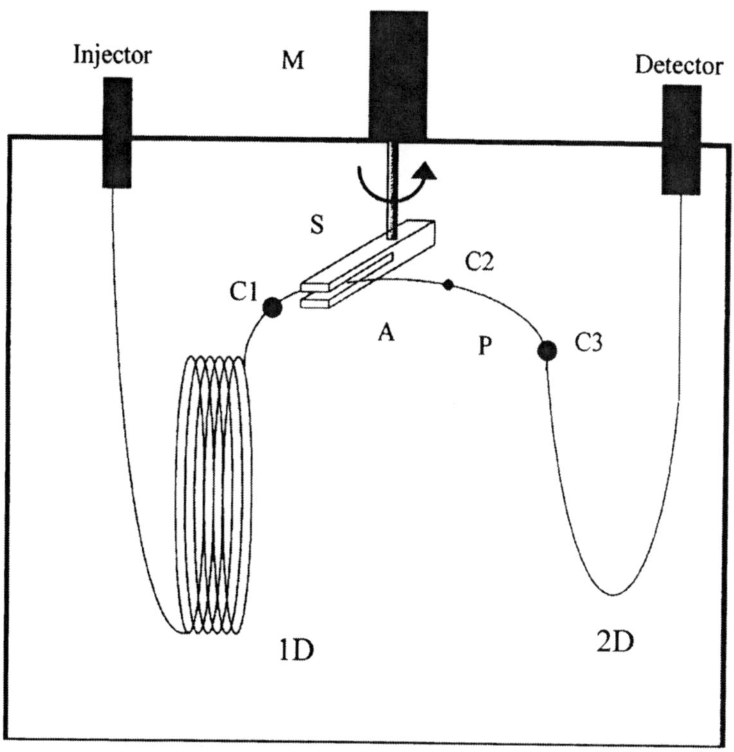

Fig. 10.3. Scheme of the "sweeper" modulator. Reproduced with permission from Marriot (2002).

In spite of the several applications carried out with this kind of modulator (Beens et al., 1998a; Beens et al., 1998b; Frysinger & Gaines, 1999; Frysinger et al., 1999; Phillips et al., 1999), it was abandoned due to a series of disadvantages. The correct alignment of all the modulator parts, to avoid contact with the moving heater, was a critical issue; the system was not suitable for highly volatile compounds, and the maximum oven temperature had to be limited to 100°C below the maximum operating temperature of the modulator stationary phase, thus restricting the range of applications.

Several other thermal modulators have been developed based on the first experiment carried out by Phillips (Liu & Phillips, 1991). Górecki and co-workers developed both single- and dual-stage modulators (Harynuk & Górecki, 2002; Goldestein et al., 2008) by using deactivated Silcosteel® packed with micro sorbent traps to increase the trapping capacity. The trapping capillary was located outside the GC oven and subjected to continuous forced air cooling and resistively heated to re-mobilize the entrapped compounds. The main disadvantage was the limited thermal stability of the sorbent, restricting the use to highly volatile compounds. In 2009, the same group introduced a modification of the previously described modulator using a single-stage thermal modulator and a temperature-programmed air vortex cooler (Górecki et al., 2009).

Cryogenic Modulators

The first cryogenic modulator was developed by Marriott and Kinghorn (1998). The longitudinally modulated cryogenic system (LMCS) consisted of a small CO_2 cryogenic trap, characterized by a hollow sleeve configuration which oscillated across a segment of column located at the head of the second dimension (Fig. 10.4). The primary column eluate was trapped and focused by the liquid CO_2-cooled trap then the trap was moved along the capillary column exposing the previously-cooled spot to the GC oven and thus ejecting the entrapped compounds in a narrow band onto the second column (estimated on the order of 50 msec) (Kinghorn & Marriott, 1999). Analyte desorption occurred by moving the trap into the downstream position in order to avoid breakthrough. Compared to heat-based modulators, the LMCS entrapped highly volatile compounds and the final oven temperature was related to the specific columns employed. The LMCS operating temperature is an important optimization parameter. A temperature of *circa* 100°C lower than the GC elution temperature is generally sufficient for effective entrapment of most compounds. However, excessively low temperatures can retard analyte release (especially for high MW compounds), while intense cooling (-120/-140°C) is required for entrapment of highly volatile compounds (i.e., C_4-C_6 alkanes). The use of CO_2 as the cryogenic agent does not generate sufficiently cold temperatures to entrap more volatile compounds such as C_3-C_4 alkanes. The main disadvantage of the LMCS modulator is the "wear and tear" caused by the continuous movement.

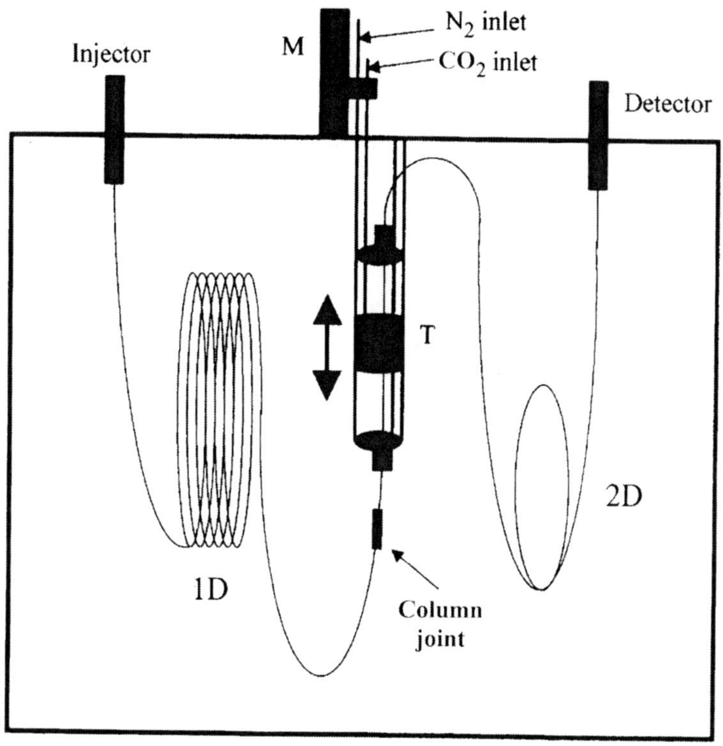

Fig. 10.4. The LMC system. T: cryogenic trap; M: pneumatic or motor driven device that moves the trap up and down. Reproduced with permission from (Marriott, 2002).

Shortly after the introduction of the first cryogenic modulator, several other approaches were developed with the aim of avoiding moving instrumental parts. Ledford (2000) presented a dual-stage modulator, formed by two cold jets and two hot jets, hence the name quad-jet modulator (Fig. 10.5). After focusing the first-dimension eluate at point A, cold jet 1 is switched off, while hot jet 1 and cold jet 2 are simultaneously activated, removing the trapped band and then compressing it at point B together with any effluent that passes through spot A during this step. The final modulation step occurs when hot jet 1 and cold jet 2 are switched off while cold jet 1 and hot jet 2 are activated, thus trapping the primary column eluate and re-injecting the trapped band in spot B onto the second dimension.

The quad-jet modulator has been used both with CO_2 or liquid nitrogen; the version commercialized by LECO Corporation uses the latter.

Beens and co-workers (2001) simplified the quad-jet configuration by removing the hot jets. The remobilization of the entrapped compounds was achieved by exploiting the heated oven air. The "Beens" version has been

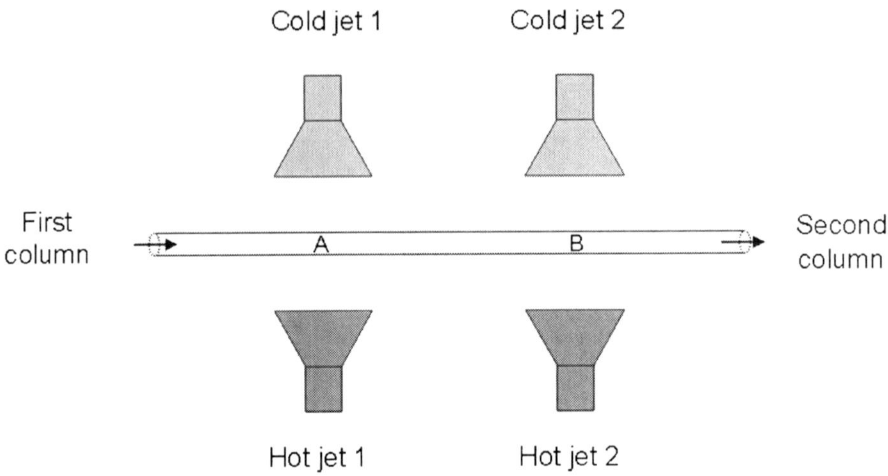

Fig. 10.5. Scheme of the quad-jet modulator.

commercialized by Thermo Scientific. Following such a trend, Adahchour and co-workers (2003) presented a single-stage cryogenic modulator consisting of a single cold jet. As for any one-stage modulator, such a configuration had to be very carefully tuned to avoid breakthrough.

The most successful simplification of the original quad-jet modulator was developed by Ledford and co-workers (2002). The simple introduction of a delay loop enabled the performance of dual-stage modulation with a single cold jet which cooled two different segments of the trapping capillary and a hot jet positioned perpendicularly that assured effective removal of the entrapped compounds (Fig. 10.6). The cold jet works continuously, trapping the primary column effluent; when the hot jet is activated, the cold jet is deviated, the first cold segment is quickly warmed up, and the analyte band moves forward into the delay loop along with any breakthrough from the first column. When the hot jet is turned off, compounds in the loop are re-trapped in the second segment and eluent from the first column is entrapped again in the first segment. The following activation of the hot jet leads to the injection of both the analyte plug, located at the second segment onto the second dimension, and the analyte band, located at the first segment into the delay loop, thus restarting the cycle. The main drawback of this modulator is the high consumption of liquid nitrogen. The loop-type modulator is commercialized by Zoex Corporation.

Riekkola and co-workers developed three versions of a dual-stage modulator (Kallio et al., 2003a; Hyötyläinen et al., 2002; Kallio et al., 2008) based on a continuous flow of CO_2 into a semi-rotating trap. In the first version, CO_2 passed into a rotating tube, equipped with two nozzles (upper and lower) at

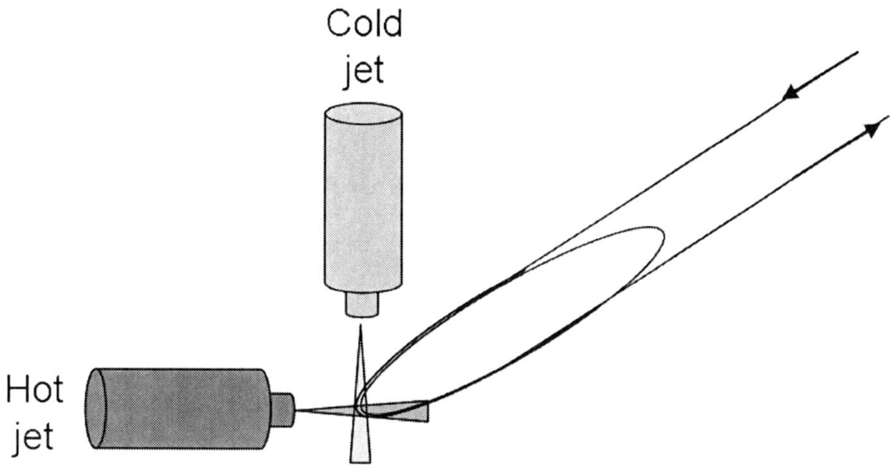

Fig. 10.6. Scheme of a dual-stage loop-type modulator.

an angle of 45°, and was directed onto the capillary in an alternating mode. The cold spots were remobilized by electrical heating. Further systems were characterized by a column holder which clenched the capillary between septa (allowing easy installation and removal) and by a single nozzle incorporated into a semi-rotating disc which directed CO_2 to the column at two trap points located 30 mm apart. Despite the moving parts, column breakage was prevented thanks to the particular configuration.

Flow Modulators

The development of cryogenic-free modulators has always been highly desirable considering the high costs of cooling gases. The first attempt in this direction was made by Bruckner and co-workers (1998), who employed a multiport diaphragm valve. The primary column effluent was directed to an exhaust line for most of the modulation period while the valve was rapidly switched only for a short time, directing the effluent to the head of the secondary column pushed by a relatively high auxiliary flow (15 mL/min). Although interesting results were obtained exploiting such an approach (Fraga et al., 2000; Prazen et al., 2001), a series of drawbacks limited the applicability of the valve system, namely the great loss in sensitivity, due to the low duty cycle and the temperature constraints because the diaphragm valve (max. operational temperature: 175°C) was situated inside the oven. The latter problem was partially resolved by locating the sensitive parts of the valve outside the oven (Seeley et al., 2000; Sinha et al., 2003). Several other works (Hamilton et al., 2003; Mohler et al., 2006) have reported the employment of valve devices, but the use of such systems has been rather limited.

Bueno and Seeley (2004) proposed a novel differential-flow device, with no diaphragm valve (Fig. 10.7). The system was composed of two sample loops connected through four T-unions. The direction of the primary column effluent, which entered the system at point A, was controlled by an auxiliary flow regulated by a three-port solenoid valve located outside the oven. The flow was directed alternately to one loop and to the other. Using such a configuration no loss of sensitivity occurred since all the primary column effluent was re-injected onto the second column, and the temperature limitations were surpassed.

Seeley and co-workers (2006) modified their first device and developed a simple in-line fluidic modulator consisting of a single collection channel constructed of deactivated fused silica (Fig. 10.8). The 15 cm × 0.45 mm i.d. collection capillary segment was connected to an auxiliary flow through two T-unions, and linked on one side to the primary column outlet and on the other side to the secondary column inlet. The high auxiliary flow was directed in an alternating mode onto the two branches by activation of a solenoid valve generating pulsed injections of the chromatography band collected in the accumulation channel.

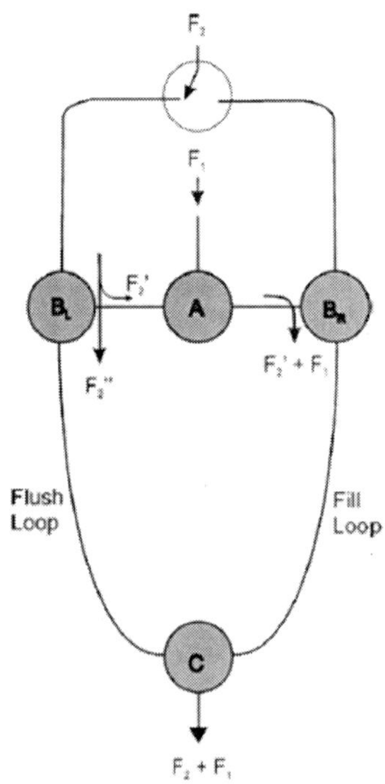

Fig. 10.7. Scheme of the flow modulator developed by Bueno and Seeley. Reproduced from Bueno et al., 2004).

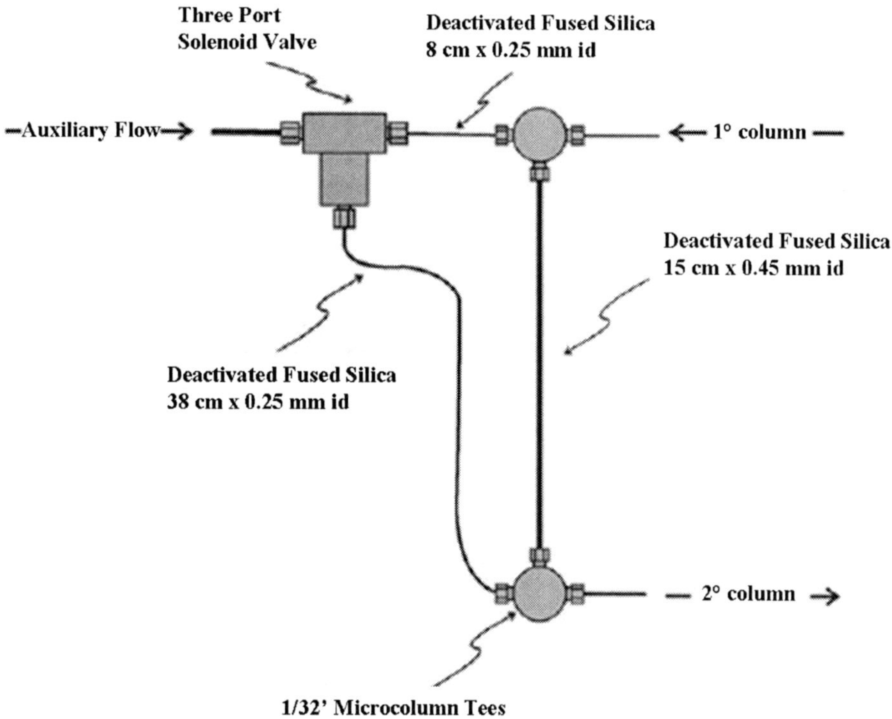

Fig. 10.8. Scheme of the in-line fluidic modulator. Reproduced with permission from (Seeley, 2006).

The in-line fluidic modulator has been incorporated into a metallic plate with internal channels created by photolithographic etching greatly reducing dead volumes. This simple and interesting flow modulator, developed with the support of Agilent Technologies, was the first to be commercialized (Quimby et al., 2007).

Other authors have verified the effectiveness of flow modulation by designing their own lab-made devices (Poliak et al., 2008b; Wang, 2007; Wang, 2008) generally with no significant differences from the principles introduced by Seeley.

Comprehensive 2D GC Method Optimization

The construction of a GC × GC system is a rather simple task because it can be achieved essentially using any commercially available GC instrument by installing a modulator between two different columns. Although the hardware configuration is quite straightforward, method optimization can be a cumbersome issue and is a factor inhibiting the widespread employment of GC × GC. Indeed, solid knowledge of chromatographic basic theory and experience in different branches of the GC field, such as conventional, classical multidimensional, fast micro-bore column, and high-speed mega-bore column low pressure GC are of great help for rapid and effective GC × GC optimization. The main parameters to be considered are modulation parameters (temperature and period), stationary phase chemistry, capillary column dimensions, gas flow, temperature program(s), outlet pressure conditions, and detector settings.

Modulation Parameters
As briefly described at the beginning of the Modulators section, modulation parameters are very important and must be carefully tuned in relation to each specific application and type of modulator. Heat-based modulators will be not considered since they are rarely employed, and flow modulation will only be briefly mentioned.

The first dimension peaks have to be sampled 3–4 (Murphy et al., 1998) times to preserve first-dimension resolution and sensitivity. Other important parameters are the entrapment and the releasing temperatures and periods, all tightly related to the specific cryogenic modulator. However, it is generally accepted that a temperature *circa* 100°C lower than that of the GC oven is sufficient for effective entrapment of most compounds; lower temperatures are needed for very volatile analytes with liquid N_2 often used for this purpose. With regards to the re-injection process, a distinction between oven temperature and hot jet remobilization is necessary. In the former case, the entrapment temperature has to be carefully tuned to avoid a delay in analyte release, particularly for high MW compounds. The re-injection process can be supported by trapping on a segment of uncoated column and/or extending the release time.

An interesting work, focused on the thermal requirements of the loop-type modulator, was reported by Gaines and Frysinger (2004). The authors found that a temperature of as much as 120–140°C lower than the first-dimension elution temperature is optimal for trapping in most applications; lower temperatures cause peak tailing due to incomplete re-mobilization. The hot jet pulse should be at least 40°C higher than the elution temperatures, and the pulse duration should be a compromise between the reduction of analyte *k* values below a specific level and the necessity to avoid breakthrough. An optimum pulse duration of 300 msec was defined by the authors. Although not reported in the aforementioned paper, the length of the delay loop is a further important issue; indeed, if the loop is too short, breakthrough can occur during hot jet operation. While on the other hand, if the loop is too long, then analytes are not sufficiently re-focused at the second cold segment or, even worse, can cause alternate "misses" in second-dimension analysis.

Speaking of flow modulation, the present chapter authors' opinion is that the devices described by Seeley et al. (2006) and by Tranchida et al., at the 7th GC × GC Symposium held in Riva del Garda (Italy) in 2010 (2010a) are characterized by the requirements to obtain a performance similar to that of a cryogenic modulator: i) an accumulation step, ii) a high-pressure injection step (compression), iii) a waste branch to optimize gas flows, iv) a duty cycle equal to one, v) no GC thermal restrictions, vi) a "normal" modulation period (3–6 sec range), vii) independent (differential) flow in each dimension. Method optimization using a differential-flow modulator is even more challenging than thermally-modulated GC × GC. The sampling time and re-injection period have to be carefully tuned as well as first- and second-dimension flows. In fact, the sampling time is highly dependent on the first-dimension gas flow and the collection loop volume, where over-filling must be avoided during the collection step. The re-injection process, which lasts up to 100–200 msec, needs an auxiliary flow at least 20 times higher (10–20 mL/min) than that of the first column (Shellie, 2008). Such high flows generate far-from-ideal linear velocities, reducing the second-dimension efficiency; therefore, to obtain a reasonable separation power the secondary column is rather long (e.g., 5 m).

Stationary Phase Combinations

Knowledge of the mixture to be analyzed is of great help for GC × GC column selection, though this is seldom the case. Natural samples are generally very complex, containing compounds with a wide range of volatilities, different chemical structures and polarities; therefore, stationary phase optimization is often obtained through trial and error testing.

The main aim of GC × GC method development is to maximize the amount of exploitable separation space, which is mainly achieved by choosing a proper stationary phase combination. The separation mechanisms of the two columns must be different to obtain a so-called "orthogonal" separation. Theoretically, considering two completely independent column selectivities and a fully optimized separation, the peak capacity of a GC × GC system should be equal to the product of the peak capacity value relative to each column. However, such a result is an excessive estimation since both of the aforementioned conditions are never fully achieved. Indeed, entirely dissimilar separation mechanisms do not exist since analyte vapour pressures play a major role in all GC separation processes. Furthermore, GC × GC optimization is the result of a compromise between the best conditions for each dimension.

The selection of the column set is related to the primary objectives of a specific application. The most popular and orthogonal combination employs a non-polar column (e.g., 100% dimethylpolysiloxane, 5% diphenyl + 95% dimethylpolysiloxane) as first dimension and a polar (e.g., 100% polyethylene glycol, 50% diphenyl + 50% dimethyl polysiloxane) one in the second dimension. In such a case, the primary column elution order occurs according to increasing boiling points, while the secondary column separation is dependent on specific polarity-based interactions (H-bond, dipole-dipole, dispersion forces, etc.).

Many sample-types are characterized by the presence of homologous series of constituents. If the right columns are selected, chemically-similar compounds can form bi-dimensional patterns which are a very useful tool for peak identification. Using an orthogonal set, the non-polar compounds are located in the lower parts of the 2D chromatogram while the more polar compounds are more retained in the second dimension, and thus are present in the upper part of the contour plot. The GC × GC analysis of fatty acid methyl esters (FAMEs) is a perfect example to illustrate group-type order. Fig. 10.9 shows a complex human plasma fatty acid profile. Homologous compounds are situated in a grid, according to their chemical characteristics. In particular, saturated FAMEs are in the lower part of the 2D chromatogram while an increase in the number of double bonds (DB) in the fatty acid chain increases retention in the second dimension. FAMEs are also separated according to the DB position. Compounds with the DB in the same ω position are aligned in parallel diagonal lines, with the higher ω positions eluting before the lower ones (ω6 FAMEs elute before ω3 ones).

Often, a so called "reversed" set (polar × non-polar) can better satisfy the aims of a specific research objective. Adahchour and co-workers (2004) studied and compared an orthogonal and non-orthogonal set for two different

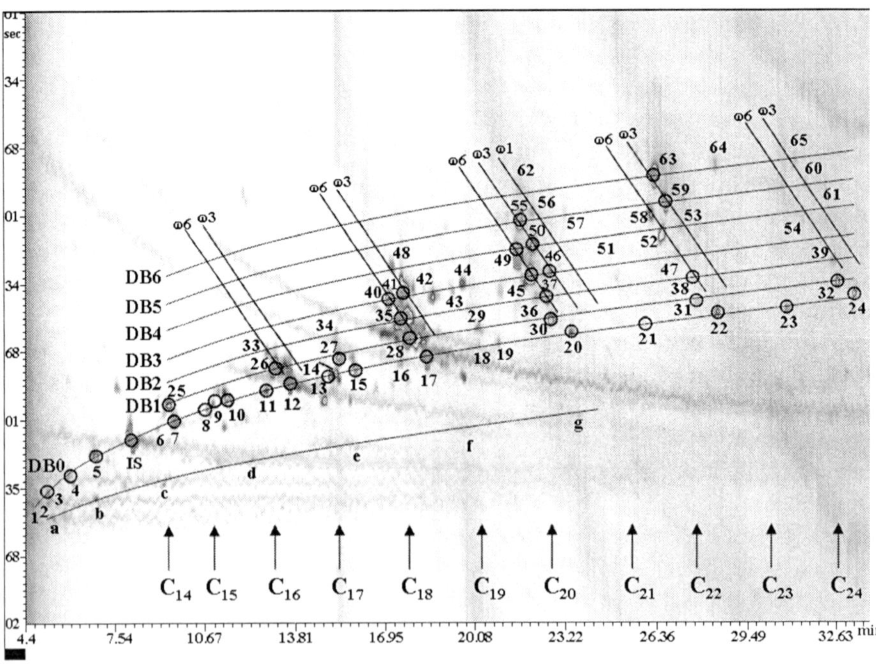

Fig. 10.9. GC × GC analysis of human plasma fatty acids. For peak identification refer to (Tranchida et al., 2008b).

complex matrices, namely diesel oil and food flavors. For the former sample-type, a completely reverse-ordered structure was obtained with the different classes (alkanes, mono-aromatics, di-aromatics, etc.) grouped tightly together giving an advantage when group-type determination is required. When food flavor samples were analyzed, the non-orthogonal approach improved the peak shape of polar compounds such as aliphatic acids and alcohols, which also improved the ordered structure of the chromatogram. Consequently, the conventional and reversed set must both be considered for the determination of target and unknown compounds in complex matrices.

Carrier Gas Linear Velocities

The most employed column set uses a conventional 25–30 m × 0.25 mm ID capillary in the first dimension and a 1–1.5 m × 0.1 mm ID one in the second dimension. Using such a configuration, the full potential of GC × GC is hardly ever expressed, mainly due to the generation of non-optimum gas linear velocities in both dimensions. In thermally modulated systems, the carrier gas velocities in both dimensions are dependent on the primary column inlet pressure. The main consequence is that most GC × GC applications occur at gas velocities near-to-ideal in the first dimension (usually slightly slower) and far higher in the second dimension (Shellie et al., 2004; Beens et al., 2005; Tranchida et al., 2007). The potential gain in efficiency is reduced by such far-from-optimum conditions. Different options exist to overcome the aforementioned drawback: i) a wider-bore ID second column (0.15–0.18 mm ID) could be used, generating lower linear velocities, but it would cause a loss in efficiency; ii) reducing the inlet pressure would generate decreased linear velocities both in the first and second column. This solution would have a negative impact on first dimension resolution (lower column efficiency), as well as on that of the second dimension, since the benefits deriving from a lower linear velocity in the second column are nullified by the higher elution temperatures; iii) a longer secondary column would reduce linear velocities; but the occurrence of wrap-around would increase, requiring extended modulation periods, causing losses in first dimension resolution.

The best solution to the gas flow problem was reported along with the invention of thc technique (Liu & Phillips, 1991; Venkatramani & Phillips, 1993), but it later disappeared in further GC × GC investigations until Tranchida and co-workers (2007) described a method defined as split-flow GC × GC in 2007. In those experiments, about 30% of the primary column flow was diverted to waste by connecting the two analytical columns to a short capillary segment. The latter was connected to a manually-operated needle valve to split the effluent from the primary column and hence regulating gas flows in both dimensions (Fig. 10.10).

Tranchida and co-workers (2007a) used an LMCS and a single GC oven, equipped with a flame ionization detector (FID). Initially, cod liver oil fatty

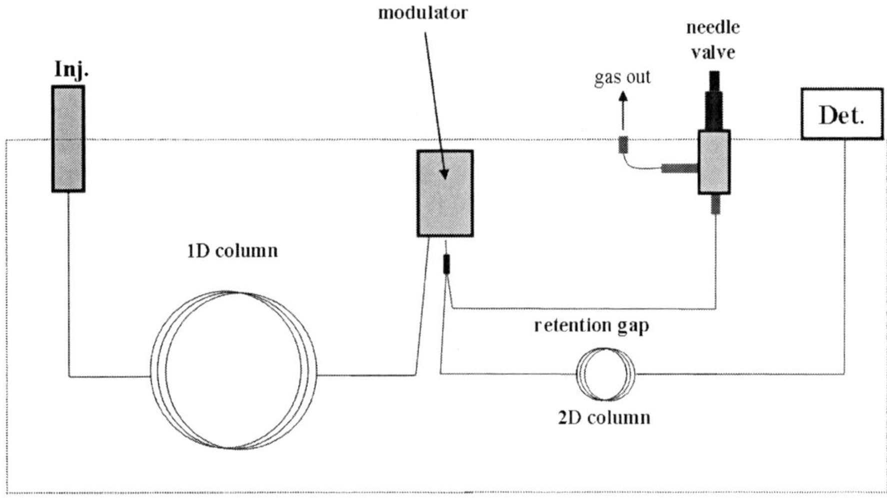

Fig. 10.10. General split-flow comprehensive 2D GC scheme.

acid methyl esters were analyzed with the split-valve closed, that is, using traditional GC × GC. A hydrogen head pressure of 194.9 kPa was applied which generated a linear velocity of about 35 cm/s and 333 cm/s in the first and second dimension, respectively. A well-structured chromatogram was obtained under such conditions, though a large amount of 2D space remained unoccupied. By employing the split-flow system, the best results were obtained using 35 and 213 cm/s in the first and second dimension, respectively. Occupation of the 2D space underwent an increase of about 32%, while resolution was enhanced by +50% for many FAMEs. It must be added that second dimension overloading is also avoided by using a split-flow configuration.

Other split-flow studies have been carried out by the same research group to confirm the effectiveness of the approach: a series of experiments were carried out using a loop-type dual-stage modulator and a double GC oven coupled with FID or MS detection (Tranchida et al., 2009a; Tranchida et al., 2009b; Tranchida et al., 2010b; Purcaro et al., 2010). In particular, the potential of a 50 μm ID capillary column was compared to a much more popular 100 μm ID one (Tranchida et al., 2009b). An apolar 30 m × 0.25 mm ID column was connected through a Y union to a 1 m × 0.05 mm ID polar one and to a 20 cm × 0.05 mm ID uncoated column segment, the latter linked to a manually-operated split valve. The 0.25 mm + 0.05 mm ID column combination operates poorly without flow splitting because excessively high gas linear velocities are generated in the second dimension. Furthermore, overloading of the secondary column can occur easily. The best operational conditions were obtained with the split valve completely opened; 83.4% of the effluent was diverted to the waste, obtaining linear velocities of *circa* 20 and 110 cm/s in the first and second dimension, respectively. The results obtained using this configuration were compared with those derived by employing an optimized system using the same first column, connected to a 1 m × 0.1 mm ID polar one and to a 30 cm × 0.1 mm ID uncoated segment. Fig. 10.11 compares expansions of a 2D chromatogram of diesel obtained employing the two sets. Separation power greatly increased by using a 50 μm ID capillary secondary column, and resolution almost doubled compared to a 100 μm ID capillary.

Detectors

The performance of the detector cannot be overlooked if an effective chromatography analysis is desired. In particular, the enhanced resolution obtained with the GC × GC must not to be impaired by improper detector coupling. Modulated peaks have baseline widths generally in the 100–600 ms range (even less in some applications), therefore dedicated detectors are required. A suitable GC × GC detector has to be characterized by: i) negligible internal volumes to avoid band broadening of the narrow chromatographic peaks, ii) rapid rise times, iii) fast acquisition rates for reliable peak reconstruction and to maintain resolution. The minimum number of data points per peak necessary for a correct reconstruction has been widely discussed in the literature, and a variety of opinions have been reported (Adahchour et al., 2005; Holland et al, 1983; Hinshaw, 2003; Dyson, 1999; Poole, 2003). For example, Hinshaw (2003) stated that 10 data points across the width at half height are required, while Dyson (1999) concluded that at least 15–20 data points/peak are needed. Dallüge and co-workers (2002) and Adahchour and co-workers (2005) are less restrictive, concluding that 5–6 and 7 data points per peak, respectively, are sufficient.

A brief discussion about the most commonly employed detectors in the GC × GC field will follow. For a thorough discussion on this topic the reader can refer to the recent review by von Mühlen and co-workers (2006).

The most commonly employed detector in GC × GC has been the flame ionization detector which is capable of very high acquisition rates easily above 100 Hz. The FID is a universal detector with a response proportional to the compound's carbon content, which is suitable for most quantitative applications. For instance, in petrochemical applications the FID is the most suitable detector (Phillips & Xu, 1995) since such samples are formed of far too many compounds that are not available as pure standards for quantification purposes using other detectors such as the mass spectrometer (MS). Indeed, the FID can be easily employed even for identification when group-type 2D chromatograms are obtained. The group type is assigned according to retention time correspondence with even one standard, and then FID quantification can easily be done. Biedermann and Grob (2009) exploited GC × GC-FID and group-type pattern formation for the quantification of mineral oil constituents

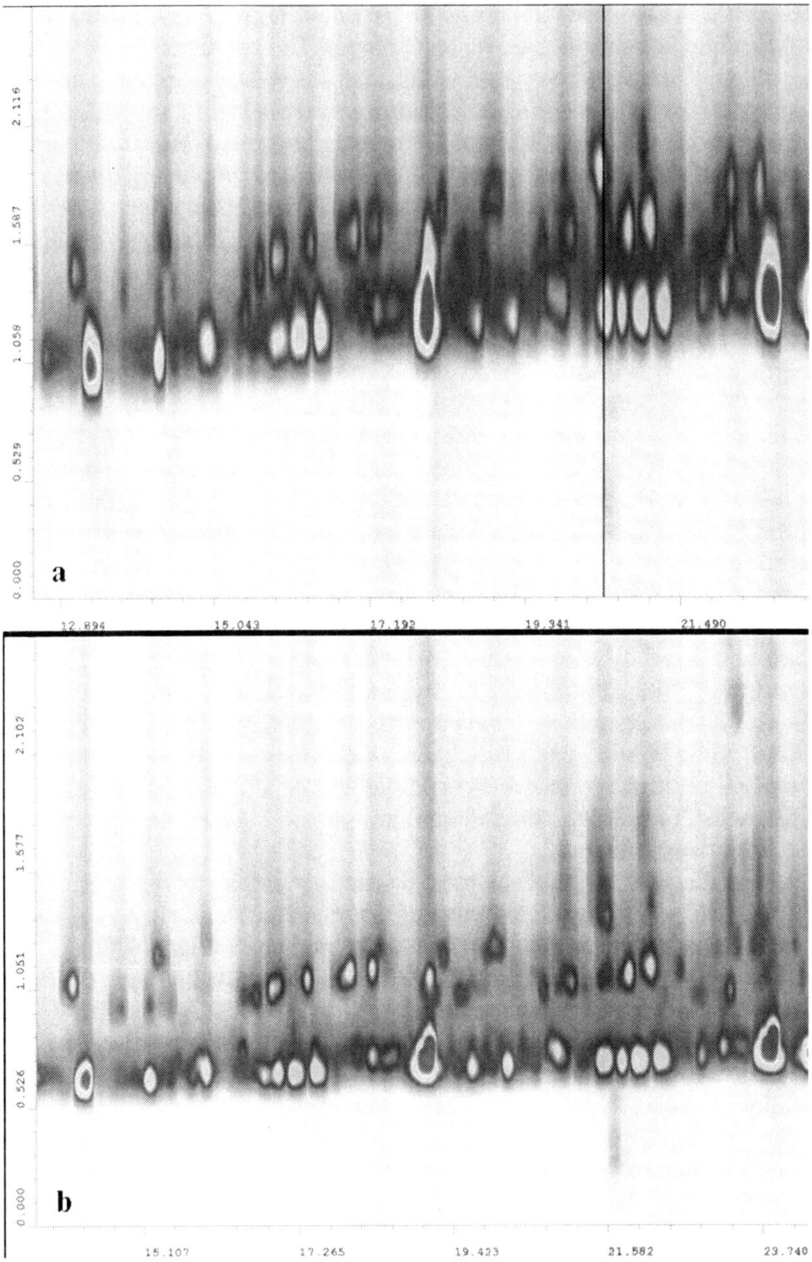

Fig. 10.11. Expansion of the same bidimensional analyte elution range derived from a diesel sample analysed using the same 1D column and a (a) 0.1 mm ID and (b) 0.5 mm ID 2D column. Reproduced with permission from (Tranchida et al., 2009b).

contained in contaminated sunflower oil. GC × GC- FID has been employed in several fields such as environmental analysis (Reddy et al., 2002; Frysinger et al., 2003; Slater et al., 2005), industrial applications (Vendeuvre et al., 2004; Vendeuvre et al., 2005), and food analysis (Tranchida et al., 2008a; Vlaeminck et al., 2007).

More selective detectors are useful in trace analysis since sensitivity can be greatly enhanced. The electron capture detector (ECD) is characterized by a high sensitivity for organic molecules containing electronegative functional groups (halogens, phosphorous and nitro-groups). The main concern in coupling such a detector has been related to a rather high internal volume which can cause band broadening. Recently, Agilent Technologies has presented a newly-designed ECD, the micro-ECD (μ-ECD), with a miniaturized internal volume (150 μL). The μ-ECD responds more rapidly to peak flux change and is compatible with reduced column flows. The performance of the μ-ECD was evaluated by Klee and co-workers (1999) in a monodimensional GC system.

Internal dead-volumes play an important but not exclusive role in band broadening. In fact, careful optimization of the GC × GC system is also fundamental, as demonstrated by Kristenson and co-workers (2003), who evaluated and compared the performance of six modulator types (sweeper, LMCS, quad-jet, dual-jet CO_2, semi-rotating cryo, loop) coupled with three commercially-available ECDs. The detectors were characterized by extremely different internal volumes, namely 150 μL, 450 μL, and 1.5 mL for the Agilent, Thermo Finnigan, and Shimadzu ECD, respectively. The authors demonstrated that peak broadening can be greatly limited by adjusting the make-up flow properly. Several applications have been published using GC × GC-ECD to improve detection limits in the trace analysis of organochlorine (OCs), organophosphorus (OPs), and polychlorinated biphenyl (PCBs) contaminants (Zrostlikova et al., 2002; Quan et al., 2002; Conka et al., 2005; Aybar et al., 2005; Gomara et al., 2002; Kusakabe et al., 2003).

The nitrogen-phosphorous detector (NPD) is structurally similar to the FID except for a ceramic bead above the jet containing an alkali metal salt which catalyses the electron transfer to nitrogen and phosphorus compounds, forming negative ions in a flameless gaseous environment. Ions are then directed to a collector electrode. NPD temperature and gas flow modification can reduce the peak tailing that often occurs due to the temporary adsorption of decomposition products on the ceramic bead (Poole, 2003). The NPD has been used for pesticides analysis in GC × GC (Khummueng et al., 2006) and in gas-oil samples (von Muehlen et al., 2007).

The sulphur chemiluminescence (SCD) and nitrogen-chemiluminescence detector (NCD) are S-and N-compound selective detectors. Oxidation and combustion, under hydrogen-rich conditions, generate the corresponding oxide compound, which reacts with ozone to form an electronically excited dioxide derivative. The latter generates light energy when returning back to ground state, SO_2 in the blue region of the spectrum and NO_2 in the red and infra-red regions. Blomberg and co-workers (2004) analyzed dibenzothiophene using SCD and FID simultaneously. The authors found that system electronics, rather than the detector dead volume, caused band broadening. Once the problem was resolved by using a modified electrometer, sulphur compounds were analyzed in a series of fuel products. The NCD has been employed to characterize N-compounds in a diesel sample, specifically carbazoles and indoles (Wang et al., 2004).

The atomic emission detector (AED) is a multi-element system with the possibility to measure up to 23 elements. The capillary GC column ends in a microwave-induced helium plasma chamber where all the elements contained in a sample are atomized and excited. Characteristic atomic emission spectra pass through a diffraction grating and are detected by a photodiode array. Although the AED is characterized by a rather limited acquisition frequency, namely 10 Hz, it was coupled to a GC × GC system for pesticides analysis (van Stee et al., 2003). To meet the low sampling rate of the detector, and with the aim of obtaining quantitative data, the authors induced band broadening by connecting a 0.7 m × 0.25 mm ID uncoated column to the end of the 0.6 m × 0.10 mm ID polar second dimension.

A separate discussion must be provided regarding mass spectrometry, the most informative detection system, considered as an additional third dimension by GC × GC users. The coupling of an MS instrument to a GC × GC system has been highly desirable since the birth of comprehensive 2D GC to add structural information to the great separation power. The first attempt was made by Frysinger and Gaines (1999), who used a quadrupole MS with a very low acquisition speed (2.43 scan/s). The authors intentionally broadened the 2D-peaks to a minimum 1-s baseline width by using a thick second dimension stationary phase and by slowing down the heating rate (0.5°C/min). Under such conditions, less than 3 data points per peak were obtained. The authors concluded that a time-of-flight (ToF)-MS would be the ideal solution to satisfy the requirements of GC × GC narrow peaks. A year later, van Deursen and co-workers (2000) described the first GC × GC-ToF-MS application, focused on kerosene analysis. The ToF-MS process generates a complete spectrum from every ion pulse from the ion source. Up to 500 spectra/s can be produced, thus a huge quantity of information needs collection. However, to optimize the acquisition rate of a ToF-MS method, a compromise among different parameters must be made, namely i) data points per peak, which have to be sufficient for a reliable peak reconstruction; ii) sensitivity, which decreases with an increase of the spectral acquisition rate; iii) data file size, because higher acquisition rates generate large data files, handled with difficulty by normal computers. Consequently, in GC × GC-ToF-MS experiments, an acquisition rate of 50 Hz is a frequent compromise. The acquisition of complete spectra and the absence of peak skewing enable the exploitation of deconvolution, a useful tool to unravel peak overlapping. Besides these prerequisites, slight resolution (at least 5 ms) and a sufficient number of data points (at least 4) are required to deconvolute peaks. The main software employed for deconvolution purposes are the LECO ChromaTOF and AMDIS.

Although ToF-MS fully satisfies GC × GC requirements, researchers and instrument developers have made substantial efforts to use quadrupole (q) MS, by carefully tuning the qMS parameters in relation to the specific GC × GC application, and by improving the scan cycle, respectively. In a quadrupole analyzer, the scan data acquisition rate (Hz) is defined as 1000 divided by the scan time (ms) and the inter-scan delay (ms) (Eq.1), where the scan time depends on the mass range (set by the operator) and on the *q* scan speed (instrument parameter) (Eq. 2).

$$\text{Data acquisition rate (Hz)} = \frac{1000}{\text{scan time (ms)} + \text{inter-scan delay (ms)}} \qquad \text{(Eq. 1)}$$

$$\text{Scan time (s)} = \frac{\text{Mass range scanned (amu)}}{\text{Scan speed (amu/s)}} \qquad \text{(Eq. 2)}$$

The inter-scan delay is defined as the time required to reset to the initial voltage, prior to the next scan. Figure 10.12 shows a graph illustrating the quadrupole MS acquisition process.

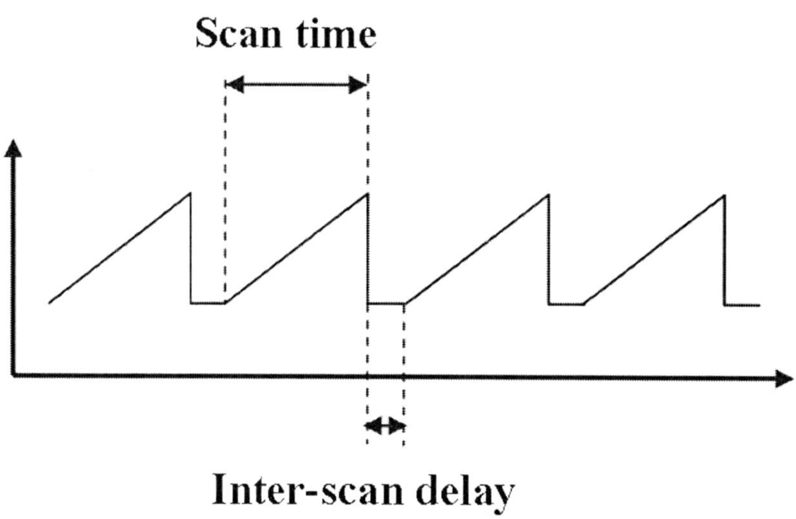

Fig. 10.12. Graph illustrating the qMS acquisition process.

Several authors tried to push the performance of the qMS by playing on the scanned mass range (Shellie et al., 2003a; Shellie & Marriott, 2003; Kallio et al., 2003; Cordero et al., 2007; Mondello et al., 2005). Shellie and co-workers (Shellie et al., 2003a; Shellie & Marriott, 2003) attained between 18.5 and 20 scan/s, using a 4000 amu/s qMS scan speed and by reducing the mass range, in essential oil and environmental applications. Cordero and co-workers (2007), although using a qMS with a high scan speed (11,111 amu/s) in allergens analysis, had to reduce the mass range scanned (40–240 m/z), to achieve 18.52 scan/s, due to the relatively high inter-scan delay (30 ms). Mondello and co-workers (2005) analysed a commercial perfume, using GC × GC in combination with a rapid-scanning qMS (10,000 amu/s). Thanks to the relatively low inter-scan delay (calculated to be 14 msec), an acquisition frequency of 20 Hz was achieved using a "normal" mass range (40–400 *m/z*). Considering its limitations, the qMS has been considered suitable only for qualitative purposes. The selected ion monitoring (SIM) mode has been often used to meet the requirements for quantitative analysis, since higher acquisition rates can be attained. In the SIM mode, acquisition speed is composed of the dwell time, which is the time spent acquiring the specific ion, and the inter-channel delay, the time to set the voltage for the next acquisition level.

Recently, the performance of a novel very-rapid qMS, with a 20,000 amu/s scan speed and a very low inter-scan delay (5 ms), has been reported (Purcaro et al., 2010a). Accurate GC × GC-MS quantification of perfume allergens was described: an average of more than 15 data points per peak were obtained, through the application of a 50 Hz acquisition rate and a 40–340 *m/z* mass range. The availability of such a very rapid qMS represents an important step towards the expansion of quantification methods in the GC × GC field.

Applications

GC × GC techniques have been applied successfully in many areas. Petrochemical samples, among the most complex sample types known to analytical chemists, have been the most widely studied by using GC × GC. Fuels are characterized by several fractions, mainly hydrocarbons and heteroatom constituents (containing S, N, O, etc.). The former are subdivided into different groups, namely paraffins, isoparaffins, naphthanes, olefins, and aromatics. These petrochemical constituents form bidimensional group-type patterns, generating well-structured GC × GC chromatograms. In many cases, emphasis has been directed to the characterization of a specific group type, (Blomberg et al., 1997; Gaines et al., 1999; Schoenmakers et al., 2000;), though target compounds (Frysinger et al., 1999; Frysinger & Gaines, 2000), and biomarkers (Frysinger & Gaines, 2001) have been investigated as well.

Different sources of environmental and food contamination have been extensively studied by GC × GC. Oil spill profiles, and related weathering patterns, were characterized by Diez and co-workers (2007), while Arey et al., (2007) identified chemical markers deriving from weathered oils to easily assess the source and the occurrence of contamination in wildlife. Dioxin and furan contamination, in crude fly ash and flue gas from a municipal waste incinerator, were studied by using a GC × GC-HR-ToF-MS (Shunji et al., 2008). Other pollutants have been analyzed, such as target contaminants in waste water (Semard et al., 2008;), ambient volatile organic compounds (VOCs) (Bartenbach et al., 2007; Vogt et al., 2007), polycyclic aromatic hydrocarbons (PAHs) in atmospheric particulates (Amador-Munoz et al., 2008; Ochiai et al., 2007) and edible oils (Purcaro et al., 2007).

Recently, the number of GC × GC papers dealing with biological and biota samples has increased. Different sample types, such as blood, urine, human breath, mouse organic tissue, yeast, plants and bacteria, have been studied in relation to drugs, organic acid metabolites, VOCs, metabolomics, polybrominated diphenyl ethers (PBDEs), etc. (Song et al., 2004a; Song et al., 2004b; Pierce et al., 2006; Sanchez et al., 2006; Libardoni et al., 2006; Welthagen et al., 2005; Shellie et al., 2005; Mohler et al., 2006b; Robrock et al., 2008). In a GC × GC-ToFMS investigation focused on the analysis of spleen extract of obese and lean mice, about 538 peaks and more than 1220 peaks were counted in the monodimensional and bidimensional chromatograms, respectively, after deconvolution and removal of known artifacts (e.g., polysiloxane, phthalates or column bleed) (Welthagen et al., 2005).

GC × GC has also proved to be a useful tool to unravel the complexity of foods and fragrances, in particular coupled to an MS detector. Zhu and co-workers (2007) identified more than 500 compounds in Chinese liquor, using ToF-MS spectra combined with structured 2D chromatograms and retention indices. Thorough information on volatile and semi-volatile components, contained in different food samples, such as coffee beans, yerba mate tea, wine, beer, and honey, has been reported (Mondello et al., 2004; Purcaro et al., 2009; Mayadunne et al., 2005; Ryan et al., 2004; Ryan et al., 2005); essential oil profiles (Ma et al., 2007; Shellie et al., 2004b) and allergens in fragrance products (Cordero et al., 2007) have also been widely investigated.

The number of applications has been continuously increasing during the last 20 years, and several GC × GC reviews have been published. A brief *excursus* of the possible fields of GC × GC application has been herein reported; for more thorough information, a series of recent reviews can be consulted (Adahchour et al., 2006b; Adahchour et al., 2006c, Adahchour et al., 2008; Tranchida et al., 2004; Cortes et al., 2009).

Lipids Analysis

Lipids are a large class of organic compounds, essential for the structure and function of living cells. Their main biological functions are energy storage, structural components of cell membranes, and important signalling molecules. Lipids, as fats and oils, provide an important part of the human dietary intake, both as natural constituents of foods and as ingredients. They are an important source of vitamins and they greatly affect the physical structure of foods and the solubility of taste and aroma constituents (Christie, 2003; Gurr et al., 2002). Lipids comprise a broad range of structures, characterized by non-polar and hydrophobic skeletons, but some of them present a polar or hydrophilic group, giving them an amphiphilic character. A general classification divides lipids into saponifiable and non-saponifiable fractions. The saponifiable compounds are fatty acid derivatives, mainly acylglycerols (tri-, di-, and monoacylglycerols), phospholipids, glycolipids, waxes, and sterol esters. The non-saponifiable constituents include mainly free fatty acids (FAs), isoprenoid lipids (steroids, carotenoids, monoterpenes), and tocopherols.

Comprehensive 2D GC is certainly a valuable tool for lipid analysis, considering the complexity of many samples, and the high importance of thoroughly investigating fat and oil profiles. GC × GC applications on the main classes of lipids are summarized in Table 10.B and will be hereafter discussed.

Fatty Acid Analysis

Natural-occurring fatty acids, in general, are straight chains containing (usually) an even number of carbon atoms, with a carboxyl group at one extremity. The carbon chain, usually within the C_4-C_{24} length, can be totally saturated or with double bonds (generally in the 1–6 range), mainly in the *cis* (or *Z*) configuration. Occasionally, higher carbon numbers can be found in specific sample types, such as microorganisms or algae. In the former, odd C-number chains can also be present, as well as a great variety of functional groups, such as *trans* double bonds, acetylenic bonds, epoxyl, hydroxyl, keto, or ether groups, and cyclopropene, cyclopropane, and cyclopentane rings. Hydroxy fatty acids are synthesized in some animal tissues as well, but usually derive from food processing (Christie, 2003). In order to analyze these lipid constituents, it is usually necessary to pre-separate a specific lipid class, and then transform the fatty acids into less polar and more volatile analytes, mainly methyl esters (FAMEs).

Conventional GC, with polyethylene glycol-type stationary phases, has been widely employed to separate the most important FAs. However, to fully resolve geometrical and conjugated FA isomers, rather long (up to 100 m) and high-polarity columns are required, such as the bis-cyanopropyl polysiloxane phase (Christie, 2003) or the novel ionic liquid phase (Ragonese et al., 2009).

The application of GC × GC to the analysis of FAs has highlighted the unsuspected complexity of some of these sample-types. Furthermore, the formation of GC × GC chemical class patterns is a very helpful tool for the assignment of FA "unknowns". In fact, although the use of the MS detector with the support of linear retention indices (LRI) has proven useful for peak identification, the confirmation of FA identity can still be a cumbersome task, due to the very similar mass spectra of the isomers. Finally, the increase of analytical sensitivity has revealed the presence of minor components, such as odd carbon-number FAs.

Table 10.B. GC × GC Applications in the Lipidic Field.

Compounds	Detector	1D column	2D column	Observation	Reference
FAMEs in edible oils	FID	HP-1 9m x 0.2mm ID x 0.33 μm	CP-wax-52 0.3m x 0.1mm ID x 0.2μm	First FAMEs separation in GC x GC	de Geus et al., 2001
FAMEs in natural fats and oils	FID	BPX-5MS 30m x 0.25mm ID x 0.25 μm Supelcowax-10 30m x 0.25mm ID x 0.25 μm	Supelcowax-10 1m x 0.1mm ID x 0.1μm SPB-5 1m x 0.1mm ID x 0.1μm	Use of orthogonal and non-orthogonal set	Mondelloet al., 2003
FAMEs in olive oils	FID	CPSil 8 10m x 0.25 mm ID x 0.25 μm BP20 wax 25m x 0.32 mm ID x 0.25 μm	BP20 wax 1m x 0.1 mm ID x 0.1 μm BPX35 1m x 0.1 mm ID x 0.1 μm	Twin-GCxGC splitted into orthogonl and non-orthogonal sets	Adahchouret al., 2005b
FAMEs in olive oil	FID	SP-2560 100m x 0.25mm ID x 0.20 μm	Equity-1 0.9m x 0.1mm ID x 0.1μm	Identification of odd-C number FAMEs	Tranchida et al., 2008a
FAMEs in milk	FID	CP 7420 100m x 0.25mm ID x 0.25 μm CP 7420 100m x 0.25mm ID x 0.25 μm HP-1 30m x 0.25mm ID x 0.25 μm HP-1 30m x 0.25mm ID x 0.25 μm	HP-5MS 1.5m x 0.1mm ID x 0.1μm HP-1 1m x 0.1mm ID x 0.1μm cyano col 1m x 0.1mm ID x 0.1μm DB-wax 1m x 0.1mm ID x 0.1μm	Four different sets were studied	Hyötyläinen et al., 2004
FAMEs in milk	FID	BPX-5MS 30m x 0.25mm ID x 0.25 μm BPX-80 30m x 0.25mm ID x 0.25 μm	BP-20 0.85m x 0.1mm ID x 0.2μm BP-35 0.25m x 0.1mm ID x 0.1μm	Identified 27 FAMEs	Vlaeminck et all., 2007
FAMEs in animal fat	ToF-MS	SLB-5MS 10m x 0.1mm ID x 0.1 μm	DB-wax 0.5m x 0.1mm ID x 0.1μm	Fast separation	Chinet al., 2009
FAMEs in bovine fats and milk	FID	SLB-IL100 60m x 0.2mm ID x 0.2 μm	BPX-50 3m x 0.1mm ID x 0.1 μm	Study of the cis/trans separation on five different sets	Villegas et al., 2010
FAMEs in animal fats	ToF-MS	DB-5ms 30m x 0.25mm ID x 0.25 μm	DB-wax 1m x 0.1mm ID x 0.1μm	Identified 51 FAMEs	Indrasti et al., 2010b
FAMEs in anchovy's brain and vegetable oil	qMS	ZB-5 30m x 0.25mm ID x 0.25 μm	ZB-50 1m x 0.1mm ID x 0.1μm		Lucci et al., 2009
FAMEs in cod liver oil	FID	BPX5 30m x 0.25mm ID x 0.25 μm	BP20 0.8m x 0.1mm ID x 0.1 μm	Study of sample dimensionality	Harynuk et al., 2006
FAMEs in cod liver oil	FID	Equity-5 30m x 0.25mm ID x 0.25 μm	Supelcowax-10 1m x 0.25mm ID x 0.25μm	GC x GC separation used to support peak identification in fast GC.	Mondello et al., 2006
FAMEs in cod liver oil	FID	Equity-5 30m x 0.25mm ID x 0.25 μm	Supelcowax-10 1m x 0.25mm ID x 0.25μm	First use of the split-flow system	Tranchida et al., 2007a
FAMEs in marine oil	FID	BPX-5 30m x 0.25mm ID x 0.25 μm BP 1 25m x 0.22mm ID x 0.2 μm	BP 20 0.8m x 0.1mm ID x 0.1 μm BPX 70 1m x 0.1mm ID x 0.2 μm	Study of isomers separation in two sets of column	Western et al., 2002
FAMEs in human plasma	FID	Equity-1 30m x 0.25mm ID x 0.25 μm	Supelcowax-10 0.95m x 0.1mm ID x 0.1μm	Identified 65 FAMEs.	Tranchida et al., 2008b
FAMEs in biota	ToF-MS	HP-5MS 30m x 0.25mm ID x 0.25 μm	BPX 50 1m x 0.1mm ID x 0.1 μm	Direct-thermal-desorption (DTD) interface was employed to inject whole cell suspension	Akoto et al., 2008
Bacteria FAs (BAMEs)	FID	HP-5MS 30m x 0.25mm ID x 0.25 μm	BPX 70 1m x 0.1mm ID x 0.1 μm	The method was compared to a reference one for taxonomic identification (Sherlock MIDI)	David et al., 2008

(continued on next page)

Table 10.B. *(continued)*

Compounds	Detector	1D column	2D column	Observation	Reference
Bacteria FAs (BAMEs)	FID, qMS	HP-5MS 30m x 0.25mm ID x 0.25 μm	BPX 70 4m x 0.25mm ID x 0.25μm	A first commercialized flow modulator was employed. After the second dimension separation the flow was splitted between an FID and a MS detector.	Gu et al., 2010
Bacteria FAs (BAMEs)	qMS	SLB-5MS 11.4m x 0.1mm ID x 0.1 μm	Supelcowax-10 1m x 0.05mm ID x 0.05μm	Two dedicated libraries were built, namely a GC x GC BAMEs library and Bacteria library.	Purcaro et al., 2010b
FAMEs in motor oil	FID, ToF-MS	Rtx-1ms 19m x 0.2mm x 0.2 μm	BPX-50 1m x 0.1mm ID x 0.1μm	Detect leakage of biodisel into motor oil	Peacock et al., 2010
FAMEs in biodisel	FID, ToF-MS	Solgel wax 30m x 0.25mm ID x 0.25 μm	DB1 1m x 0.1 mm ID x 0.1 μm	Several type of first and second dimension stationary phases were compared to maximize the FAs and alkanes resolution	Adam et al., 2008
FAMEs in biodiesel	FID	BPX-5 15m x 0.25mm ID x 0.25 μm	BP 20 1m x 0.1mm ID x 0.1 μm	Useful method to differentiate the origin and type of biodiesel.	Tiyapongpattana et al., 2008
Mono- and di-glyceride in fats and oils	ToF-MS	DB17ht 6m x 0.1mm ID x 0.1 μm	SLB-5MS 0.6m x 0.1mm ID x 0.1 μm		Indrasti et al., 2010a
TAGs edible oils and fats	FID	CP-wax 30m x 0.25mm ID x 0.25 μm	VF-23 3m x 0.1mm ID x 0.1 μm	TAGs previously separated in Ag-LC	de Koning et al., 2006
waxs in lanolin	ToF-MS	XTI-5 10m x 0.25mm ID x 0.25 μm	BPX-50 1m x 0.1mm ID x 0.1μm	Five different set were tested	Jover et al., 2005
wax in edible oils	FID	PS-255 20m x 0.25mm ID x 0.12 μm	SOP-50 1.5m x 0.15mm ID x 0.075μm	Use of a LC preseparation	Biedermann et al., 2008
Sterols in lip mussels and cow faecal mixtures	FID	BPX-5 30m x 0.25mm ID x 0.25 μm	BP 50 2m x 0.1mm ID x 0.1 μm	Separation of critical couples	Truong et al., 2003
steroids (standards)	IRMS	DB-5ms 30m x 0.25mm ID x 0.25 μm	DB-17 1m x 0.1mm ID x 0.1μm	Optimization of the IRMS coupling	Tobias et al., 2008
Steroids in urine	FID, ToF-MS	BPX-5 30m x 0.25mm ID x 0.25 μm BPX-50 30m x 0.25mm ID x 0.25 μm	BPX 50 1m x 0.1mm ID x 0.1 μm BPX 5 1m x 0.1mm ID x 0.1 μm	Highlighted the necessity of dedicated ToF-MS library	Mitrevski et al., 2008
Steroids in urine	FID	BPX-5 18m x 0.18mm ID	BPX 50 1m x 0.1mm ID	Stationary phase thickness not reported	Mitrevski et al., 2007
Steroids in urine	FID, ToF-MS	BPX-5 30m x 0.25mm ID x 0.25 μm	BPX 50 1m x 0.1mm ID x 0.2 μm	Six anabolic agents listed by World Anti-Doping Agency (WADA) were investigated.	Mitrevski et al., 2010
Steroids in urine	ToF-MS	HP-1 17m x 0.2mm ID x 0.11 μm	OV-1701 1m x 0.1mm ID x 0.1μm	Method validated for five steroid-doping compounds	Silva et al., 2009
Steroids in nutritional supplements	ToF-MS	DB-5ms 30m x 0.25mm ID x 0.25 μm	BPX-50 2m x 0.1mm ID x 0.1μm	Twenty-five steroids were analysed	Stepan et al., 2008

FAMEs in Food Samples

The first FA investigation in the GC × GC field was carried out in 2001 by de Geus and co-workers (2001), using a thermal sweeper (5-s modulation period), a 9.0 m × 0.2 mm ID × 0.33 μm HP-1 (dimethylpolysiloxane) column in the first dimension and a 0.30 m × 0.1 mm ID × 0.2 μm CP-WAX-52 (polyethylene glycol) as the second column. The FA profiles of several vegetable and fish oil samples were visualized using an FID detector. The formation of chemical class patterns was highlighted by a series of curves drawn on the 2D plot, linking compounds with the same number of double bonds. A series of peaks were tentatively identified (marked with a question mark in Fig. 10.13) according to their specific position in the 2D chromatogram. Furthermore, sensitivity enhancement enabled the identification of several odd carbon-number FAs. Fig. 10.13 shows the 2D chromatogram of a herring oil sample.

Marriott and co-workers (Western et al., 2002) confirmed the advantages of GC × GC in FA analysis, by employing a cryogenic modulator, namely the LMCS, and testing two sets of columns: BPX5 (5% phenyl polysilphenylene siloxane phase) × BP20 (polyethylene glycol phase) and BP1 (100% polydimethylsiloxane) × BPX70 (cyanopropyl polysilphenylene siloxane equivalent to 70% cyanopropyl siloxane). With the latter set, about 70 peaks were separated; among these, 49 peaks were identified by exploiting group-pattern formation. Harynuk and co-workers (2006) studied what they called "sample dimensionality", in the case of FAMEs. It is noteworthy that, in the case of FAME analysis, the universally accepted Kovats' linear retention index (LRI) is often replaced by analogous parameters known as equivalent chain length (ECL) and fractional chain-length (FCL). The former is calculated by plotting the logarithms of the retention times of homologues series of straight-chain saturated FAMEs *vs.* the number of carbon atoms in the aliphatic chain. The FCL is simply the fractional increment in the ECL value, which is dependent on the number and position of double bonds and the presence of substituent groups (methyl branches, ring system and oxygenated moieties). In GC × GC separations, group pattern formation has been usually discussed in terms of primary and secondary retention times (${}^1t_R$ and ${}^2t_R$), though using such parameters, the relationship is strictly limited to the specific instrumental set-up. Therefore, the authors explored the relationship between "sample dimensionality" and "separation dimensionality". Dimensionality is defined by Giddings as "the number of independent variables that must be specified to identify the components of the sample" (Giddings, 1995). Thus "sample dimensionality" comprises chemical characteristics, such as carbon-chain length, number of double bonds, branch position, functional groups, etc. "Separation dimensionality", on the other hand, is the number of different, isolated separation mechanisms to which the sample is subjected. To achieve a better understanding of the FAMEs analyzed, the plot of ${}^1t_R$ *vs.* ${}^2t_R$ was compared with several other plots using basic chemical information, such as ${}^2k$ *vs.* the chain length, ${}^2t_R$ *vs.* chain length,

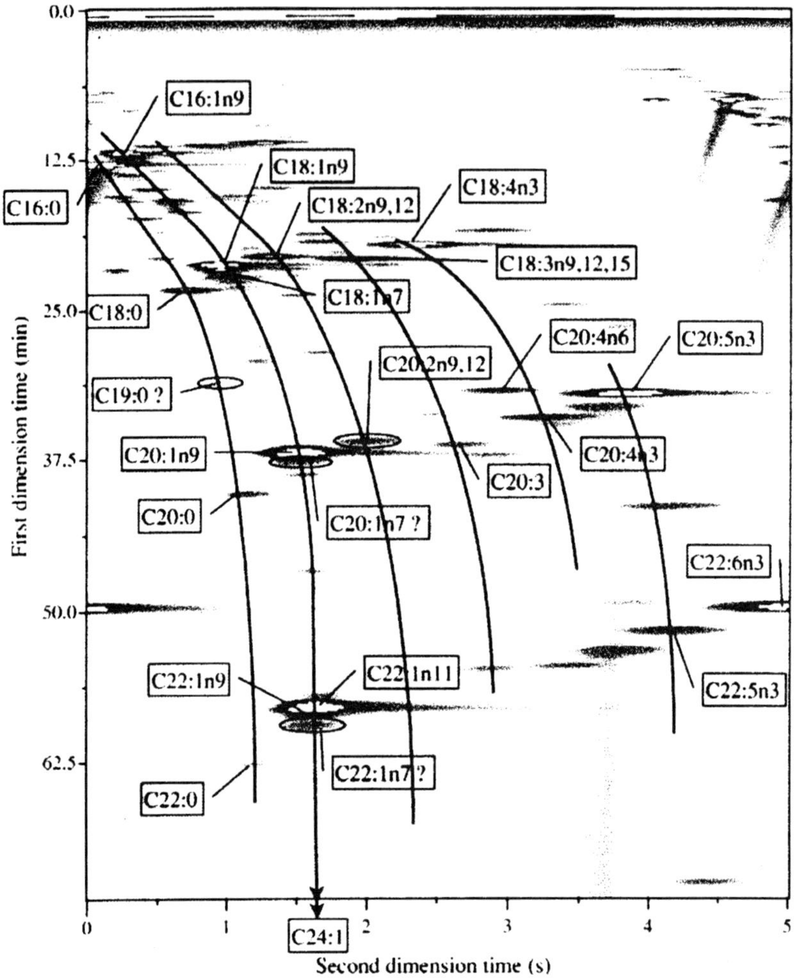

Fig. 10.13. GC × GC-FID chromatogram of herring oil FAMEs. (Reproduced with permission from de Geus et al., 2001.)

and FCL *vs.* number of double bonds. Such information can be used to predict the separation of a known sample in the 2D space and to extract basic chemical information of "unknowns" from the GC × GC chromatogram.

Vlaeminck and co-workers (2007) studied the effects of a cow diet, supplemented with marine algae, on the milk FA profile; the amount and the number of unsaturated FAs greatly increased. The resolution power of GC × GC appeared to be insufficient for a series of fatty acids, such as *trans-* and *cis-* $C_{18:1}$

isomers. The authors suggested the use of a longer and more selective first dimension column to optimize resolution.

The performances of four column combinations, to maximize FAME resolution, were studied in the analysis of milk FAs (Hyötyläinen *et al.*, 2004). A very long (100 m) cyano column (CP7420) coupled with two different non-polar columns (HP-1 and HP-5) were compared to a traditional 30 m non-polar column (HP-1) coupled with a 1 m × 0.1 mm ID × 0.1 μm cyano capillary and with a 0.35 m × 0.05 mm ID × 0.1 μm DB-Wax column. The latter column combination resulted in the best choice, even if the high backpressure caused by the micro-bore column enabled the use of only a short segment. Villegas and co-workers (2010) tested five different sets of columns to optimize the resolution of *cis/trans* C_{18} isomers, to quantify *trans*-vaccenic acid in bovine fat without any interference. The best results were obtained using an SLB-IL 100 column (60 m × 0.2 mm × 0.2 μm) as the first dimension and a medium polarity column in the second dimension (BPX-50 3 m × 0.1 mm × 0.1 μm).

Indrasti and co-workers (2010b) used a GC × GC system, coupled with a ToF-MS detector, to study the FAME profile of lard in comparison with goat, chicken, and cattle fat; the objective of the investigation was to evaluate the possibilities to determine lard adulteration. It is worth noting that the FAME composition was statistically different ($p<0.05$) among the various species of animals. In total, fifty-one components were identified, and in particular the authors concluded that three FAs can be used as markers for lard, namely $C_{18:3n3t}$, $C_{20:3n3t}$, and $C_{20:2n6}$.

Mondello and co-workers (2003) applied GC × GC-FID to the analysis of vegetable and fish oils, using both a non-polar × polar and polar × non-polar column combinations. FA-class pattern formation was thoroughly studied by deriving log curves for each series of homologues. Identification was also supported by using GC-MS and LRI data. The same group (Mondello *et al.*, 2006) highlighted the degree of order within the C20 carbon-number group, in an analysis of cod liver oil (Fig. 10.14). It was observed that esters with the same double-bond number aligned themselves along a slanting horizontal line, while FAMEs with the same ω value were located along a slanting vertical line, forming a grid with specific esters located at precise intersection points. The cod liver oil was found to be highly complex, with more than 200 peaks counted on the 2D plane.

A non-orthogonal set with a long polar column in the first dimension (100 m) was employed by Tranchida and co-workers (2008a) to optimize the separation of critical isomer couples in olive and hazelnut oils. Thanks to the increased sensitivity, several odd C-number FAMEs were detected for the first time in such samples; furthermore, a series of FAMEs were identified only in hazelnut oil. In particular $C_{16:1\omega5}$ was detected at relatively high concentrations (0.007%), enabling the detection of the fraudulent addition of 5% of hazelnut oil into olive oil.

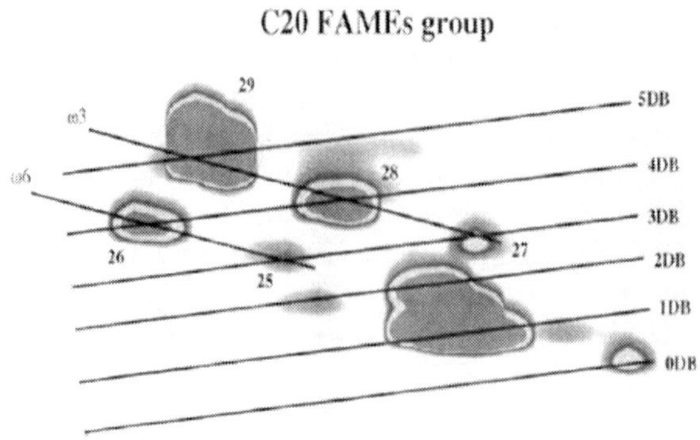

Fig. 10.14. Expansion of a GC × GC cod liver oil FAMEs (C_{20} group) chromatogram. Peak identification: (25) $C_{20:3\omega6}$; (26) $C_{20:4\omega6}$; (27) $C_{20:3\omega3}$; (28) $C_{20:4\omega3}$; (29) $C_{20:5\omega3}$. (Reproduced with permission from Mondello et al., 2006.)

It has been previously discussed that both orthogonal and non-orthogonal column sets can be useful to unravel the profiles of complex samples. In 2005, Adahchour and co-workers (2005b) analyzed olive oil FAMEs by combining two methods in one system, by using a column-entrance-split, which diverted the sample (50:50) onto two distinct sets; therefore, all the desired information was obtained in the time of a single run. Retention time repeatability was evaluated using both configurations, injecting a standard mixture five consecutive times; values lower than 0.3 and 1.7% were obtained on the first and second columns, respectively. One drawback of the reverse set was the lower thermal stability of the wax phase, which generated bands of modulated bleed material.

In a study on GC × GC optimization, Tranchida and co-workers (2007a) emphasized that partial exploitation of the bidimensional chromatogram space is due to: a) partial correlation between the two stationary phases, and b) non-ideal gas flow conditions, which affects mainly the secondary column. Indeed, considering a 30 m × 0.25 mm ID non-polar × 1 m × 0.1 mm ID polar column combination, with a 194.9 kPa inlet pressure, a near-to-ideal 1D linear velocity (*circa* 35 cm/s) and a far-too-high 2D linear velocity (*circa* 333 cm/s) were generated. The use of a Y press-fit between the two columns, with a 30 cm × 0.1 mm ID branch (uncoated column) connected to a manual split valve, enabled fine tuning of the flow entering the second column, thus reducing the linear velocity. The optimum conditions were reached with 65 % of the primary column flow diverted to the split valve; under such conditions, the 1D linear velocity was maintained, while a second-dimension velocity of 213 cm/s was generated. The authors "proved their point" in cod liver oil applications: exploitation of the 2D space was increased by more than 70% (from 22 to 38.4% of the occupied space), as shown in Fig. 10.15.

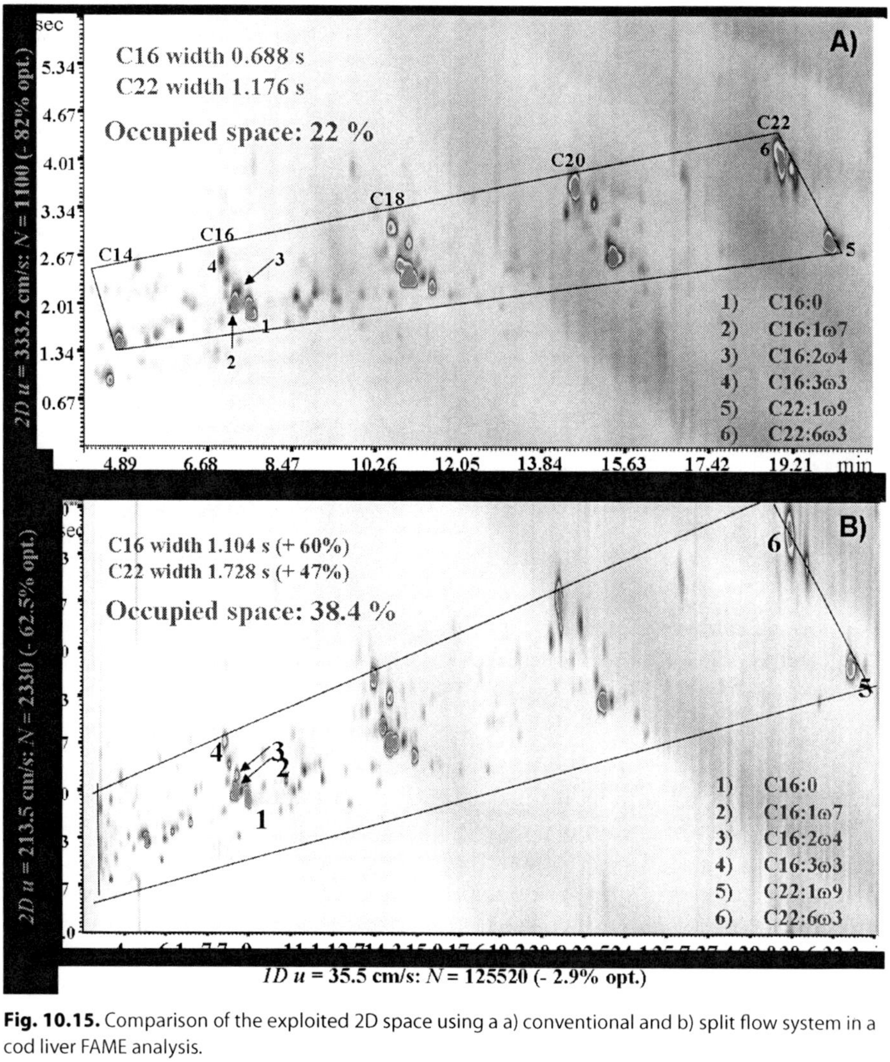

Fig. 10.15. Comparison of the exploited 2D space using a a) conventional and b) split flow system in a cod liver FAME analysis.

Under split-flow conditions, the second dimension separation improved greatly, as shown in Fig. 10.16, which illustrates the C_{16} FAME group of a cod liver oil. It was observed that resolution between $C_{16:2\omega4}$ and $C_{16:2\omega7}$ increased from 0.8, calculated using a traditional GC × GC system, to a value of 1.2 using the split-flow GC × GC system, meaning a 50% increase.

FAs are usually bonded to glycerol, forming triacylglycerols (TAGs), the main constituents of fats and oils (up to 98%). TAGs are high boiling-point compounds and, therefore, are usually analyzed intact by means of HPLC,

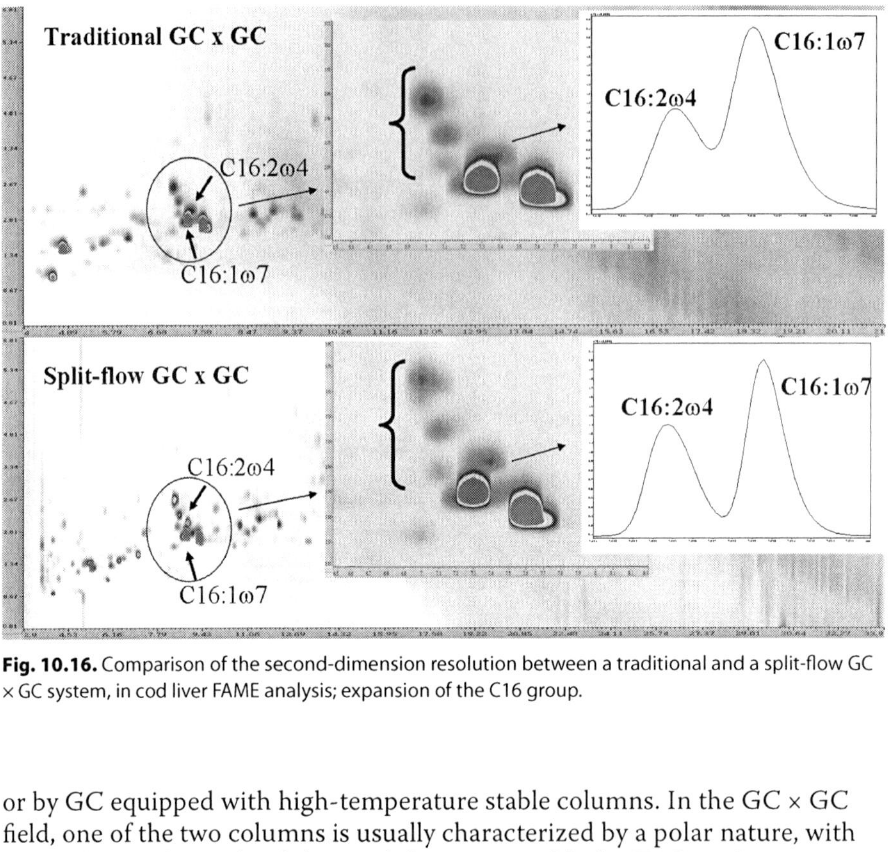

Fig. 10.16. Comparison of the second-dimension resolution between a traditional and a split-flow GC × GC system, in cod liver FAME analysis; expansion of the C16 group.

or by GC equipped with high-temperature stable columns. In the GC × GC field, one of the two columns is usually characterized by a polar nature, with relatively low thermal stability, and thus not suitable for TAG analysis. An interesting three-step approach was proposed to characterize TAG profiles: after a silver-phase LC (AgLC) separation, trans-methylation of the TAGs was achieved, then followed by GC × GC analysis (de Koning *et al.*, 2006). A CP-Wax column was used as first dimension, in order to separate FAMEs according to chain length and the number and position of the double bonds, while a VF-23ms capillary was employed as second dimension, for improved *cis/trans* isomer isolation. Method optimization was carried out with a relatively simple sample of hydrogenated vegetable oil, with a total analysis time of 2 h.

FAMEs in Biological and Biota Matrices

All the five advantages of GC × GC, namely I) separation power, II) sensitivity, III) selectivity, IV) structured chromatogram, V) speed (as number of peaks per minute), were highlighted in a GC × GC investigation on human plasma FAMEs (Tranchida *et al.*, 2008b). Indeed, despite the high complexity of such a matrix, 65 compounds were identified using pure standard compounds and according to specific positions in highly structured chromatograms

(Fig. 10.10). Investigation of the plasma FA profile is of high interest, in particular the amount of saturated and polyunsaturated FAs, since it can be related to a series of diseases, such as hypertension, diabetes, coronary heart disease, inflammatory and autoimmune disorders and cancer.

The FA profile is important to classify living organisms, such as micro-algae, meiofauna species and microorganisms. Akoto and co-workers (2008) used a GC × GC –ToF MS system, equipped with a direct thermal-desorption (DTD) injection interface, for the analysis of whole/intact phytoplankton and zooplankton samples, by simultaneously performing a thermally-assisted hydrolysis and methylation. Thanks to the increased sensitivity and to the improved separation, the authors found a series of FAs never mentioned before in several types of micro-algae and plankton.

The use of FA profiles in the taxonomic classification of microorganisms has been described since 1963 (Abel et al., 1963). The FA composition is markedly affected, both in qualitative and quantitative terms, by the nature of the medium, by the growth conditions, and by the age of the culture when harvested. Bacterial lipids are characterized by uncommon FAs, such as odd-chain, branched-chain (mainly iso- and anteiso-), cyclopropane and hydroxyl FAs, as well as positional isomers of unsaturated FAs. Therefore, the structured chromatograms obtained through GC × GC analysis are very useful to elucidate the composition of such sample-types. David and co-workers (2008) compared the data obtained from the well-established Sherlock MIDI analysis method (M. Sasser, MIDI Technical Note 101, 1990, www.midi-inc.com), with that derived by using loop-modulated GC × GC–FID. Two parameters were exploited for peak assignment, namely equivalent chain length in the first dimension and relative polarity in the second dimension. The latter was calculated as the ratio between retention time of the solute over the average retention times of preceding and proceeding saturated linear methyl esters. The same group (Gu et al., 2010) compared the separations obtained, for the same bacteria fatty acid sample, using a cryogenic modulator and a flow modulator, as shown in Fig. 10.17. Furthermore, a flow-modulated method was optimized using parallel detection. A Y-splitter was located at the end of the second column, diverting the flow between a 1.2 m × 0.25 mm ID uncoated tube connected to an FID and a 0.5 m × 0.1 mm ID capillary linked to a qMS detector. Using such an instrumental configuration, the MS can be used to identify unknown compounds and the FID for quantification purposes.

Purcaro and co-workers (2010b) used a fast one-step sample preparation method for extraction and methylation of bacteria fatty acids, requiring 2 minutes, and a relatively rapid split-flow twin-oven comprehensive 2D GC-qMS separation (25 min). The authors used the linear retention index approach, with the support of novel GC × GC software, to unambiguously identify each compound. Dedicated MS libraries, created using both comprehensive and monodimensional GC systems, were employed. The library

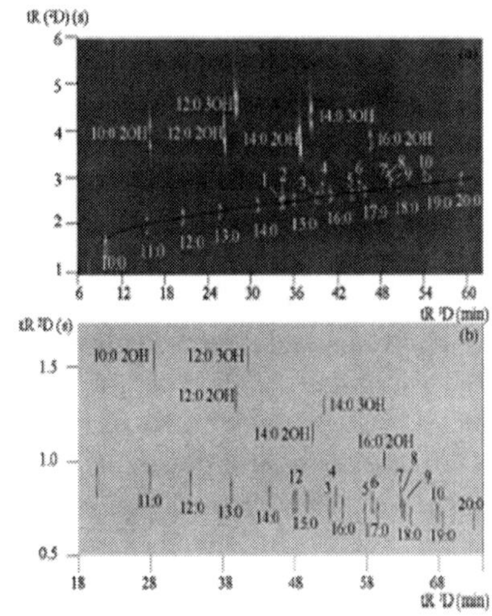

Fig. 10.17. Comparison of a BAMEs GC × GC chromatograms obtained by using (a) thermal modulation and (b) flow-modulation. Peak identification: 1, C15:0 iso; 2, C15:0 anteiso; 3, C16:0 iso; 4, C16:1 w9c; 5, C17:0 iso; 6, C17:0 cyclo; 7, C18:2 w9w12 cc; 8, C18:1 w9c; 9, C18:1 w9t and 10, C19:0 cyclo. (Reproduced with permission from Gu & Sandra, 2010.)

LRIs, calculated under monodimensional GC conditions, were comparable to the LRI values obtained using the 2D GC system, equipped with the same non-polar column (SLB-5ms) in the first dimension. Furthermore, the authors proposed a so-called "Bacteria Library" containing single averaged spectra relative to specific bacterium samples, comprising all the methyl esters in the retention time range and then subtracting the compressed chemical noise at three points across the chromatogram. The data derived can be considered as a sample fingerprint, comparable to a direct MS injection. The idea of the authors was to first give a name to the bacteria analyzed, and then investigate the FA profile.

FAMEs in Petrochemical Matrices

The characterization of the FA profile in biodiesel is important for both quality control and forensic reasons. Usually, biodiesel is mixed with conventional diesel and these blends are designed as BX fuels (where X is the percentage of pure biodiesel). The unravelling of the compositions of such mixtures is a challenging task. Several stationary phases were investigated by Adam and co-workers (2008) to maximize the resolution of three different series of homologues (alkanes, FAMEs, n-alkylaromatic compounds) in both the first and second dimensions. Six types of columns, having the same dimensions (30 m × 0.25 mm × 0.25 μm), were tested as the first dimension, namely SPB 5,

DB 1701, UB Wax, Solgel Wax, BPX 50, and BPX 90. Poor resolution between alkanes and FAMEs, as well as between FAMEs with different double bond numbers, was attained using an apolar phase (DB-1). Solgel Wax and UB Wax showed the best performance, with the former the prime choice because of the maximum operating temperature (300°C *vs.* 260°C). Five different stationary phases were evaluated in the second dimension (1 m × 0.1 mm × 0.1 μm), namely DB 1, DB 1701, BPX 50, BPX 70, and BPX 90. Obviously, poor resolution results were reported using either the BPX 70 or BPX 90 column, since the combination with the Solgel Wax was excessively correlated. The best separation between FAs and aromatics was obtained, as expected, by coupling a secondary non polar column (DB 1). The optimized set was employed to analyze two blends of commercial petroleum diesel, with transesterified coprah oil and rapeseed oil; a ToF-MS detector was employed for identification, while an FID was exploited for quantification purposes. The quantitative results obtained were in good agreement with the reference European method [European norm EN 14331. Liquid petroleum products, separation and characterization of fatty acid methyl esters (FAME) from middle distillates, 1 April 2004], except for minor components, where slight differences were highlighted. It was the authors' opinion that such inconsistency was caused by the increased sensitivity and resolution obtained by using GC × GC, compared to monodimensional GC.

Studies on diesel and gasoline engines have highlighted an accumulation of toxic compounds, in particular polycyclic aromatic hydrocarbons (PAHs), in used motor oil, derived from a direct leakage of fuel and incomplete combustion products (Grimmer et al., 1981; Pruell et al., 1988). Since a series of biodiesel studies have reported increased toxicity of exhaust emission particles (cytotoxicity or mutagenicity) compared to fossil diesel (Swanson et al., 2009; Krahl et al., 2009), Peacock and co-workers (2010) investigated whether motor oils from modern biodiesel cars accumulate toxic compounds. GC × GC coupled both with FID and ToF-MS detection was employed for the study. Low PAH levels were detected in the motor oil, with no other toxic compounds identified. A 0.5% amount of biodiesel-derived FAMEs was determined, which was of no concern since fatty acids are harmless and easily biodegraded.

Sterols and Wax Analysis

Fewer applications, compared to FA analysis, have been carried out on minor lipid components. Sterols are high boiling-point compounds, also known as steroid alcohols, and are a sub-group of steroids based on the same carbon structure, namely 1,2-cyclopentanoperhydrophenanthrene (sterane). Sterols are characterized by a hydroxyl group in position 3 of the A-ring, and an aliphatic side chain, of eight or more carbon atoms, attached to position 17, on the fourth ring. Sterols are found both in the free state and in the esterified form. GC × GC has been applied to the analysis of sterols contained in environmental water samples (Truong et al., 2003); the ratio between faecal sterols and their hydrogenated products (stanols) is a marker of faecal pollution. Slightly polar and moderately polar thermally-stable columns were employed, due to the low volatility of the analytes of interest. A near-to-baseline separation of 24-ethylcoprostanol, 24-methylenecholesterol, campesterol, and 24-ethyl-epi-coprostanol, which overlapped in 1D analysis, was obtained. Several GC × GC-ToF-MS works have been focused on the study of steroids in urine for anti-doping purposes (Mitrevski et al., 2008; Mitrevski et al., 2007; Mitrevski et al., 2010, Silva et al., 2009). In particular, Mitrevski et al. (2008) derivatized urine samples by using N,O-bis(trimethylsilyl)trifluoroacetamide (BSTFA), prior to injection in a GC × GC system equipped both with FID and ToF-MS detectors. Peak shape was thoroughly studied in relation with the modulation trapping temperature; it was observed that excessively low temperatures generated peak tailing, while, on the other hand, higher-than-optimum temperatures caused peak fronting, as shown in Fig. 10.18.

An interesting aspect of the investigation concerned ToF-MS spectra similarity matching, carried out using a commercial quadrupole-based MS database (NIST05) and a laboratory-made ToF-MS database. Although the mass spectra appeared very similar, the 19 standard sterols evaluated were positively identified, with an average match quality of 950, when the laboratory-made database was employed; on the other hand, only four sterols were positively identified, setting a match threshold of 800, using the NIST05 library. It can be concluded that dedicated MS libraries would be needed when using ToF-MS detection.

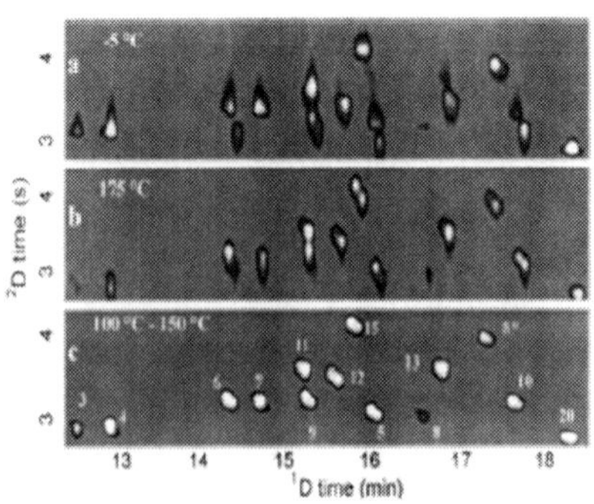

Fig. 10.18. Comparison of the effect of the modulation temperature (T_M) on the chromatographic performance in sterols analysis. (a) low T_M (-5°C); (b) high T_M (175°C); (c) optimized T_M (programmed from 100 to 150°C) (Reproduced with permission from Mitrevski et al., 2008.)

Stepan and co-workers (2008) optimized a GC × GC-ToF-MS method for the analysis of 25 anabolic steroids contained in nutritional supplements. Ethyl acetate, employed as extraction solvent, did not eliminate several interfering compounds, namely fatty acids (palmitic acid and stearic acid), saccharide-related compounds (1,6-anhydro-D-glucopyranose and 1,6-anhydro-D-galactofuranose), and aromatic compounds (vanillin and 4-hydroxy-2-methoxycinnamaldehyde). Such compounds may pose a problem for the chromatography performance, by deposition in the injection liner and in the first part of the capillary column; therefore, a dispersive solid-phase extraction was performed as a clean-up step, by using primary secondary amine (PSA) as a sorbent. The validated method showed good results except for the steroid oxymetholone, which due to the particular chemical structure, interacted more strongly with PSA (through hydrogen bonding), causing low recovery.

Tobias and co-workers (2008) coupled a GC × GC system with an isotope ratio MS (IRMS), for the measurement of steroid $^{13}C/^{12}C$ ratios. In carbon-13 IRMS analyses, all the organic compounds are transformed into CO_2, isotopic abundance measurements are achieved, and thus fragmentation profiles are not generated. The main problem in a GC × GC and IRMS system is excessive band broadening caused by the combustion chamber (about 1 sec), and a slow detector response (again, about 1 sec). Therefore, the instrument was carefully modified, in terms of solvent elimination process, combustion reactor, transfer lines, water trap, and the open split. The description of such modifications is outside of the aim of this chapter; however band broadening was greatly reduced. The final acquisition frequency of the IRMS was 25 Hz.

Waxes are esters of long-chain fatty acids with long-chain linear or branched alcohols. However, the so-called "wax fraction" includes a broadened range of compounds, such as di-esters and hydroxyl esters of high molecular weight alcohols (aliphatic and steroidal), high molecular-weight acids, free FAs and free fatty alcohols. Wax has a large realm of applications in daily life (candles, polishes, wax paper, etc.), as well as in foodstuffs, such as coating for many cheeses, and glazing agents (e.g., E913). Jover and co-workers (2005) analyzed lanolin, the wool wax secreted by the sebaceous glands of sheep, by GC × GC-ToF-MS without pre-fractionation from the other various classes of lipids. Various derivatization processes (methylation, silylation, and methylation/silylation) and column combinations were tested. Although the double derivatization process (methylation for acidic groups and silylation for alcoholic groups) was the most time consuming, it also proved to be the most effective in reducing peak tailing. In fact, the second-dimension separation was maximized when the polarity of both acidic and alcoholic groups were reduced. The column set chosen consisted of an XTI-5 10 m × 0.25 mm ID × 0.25 μm first dimension and a BPX-50 1 m × 0.1 mm ID × 0.1 μm second dimension. A well-structured chromatogram was obtained, highlighted by examining extracted ion chromatograms (Fig. 10.19); a series of horizontal analyte bands were observed, formed by compounds of the same chemical class, namely FAs, fatty alcohols, hydroxyl acids and diols. Despite the large number of compounds reported by the ToF-MS software, only 30 compounds were identified with an acceptable spectral similarity, that is, over 800 (the similarity scale varied in the 0–999 range). According to the authors, such a poor result was due to a rather incomplete MS library, especially in the high MW range, and to poor quality mass spectra, characterized by a considerable reduction of the high MW ion intensity.

Biedermann and co-workers (2008) investigated the wax ester fraction of various plant oils, by using LC-GC-MS and LC-GC × GC-FID. The normal-phase LC (NPLC) step achieved pre-separation of the triacylglycerols, before transferring the wax fraction to the GC system through an on-column interface, by using the concurrent eluent evaporation technique. The LC fraction of interest was collected on a laboratory-coated precolumn (OV-1701-OH, 40 cm × 0.53 mm ID × 0.03 μm film thickness) connected to a vapour exit. Then, the precolumn was dismounted and attached to the inlet of the first column in the GC × GC instrument, where the analysis was performed. The authors highlighted a certain degree of degradation due to thermal stress, in particular of diterpene esters. Such an effect was aggravated by using GC × GC, since the flow resistance generated by the second micro-bore column retarded elution from the first column, thus increasing the elution temperatures. Compound losses were maintained at a moderate level by using a high flow rate (2 ml/min), using hydrogen as carrier gas, shortening columns, and carefully selected the geometry and film thicknesses of the stationary phases (1D: PS-255 20 m × 0.25 mm ID × 0.12 μm film thickness; 2D: SOP-50 1.5m × 0.15 mm ID × 0.075 μm film thickness).

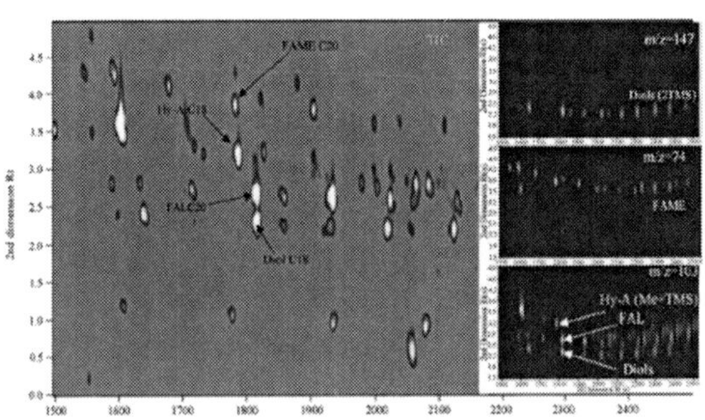

Fig. 10.19. GC × GC-ToFMS TIC-chromatogram of a methylated and silylated lanolin sample (on the left). Inserted windows of extracted ions (m/z 74, 103, 147). Abbreviations: FAL, fatty alcohol; Hy-A, hydroxyl acids; ME, methyl derivative; TMS, trimethylsilyl derivative. (Reproduced with permission from Jover et al., 2005.)

Final Remarks

Comprehensive two-dimensional gas chromatography, especially if combined with mass spectrometry, appears to be a perfect technique for the analysis of volatile and semi-volatile lipid molecules. In fact, any lipid analyst should be happy with the heightened selectivity (three separation dimensions, related to volatility, polarity, and mass), increased peak capacity and sensitivity, as well as enhanced speed, generated by the 3D approach. Moreover, the benefits of structured chromatograms have been illustrated in the present contribution.

It can be anticipated that future research in the field of GC × GC lipid analysis will be focused especially on adulteration in the lipid field, an issue which is not only the cause of huge economic loss, but also is often very difficult to determine. In this respect, GC × GC could be very useful in the minimization of the sample preparation process, or to avoid it completely, with the aim of attaining a fingerprint of the specific lipid mixture analyzed. The result of such a process, namely a highly detailed and sample-specific chromatogram could be exploited for the evaluation of quality, or the discovery of frauds related to the mixing of different kinds of oils.

Another important topic, which has been only partially exploited, concerns the biological lipidome, in particular related to metabolic diseases and hormonal dysfunctions. It is obvious that both adulteration and lipidomic studies would require the use of dedicated chemometric tools.

In the near future, considering the high price of cryogenic fluids, an important effort will be dedicated to the improvement of flow modulators in place of cryogenic modulators. However, it is the authors' impression that many described GC × GC applications, not only in the lipid field, could have also been carried out by using a less powerful technique.

References

Abel, K.; De Schmertzing, H.; Peterson, J.I. Classification of micro-organisms by analysis of chemical composition. I. Feasibility of gas-chromatography, J. Bacteriol. 1963, 85, 1039-1044.

Adahchour, M.; Beens, J.; Brinkman, U.A.Th. Analyst 2003, 128(3), 213-216.

Adahchour, M.; Beens, J.; Vreuls, R.J.J.; Brinkman, U.A.Th. Recent developments in comprehensive two-dimensional gas chromatography (GC × GC) I. Introduction and instrumental set-up. TrAC 2006a, 25(5), 438-454.

Adahchour, M.; Beens, J.; Vreuls, R.J.J.; Brinkman, U.A.Th. Recent developments in comprehensive two-dimensional gas chromatography (GC × GC) IV. Further applications, conclusions and perspectives. TrAC 2006c, 25(8), 821-840.

Adahchour, M.; Jover, E.; Beens, J.; Vreuls, R.J.J.; Brinkman, U.A.Th. Twin comprehensive two-dimensional gas chromatographic system: Concept and applications. J. Chromatogr. A 2005b, 1086(1-2), 128-134.

Adahchour, M.; Beens, J.; Vreuls, R.J.J.; Brinkman, U.A.Th. Recent developments in comprehensive two-dimensional gas chromatography (GC × GC) III. Applications for petrochemicals and organohalogens. TrAC 2006b, 25(7), 726-741.

Adahchour, M.; Beens, J.; Vreuls, R.J.J.; Batenburg, A.M.; Brinkman, U.A.Th. Comprehensive two-dimensional gas chromatography of complex samples by using a "reversed-type" column combination: Application to food analysis. J. Chromatogr. A, 2004, 1054, 47-55.

Adahchour, M.; Brandt, M.; Baier, H-U.; Vreuls, R.J.J.; Batenburg, A.M.; Brinkman, U.A.Th. Comprehensive two-dimensional gas chromatography coupled to a rapid-scanning quadrupole mass spectrometer: Principles and applications. J. Chromatogr. A 2005a, 1067, 245-254.

Adam, F.; Bertoncini, F.; Coupard, V.; Charon, N.; Thiébaut, D.; Espinat, D.; Hennion, M.-C. Using comprehensive two-dimensional gas chromatography for the analysis of oxygenates in middle distillates. I. Determination of the nature of biodiesels blend in diesel fuel. J. Chromatogr. A 2008, 1186(1-2), 236-244.

Akoto, L.; Stellaard F.; Irth H.; Vreuls, R.J.J.; Pel, R. Improved fatty acid detection in micro-algae and acquatic meiofauna species using a direct thermal desorption interface combined with comprehensive gas chromatography-time-of-flight mass spectrometry. J. Chromatogr. A 2008, 1186, 254-261.

Amador-Munoz, O.; Villalobos-Pietrini, R.; Aragon-Pina, A.; Tran, T.C.; Morrison, P.D.; Marriott, P.J. Quantification of polycyclic aromatic hydrocarbons based on comprehensive two-dimensional gas chromatography-isotope dilution mass spectrometry. J. Chromatogr. A 2008, 1201, 161-168.

Arey, J.S.; Nelson, R.K.; Plata, D.L.; Reddy, C.M. Disentangling oil weathering using GC×GC. Part 2. Mass transfer calculations. Environ. Sci. Technol. 2007, 41, 5747-5755.

Aybar-Munoz, J; Fernandez-Gonzalez, E.; Garcia-Ayuso, L.E.; Gonzalez-Casado, A.; Cuadros-Rodriguez, L. Semiqualitative method for detection of pesticide residues over established limits in vegetables by use of GC-uECD and GC-(EI)MS. Chromatographia 2005, 61, 505-513.

Bartenbach, S.; Williams, J.; Plass-Duelmer, C.; Berresheim, H.; Lelieveld, J. In-situ measurement of reactive hydrocarbons at Hohenpeissenberg with comprehensive two-dimensional gas chromatography (GCxGC-FID): use in estimating HO and NO3, Atmos. Chem. Phys. 2007, 7, 1-14.

Beens, J.; Adahchour, M.; Vreuls, R.J.J.; van Altena K.; Brinkman U.A.Th. Simple, non-moving modulation interface for comprehensive two-dimensional gas chromatography. J. Chromatogr. A 2001, 919(1), 127-132.

Beens, J.; Boelens, H.; Tijssen, R.; Blomberg, J. Simple, non-moving modulation interface for comprehensive two-dimensional gas chromatography. J High Resolut. Chromatogr. 1998a, 21(1), 47-54.

Beens, J.; Tijssen, R.; Blomberg, J. Comprehensive two-dimensional gas chromatography (GC × GC) as a diagnostic tool. J. High Resolut. Chromatogr. 1998b, 21(1), 63-64.

Beens, J.; Janssen, H.-G.; Adahchour, M.; Brinkman, U.A.Th. Flow regime at ambient outlet pressure and its influence in comprehensive two-dimensional gas chromatography. J. Chromatogr. A 2005, 1086, 141-150.

Biedermann, M.; Grob, K.; Comprehensive two dimensional GC after HPLC preseparation for the characterization of aromatic hydrocarbons of mineral oil origin in contaminated sunflower oil. J. Sep. Sci. 2009, 32, 3726-3737.

Biedermann, M.; Hasse-Aschoff, P.; Grob, K. Wax ester fraction of edible oils: Analysis by on-line LC-GC-MS and GC × GC-FID. Eur. J. Lipid Sci. Technol. 2008, 110, 1084-1094.

Blomberg, J.; Riemersma, T.; van Zuijlen, M.; Chaabani H. Comprehensive two-dimensional gas chromatography coupled with fast sulphur-chemiluminescence detection: implications of detector electronics. J. Chromatogr. A 2004, 1050, 77-84.

Blomberg, J.; Schoenmakers, P.J.; Beens, J.; Tijssen, R. Comprehensive two-dimensional gas chromatography (GC×GC) and its applicability to the characterization of complex (petrochemical) mixtures. J. High Resolut. Chromatogr. 1997, 20, 539-544.

Bruckner, C.A.; Prazen, B.J.; Synovec, R.E. Comprehensive two-dimensional high-speed gas chromatography with chemometric analysis. Anal. Chem. 1998, 70(14), 2796-2804.

Bueno, P.A.; Seeley J.V. Flow-switching device for comprehensive two-dimensional gas chromatography. J. Chromatogr. A 2004, 1027, 3-10.

Chin, S.-T.; Che Man, Y.B.; Tan, C.P.; Hashim, D.M. Rapid profiling of animal-derived fatty acids using fast GC × GC coupled to Time-of-flight mass spectrometry. J. Am. Oil Chem. Soc. 2009, 86(10), 949-958.

Christie W.W. In Lipid Analysis, Oily Press: Bridgwater, England, 2003.

Conka, K.; Drobna, B.; Kocan, A.; Petrik, J. Simple solid-phase extraction method for determination of polychlorinated biphenyls and selected organochlorine pesticides in human serum. J. Chromatogr. A 2005, 1084, 33-38.

Cordero, C.; Bicchi, C.; Joulain, D.; Rubiolo, P. Identification, quantitation and method validation for the analysis of suspected allergens in fragrances by comprehensive two-dimensional gas chromatography coupled with quadrupole mass spectrometry and with flame ionization detection. J. Chromatogr. A 2007, 1150, 37-49.

Cortes, H.J.; Winniford, B.; Luong, J.; Pursch, M. Comprehensive two dimensional chromatography review. J. Sep. Sci. 2009, 32, 883-904.

Dallüge, J.; Vreuls, R.J.J.; van Iperen, D.J.; van Rijn, M.; Brinkman, U.A.Th. Resistively heated gas chromatography coupled to quadrupole mass spectrometry. J. Sep. Sci. 2002, 25, 608-614.

David, F.; Tienpont, B.; Sandra, P. Chemotaxonomy of bacteria by comprehensive GC and GC-MS in electron impact and chemical ionisation mode. J. Sep. Sci. 2008, 31, 3395-3403.

de Geus, H.-J.; Aidos, I.; de Boer, J.; Luten, J.B.; Brinkman, U.A.Th. Characterisation of fatty acids in biological oil samples using comprehensive multidimensional gas chromatography. J. Chromatogr. A 2001, 910, 95-103.

de Koning, S.; Janssen, H.-G.; Brinkman, U.A.Th. Characterization of triacylglycerides from edible oils and fats using a single and multidimensional techniques. LC-GC Eur. 2006, 19(11), 590-600.

Diez, S.; Jover, E.; Bayona, J.M.; Albaiges, J. Prestige oil spill. III. Fate of a heavy oil in the marine environment. Environ. Sci. Technol. 2007, 41, 3075-3082.

Dyson, N. Peak distortion, data sampling errors and the integrator in the measurement of very narrow chromatographic peaks. J. Chromatogr. A 1999, 842, 321-340.

Fraga, C.G.; Prazen, B.J.; Synovec, R.E. Comprehensive two-dimensional gas chromatography and chemometrics for the high-speed quantitative analysis of aromatic isomers in a jet fuel using the standard addition method and an objective retention time alignment algorithm. Anal. Chem. 2000, 72, 4154-4162.

Frysinger, G.S.; Gaines, R.B. Determination of oxygenates in gasoline by GC × GC. J. High Resolut. Chromatogr. 2000, 23, 197-201.

Frysinger, G.S.; Gaines, R.B. Separation and identification of petroleum biomarkers by comprehensive two-dimensional gas chromatography. J. Sep. Sci. 2001, 24, 87-96.

Frysinger, G.S.; Gaines, R.B. Comprehensive two-dimensional gas chromatography with mass spectrometric detection (GC×GC/MS) applied to the analysis of petroleum J. High Resolut. Chromatogr. 1999, 22(5), 251-255.

Frysinger, G.S.; Gaines, R.B.; Ledford, E.B. Quantitative determination of BTEX and total aromatic compounds in gasoline by comprehensive two-dimensional gas chromatography (GC×GC) J. High Resolut. Chromatogr. 1999, 22(4), 195-200.

Frysinger, G.S.; Gaines, R.B.; Xu, L.; Reddy, C.M. Resolving the unresolved complex mixture in petroleum-contaminated sediments. Environ. Sci. Technol. 2003, 37, 1653-1662.

Gaines, R.B.; Frysinger, G.S.; Hendrick-Smith, M.S.; Stuart, J.D. Oil spill source identification by comprehensive two-dimensional gas chromatography. Environ. Sci. Technol. 1999, 33, 2106-2112.

Gaines, R.B.; Frysinger, G.S. Temperature requirements for thermal modulation in comprehensive two dimensional gas chromatography. J. Sep. Sci. 2004, 27, 380-388.

Giddings, J.C. Sample dimensionality: A predictor of order-disorder in component peak distribution in multidimensional separation. J. Chromatogr. A 1995, 703, 3-15.

Golay, M.J.E. Gas chromatography. Academic Press: New York, 1958.

Goldstein, A.H.; Worton, D.R.; Williams, B.J.; Hering, S.V.; Kreisberg, N.M.; Panic, O.; Górecki, T. Thermal desorption comprehensive two-dimensional gas chromatography for in-situ measurements of organic aerosols J. Chromatogr. A 2008, 1186(1-2), 340-347.

Gomara, B.; Ramos, L.; Gonzalez, M.J. Determination of polychlorinated biphenyls in small-size serum samples by solid-phase extraction followed by gas chromatography with micro-electron-capture detection. J. Chromatogr. B 2002, 766, 279-287.

Górecki, T.; Goldstein, A.H.; Kreisberg, N.M.; Hering, S.V.; Teng, A.P.; Worton, D.R.; Mcneish, C. editor. International Network of Environmental Forensics Conference; 2009 August 31 - September 2; Calgary (AB). Abstracts p. 30.

Grimmer, G.; Jacob, J.; Naujack, K.W.; Dettbarn, G. Profile of the polycyclic aromatic hydrocarbons from used engine oil. Inventory by GC/MS-PAH in environmental materials. Fresenius J. Anal. Chem. 1981, 309, 13-19.

Gu, Q.; David, F.; Lynen, F.; Rumpel, K.; Xu, G.; De Vos, P.; Sandra, P. Analysis of bacteria fatty acids by flow modulated comprehensive two-dimensional gas chromatography with parallel flame ionization detector/mass spectrometry. J. Chromatogr. A 2010, 1217, 4448-4453.

Gurr, M.I.; Harwood, J.L.; Frayn, K.N. Lipid Biochemistry W.H. Freeman Publishers: New York, 2002.

Hamilton, J.F.; Lewis, A.C.; Bartle, K.D. Peak amplitude and resolution in comprehensive gas chromatography using valve modulation. J. Sep. Sci. 2003, 26, 578-584.

Harynuk, J.; Górecki, T. Design considerations for a GC×GC system. J. Sep. Sci. 2002, 25(5/6), 304-310.

Harynuk, J.; Vlaeminck, B.; Zaher, P.; Marriott, P.J. Projection of multidimensional GC data into alternative dimensions-exploiting sample dimensionality and structured retention patterns. Anal. Bioanal. Chem. 2006, 386(3), 602-613.

Hinshaw, J.V. Handling fast peaks. LCGC North Am. 2003, 21, 268-272.

Holland, J.F.; Enke, C.G.; Allison, J.; Stults, J.T.; Pinkston, B.; Newcome, J.T. Mass spectrometry on the chromatographic time scale: Realistic expectations. Anal. Chem. 1983, 55, 997A-1012A.

Hyötyläinen, T.; Kallio, M.; Hartonen, K.; Jussila, M.; Palonen, S.; Riekkola, M. Modulator design for comprehensive two-dimensional gas chromatography: quantitative analysis of polyaromatic hydrocarbons and polychlorinated biphenyls. Anal. Chem. 2002, 74(17), 4441-4446.

Hyötyläinen, T.; Kallio, M.; Lehtonen, M.; Lintonene, S.; Peräjoki, P.; Jussila, M.; Riekkola M.-L. Comprehensive two-dimensional gas chromatography in the analysis of dietary fatty acids. J. Sep. Sci. 2004, 27, 459-467.

Indrasti, D.; Che Man, Y.B.; Chin, S.T.; Mustafa, S.; Mat Hashim, D.; Abdul Manaf, M. Regiospecific analysis of mono- and diglycerides in glycerolysis products by GC × GC-TOF-MS J. Am. Oil Chem. Soc. 2010a, 87(11), 1255-1262.

Indrasti, D.; Che Man, Y.B.; Mustafa, S.; Hashim, D.M. Lard detection based on fatty acids profile using comprehensive gas chromatography hyphenated with time-of-flight mass spectrometry. Food Chem. 2010b 122(4), 1273-1277.

James, A.T.; Martin, A.J.P. Gas-liquid partition chromatography: the separation and micro-estimation of volatile fatty acids from formic acid to dodecanoic acid. Biochem. J. 1952, 50, 679-690.

Jover, E.; Adahchour, M.; Bayona, J.M.; Vreuls, R.J.J.; Brinkman, U.A.Th. Characterization of lipids in complex samples using comprehensive two-dimensional gas chromatography with time-of-flight mass spectrometry. J. Chromatogr. A 2005, 1086, 2-11.

Kallio, M.; Hyötyläinen, T.; Jussila, M.; Hartonen, K.; Palonen, S.; Shimmo, M.; Riekkola, M. Semi-rotating cryogenic modulator for comprehensive two-dimensional gas chromatography. Anal. Bioanal. Chem. 2003a, 375(6), 725-731.

Kallio, M.; Hyötyläinen, T.; Lehtonen, M.; Jussila, M.; Hartonen, K.; Shimmo, M.; Riekkola, M.-L. Comprehensive two-dimensional gas chromatography in the analysis of urban aerosols. J. Chromatogr. A 2003b, 1019, 251-260.

Kallio, M.; Jussila, M.; Raimi, P.; Hyötyläinen, T. Modified semi-rotating cryogenic modulator for comprehensive two-dimensional gas chromatography. Anal. Bioanal. Chem. 2008, 391(6), 2357-2363.

Khummueng, W.; Trenerry, C.; Rose, G.; Marriott, P.J. Application of comprehensive two-dimensional gas chromatography with nitrogen-selective detection for the analysis of fungicide residues in vegetable samples. J. Chromatogr. A 2006, 1131, 203-214.

Kinghorn, R.M.; Marriott, P.J. Comprehensive two-dimensional gas chromatography using a modulating cryogenic trap J. High Resolut. Chromatogr. 1998, 21(11), 620-622.

Kinghorn, R.M.; Marriott, P.J. High Speed Cryogenic Modulation – A technology enabling comprehensive multidimensional gas chromatography J. High Resolut. Chromatogr. 1999, 22(4), 235-238.

Klee, M.S.; Williams, M.D.; Chang, I.; Murphy, J. Superior ECD performance through design and application. J. High Res. Chromatogr. 1999, 22, 24-28.

Krahl, J.; Knothe, G.; Munack, A.; Ruschel, Y.; Schroder, O.; Hallier, E.; Westphal, G.; Bunger, J. Comparison of exhaust emissions and their mutagenicity from the combustion of biodiesel, vegetable oil, gas-to-liquid and petrodiesel fuels. Fuel 2009, 88, 1064-1069.

Kristenson, E.M.; Korytár, P.; Danielsson, C.; Kallio, M.; Brandt, M.; Mäkelä, J.; Vreuls, R.J.J.; Beens, J.; Brinkman, U.A.Th. Evaluation of modulators and electron-capture detectors for comprehensive two-dimensional GC of halogenated organic compounds. J. Chromatogr. A 2003, 1019, 65-77.

Kusakabe, T.; Saito, T.; Takeichi, S. Headspace solid-phase micro-extraction and gas chromatography with micro-electron capture detection for the measurement of p,p'-DDE in rat whole blood and hair. Tokai J. Exper. Clin. Med. 2003, 28, 131-138.

Ledford, E.B.; Billesbach, C.; Termaat, J. Pittcon 2002, March 17-22, New Orleans, LA. contribution #2262P.

Ledford, E.B. Presented at the 23rd Symposium on Capillary Gas Chromatography, Riva del Garda, June 2000.

Libardoni, M.; Stevens, P.T.; Hunter Waite, J.; Sacks, R. Analysis of human breath samples with a multi-bed sorption trap and comprehensive two-dimensional gas chromatography (GC×GC). J. Chromatogr. B 2006, 842, 13-21.

Liu, Z.; Phillips, J.B. Comprehensive two-dimensional gas chromatography using an on-column thermal modulator interface. J. Chromatogr. Sci. 1991, 29(6), 227-231.

Liu, Z.; Phillips, J.B. Sensitivity and detection limit enhancement of gas chromatography detection by thermal modulation. J. Microcolumn Sep. 1994, 6(3), 229-235.

Lucci, P.; Pacetti, D.; Boselli, E.; Frega, N.G. Comprehensive multidimensional gas chromatography in the analysis of food lipids. Progress Nutr. 2009, 11(3), 143-148.

Ma, C.; Wang, H.; Lu, X.; Li, H.; Liu, B.; Xu, G. Analysis of Artemisia annua L. volatile oil by comprehensive two-dimensional gas chromatography time-of-flight mass spectrometry. J. Chromatogr. A 2007, 1150, 50-53.

Marriott, P.J. In: Multidimensional Chromatography. L. Mondello, A.C. Lewis, K.D. Bartle, Eds.; John Wiley and Sons, New York, 2002, 77-108.

Mayadunne, R.; Nguyen, T.-T.; Marriott, P.J. Amino acid analysis by using comprehensive two-dimensional gas chromatography. Anal. Bioanal. Chem. 2005, 382, 836-847.

Mitrevski, B.S.; Brenna, J.T.; Zhang, Y.; Marriott, P.J. Application of comprehensive two-dimensional gas chromatography to sterols analysis. J. Chromatogr. A 2008, 1214(1-2), 134-142.

Mitrevski, B.S.; Brenna, J.T.; Zhang, Y.; Marriott, P.J. A new paradigm in drugs analysis: Comprehensive two-dimensional gas chromatography for steroid analysis. Chem. Aust. 2007, 74, 3-5.

Mitrevski, B.S.; Wilairat, P.; Marriott, P. J. Comprehensive two-dimensional gas chromatography improves separation and identification of anabolic agents in doping control. J. Chromatogr. A 2010, 1217(1), 127-135.

Mohler, R.E.; Prazen, B.J.; Synovec, R.E. Total-transfer, valve-based comprehensive two-dimensional gas chromatography. Anal. Chim. Acta 2006a, 555(1), 68-74.

Mohler, R.E.; Dombek, K.M.; Hoggard, J.C.; Young, E.T.; Synovec, R.E. Comprehensive two-dimensional gas chromatography time-of-flight mass spectrometry analysis of metabolites in fermenting and respiring yeast cells. Anal. Chem. 2006b, 78, 2700-2709.

Mondello, L.; Casilli, A.; Tranchida, P.Q.; Dugo, G.; Dugo, P. Comprehensive two-dimensional gas chromatography in combination with rapid scanning quadrupole mass spectrometry in perfume analysis. J. Chromatogr. A 2005, 1067, 235-243.

Mondello, L.; Casilli, A.; Tranchida, P.Q.; Dugo, P.; Costa, R.; Festa, S.; Dugo, G. Comprehensive multidimensional GC for the characterization of roasted coffee beans. J. Sep. Sci. 2004, 27, 442-450.

Mondello, L.; Casilli, A.; Tranchida, P.Q.; Dugo, P.; Dugo, G. Detailed analysis and group-type separation of natural fats and oils using comprehensive two-dimensional gas chromatography. J. Chromatogr. A 2003, 1019, 187-196.

Mondello, L.; Casilli, A.; Tranchida, P.Q.; Lo Presti, M.; Dugo, P.; Dugo, G. Comprehensive gas chromatography coupled to mass spectrometry for the separation of pesticides in a very complex matrix. Anal. Bioanal. Chem. 2007, 389, 1755-1763.

Mondello, L.; Casilli, A.; Tranchida, P.Q.; Lo Presti, M.; Dugo, P.; Dugo G. Comprehensive two-dimensional gas chromatography-mass spectrometry: A review. Mass Spectrom. Rev. 2008, 27(2), 101-124.

Mondello, L.; Tranchida, P.Q.; Dugo, P.; Dugo, G. Rapid, micro-scale preparation and very fast gas chromatographic separation of cod liver oil fatty acid methyl ester. J. Pharm. Bio. Anal. 2006, 41, 1566-1570.

Murphy, R.E.; Schure, M.R.; Foley, J.P. Effect of sampling rate on resolution in comprehensive two-dimensional liquid chromatography. Anal. Chem. 1998, 70(8), 1585-1594.

Ochiai, N.; Ieda, T.; Sasamoto, K.; Fushimi, A.; Hasegawa, S.; Tanabe, K.; Kobayashi, S. Comprehensive two-dimensional gas chromatography coupled to high-resolution time-of-flight mass spectrometry and simultaneous nitrogen phosphorous and mass spectrometric detection for characterization of nanoparticles in roadside atmosphere. J. Chromatogr. A 2007, 1150, 13-20.

Peacock, E.E.; Arey, J.S.; DeMello, J.A.; McNichol, A.P.; Nelson, R.K.; Reddy, C.M. Molecular and isotopic analysis of motor oil from a biodiesel-driven vehicle. Energy Fuels, 2010, 24(2), 1037-1042.

Phillips, J.B.; Ledford, E.B. Thermal Modulation: A Chemical instrumentation component of potential value in improving portability. Field Anal. Chem. Technol. 1996, 1(1), 23-29.

Phillips, J.B; Xu, J. Comprehensive multi-dimensional gas chromatography. J. Chromatogr. A 1995, 703(1-2), 327-334.

Phillips, J.B.; Gaines, R.B.; Blomberg, J.; Van Der Wielen, F.W.M.; Dimandja, J.; Green, V.; Granger, J.; Patterson, D.; Racovalis, L.; De Geus H.; et al. A robust thermal modulator for comprehensive two-dimensional gas chromatography. J. High Resolut. Chromatogr. 1999, 22(1), 3-10.

Pierce, K.M.; Hoggard, J.C.; Hope, J.L.; Rainey, P.M.; Hoofnagle, A.N.; Jack, R.M.; Wright, B.W.; Synovec, R.E. Fisher ratio method applied to third-order separation data to identify significant chemical components of metabolite extracts. Anal. Chem. 2006, 78, 5068-5075.

Poliak, M.; Fialkov, A.B.; Amirav, A. Pulsed flow modulation two-dimensional comprehensive gas chromatography-tandem mass spectrometry with supersonic molecular beams. J. Chromatogr. A 2008a, 1210, 108-114.

Poliak, M.; Kochman, M.; Amirav, A. Pulsed flow modulation comprehensive two-dimensional gas chromatography. J. Chromatogr. A 2008b, 1186, 189-195.

Poole, C.F. In: The Essence of Chromatography. Elsevier: Amsterdam, 2003, 66-67.

Prazen, B.J.; Johnson, K.J.; Weber, A.; Synovec, R.E. Two-dimensional gas chromatography and trilinear partial least squares for the quantitative analysis of aromatic and naphthene content in naphtha. Anal. Chem. 2001, 73, 5677-5682.

Pruell, R.J.; Quinn, J.G. Accumulation of polycyclic aromatic hydrocarbons in crankcase oil. Environ. Pollut. 1988, 49, 89-97.

Purcaro, G.; Tranchida, P.Q.; Assis Jacques, R.; Bastos Caramão, E.; Moret, S.; Conte, L.; Dugo, P.; Dugo, G.; Mondello, L. Characterization of the yerba mate (Ilex Paraguariensis) volatile fraction by using solid-phase microextraction comprehensive two dimensional chromatography-quadrupole mass spectrometry. J. Sep. Sci. 2009, 32, 3755-3763.

Purcaro, G.; Morrison, P.; Moret, S.; Conte, L.S.; Marriott, P.J. Determination of polycyclic aromatic hydrocarbons in vegetable oils using solid phase microextraction comprehensive two-dimensional gas chromatography coupled with time of flight mass spectrometry. J. Chromatogr. A 2007, 116, 284-291.

Purcaro, G.; Tranchida, P.Q.; Ragonese, C.; Conte, L.S.; Dugo, P.; Dugo, G.; Mondello, L. Evaluation of a novel rapid-scanning quadrupole mass spectrometer in an apolar × ionic-liquid comprehensive two-dimensional gas chromatography system. Anal. Chem. 2010a, 82, 8503-8590.

Purcaro, G.; Tranchida, P.Q.; Dugo, P.; La Camera, E.; Bisignano, G.; Conte, L.; Mondello, L. Characterization of bacterial lipid profiles by using rapid sample preparation and fast comprehensive two-dimensional gas chromatography in combination with mass spectrometry. J. Sep. Sci. 2010b, 33, 2334-2340.

Quan, X.; Chen, S.; Platzer, B.; Chen, J.; Gfrerer, M. Simultaneous determination of chlorinated organic compounds in environmental samples using gas chromatography coupled with a micro electron capture detector and micro-plasma atomic emission detector. Spectrochim. Acta, Part B 2002, 57B, 189-199.

Quimby, B.; McCurry, J.; Norman, W. Chromatography: Reinvigorating a mature analytical discipline. LC GC The Peak 2007, April, 7-15.

Ragonese, C.; Tranchida, P.Q.; Dugo, P.; Dugo, G.; Sidisky, L.M.; Robillard, M.W.; Mondello, L. Evaluation of use of a dicationic liquid stationary phase in the fast and conventional gas chromatographic analysis of health-hazardous C-18 cis/trans fatty acids. Anal. Chem. 2009, 81(13), 5561-5568.

Reddy, C.M.; Eglinton, T.I.; Hounshell, A.; White, H.K.; Xu, L.; Gaines, R.B.; Frysinger, G.S. The west falmouth oil spill after thirty years: the persistence of petroleum hydrocarbons in marsh sediments. Environ. Sci. Technol. 2002, 36, 4754-4760.

Robrock, K.R.; Korytar, P.; Alvarez-Cohen, L.; Pathways for the anaerobic microbial debromination of polybrominated diphenyl ethers. Environ. Sci. Technol. 2008, 42, 2845-2852.

Ryan, D.; Watkins, P.; Smith, J.; Allen, M.; Marriott, P. Analysis of methoxypyrazines in wine using headspace solid phase microextraction with isotope dilution and comprehensive two-dimensional gas chromatography. J. Sep. Sci. 2005, 28, 1075-1082.

Ryan, D.; Shellie, R.; Tranchida, P.; Casilli, A.; Mondello, L.; Marriott, P. Analysis of roasted coffee bean volatiles by using comprehensive two-dimensional gas chromatography–time-of-flight mass spectrometry. J. Chromatogr. A 2004, 1054, 57-65.

Sanchez, J.M.; Sacks, R.D. Development of a multibed sorption trap, comprehensive two-dimensional gas chromatography, and time-of-flight mass spectrometry system for the analysis of volatile organic compounds in human breath. Anal. Chem. 2006, 78, 3046-3054.

Schoenmakers, P.J.; Oomen, J.L.M.M.; Blomberg, J.; Genuit, W.; van Velzen, G. Comparison of comprehensive two-dimensional gas chromatography and gas chromatography – mass spectrometry for the characterization of complex hydrocarbon mixtures. J. Chromatogr. A 2000, 892, 29-46.

Schoenmakers, P.; Marriott, P.; Beens, J. Nomenclature and Conventions in Comprehensive Multidimensional Chromatography, LCGC Europe, 2003, June, 1-4.

Seeley, J.V.; Kramp, F.; Hicks, C.J. Comprehensive two-dimensional gas chromatography via differential flow modulation. Anal. Chem. 2000, 72, 4346-4352.

Seeley, J.V.; Micyus, N.J.; McCurry, J.D.; Seeley, S.K. Comprehensive two-dimensional gas chromatography with a simple fluidic modulator. Amer. Lab. 2006, 38, 24-26.

Semard, G.; Bruchet, A.; Cardinael, P.; Bouillon, J.P. Use of comprehensive two-dimensional gas chromatography for the broad screening of hazardous contaminants in urban wastewaters. Water Sci. Technol. 2008, 57, 1983-1989.

Shellie, R.; Marriott, P.; Morrison, P.; Mondello, L. Effects of pressure drop on absolute retention matching in comprehensive two-dimensional gas chromatography. J. Sep. Sci. 2004a, 27, 504-512.

Shellie, R. Comprehensive 2D GC-QMS with differential flow modulation. LC GC Eur. 2008, 21, 572-578.

Shellie, R.; Marriott, P.; Morrison, P. Comprehensive two-dimensional gas chromatography with flame ionization and time-of-flight mass spectrometry detection: Qualitative and quantitative analysis of west Australian sandalwood oil. J. Chromatogr. Sci. 2004b, 42, 417-422.

Shellie, R.A.; Marriott, P.J. Comprehensive two-dimensional gas chromatography-mass spectrometry analysis of Pelargonium graveolens essential oil using rapid scanning quadrupole mass spectrometry. Analyst 2003, 128, 879-883.

Shellie, R.A.; Welthagen, W.; Zrostlikova, J.; Spranger, J.; Ristow, M.; Fiehn, O.; Zimmermann, R. Statistical methods for comparing comprehensive two-dimensional gas chromatography–time-of-flight mass spectrometry results: Metabolomic analysis of mouse tissue extracts. J. Chromatogr. A 2005, 1086, 83-90.

Shellie, R.A.; Marriott, P.J.; Huie, C.W. Comprehensive two-dimensional gas chromatography (GC x GC) and GC x GC- quadrupole MS analysis of Asian and American ginseng. J. Sep. Sci. 2003a, 26, 1185-1192.

Shunji, H.; Yoshikatsu, T.; Akihiro, F.; Hiroyasu, I.; Kiyoshi, T.; Yasuyuki, S.; Masa-aki, U.; Akihiko, K.; Kazuo, T.; Hideyuki, O.; Katsunori, A. Quantification of polychlorinated dibenzo-p-dioxins and dibenzofurans by direct injection of sample extract into the comprehensive multidimensional gas chromatography/high-resolution time-of-flight mass spectrometer. J. Chromatogr. A 2008, 1178, 187-198.

Silva, A.I.; Pereira, H.M.G.; Casilli, A.; Conceicao, F.C.; Neto, F.R.A. Analytical challenges in doping control: Comprehensive two-dimensional gas chromatography with time of flight mass spectrometry, a promising option. J. Chromatogr. A 2009, 1216, 2913-2922.

Sinha, A.E.; Johnson, K.J.; Prazen, B.J.; Lucas, S.V.; Fraga, C.G.; Synovec, R.E. Comprehensive two-dimensional gas chromatography of volatile and semi-volatile components using a diaphragm valve-based instrument. J. Chromatogr. A 2003a, 983, 195-204.

Slater, G.F.; White, H.K.; Eglinton, T.I.; Reddy, C.M. Determination of microbial carbon sources in petroleum contaminated sediments using molecular 14C analysis. Environ. Sci. Technol. 2005, 39, 2552-2558.

Song, S.M.; Marriott, P.J.; Kotsos, A.; Drummer, O.H.; Wynne, P. Comprehensive two-dimensional gas chromatography with time-of-flight mass spectrometry (GC × GC-TOFMS) for drug screening and confirmation. Forens. Sci. Int. 2004b, 143, 87-101.

Song, S.M.; Marriott, P.J.; Wynne, P. Comprehensive two-dimensional gas chromatography—quadrupole mass spectrometric analysis of drugs. J. Chromatogr. A 2004a, 1058, 223-232.

Stepan, R.; Cuhra, P.; Barsova, S. Comprehensive two-dimensional gas chromatography with time-of-flight mass spectrometric detection for the determination of anabolic steroids and related compounds in nutritional supplements, Food Add. Contam. Part A 2008, 25(5), 557-565.

Swanson, K.L.; Kado, N.Y.; Funk, W.E.; Pleil, J.D.; Madden, M.C.; Ghio, A.J. Release of the Pro-inflammatory Markers by BEAS-2B Cells Following in vitro Exposure to Biodiesel Extracts. Open Toxicol. J. 2009, 3, 8-15.

Tiyapongpattana, W.; Wilairat, P.; Marriott, P.J. Characterization of biodiesel and biodiesel blends using comprehensive two-dimensional gas chromatography. J. Sep. Sci. 2008, 31, 2640-2649.

Tobias, H.J.; Sacks, G.L.; Zhang, Y.; Brenna, J.T. Comprehensive two-dimensional gas chromatography combustion isotope ratio mass spectrometry. Anal. Chem. 2008, 80, 8613-8621.

Tranchida, P.Q.; Purcaro, G.; Dugo, P.; Dugo, G.; Dawes, P.; Mondello, L. Development of a flexible chip-based pneumatic modulator for comprehensive two-dimensional gas chromatography, the 34th International Symposium on Capillary Chromatography (ISCC) and the 7th GC×GC Symposium, Riva del Garda, May 30-June 4, 2010a (oral presentation).

Tranchida, P.Q.; Costa, R.; Donato, P.; Sciarrone, D.; Ragonese, D.; Dugo, P.; Dugo, G.; Mondello, L. Acquisition of deeper knowledge on the human plasma fatty acid profile exploiting comprehensive 2-D GC. J. Sep. Sci, 2008b, 31, 3347-3351.

Tranchida, P.Q.; Casilli, A.; Dugo, P.; Dugo, G.; Mondello, L. Generation of improved gas linear velocities in a comprehensive two-dimensional gas chromatography system. Anal. Chem. 2007a, 79, 2266-2275.

Tranchida, P.Q.; Donato, P.; Dugo, P.; Dugo, G.; Mondello, L. Comprehensive chromatographic methods for the analysis of lipids. Trends Anal. Chem. 2007b, 26, 191-205.

Tranchida, P.Q.; Purcaro, G.; Conte, L.; Dugo, P.; Dugo, G.; Mondello, L. Optimized use of a 50 μm internal diameter secondary column in a comprehensive two-dimensional gas chromatography System. Anal. Chem. 2009b, 81, 8529-8537.

Tranchida, P.Q.; Purcaro, G.; Conte, L.; Dugo, P.; Dugo, G.; Mondello, L. Higher resolution comprehensive two-dimensional gas chromatography applied to the analysis of roasted coffee volatiles. J. Chromatogr. A 2009a, 1216, 7301-7306.

Tranchida, P.Q.; Purcaro, G.; Fanali, C.; Dugo, P.; Dugo, G.; Mondello, L. Optimized use of a 50 μm ID secondary column in a comprehensive two-dimensional gas chromatography-mass spectrometry system. J. Chromatogr. A 2010b, 1217, 4160-4166.

Tranchida, P.Q; Giannino, A.; Mondello, M.; Sciarrone, D.; Dugo, P.; Dugo, G.; Mondello, M. Elucidation of fatty acid profiles in vegetable oils exploiting group-type patterning and enhanced sensitivity of comprehensive two-dimensional gas chromatography. J. Sep. Sci. 2008a, 31, 1797-1802.

Tranchida, P.Q.; Dugo, P.; Dugo, G.; Mondello, L. Comprehensive two-dimensional chromatography in food analysis. J. Chromatogr. A 2004, 1054(1-2), 3-16.

Truong, T.T.; Marriott, P.J.; Porter, N.A.; Leeming, R. Application of comprehensive two-dimensional gas chromatography to the quantification of overlapping faecal sterols. J. Chromatogr. A 2003, 1019, 197-210.

van Deursen, M.; Beens, J.; Reijenga, J.; Lipman, P.; Cramers, C.; Blomberg, J. Group-type identification of oil samples using comprehensive two-dimensional gas chromatography coupled to a time-of-flight mass spectrometer (GC x GC-TOF). J. High Resol. Chromatogr. 2000, 23, 507-510.

Van Stee, L.L.P.; Beens, J.; Vreuls, R.J.J.; Brinkman, U.A.T. Comprehensive two-dimensional gas chromatography with atomic emission detection and correlation with mass spectrometric detection: Principles and application in petrochemical analysis. J. Chromatogr. A 2003, 1019, 89-99.

Vendeuvre, C.; Bertoncini, F.; Duval, L.; Duplan, J.L.; Thiébaut, D.; Hennion, M.-C. Comparison of conventional gas chromatography and comprehensive two-dimensional gas chromatography for the detailed analysis of petrochemical samples. J. Chromatogr. A, 2004, 1056, 155-162.

Venkatramani, C.J.; Phillips, J.B. Comprehensive two-dimensional gas chromatography applied to the analysis of complex mixtures. J. Microcol. Sep. 1993, 5(6), 511-516.

Villegas, C.; Zhao, Y.; Curtis, J.M. Two methods for the separation of monounsaturated octadecenoic acid isomers. J. Chromatogr. A 2010, 1217, 775-784.

Vlaeminck, B.; Harynuk, J.; Fievez, V.; Marriott, P. Comprehensive two-dimensional gas chromatography for the separation of fatty acids in milk. Eur. J. Lipid Sci. Technol. 2007, 109, 757-766.

Vogt, L.; Groeger, T.; Zimmermann, R. Automated compound classification for ambient aerosol sample separations using comprehensive two-dimensional gas chromatography–time-of-flight mass spectrometry. J. Chromatogr. A 2007, 1150, 2-12.

von Mühlen, C.; Khummueng, W.; Zini, C.A.; Caramão, E.B.; Marriott, P.J. Detector technologies for comprehensive two-dimensional gas chromatography. J. Sep. Sci. 2006, 29, 1909-1921.

von Muehlen, C.; de Oliveira, E.C.; Morrison, P.D.; Zini, C.A.; Caramao, E.B.; Marriott, P.J. Qualitative and quantitative study of nitrogen-containing compounds in heavy gas oil using comprehensive two-dimensional gas chromatography with nitrogen phosphorus detection. J. Sep. Sci. 2007, 30, 3223-3232.

Wahl, J.H.; Riechers, D.M.; Vucelick, M.E.; Wright, B.W. A portable multi-dimensional gas chromatographic system for field applications. J. Sep. Sci. 2003, 26, 1083-1090.

Wang, F.C.Y. New valve switching modulator for comprehensive two-dimensional gas chromatography. J. Chromatogr. A 2008, 1188, 274-280.

Wang, F.C.Y. PatentsWO 2007106505 20070920, PCT Int. Appl. (2007), US 2007/0214866 A1.

Wang, F.C. Two Valve Switching Modulator for Comprehensive Two-dimensional Gas Chromatography. U.S. Patent 7,779,670. Filed March 9, 2007, issued August 24, 2010.

Wang, F.C.Y.; Robinson, W.K.; Greaney, M.A. Speciation of nitrogen-containing compounds in diesel fuel by comprehensive two-dimensional gas chromatography. J. Sep. Sci. 2004, 27, 468-472.

Welthagen, W.; Shellie, R.A.; Spranger, J.; Ristow, M.; Zimmermann, R.; Fiehn, O. Comprehensive two-dimensional gas chromatography–time-of-flight mass spectrometry (GC x GC-TOF) for high resolution metabolomics: Biomarker discovery on spleen tissue extracts of obese NZO compared to lean C57BL/6 mice. Metabolomics 2005, 1, 65-73.

Western, R.J.; Lau, S.S.G.; Marriott, P.J.; Nichols, P.D. Positional and geometric isomer separation of FAME by comprehensive 2-D GC, Lipids 2002, 37(7), 1-10.

Zhu, S.; Lu, X.; Ji, K.; Guo, K.; Li, Y.; Wu, C.; Xu, G. Characterization of flavor compounds in Chinese liquor Moutai by comprehensive two-dimensional gas chromatography/time-of-flight mass spectrometry. Anal. Chim. Acta 2007, 597, 340-348.

Zrostlikova, J.; Lehotay, S.J.; Hajslova, J. Simultaneous analysis of organophosphorus and organochlorine pesticides in animal fat by gas chromatography with pulsed flame photometric and micro-electron capture detectors. J. Sep. Sci. 2002, 25, 527-537.

11

Thin-layer Chromatography with Direct Mass Spectrometric Analyte Detection

Beate Fuchs and Jürgen Schiller
University of Leipzig, Medical Department, Institute of Medical Physics & Biophysics, Härtelstr. 16/18, D-04107 Leipzig, Germany

Abstract

Thin-layer chromatography (TLC), and particularly high-performance thin-layer chromatography (HPTLC), is a widely used, fast, and inexpensive separation method. Even minor analytes of complex mixtures often can be separated from bulk constituents and monitored with high sensitivity and resolution. However, unambiguous assignment of the obtained spots to defined compounds often is difficult, because the application of common TLC staining methods reveals a substance class but not a specific compound. In the case of phospholipids, the lipid class, but not the detailed fatty acyl composition, can be determined. Combining TLC directly with spectroscopic methods, e.g., nuclear magnetic resonance (NMR) spectroscopy is an option. However, limited sensitivity of such methods deters their use. Today, mass spectrometry (MS) is the most suitable method of analyte characterization, due to its high sensitivity and the detailed mass and fragment information provided. Invention of soft ionization methods, such as matrix-assisted laser desorption and ionization time-of-flight (MALDI-TOF) MS and electrospray ionization (ESI), made this possible; today, other desorption and atmospheric pressure ionization techniques are also available.

This chapter covers the direct combination of TLC with MS, or TLC/MS, and overviews different molecule classes that have been analyzed by this approach. The significant progress from 2005–2010 will be emphasized.

Chromatographic Methods and Product Identification

Planar chromatography (in particular thin-layer or high-performance thin-layer chromatography) and high-performance liquid chromatography (HPLC) are very important tools of modern analytical chemistry (Hahn-Deinstrop, 2006).

TLC is a frequently used separation technique in (synthetic) organic chemistry and natural product chemistry, where normally only small amounts of the samples are available. HPLC can also be used for the analysis of large biomolecules, such as proteins or polysaccharides, which cannot be appropriately characterized by TLC. When compared with HPLC, TLC offers the following important advantages, in addition to the simple performance and the relatively low costs (Hahn-Deinstrop, 2006; Fuchs et al., 2009):

- TLC is affected to a much lesser extent than HPLC by impurities in the sample. Therefore, TLC may be used for the analysis of "suspicious" samples (for instance, crude tissue or food extracts) that might easily damage or plug the column of an HPLC system.
- TLC is less expensive, and particularly less harmful, as well as more environmentally friendly, because the consumption of solvents is much lower than in the case of HPLC.
- TLC does not provide any "memory" effects, since a completely new stationary phase is used in all cases. HPLC columns are normally expensive and can be reused after careful washing to remove previous analytes, but there can be risk of some residual material on the column. Thus, it is often more difficult to certify an HPLC assay than a TLC assay.
- Nowadays there are "multiplexing" HPLC systems available (Woodbury et al., 1995) that allow the simultaneous analysis of several samples in parallel. Note, the simple analysis of a multitude of samples is an obvious advantage of TLC.
- Quantitative data analysis using specific dyes that bind to a certain analyte is possible. This is simpler than in HPLC, where post-column derivatization of non-chromophores is often difficult.
- TLC plates can be stored after development and later re-analyzed.

This chapter does not cover improvements of the achievable separation quality, done either by optimizing the stationary phase or the solvent system or sample application, because there is hardly a class of small molecules that has not yet been analyzed by TLC. Separation is one important step of successful analysis, however; detection of the different analytes on the TLC plate and their structural elucidation also is very important. The retardation factor (R_f value) by which a compound is characterized, is influenced by many parameters, such as the solvent composition, the TLC plates used, the relative humidity, the temperature, etc. Therefore, it is not easily reproducible. If unknown natural products must be analyzed, analyte identification is problematic if no reliable reference compounds are available (Hahn-Deinstrop, 2006; Marderosian & Chao, 1974).

The following approaches can be used for identification of TLC-separated spots (Geiss, 1987).

(a) Subsequent to the re-elution of individual compounds from the TLC plate
This method is tedious, time-consuming, and risks losing large amounts of the analytes, yet seems most straightforward, applicable to all compounds, and all frequently used spectroscopic methods (such as NMR or MS) can be applied. A very useful tool is commercially available from CAMAG that helps to extract samples from a TLC plate in a very convenient and highly reproducible manner. It is based on an interface for hyphenation of planar chromatography with mass spectrometry (Luftmann et al., 2007). Nevertheless, it was recently shown in the context of lipids that TLC may be used without any risk of major sample alterations. Although there were significant solvent-dependent losses in the absolute amount of each lipid, the relative composition of the mixture remained constant (Teuber et al., 2010).

(b) Directly on the TLC plate by spectroscopic analysis
In contrast to (a) this approach does not suffer from potential losses upon the re-extraction process of the analyte and retains the achieved separation quality completely (components with very similar R_f values cannot be very accurately separately re-eluted and analyzed by the method described previously). Several different spectroscopic techniques have been applied successfully and will be discussed in this chapter.

(c) Without any derivatization directly on the TLC plate
This method is limited to molecules such as colored compounds (e.g., plant or synthetic dyes) or molecules with an intense UV absorption or fluorescence. The analyte can be directly visualized, is not destroyed by any harsh staining process, and afterwards can be re-analyzed by any suitable analytical method. Unfortunately, most compounds cannot be detected by this simple approach, because they are not colored or lack UV absorbance, and the achievable sensitivity is normally quite low.

d) Subsequent to specific derivatization directly on the TLC plate
Characteristic dyes commonly are used to monitor specific compounds that react with the dye due to the presence of defined chemical functional groups (Nakamura & Pisano, 1976). For example, ninhydrin reacts specifically with amino groups, whereas dinitrophenylhydrazine (DNPH) is a well-known and useful derivatization agent of carbonyl groups present in aldehydes or ketones. The use of moderately concentrated sulfuric acid and subsequent heating ("charring") is a commonly used but completely nonspecific approach (Hahn-Deinstrop, 2006). Under these conditions, the resulting black spots allow the identification of colorless compounds directly on the TLC plate, but virtually all organic compounds are destroyed. This is a frequently used approach in the field of lipid or fatty acid analysis, i.e., if compounds with relatively high carbon contents are of interest (Klein et al., 1998).

Mass Spectrometry

Mass spectrometric (MS) techniques offer high sensitivities and are used most often to evaluate TLC-separated compounds (for a more detailed survey of different MS techniques, see Morlock & Schwack, 2010, 2010a). We will focus here on the two most common approaches. One of these two techniques characterizes the sample by MS after re-elution from the TLC plate. For the sample evaluation, normally electrospray ionization (ESI), MS is used because it seems the most adaptable ionization technique. For example, ESI MS enables the ionization of nearly all compounds, even rather labile ones. The second MS technique is based on a direct generation of the ions from a solid surface.

MS Methods Based on Previous Extraction of the Analytes

For quantitative TLC/ESI MS analysis on HPTLC silica gel plates, a plunger-based extraction interface (commercially available as the "ChromXtract" from CAMAG) is used with an HPLC pump. For a selected model compound (harmane, a heterocyclic aromatic amine), these MS spectra provide good results regarding repeatability, the achievable limit of detection (LOD), and the limit of quantification (LOQ) (Jautz & Morlock, 2006). Complete and highly reproducible extraction of different analytes from glass-backed as well as aluminum-backed TLC and HPTLC plates, with layer thicknesses up to 100 μm and different stationary phases, were enabled by slight modifications of the device (Alpmann & Morlock, 2006).

In a solid pharmaceutical formulation, the "ChromXtract" was also used for the simultaneous determination of caffeine, ergotamine, and metamizole by HPTLC [silica gel 60F layer, ethyl acetate/methanol/ammonia (90:15:1, *v/v/v*) as the mobile phase], with UV absorption densitometry for caffeine and metamizole and fluorescence densitometry for ergotamine. By single quadrupole [a common mass analyzer (Hillenkamp & Peter-Katalinić, 2007; Fuchs et al., 2008) MS] additional mass confirmation was performed in the positive ion ESI full scan mode for caffeine and ergotamine and the negative ion mode for metamizol (Aranda & Morlock, 2007). A caffeine standard with (deuterated) caffeine-d_3 was used for the correction of the ChromXtractor plunger positioning; this resulted in considerably enhanced data accuracy.

This is a highly versatile approach. For more information on TLC/MS by re-extraction of the analyte, see Morlock and Sherma (2008, 2009). The following sections focus on the use of desorption techniques.

MS Desorption Methods

Of the many desorption MS techniques in use, we focus on two of them: (1) matrix-assisted laser desorption ionization (Hillenkamp & Peter-Katalinić, 2007) and (2) desorption electrospray (DESI) (Miao & Chen, 2009). MS. MALDI MS will be discussed in more detail, but a detailed discussion of DESI MS is beyond the scope of this chapter (see Chen et al., 2005).

A very new and promising method is HPTLC/Direct Analysis in Real Time DART-TOF MS, which is commercially available from the IonSense company. Isopropylthioxanthone (ITX) as a selected example could be successfully identified directly from a TLC plate with detection limits in the low ng/zone range. ITX is used as a photoinitiator in offset printing and was previously found as an impurity in baby milk products. By using stable isotope labeled standards, the quantitative analysis could be notably improved. By using an in-house plate holder system, the spatial resolution was less than 3 mm (Morlock & Ueda, 2007).

This technique can characterize TLC plates and native tissue samples (Schäfer et al., 2009).

For example, this method would be very useful for the differentiation between healthy and cancer tissue by the presence or absence of certain marker molecules. The rapid performance of this technique ["Direct Analysis in Real Time" ("DART")] can make its applications during surgical procedures very likely in the near future.

NMR Spectroscopy

Nuclear magnetic resonance spectroscopy is not very sensitive, yet it is a powerful analytical tool. However, most of the available data are based on ^{1}H NMR because ^{1}H is the most sensitive nucleus (Berger & Braun, 2004). Unfortunately, there are yet only a very few papers about this topic: Organic reaction products could be identified directly from TLC spots by high-resolution magic angle spinning (HR MAS) NMR spectroscopy (Bradley & McLaughlin, 2007; Wilson et al., 1997).

HR MAS uses fast sample rotation (a few thousand Hertz) at the magic angle (54.7°) to reduce the line-widths of the individual resonances. An advantage is that the less complex pulse sequences of high resolution NMR can be used instead of the complex pulse sequences of solid state NMR (Schiller et al., 2004).

- The spot of interest is scraped from the TLC plate and transferred to an HR MAS sample rotor (available from nearly all manufacturers of NMR spectrometers).
- For the measurements, a suspension in a small amount of deuterated solvent is required and the use of D_2O instead of H_2O helps to suppress the solvent signal.
- A special reduced-volume insert of only about 20 μl in the rotor reduces the amount of required sample.

One drawback of this method is that the stationary phase, i.e., the -OH groups of the silica gel give a broad background signal near about 4.6 ppm. This broad resonance may dominate the spectra and reduce the sensitivity to detect resonances of interest. Nevertheless, the intensities of broad resonances can be significantly reduced by the Carr-Purcell-Meiboom-Gill (CPMG) pulse sequence that eliminates compounds with short relaxation times that would lead to broad resonances (Berger & Braun, 2004). One-dimensional ^{1}H NMR spectra with a good signal-to-noise ratio (S/N) require only a few minutes in contrast to two-dimensional ^{1}H NMR that require several hours. All TLC/NMR applications were exclusively based on ^{1}H NMR, and no attempts were made to record NMR spectra of less sensitive nuclei such as ^{13}C (Berger & Braun, 2004).

IR and Raman Spectroscopy

Infrared (IR) and Raman spectroscopy offer complementary information but IR is the more common technique (Günzler & Gremlich, 2002). IR spectroscopy is particularly suitable to detect polar functional groups such as C=O or O—H groups. In contrast, Raman spectroscopy is more useful for the characterization of relatively apolar groups such as C—H groups. IR irradiation has a longer wavelength and thus a lower energy than UV light (that is very common for monitoring spots of aromatic or unsaturated compounds on a TLC plate) and penetrates therefore deeper into the TLC sample plate. This reduces significantly the risk of "loosing" compounds at the inner portion of the silica gel layer.

By HPTLC and surface-enhanced Raman spectrometry (SERS) the ingredients from the small berry fruit of *Evodia rutaecarpa* (known as "Wu-Zhu-Yu" in Chinese traditional medicine with anti-inflammatory activity) were analyzed at the microgram level. The achieved absorptions could be directly used and no standard samples were needed (Zhang et al., 2007). Recently, amino acids such as L-lysine were quantitatively determined directly from the TLC plate by near IR (NIR) spectroscopy. Commercially available apple juice was used as the sample and cellulose TLC with n-butanol/acetic acid/water (8:4:2, *v/v/v*) was used to separate the analytes of interest. The detection level of amino acids under these conditions was 10 μg/mL (Heigl et al., 2006).

Functionality of MALDI-TOF MS and Combination with TLC

Only a few applications of the combined TLC/MALDI technique have been described, even though this technique is commercially available from Bruker Daltonics (Fuchs et al., 2009). This topic is currently under intense research and a "hot topic" in the TLC community and will be discussed in more detail in this chapter. MALDI MS increasingly is used as an "imaging" technique for thin-tissue slices and much research work has been performed to improve the quality and reproducibility of the achievable images (Cornett et al., 2007). The MALDI MS "imaging" technique, where the analyte of interest is dispersed in a "matrix" of other molecules, uses thin-tissue sample slices subsequent to dehydration. As a TLC plate and a thin-tissue slide are similar samples, some of the methods developed for tissue imaging may also be used to analyze a lane on a TLC plate and to obtain a "scan" of the TLC surface.

As previously mentioned, directly combined TLC/MALDI MS presents some important advantages: (a) the mass spectrum of a given compound is much more indicative than the corresponding, rather difficult to reproduce Rf value and (b) significant spatial resolution may be obtained by MALDI MS as the laser beam has a focal spot of the order of 100 μm or even smaller (Hillenkamp et al., 1975).

In this way, even spots that cannot be differentiated by visual inspection of the TLC plate or by scraping off the spots of interest can be identified. For an increase in the achievable resolution, a further decrease of the laser spot size would be highly desirable. However, also consider that the spot size correlates directly with the achievable sensitivity, because ions are directly desorbed from the area irradiated by the laser beam. Thus, 100 μm is a good compromise between resolution and sensitivity.

MALDI MS was developed at the end of the 1980s and today is widely used. A detailed survey of the different applications of MALDI-TOF MS is provided in the timely book by Franz Hillenkamp and Jasna Peter-Katalinić (2007). This chapter focuses exclusively on the MALDI method, although other soft-ionization methods such as electrospray (ESI) are also available. Furthermore, we will focus here nearly exclusively on the combination with a time-of-flight (TOF) mass analyzer (Fuchs et al., 2008), although there are other mass analyzers to combine with the MALDI ion source. The majority of commercially available MALDI MS devices is equipped with a TOF mass analyzer and, thus, available to the majority of MS spectroscopists.

In Fig. 11.1, a schema of ion generation and mass analysis by MALDI-TOF MS is provided. MALDI-TOF MS utilizes in most cases an ultraviolet-absorbing matrix. There are also IR lasers in use that require completely different matrix compounds such as glycerol that absorbs intensively in the IR range (Fuchs & Schiller, 2009). However, UV matrices primarily are discussed because most commercially available MALDI devices are equipped so far with N_2 lasers emitting at $\lambda = 337$. The matrix fulfills two important tasks: (a) it absorbs the energy emitted by the laser and (b) prevents the aggregation of the analyte molecules that would otherwise result in cluster ion formation and seriously complicate the spectra (Schiller et al., 2007).

The sample on the MALDI sample plate (the "target") contains primarily co-crystals between the matrix and the analyte, in which the matrix is present in a 50- to 1000-fold excess over the analyte of interest. When the pulsed nitrogen laser hits the sample, the matrix absorbs its energy. Consequently, the matrix is vaporized, and carries intact analyte molecules into the vapor phase. Ions such as H^+ and Na^+ or other alkali metal ions are exchanged between the matrix and the analyte during the expanding process of this gas

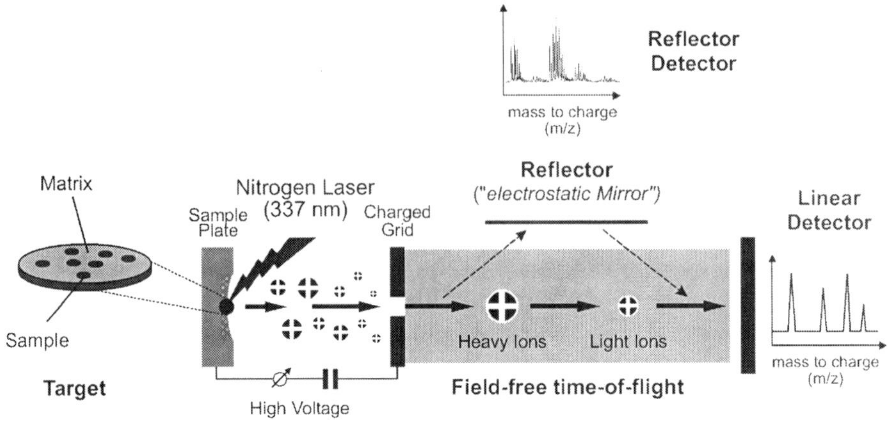

Fig. 11.1. Schema of the events occurring in a conventional MALDI-TOF mass spectrometer: When the laser beam hits the spot on the sample plate the analyte–matrix mixture is evaporated and ionization occurs by collisions with ions present in the matrix analyte mixture. These analyte ions are accelerated in an electric field, pass a charged grid and afterwards are allowed to drift in a field-free space. A separation between low-mass and high-mass ions occurs over the "flying" distance. Therefore "time-of-flight" (t) depends on the flight path (L), the mass of the ion (m), its energy (e U) and the number of charges (z). Reprinted with modifications and with permission from Elsevier (Schiller et al., 2004a).

cloud. This leads to the formation of charged analyte molecules, which are called "adducts" or "quasimolecular ions". Afterwards, ions drift freely over a field-free pathway and mass separation is achieved. Low-mass ions arrive at the detector in a shorter time than high mass ions (Schiller et al., 2004).

MALDI is often used for the detection of large molecules. Therefore, the TOF detector is very popular since it has a nearly unlimited mass range. An additional reason for the popularity of the TOF detector is the pulsed ion generation of MALDI that is most suitable for this detector (Hillenkamp & Peter-Katalinić, 2007). Besides the described cation generation, anion generation may also be achieved in the MALDI process by abstracting H^+ or Na^+ ions from the analyte. Positive and negative ions can be simply differentiated by inverting the direction of the applied electric field. The ratio between cations and anions is determined by the gas phase acidities of the analyte and the matrix (Fuchs & Schiller, 2009).

MALDI-TOF MS also can be used for small molecules, although the most popular use is the analysis of molecules with higher masses in the range of several kilodaltons such as proteins. The analysis of small molecules is covered in this chapter. However, the analysis of small molecules has one major drawback: The matrix tends to undergo photochemical reactions and the resulting products may interfere with the analytes of interest in the low mass range (Peterson, 2007).

This is called matrix interference. Some approaches suggest using unconventional matrices such as pencil lead instead of conventional small organic molecule matrices. However, currently these alternative approaches are not widely used. Peterson (2007) provides more details on these approaches. The chemical structures of some selected important small organic molecule matrices are shown in Fig. 11.2.

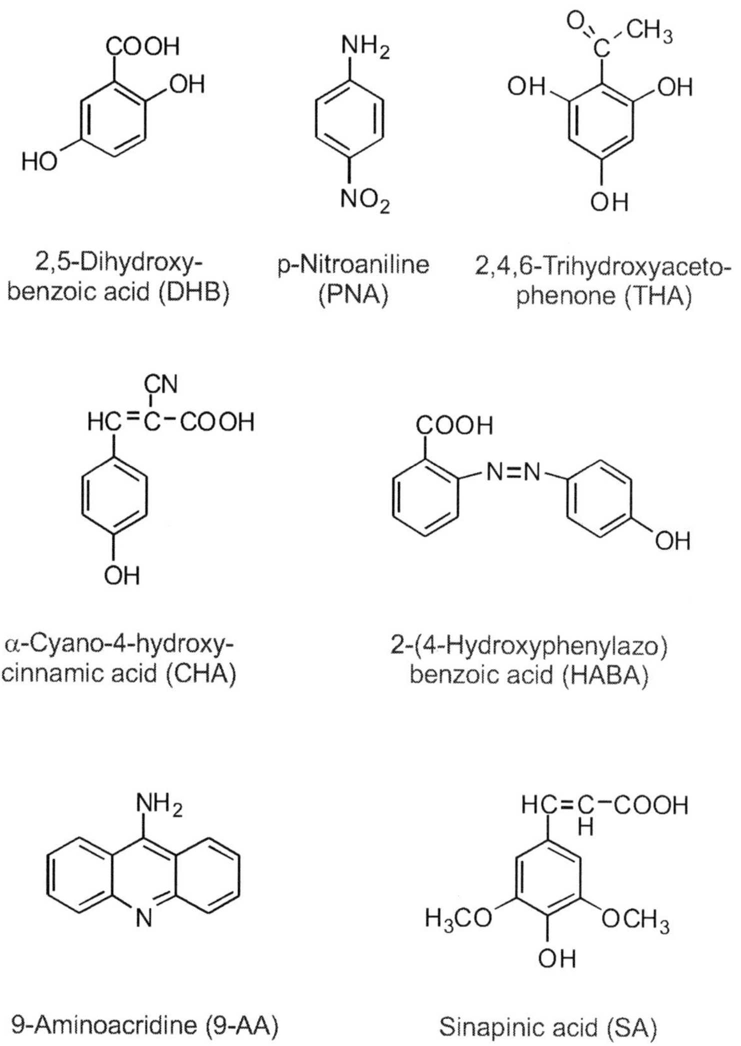

Fig. 11. 2. Chemical structures of important, frequently used UV MALDI matrix compounds (small organic molecules). Please note that 9-AA, THA and PNA are particularly useful for recording negative ion spectra. Other matrices such as DHB can be used in the positive as well as in the negative ion mode but give normally better quality positive ion spectra due to the presence of the carboxylic acid.

Quantitative Aspects

Absolute quantitative data cannot usually be provided by MALDI-TOF MS, because the detection limits of different compounds may differ significantly. For example, the detection limit of phospholipids is of the order of 20 picograms (Gellermann et al., 2006) corresponding to about 25 femtomole if 800 Da is assumed as a typical molecular mass of a phospholipid. However, the presence of salts, detergents, other lipids, etc., may significantly increase the detection levels (Zschörnig et al., 2006). This means the mentioned detection limit for phospholipids is valid only for an isolated lipid in nearly salt-free organic solvents in the absence of impurities and only for the use of 2,5-dihydroxybenzoic acid (DHB) as matrix. These considerations do not hold exclusively for phospholipids, but for other substances, too. Furthermore, the accuracy of quantitative information available from MALDI-TOF mass spectra is often questioned due to the irregularities of the co-crystals between the matrix and the analyte. Regarding quantitative data analysis, chromatographic and mass spectrometric methods (LC/MS) that make use of solutions provide obviously more accurate data. By definition, solutions always show a homogeneous distribution of the analyte and there are no inhomogeneities. Nevertheless, there is an increasing number of papers showing that carefully recorded MALDI mass spectra may be quantitatively analyzed (Duncan et al., 2008). The different, structure-dependent detectabilities of individual compounds (related to the pK values) cause additional problems if complex mixtures are analyzed (Schiller et al., 2007a). A previous separation of the crude mixture into the individual compounds or at least compound groups is normally required in order to obtain reliable data (Schiller et al., 2007a).

Combining TLC with MALDI

The direct combination of TLC with MS detection is driven by the aim of obtaining a detailed TLC "image" without compromising TLC resolution and MS sensitivity. This effort has been pursued for about 30 years. Several different approaches to combine TLC with MS detection have been suggested.

1. Scraping the spot of the analyte from the TLC plate and extracting the compound of interest into a suitable solvent. Afterwards the extracted analyte can be analyzed by MS in the conventional manner; but compounds with very similar Rf values cannot be readily differentiated under these conditions. This approach also is very tedious and time-consuming, yet very simple to use and provides generally the desired information.
2. A small piece of the entire TLC plate is cut off and analyzed by MALDI MS: Thus, the spot of interest is analyzed in the presence of the stationary phase-silica gel in the majority of cases. This method has the

advantage that irregularities of the TLC plate play the smaller roles, the smaller the size of the TLC plate.

3. The entire TLC plate is placed into the mass spectrometer for direct analysis of the entire TLC plate so that two-dimensional scanning can be achieved. This is unequivocally the most direct approach.

When the spot of interest is analyzed in the presence of the silica gel on the TLC plate, two important requirements are indispensable for successful recording MALDI mass spectra: the UV-absorbing matrix (if using the most common devices equipped with UV lasers) must be applied onto the TLC plate and the analyte must be extracted from the inner portion of the sample plate to the TLC surface, because UV laser irradiation does not penetrate deeply into the TLC plate. Furthermore, major spreading of the analyte during the addition of the matrix and the solvent must be avoided as far as possible. Otherwise, this would significantly compromise the achieved quality of chromatographic separation. Both important prerequisites are discussed in more detail in (Mehl et al., 1997).

Thus, the application of the MALDI matrix solution is an important prerequisite of the TLC/MALDI measurement (Wilson, 1999). Successful methods to apply the matrix can be summarized as follows:

(a) Adding small droplets of the matrix solution onto the (previously marked) spots of interest by a common pipette.
(b) Using a solvent or a solvent mixture with a relatively high surface tension resulting in rather small spots when applied to the TLC plate. Mixtures of water and acetonitrile were already successfully used in this context (Nakamura et al., 2006).
(c) Spraying the matrix solution onto the TLC plate (e.g., by using electrospray equipment). Instruments based on this approach are nowadays already commercially available in the context of MS imaging (for instance the "ImagePrep" device from Bruker Daltonics).
(d) Brushing the TLC plate with a "supersaturated" solution of the matrix (Mowthorpe et al., 1999).
(e) "Pressing" matrix crystallized on a special surface (e.g., a stainless steel block) onto the TLC plate.
(f) In order to overcome problems with surface irregularities of the TLC plate (and the resulting mass inaccuracies), the use of a "hybrid plate" was also suggested (Mehl & Hercules, 2000). After TLC separation, the analyte spots were eluted from the TLC silica layer to the MALDI target. Although this is a rather indirect approach, it was shown that the detection limits (in the femtomole range) could be significantly improved under these conditions (Mehl & Hercules, 2000).

After application of the matrix, the TLC plates need to be attached to the MALDI plates and the most obvious and simple approach is to fix a common TLC plate with conductive adhesive tape onto a standard MALDI target and to load it directly into the mass spectrometer. However, the back of the TLC plate is very important. As electric conductivity is required in order to allow desorption of ions from the TLC plate, glass backs may not be used as they are not conductive. The simplest way to overcome this problem is to use commercially available TLC plates with aluminum backs.

Summaries of the specifics of the coupling between MALDI and planar as well as column chromatography can be found in Gusev (2000) and Busch (2004).

The following is an overview of compounds that were recently investigated by combined TLC/MALDI. Some selected examples are chosen for illustration of the capabilities and limitations of this approach, although it is not our aim to give a complete survey. The focus of this survey will be on smaller molecules because TLC is particularly useful for the separation of such molecules (Sherma, 2008).

Characterizing of DNA and Nucleic Acids by Direct TLC/MALDI Coupling
The MS analysis of DNA is challenging because of the phosphate backbone and the resulting significant negative charge density (Douthwaite & Kirpekar, 2007). Charged compounds are much more resistant to MALDI MS analysis than neutral molecules with a comparable molecular weight (Hillenkamp & Peter-Katalinić, 2007; Fuchs et al., 2008). Thus, for charged molecules much higher amounts of samples are needed because of a much lower sensitivity. Attempts to protonate the phosphate group to screen its charge by performing the MS analysis under sufficiently acidic conditions should be regarded with great caution as cleavage of the phosphate residues might easily occur. However, the significant negative charge density of DNA can be used to record negative ion spectra: Phosphorylated inositols show that each phosphate residue aggravates the MS detection. This particularly holds for positive ion detection, but also to a minor extent for negative ion detection (Müller et al., 2001).

Detection of smaller oligonucleotides (obtained by enzymatic digestion of native DNA prior to TLC) is possible without major problems (Isbell et al., 1999). The detection limits of small oligonucleotides are in the range of about 10 pg when 2,4,6-trihydroxyacetophenone is used as the matrix and ammonium hydroxide/methanol as solvent system (Isbell et al., 1999).

Characterizing of Small Proteins and Peptides by Direct TLC/MALDI Coupling
The analysis of proteins and peptides ("proteomics") is the traditional focus of MALDI (Yates et al., 2009). Therefore, peptides were historically used as the first components in order to evaluate the capabilities and limitations of combined TLC/MALDI. Gusev and coworkers (1995) investigated in 1995

the suitability of various MALDI matrices for the detection of smaller proteins (bradykinin, angiotensin, and enkephalin that possess molecular masses between about 500 and 1500 Da) on silica gel and cellulose TLC plates. These peptides are particularly important because they are widely used as mass calibrants for MALDI-TOF mass spectra (Hillenkamp & Peter-Katalinić, 2007).

For the direct investigation of TLC-separated tryptic digests of phosphopeptide mixtures, detection limits in the femtomolar range and a superior separation quality in comparison to the established (electrophoretic) protocols could be obtained (Kochin et al., 2006). A new approach regarding the analysis of phosphopeptides, a combined two-dimensional planar electrochromatography / TLC with MALDI-TOF mass spectrometric detection was recently suggested (Panchagnula et al., 2007). Phosphopeptides migrate more slowly in the first dimension, based upon their anionic phosphate residues and the resulting charge, and certain predominantly acidic phosphopeptides even migrate in the opposite direction, relative to the bulk of the peptides. Phosphopeptides were further distinguished based upon their hydrophilicities in the second dimension. This permits a restricted region of the plate to be directly interrogated for the presence of phosphopeptides by MALDI-TOF MS without the interference of residual proteins.

Homemade TLC plates are sometimes more advantageously used in certain applications than commercially available ones. For example, a monolithic porous polymer layer significantly improved both the TLC separation of larger peptides, in comparison to commercially available pre-coated TLC plates, and their detection by MALDI-TOF MS (Bakry et al., 2007). These monolithic thin layers were prepared *in situ* using UV-initiated polymerization; 50- to 200-μm thin poly(butyl methacrylate-co-ethylene dimethacrylate) layers, and could be prepared directly on the stainless steel MALDI targets.

Using 2-(4-hydroxyphenylazo)benzoic acid (HABA) as matrix provided the very best homogeneity regarding peptide analytes and absolute detection limits of 2–4 ng (on the TLC plate) could be established. In contrast, ferulic acid and sinapinic acid were less sensitive but showed a higher salt tolerance. Analyte fragmentation was comparable to standard MALDI (on a stainless steel target). The problem of analyte spreading upon applying the matrix solution to the TLC plate was solved in an interesting way (Gusev et al., 1995). A matrix layer was first generated on a stainless steel target, from which it was transferred to the TLC plate of interest. The transfer was accomplished by pressing the matrix onto the previously wetted TLC plate. Subsequent to this preparation, the TLC plate was directly inserted into the MS device. Under these conditions, absolute detection limits in the picogram range and spatial resolutions of about 250–500 μm could be obtained even for larger molecules (Gusev et al., 1995). However, TLC/MALDI has experienced very little interest in the field of peptide analysis most probably due to the widespread establishment of LC/MS methods in this field (Müller et al., 2007).

Characterizing Phospholipids by Direct TLC/MALDI Coupling

Phospholipids (PL) are important constituents of all tissues and particularly cells. In the past they were not considered to be very important but were primarily regarded to represent the "packaging material" for other much more important molecules such as proteins. This situation has changed significantly when the important role of PLs in different pathologies and as important second messenger molecules was discovered (Fernandis & Wenk, 2007). Nowadays the term "lipidomics" has become very popular (Fuchs & Schiller, 2009a) and MALDI-TOF MS analysis of lipids, as well (Schiller et al., 1999; Harvey, 1995). Especially MALDI imaging has improved the acceptance because PLs ionize particularly well and are, thus, easily detectable under MALDI imaging conditions. A brief overview of the structures of selected sphingo- and phospholipids is given in Fig. 11.3.

The strongly varying detectabilities of the individual PL classes represent serious problems regarding mixture analyses. Compounds with quaternary

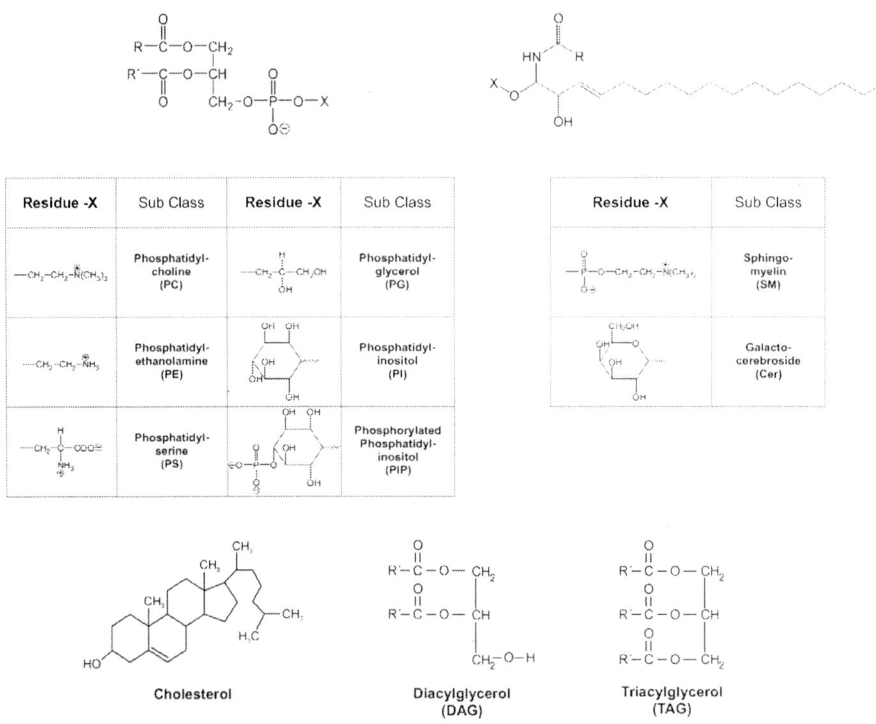

Fig. 11. 3. Overview of lipids occurring in cellular membranes. The head group structures of the relevant phospholipids are shown in the left table, whereas the structures of sphingosine-derived lipids are shown at the right. The sinuous lines indicate that stereochemical aspects of carbohydrate chemistry will not be considered. Please note that only PC and PE represent neutral (zwitterionic) phospholipids, while all others are acidic phospholipids. R and R′ represent varying fatty acyl residues. At the bottom, some apolar lipids are additionally shown.

ammonia groups [e.g., phosphatidylcholine (PC), sphingomyelin (SM), and lysophosphatidylcholine (LPC)] are detectable with high sensitivities as positive ions and may suppress further lipids that are only detectable with lower sensitivities [e.g., phosphatidylethanolamine (PE), phosphatidylinositol (PI), or phosphatidylglycerol (PG)] (Petković et al., 2001). This is illustrated in Fig. 11.4 on the selected example of an extract from hen egg yolk.

The strongly varying detectabilities may also be considered as an advantage because only some selected PL classes are detected in the mixtures and, thus, spectra are less complex. However, for a detailed analysis of complex PL mixtures chromatographic separation of the total lipid extracts is normally required.

There are many reports where MALDI MS analysis was performed subsequent to TLC separation accompanied by scraping off the spots of interest and the re-elution of the related lipids (Touchstone, 1995). For example, the lipid composition of body fluids (Sommerer et al., 2004), cells (Leßig et al., 2004), and tissues (Schiller et al., 2003) could be successfully analyzed by this approach.

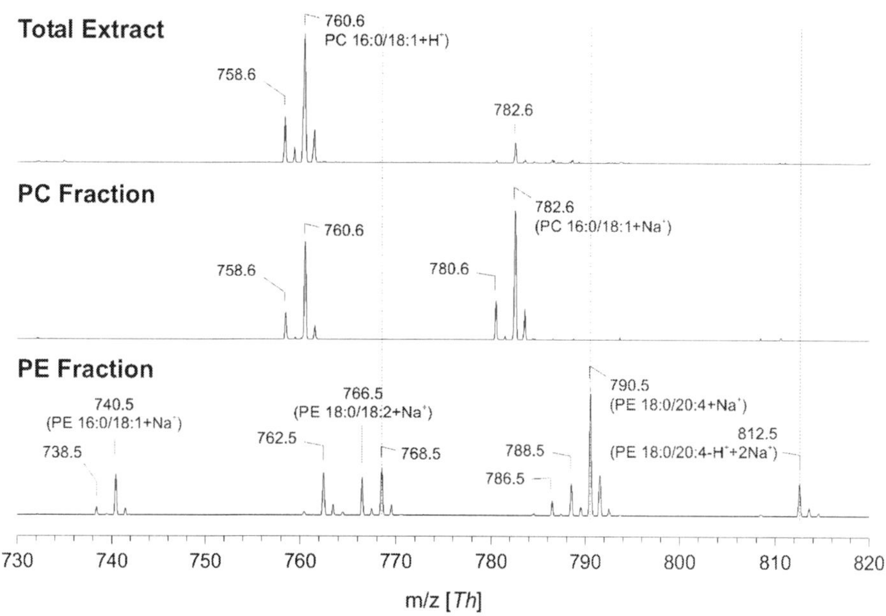

Fig. 11.4. MALDI-TOF mass spectra of an organic extract from hen egg yolk to illustrate significant differences regarding detection limits. The upper spectrum demonstrates the total extract (without chromatographic separation). For the middle and the lower spectra, separation of lipids was performed by TLC prior to MS analysis. The middle spectrum represents the isolated phosphatidylcholine (PC) fraction that is virtually identical to the spectrum of the total extract. In the lower spectrum the phosphatidylethanolamine (PE) fraction is shown. It is obvious that there is not even a small peak of the most abundant PE species (PE 18:0/20:4 +Na$^+$, m/z = 790.5) detectable in the mass spectrum of the total extract. Only the most intense peaks have been assigned in the figure.

Another important note regarding TLC analysis of phospholipids is that in addition to the "normal" diacyl-phospholipids there are also alkenyl-acyl phospholipids called "plasmalogens". For instance, spermatozoa or stem cells possess an extraordinary high plasmalogen content (Leßig & Fuchs, 2009; Schiller et al., 2003). These plasmalogens are not formed by ester condensation of the glycerol backbone with a fatty acid but with a fatty aldehyde. A comprehensive review about plasmalogens is available in (Leßig & Fuchs, 2009). The finding that plasmalogen PL fractions do often contain the corresponding lysophospholipids formed by decomposition clearly indicates that the alkenyl ether group is extremely sensitive towards even traces of acids (Schiller et al., 2003). Thus, plasmalogens are even sensitive against the acidic properties of the silica gel and are already hydrolyzed under these conditions into the corresponding lyso-compounds.

Two different methods were so far attempted for the direct analysis of phospholipids separated on a TLC plate by MALDI-TOF MS. One approach is based on the use of an IR laser and glycerol as matrix (Rohlfing et al., 2007). This approach has the significant advantage that IR radiation penetrates deeper into the TLC plate, but simultaneously confers the disadvantage that abundant glycerol adducts (and to a minor extent even NaCl adducts) with the PLs of interest are detected. Another problem is that IR lasers are commercially barely available so far.

Another approach used a commercially available nitrogen laser-equipped MALDI-TOF MS device and standard DHB as matrix (Fuchs et al., 2007). A special adapter device was used for the introduction of the TLC plate into the mass spectrometer (Fig. 11.5), whereby the TLC plate was simply mounted onto the target by conductive adhesive tape (Fig. 11.5a) (Fuchs et al., 2007; Fuchs et al., 2009a).

After a test period of the MALDI-TOF adapter target prototype used with conductive adhesive tape to fix the sample (Bruker Daltonics, Bremen, Germany) an adapter plate with mechanical fixation of the TLC plate was developed by Bruker Daltonics. The TLC plate is inserted into the centre of the target and can then be fixed mechanically (Fig. 11.5b). This adapter target is now commercially available.

In order to illustrate the achievable analytical power of this approach, one selected lane of a stem cell extract separated by TLC and some selected positive ion MALDI mass spectra that were directly recorded from the developed TLC plate are shown in Fig. 11.6.

Two facts are particularly remarkable. First, even rather minor fractions of phosphatidylglycerol (PG), e.g., can be easily analyzed and give spectra with an adequate signal-to-noise (S/N) ratio. As PG makes up only about 1.3% of the total PLs of the stem cells (Fuchs et al., 2008a). This proves the significant dynamic range of more than two orders of magnitude. Second, the lower and the upper part of the PC spot provide significantly different mass spectra,

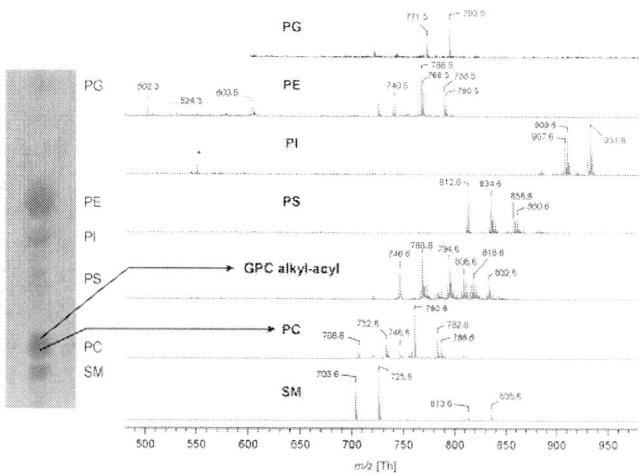

Fig. 11.5. Expanded region of a TLC-separated human mesenchymal stem cell extract and the corresponding positive ion MALDI-TOF mass spectra recorded directly from the indicated positions on the plate. More detailed mass assignments are available in (Fuchs et al., 2008a). Reprinted with modification and kind permission by Springer (Fuchs et al., 2008a).

a) b)

Fig. 11.6. Photographs of the MALDI-TOF/TLC adapter target prototypes used in these studies (Bruker Daltonics, Bremen, Germany). The TLC plate was simply mounted onto the target by conductive adhesive tape (a) or inserted into the centre of the target and then fixed mechanically (b).

whereby alkyl-acyl glycerophosphocholines (GPC) are found in the upper and diacyl PCs in the lower part of the spot. This is a clear indication that even the separation quality obtained under normal phase conditions [that primarily results in separation by polarity differences caused by different polar head-groups Fuchs et al., in press)] is sufficient to allow the differentiation between alkyl-ac(yl and acyl-acyl GPCs if MS is used for subsequent analysis (Fuchs et al., 2009a).

Regarding detection limits, the UV and the IR MALDI approaches (Rohlfing et al., 2007; Fuchs et al., 2007) provided comparable results (about 400 pmol); therefore, both approaches might be useful for routine lipid analysis. This particularly holds true as surprisingly good mass accuracies (about 100 ppm) and mass resolutions (about 3000) could be obtained that are absolutely sufficient for the differentiation of different phospholipids.

An automated routine became available very recently and was successfully applied for the investigation of the lipid composition of human mesenchymal stem cells as well as plasma PL signatures associated with respiratory disease severity in cystic fibrosis patients (Guerrera et al., 2009). This is a clear indication that MALDI mass spectra of lipids can be (at least relative to each other) quantitatively analyzed without major problems, although further attempts are clearly necessary to establish more useful matrix compounds particularly for negative ion detection (Fuchs et al., 2009b).

Characterizing Glycolipids by Direct TLC/MALDI Coupling

Glycolipids are nowadays regarded as important molecules that are involved in the pathogenesis of different diseases such as cancer. Therefore, glycolipid analysis is of increasing interest. In an early attempt native glycosphingolipids were separated on a conventional normal phase silica gel TLC plate and subsequently heat-transferred to a different membrane to get rid of impurities that potentially came from the TLC material and because spectral quality, and particularly the achievable sensitivity, can be highly improved under these conditions. The very best results were provided by polyvinylidene difluoride (PVDF) membranes (Guittard et al., 1999).

By the "classical" approach, i.e. by scraping off the individual bands from the TLC plate and their subsequent MALDI analysis, lipopolysaccharides that experience currently significant biochemical interest were primarily analyzed. However, when using more sophisticated equipment [Fourier Transform (FT) MS] (Guittard et al., 1999), the analysis of gangliosides (that primarily occur in brain tissues and are assumed to be indicative of Alzheimer disease) is also possible by direct TLC/MALDI without the induction of major analyte fragmentation. To enhance the stability of the generated ions, these authors used a relatively high gas pressure to allow "collisional cooling" of the generated ions (O'Connor et al., 2004). Using very simple equipment, i.e., a commercially available MALDI-TOF MS device and a normal nitrogen laser, glycolipids from brain also can be analyzed (Fuchs, 2008b).

Most of the TLC/MALDI studies of glycolipids used UV lasers. In a few studies, infrared lasers have been applied. These primarily are available on home-built MALDI devices but commercially only on special request. Er:YAG (Erbium-doped yttrium-aluminum-Garret) IR lasers normally are used and, as stated previously, have the advantage that IR radiation penetrates deeper into the sample than UV radiation. Therefore, bringing the complete analyte from the inner portion of the plate to the TLC surface is less important. Besides an IR laser, Dreisewerd and coworkers (2005) used an orthogonal instead of axial system for glycolipid analysis. The orthogonal configuration has the significant advantage that irregularities of the sample do not significantly influence the achievable spectral mass accuracy. Even minor gangliosides from a complex mixture extracted from cultured Chinese hamster ovary cells can be characterized.

These authors (Dreisewerd et al., 2005) also provided convincing evidence that the fluorescent dye "primuline" that is widely used in lipid research as it binds non-covalently to the fatty acyl residues (White et al., 1998) may be used as a non-destructive, MALDI-compatible staining agent. Similar data were independently obtained by another group that additionally introduced "vibrational" cooling (Ivleva et al., 2005). The term "vibrational cooling" describes the desorption process in the pressure range, where "cooling" of the excess energy of the generated ions is achieved. Under these conditions, fragmentation of labile bonds of the analytes can be minimized. This is normally performed by introducing an inert gas (e.g., N_2) under a moderate pressure.

A combination between MALDI, TLC and antibody detection was also recently suggested (Distler et al., 2008; Souady et al., 2009; Müsken et al., 2010). The authors used the following workflow: (a) TLC separation of cancer-associated glycosphingolipids (GSLs) from human hepatocellular and pancreatic tumors as well as GSL-specific bacteria, (b) their detection with oligosaccharide-specific proteins, and (c) the in situ MS analysis of the previously protein-detected GSLs. Detection limits of less than 1 ng of immunostained GSLs could be obtained under these conditions. An advantage of this approach is that only crude lipid extracts of biological sources are required for TLC-IR-MALDI-MS and no laborious previous GSL purification is needed. A "TLC-Blot-MALDI-Imaging" method was recently also suggested (Goto-Inoue, 2009) to visualize whole lipids and individual molecular species. This method provides higher sensitivity than common staining methods and allows for visualization of all lipids within a linear range of approximately one order of magnitude. The important field of glycolipid analysis was comprehensively reviewed (Müthing & Distler, 2010).

Recently, 2,5-dihydroxybenzoic acid (DHB) in acetonitrile / water (1:1, v/v) was found to be very useful as the matrix for the analysis of glycosphingolipids. DHB is highly soluble in this solvent mixture and the solvent mixture has a relatively high surface tension, leading to reduced analyte spreading. Sensitivities between 25 and 50 pmol were obtained under these conditions (Nakamura et al., 2006). The application of highly concentrated matrix solutions is very important because a larger excess of matrix in comparison with standard MALDI is required for the direct TLC/MALDI coupling to minimize fragmentation of the analyte (Fuchs et al., 2009; Fuchs et al., 2008b).

One important problem particularly related to MS imaging of biological tissues stems from the interference of matrix peaks with the biological molecules of interest. Using graphite-assisted laser desorption/ionization (GALDI) MS to overcome this problem was suggested; this technique was already successfully applied for the analysis of cerebrosides in a complex total brain lipid extract (Cha & Yeung, 2007). Beside the lack of typical "matrix" peaks, GALDI suffers far less from suppression effects, for instance, the suppression of cerebrosides by other dominant phospholipids, particularly PC. In addition to positive ion detection, negatively charged sulfatides easily could be detected using a thin layer of graphite. Thus, graphite enables the recording of positive and negative ion spectra.

Characterizing Carbohydrates by Direct TLC/MALDI Coupling

Carbohydrates are highly resistant to the MALDI ionization process, which is why they are less frequently investigated by MALDI-TOF MS in comparison to peptides or lipids (Fuchs et al., 2008). Carbohydrates also generally give much smaller ion yields, i.e., they are detectable only if present in relatively huge amounts. This holds true for acidic oligosaccharides that are still more difficult to detect than neutral ones (Harvey, 2009)

Using oligosaccharides isolated from human and elephant milk, Dreisewerd and coworkers (2006) showed that amounts of the order of 10 pmol still can be analyzed easily by direct TLC/MALDI coupling. Since these authors used an IR laser, simple glycerol could be applied as MALDI matrix resulting in a very homogeneous wetting of the silica gel and a high reproducibility of the mass spectra. The orthogonal MALDI device (Guilhaus et al., 2000) that enables the decoupling of the ion source from the mass analyzer provided an excellent mass accuracy, independent of potential irregularities of the TLC silica gel surface.

Data obtained by the UV laser were compared with data obtained by an IR laser on the same instrument (Dreisewerd et al., 2006). As matrix for the UV laser, α-cyano-4-hydroxycinnamic acid was used. Compared to the IR-MALDI mode, the sensitivity of the UV-MALDI was about one order of magnitude lower, whereas nonspecific analyte ion fragmentation and adduct formation was significantly reduced upon UV excitation (Dreisewerd et al., 2006). This particularly holds for glycerol adducts observed with significant intensities using the IR laser and glycerol as the matrix.

Of particular interest are glycosaminoglycans (GAG) such as hyaluronan (copolymer of glucuronic acid and N-acetylglucosamine) or chondroitin sulfate (copolymer of glucuronic acid and N-acetylgalactosamine sulfated primarily in the 4- or 6-position) because they are very abundant in the human body and possess a lot of biomedical applications that are currently investigated in more detail. One drawback is their high polarity that causes problems, and highly polar solvent systems have to be applied if successful separation by TLC is planned. A solvent system that contains significant amounts of formic acid helps to separate the individual oligosaccharides that can be easily obtained by enzymatic digestion of the native polysaccharides (Zhang et al., 2007a). Furthermore, we have recently shown that the direct characterization of these highly acidic oligosaccharides is possible directly on the TLC plate by MALDI-TOF MS (Nimptsch et al., 2010).

However, there are two problems that complicate the situation. The first problem comes from the use of high amounts of formic acid in the mobile phase. Under these conditions esterification occurs (detectable by the mass shift of 28) during the chromatographic run and, thus, a certain part of the GAG is not detectable as such but only as the corresponding formyl ester. This is, however, not a major problem because the extent of formylation is only moderate. A second much more serious problem is the loss of the sulfate residue. This has occurred under all the tested MS conditions reported (Tissot et al., 2007) and is only slightly enhanced if measurements are performed directly on the TLC plate.

Direct coupling of TLC with MS was successfully applied for a TLC / matrix free material enhanced laser desorption/ionization mass spectrometry (mf-MELDI-MS) analysis of carbohydrate reference standards (mono- and oligosaccharides: glucose, sucrose, raffinose) as well as in a plant extract (*Quercus robur*) (Qureshi et al., 2010). A new material for mf-MELDI-MS is prepared by immobilizing bradykinin, a peptide, on silica gel coupled to 4-(3-triethoxysilylpropylureido) azobenzene. This modification enables the absorption of laser energy sufficient for desorption and ionization of low molecular weight molecules like carbohydrates and amino acids. This newly synthesized material delivered excellent results with respect to signal-to-noise (S/N) ratio and sensitivity.

Characterizing Pharmaceuticals and Other Small Molecules

As many companies (particularly pharmaceutical ones) need to certify their products, this is a very important aspect. Since TLC always uses a completely new stationary phase, while HPLC always bears the risk that there is some residual material on the column, a TLC-based separation method is more easily certifiable than an HPLC-based method. In a pioneering study by Crecelius and coworkers (2000), (known) impurities of certain pharmaceuticals were directly investigated by combined TLC/MALDI. α-Cyano-4-hydroxycinnamic acid was used as the matrix and was electrosprayed onto the TLC plate in order to provide maximum homogeneity. This approach gave relatively stable signals and a good reproducibility. However, these authors also noted that irregularities of the used TLC plate led to significant deviations of mass accuracy although standard MALDI-TOF MS is capable of providing mass accuracies of the order of 20 ppm (Crecelius et al., 2000). Therefore, a program was developed that recalibrates each mass spectrum acquired during a run by using matrix ions or other selected ions as "lock masses."

The same authors have shown that it is possible to record "post source decay" (PSD) spectra (a special type of MS/MS-like spectra typical of TOF devices, for details see (Hillenkamp & Peter-Katalinić, 2007) directly from the TLC sample plate. As in a typical PSD experiment, several individual mass spectra must be recorded; this is an additional confirmation of the high reproducibility of the suggested approach (Crecelius et al., 2002)

Almost simultaneously, researchers found that the application of a particle suspension matrix may help to further improve the mass spectral quality of smaller molecules (Crecelius et al., 2002a). Particles of different materials and sizes (TiN, TiO_2, silicone, and graphite) were applied as suspensions. Among them, graphite showed the lowest background and the highest peak intensities of the analyte. Using this approach, the antibiotics tetracycline and chlortetracycline were easily differentiated by TLC/MALDI, although they gave only an unresolved spot on the TLC (Crecelius et al., 2002a).

Quantitative analysis by TLC/MALDI often yields comparable results to those obtained with stainless steel targets (the standard MALDI sample plate) (Pan et al., 2007). However, the achievable detection limits normally differ for both approaches. For instance, 60 pg of cocaine were the detection limit on the TLC plate, whereas only about 10 pg were required on the stainless steel plate. This difference might be caused by (a) irregularities of the surface of the TLC plate or (b) the inability of bringing all the analyte to the surface of the TLC plate. As already indicated, UV lasers do not penetrate very deeply into the sample; therefore, only the analyte located on the surface of the TLC plate is accurately detected. This is a major drawback of the application of UV lasers. Note that an internal reference compound of known structure and concentration is mandatory if absolute amounts (instead of relative intensity ratios) are to be determined. An internal standard was also used when Crecelius and coworkers determined the absolute amounts of the drug Piroxicam by combined TLC/MALDI (Crecelius et al., 2004). A quantity of 400–800 ng of the drug were still detectable by this approach.

Comparable sensitivities as previously reported were also found upon the analysis of siderophores (Hayen & Volmer, 2005), which are low molecular weight Fe(III)-specific binding molecules that typically occur in bacteria. These authors also reported that the sensitivity directly on the sample plate is about 4–24 times lower than that of conventional stainless steel MALDI targets. These differences were assumed to be caused by the incomplete transport of the trapped analyte molecules from the deeper layers of the TLC gel to the surface (Hayen & Volmer, 2005). Thus, a reduced thickness of the stationary phase was thought to enhance the detectability of the complete analyte. Standard TLC plates were replaced by ultra-thin layer chromatography (UTLC) plates (Salo et al., 2005), and due to the thinner adsorbent layer, the monolithic UTLC plates provided 10–100 times better sensitivities.

Although the majority of commercially available MALDI devices use high vacuum conditions, atmospheric pressure (AP) MALDI is increasingly used for two reasons (a) these devices are less accident-sensitive and (b) they provide the capability to record collisionally-induced dissociation (CID) fragment ion spectra without the need of an extra collision cell. Unfortunately, AP-MALDI possesses normally lower sensitivity than common high vacuum MALDI. However, it was shown that AP-MALDI in combination with UTLC is sufficiently sensitive to detect benzodiazepines directly in human urine without the need of a major workup of the sample (Salo et al., 2007).

One interesting application is the investigation of changes of dyes during photo-aging, i.e., upon the daylight-induced bleaching process. Paintings of old masters that are a few hundred years old usually contain di- and triterpenoids as varnish resins and undergo photooxidation, particularly in the presence of atmospheric O_2. Using such compounds on a cellulosic surface it could be shown that different di- and triterpenoids as well as their characteristic oxidation products containing up to six oxygen molecules could be easily detected by direct TLC/MALDI (Scalarone, 2005). Another related work focused on the question if the age of an ink entry from a questioned document that is an important issue in forensic sciences is possible by direct MS investigations. The age-induced changes of methyl and ethyl violet (that are both commonly used as ballpoint dyes) under the influence of light were investigated by LDI and MALDI MS directly on a layer of common paper (Weyermann et al., 2006). Many different parameters (including light, wavelength of light, heat, humidity, etc.) could be identified as responsible for dye aging and could be clearly differentiated by MS. Due to the relatively small masses of the dyes, in this application, LDI is superior in comparison to MALDI because there is no matrix background seen upon LDI.

Recently, different quaternary protoberberine alkaloids (a pharmaceutically important class of isoquinoline alkaloid) were also investigated by TLC and subsequent LDI MS) (Shariatgorji et al., 2009). Successful separation of berberine and palmatine from *Berberis barandana* as well as quantitative data analysis could be accomplished. Metabolites could be detected very sensitively due to the presence of the quaternary ammonia group that lends the molecules a permanent positive charge.

Summary and the Future

TLC is a simple, economical, and convenient analytical tool for many organic molecules, particularly small ones. However, this method often fails if unknown molecules have to be analyzed and particularly if no standards of the analytes of interest are available. There are many dyes that enable the assignment of the individual analytes by their binding to characteristic functional groups of the analyte. However, an additional detection system is needed in many cases. MS analysis is widely used in combination with liquid chromatography, but there have been fewer attempts to combine TLC with MS. Combining TLC with MS desorption techniques such as DESI MS have been reported.

As the MALDI method makes use of a solid sample, this technique seems the ionization method of choice for combining TLC and MS. Thus, this chapter focused on the combination between TLC and MALDI. TLC/MALDI mass spectra are simple to interpret because primarily singly charged ions are generated by MALDI and complex adduct patterns can be easily avoided. Additionally, the achievable MALDI mass spectra exhibit a high spatial resolution provided by the small laser beam spot as well as high detection sensitivities. The most important advantage, however, is the simple coupling (the developed TLC plate is simply mounted onto a commercially available MALDI target) between commercially available MALDI devices and TLC.

The significant progress made recently does not preclude the number of problems still to be addressed:

- The matrices now in use may not provide the optimum results. Therefore, the search for other useful TLC/MALDI matrices has to be ongoing.
- UV does not penetrate deeply into the TLC layer but remains primarily on the surface. We need to evaluate how much of the analyte is actually detected if UV lasers (the commercially available devices) are applied.
- Regarding the preceding point, major efforts are necessary to shed light on the quantitative aspects of TLC/MALDI.
- The reduced mass accuracy obtained by the combined TLC/MALDI approach in comparison to standard MALDI due to irregularities of the TLC surface must be overcome. However, by adding a suitable calibration standard, by using ultra-thin layer chromatography (UTLC) plates or MALDI devices with an orthogonal geometry this problem may be reduced significantly.

We surveyed the different substance classes already investigated by TLC/MALDI. Small, relatively apolar molecules such as lipids can be most successfully analyzed by this method. In contrast, highly polar molecules such as carbohydrates, especially if they contain additional charged function groups such as phosphate or sulfate groups, are more difficult to analyze by TLC/MALDI. Despite these problems, the examples provided confirmed the significant future potential of the direct combination between TLC and MALDI and its wide applicability.

Acknowledgments

The authors wish to thank all colleagues and friends who helped them in writing this chapter. In particular we would like to thank Dr. Suckau, Dr. Schürenberg and Dr. Becker (Bruker Daltonics, Bremen) for their continuous support and many fruitful discussions.

This work was supported by the German Research Council (DFG FU 771/1-1, DFG Schi 476/7-1, Schi 476/12-1 and the TRR 67).

References

Alpmann, A.; Morlock, G. Improved online coupling of planar chromatography with electrospray mass spectrometry: extraction of zones from glass plates. Anal. Bioanal. Chem. 2006, 386, 1543-1551.

Aranda, M.; Morlock, G. Simultaneous determination of caffeine, ergotamine, and metamizol in solid pharmaceutical formulation by HPTLC-UV-FLD with mass confirmation by online HPTLC-ESI-MS. J. Chromatogr. Sci. 2007, 45, 251-255.

Bakry, R.; Bonn, G.K.; Mair, D.; Svec, F. Monolithic porous polymer layer for the separation of peptides and proteins using thin-layer chromatography coupled with MALDI-TOF-MS. Anal. Chem. 2007, 79, 486-493.

Berger, S.; Braus, S. 200 and More NMR Experiments - A Practical Course. Wiley-VCH: Weinheim, 2004.

Bradley, S.A.; McLaughlin, R.L. High-resolution magic-angle spinning NMR for the identification of reaction products directly from thin-layer chromatography spots; Magn. Reson. Chem. 2007, 45, 814-818.

Busch, K.L. Planar separations and mass spectrometric detection. JPC-J. Planar Chromatogr.-Mod. TLC 2004, 17, 398-403.

Cha, S.; Yeung, E.S. Colloidal graphite-assisted laser desorption/ionization mass spectrometry and MSn of small molecules. 1. Imaging of cerebrosides directly from rat brain tissue. Anal. Chem. 2007, 79, 2373-2385.

Chen, H.; Talaty, N.N.; Takáts, Z.; Cooks, R.G. Desorption electrospray ionization mass spectrometry for high-throughput analysis of pharmaceutical samples in the ambient environment. Anal. Chem. 2005, 77, 6915-6927.

Cornett, D.S.; Reyzer, M.L.; Chaurand, P.; Caprioli, R.M. MALDI imaging mass spectrometry: molecular snapshots of biochemical systems. Nat. Methods 2007, 4, 828-833.

Crecelius, A.; Clench, M.R.; Richards, D.S.; Evason, D.; Parr, V. Thin-layer chromatography-postsource-decay matrix-assisted laser desorption/ionization time-of-flight mass spectrometry of small drug molecules. J. Chromatogr. Sci. 2002a, 40, 614-620.

Crecelius, A.; Clench, M.R.; Richards, D.S.; Mather, J.; Parr, V. Analysis of UK-224,671 and related substances by thin-layer chromatography-matrix-assisted laser-desorption ionization-time of flight mass spectrometry. JPC – J. Planar Chromat. 2000, 13, 76-81.

Crecelius, A.; Clench, M.R.; Richards, D.S.; Parr, V. Quantitative determination of Piroxicam by TLC-MALDI TOF MS. J. Pharmaceut. Biomed. Anal. 2004, 35, 31-39.

Crecelius, A.; Clench, M.R.; Richards, D.S.; Parr, V. Thin-layer chromatography-matrix-assisted laser desorption ionisation-time-of-flight mass spectrometry using particle suspension matrices. J. Chromatogr. A 2002, 958, 249-260.

Distler, U.; Hülsewig, M.; Souady, J.; Dreisewerd, K.; Haier, J.; Senninger, N.; Friedrich, A.W.; Karch, H.; Hillenkamp, F.; Berkenkamp, S.; Peter-Katalinić, J.; Müthing, J. Matching IR-MALDI-o-TOF mass spectrometry with the TLC overlay binding assay and its clinical application for tracing tumor-associated glycosphingolipids in hepatocellular and pancreatic cancer. Anal. Chem. 2008, 80, 1835-1846.

Douthwaite, S.; Kirpekar, F. Identifying modifications in RNA by MALDI mass spectrometry. Methods Enzymol. 2007, 425, 1-20.

Dreisewerd, K.; Müthing, J.; Rohlfing, A.; Meisen, I.; Vukelić, Z.; Peter-Katalinić, J.; Hillenkamp, F.; Berkenkamp, S. Analysis of gangliosides directly from thin-layer chromatography plates by infrared matrix-assisted laser desorption/ionization orthogonal time-of-flight mass spectrometry with a glycerol matrix. Anal. Chem. 2005, 77, 4098-4107.

Dreisewerd, K.; Kölbl, S.; Peter-Katalinić, J.; Berkenkamp, S.; Pohlentz, G. Analysis of native milk oligosaccharides directly from thin-layer chromatography plates by matrix-assisted laser desorption/ionization orthogonal-time-of-flight mass spectrometry with a glycerol matrix. J. Am. Soc. Mass Spectrom. 2006, 17, 139-150.

Duncan, M.W.; Roder, H.; Hunsucker, S.W. Quantitative matrix-assisted laser desorption/ionization mass spectrometry. Brief Funct. Genomic Proteomic 2008, 7, 355-370.

Fernandis, A.Z.; Wenk, M.R. Membrane lipids as signaling molecules, Curr. Opin. Lipidol. 2007, 18, 121-128.

Fuchs, B.; Bischoff, A.; Süß, R.; Teuber, K.; Schürenberg, M.; Suckau, D.; Schiller, J. Phosphatidylcholines and -ethanolamines can be easily mistaken in phospholipid mixtures: a negative ion MALDI-TOF MS study with 9-aminoacridine as matrix and egg yolk as selected example. Anal. Bioanal. Chem. 2009b, 395, 2479-2487.

Fuchs, B.; Nimptsch, A.; Süß, R.; Schiller, J.Analysis of brain lipids by directly coupled matrix-assisted laser desorption ionization time-of-flight mass spectrometry and high-performance thin-layer chromatography. J. AOAC Int. 2008b, 91, 1227-1236.

Fuchs, B.. Schiller, J. Application of MALDI-TOF mass spectrometry in lipidomics. Eur. J. Lipid Sci. Technol. 2009a, 111, 83-98.

Fuchs, B.; Schiller, J.; Süß, R.; Nimmptsch, A.; Schürenberg, M.; Suckau, D. Capabilities and disadvantages of combined matrix-assisted laser desorption/ionization time-of-flight mass spectrometry (MALDI-TOF MS) and high-performance thin-layer chromatography (HPTLC): Analysis of egg yolk lipids. JPC – J. Planar Chromat. 2009a, 22, 35-42.

Fuchs, B.; Schiller, J.; Süß, R.; Schürenberg, M.; Suckau, D. A direct and simple method of coupling matrix-assisted laser desorption and ionization time-of-flight mass spectrometry (MALDI-TOF MS) to thin-layer chromatography (TLC) for the analysis of phospholipids from egg yolk. Anal. Bioanal. Chem. 2007, 389, 827-834.

Fuchs, B.; Schiller, J.; Süß, R.; Zscharnack, M.; Bader, A.; Müller, P.; Schürenberg, M.; Becker, M.; Suckau, D. Analysis of stem cell lipids by offline HPTLC-MALDI-TOF MS. Anal. Bioanal. Chem. 2008a, 392, 849-860.

Fuchs, B.; Schiller, J. Recent developments of useful MALDI matrices for the mass spectrometric characterization of apolar compounds. Curr. Org. Chem. 2009, 13, 1664-1681.

Fuchs, B.; Arnold, K.; Schiller, J. In Encyclopedia of Analytical Chemistry: Mass Spectrometry of Biological Molecules; R.A. Meyers, Ed.; John Wiley & Sons: Chichester, 2008; 1, 1-39.

Fuchs, B.: Süß, R.; Nimptsch, A.; Schiller, J. Matrix-assisted laser desorption and ionization time-of-flight mass spectrometry (MALDI-TOF MS) directly combined with thin-layer chromatography (TLC) - A review of the current state. Chromatographia 2009, 69(Suppl.), 95-105.

Fuchs, B.; Süß, R.; Teuber, K.; Eibisch, M.; Schiller, J. Lipid analysis by thin-layer chromatography - A review of the current state. J. Chromatogr. A in press.

Geiss, F. Fundamentals of Thin Layer Chromatography, 1; Hüthig: Heidelberg, 1987.

Gellermann, G.P.; Appel, T.R.; Davies, P.; Diekmann, S. Paired helical filaments contain small amounts of cholesterol, phosphatidylcholine and sphingolipids. Biol. Chem. 2006, 387, 1267-1274 .

Goto-Inoue, N.; Hayasaka, T.; Taki, T.; Gonzalez, M.; Setou, M. A new lipidomics approach by thin-layer chromatography-blot-matrix-assisted laser desorption/ionization imaging mass spectrometry for analyzing detailed patterns of phospholipid molecular species. J. Chromatogr. A 2009, 1216, 7096-7101.

Guerrera, I.C.; Astarita, G.; Jais, J.P.; Sands, D.; Nowakowska, A.; Colas, J.; Sermet-Gaudelus, I.; Schuerenberg, M.; Piomelli, D.; Edelman, A.; Ollero, M. A novel lipidomic strategy reveals plasma phospholipid signatures associated with respiratory disease severity in cystic fibrosis patients. PLoS One 2009, 4, e7735.

Guilhaus, M.; Selby, D.; Mlynski, V. Orthogonal acceleration time-of-flight mass spectrometry. Mass Spectrom. Rev. 2000, 19, 65-107.

Guittard, J.; Hronowski, X.P.L.; Costello, C.E.. Direct matrix-assisted laser desorption/ionization mass spectrometric analysis of glycosphingolipids on thin layer chromatographic plates and transfer membranes. Rapid Commun. Mass Spectrom. 1999, 13, 1838-1849.

Günzler, H.; Gremlich, H.U. IR-Spectroscopy - An Introduction. Wiley-VCH Weinheim, 2002.

Gusev, A.I.; Proctor, A.; Rabinovich, Y.I. Hercules D.M. Thin-layer chromatography combined with matrix-assisted laser-desorption ionization mass spectrometry. Anal. Chem. 1995, 67, 1805-1814.

Gusev, A.I. Interfacing matrix-assisted laser desorption/ionization mass spectrometry with column and planar separations. Fresenius J. Anal. Chem. 2000, 366, 691-700.

Gusev, A.I.; Vasseur, O.J.; Proctor, A.; Sharkey, A.G.; Hercules, D.M. Imaging of thin-layer chromatograms with matrix-assisted laser-desorption ionization mass spectrometry. Anal. Chem. 1995, 67, 4565-4570.

Hahn-Deinstrop, E. Applied Thin-Layer Chromatography - Best Practice and Avoidance of Mistakes, Wiley-VCH: Weinheim, 2006.

Harvey, D.J. Analysis of carbohydrates and glycoconjugates by matrix-assisted laser desorption/ionization mass spectrometry: An update for 2003-2004. Mass Spectrom. Rev. 2009, 28, 273-361.

Harvey, D.J. Matrix-assisted laser desorption/ionization mass spectrometry of phospholipids. J. Mass Spectrom. 1995, 30, 1333-1346.

Hayen, H.; Volmer, D.A. Rapid identification of siderophores by combined thin-layer chromatography/matrix-assisted laser desorption/ionization mass spectrometry, Rapid Commun. Mass Spectrom. 2005, 19, 711-720.

Heigl, N.; Huck, C.W.; Rainer, M.; Najam-Ul-Haq, M.; Bonn, G.K. Near infrared spectroscopy, cluster and multivariate analysis hyphenated to thin layer chromatography for the analysis of amino acids. Amino Acids 2006, 31, 45-53.

Hillenkamp, F.; Unsöld, E.; Kaufmann, R.; Nitsche, R. Laser microprobe mass analysis of organic materials. Nature 1975, 256, 119-120.

Hillenkamp, F.; Peter-Katalinić, J. MALDI MS - A Practical Guide to Instrumentation, Methods and Application, Wiley-VCH: Weinheim, 2007.

Isbell, D.T.; Gusev, A.I.; Taraneko, N.I.; Chen, C.H.; Hercules, D.M. Analysis of nucleotides directly from TLC plates using MALDI-MS detection. Fresenius J. Anal. Chem. 1999, 365, 625-630.

Ivleva, V.B.; Sapp, L.M.; O'Connor, P.B.; Costello, C.E. Ganglioside analysis by thin-layer chromatography matrix-assisted laser desorption/ionization orthogonal time-of-flight mass spectrometry. J. Am. Soc. Mass Spectrom. 2005, 16, 1552-1560.

Jautz, U.; Morlock, G. Efficacy of planar chromatography coupled to (tandem) mass spectrometry for employment in trace analysis. J. Chromatogr. A 2006, 1128, 244-250.

Klein, T.R.; Kirsch, D.; Kaufmann, R.; Riesner, D. Prion rods contain small amounts of two host sphingolipids as revealed by thin-layer chromatography and mass spectrometry. Biol. Chem. 1998, 379, 655-666.

Kochin, V.; Imanishi, S.Y.; Eriksson, J.E. Fast track to a phosphoprotein sketch - MALDI-TOF characterization of TLC-based tryptic phosphopeptide maps at femtomolar detection sensitivity. Proteomics 2006, 6, 5676-5682.

Leßig, J.; Fuchs, B. Plasmalogens in biological systems: their role in oxidative processes in biological membranes, their contribution to pathological processes and aging and plasmalogen analysis. Curr. Med. Chem. 2009, 16, 2021-2041.

Leßig, J.; Gey, C.; Süß, R.; Schiller, J.; Glander, H.J.; Arnhold, J. Analysis of the lipid composition of human and boar spermatozoa by MALDI-TOF mass spectrometry, thin layer chromatography and 31P NMR spectroscopy. Comp. Biochem. Physiol. B Biochem. Mol. Biol. 2004, 137, 265-277.

Luftmann, H.; Aranda, M.; Morlock, G.E. Automated interface for hyphenation of planar chromatography with mass spectrometry, Rapid Commun. Mass Spectrom. 2007, 21, 3772-3776.

Marderosian, A.D.; Chao, J.M. Identification of active drug principles in natural products. J. Chromatogr. Sci. 1974, 12, 285-292.

Mehl, J.T.; Gusev, A.I.; Hercules, D.M. Coupling protocol for thin layer chromatography matrix-assisted laser desorption ionization. Chromatographia 1997, 46, 358-364.

Mehl, J.T.; Hercules, D.M. Direct TLC-MALDI coupling using a hybrid plate. Anal. Chem. 2000, 72, 68-73.

Miao, Z.; Chen, H. Direct analysis of liquid samples by desorption electrospray ionization-mass spectrometry (DESI-MS). J. Am. Soc. Mass Spectrom. 2009, 20, 10-19.

Morlock, G.; Sherma, J. New and improved liquid chromatographic methods for food analysis. J. AOAC Int. 2009, 92, 689-690.

Morlock, G.; Schwack, W. Coupling of planar chromatography to mass spectrometry. TrAC - Trends Anal. Chem. 2010, 29 1157-1171.

Morlock, G. Schwack, W. Hyphenations in planar chromatography. J. Chromatogr. A 2010a, 1217, 6600-6609.

Morlock, G. Ueda, Y. New coupling of planar chromatography with direct analysis in real time mass spectrometry. J. Chromatogr. A 2007, 1143, 243-251.

Mowthorpe, S.; Clench, M.R.; Cricelius, A.; Richards, D.S.; Parr, V.; Tetler, L.W. Matrix-assisted laser desorption/ionisation time-of-flight thin layer chromatography mass spectrometry - A rapid method for impurity testing. Rapid Commun. Mass Spectrom. 1999, 13, 264-270.

Mueller, D.R.; Voshol, H.; Waldt, A.; Wiedmann, B.; Van Oostrum, J. LC-MALDI MS and MS/MS - an efficient tool in proteome analysis. Subcell. Biochem. 2007, 43, 355-380.

Müller, M.; Schiller, J.; Petković, M.; Oehrl, W.; Heinze, R.; Wetzker, R.; Arnold, K.; Arnhold, J. Limits for the detection of (poly-)phosphoinositides by matrix-assisted laser desorption and ionization time-of-flight mass spectrometry (MALDI-TOF MS). Chem. Phys. Lipids 2001, 110, 151-164.

Müsken, A.; Souady, J.; Dreisewerd, K.; Zhang, W.; Distler, U.; Peter-Katalinić, J.; Miller-Podraza, H.; Karch, H.; Müthing, J. Application of thin-layer chromatography/infrared matrix-assisted laser desorption/ionization orthogonal time-of-flight mass spectrometry to structural analysis of bacteria-binding glycosphingolipids selected by affinity detection. Rapid Commun. Mass Spectrom. 2010, 24, 1032-1038.

Müthing, J.; Distler, U. Advances on the compositional analysis of glycosphingolipids combining thin-layer chromatography with mass spectrometry. Mass Spectrom. Rev. 2010, 29, 425-479.

Nakamura, H. Pisano, J.J. Derivatization of compounds at the origin of thin-layer plates with fluorescamine. J. Chromatogr. 1976, 121, 33-40.

Nakamura, K.; Suzuki, Y.; Goto-Inoue, N.; Yoshida-Noro, C.; Suzuki, A. Structural characterization of neutral glycosphingolipids by thin-layer chromatography coupled to matrix-assisted laser desorption/ionization quadrupole ion trap time-of-flight MS/MS. Anal. Chem. 2006, 78, 5736-5743.

Nimptsch, K.; Süß, R.; Riemer, T.; Nimptsch, A.; Schnabelrauch, M.; Schiller, J. Differently complex oligosaccharides can be easily identified by matrix-assisted laser desorption and ionisation time-of-flight mass spectrometry directly from a standard thin-layer chromatography plate. J. Chromatogr. A 2010, 1217, 3711-3715.

O'Connor, P.B.; Budnik, B.A.; Ivleva, V.B.; Kaur, P.; Moyer, S.C.; Pittman, J.L.; Costello, C.E. A high pressure matrix-assisted laser desorption ion source for Fourier transform mass spectrometry designed to accommodate large targets with diverse surfaces. J. Am. Soc. Mass Spectrom. 2004, 15, 128-132.

Pan, C.; Xu, S.; Zhou, H.; Fu, Y.; Ye, M.; Zou, H. Recent developments in methods and technology for analysis of biological samples by MALDI-TOF-MS. Anal. Bioanal. Chem. 2007, 387, 193-204.

Panchagnula, V.; Mikulskis, A.; Song, L.; Wang, Y.; Wang, M.; Knubovets, T.; Scrivener, E.; Golenko, E.; Krull, I.S.; Schulz, M.; et al. Phosphopeptide analysis by directly coupling two-dimensional planar electrochromatography/thin-layer chromatography with matrix-assisted laser desorption/ionization time-of-flight mass spectrometry. J. Chromatogr. A 2007, 1155, 112-123.

Peterson, D.S. Matrix-free methods for laser desorption/ionization mass spectrometry. Mass Spectrom. Rev. 2007, 26, 19-34.

Petković, M.; Schiller, J.; Müller, M.; Benard, S.; Reichl, S.; Arnold, K.; Arnhold, J. Detection of individual phospholipids in lipid mixtures by matrix-assisted laser desorption/ionization time-of-flight mass spectrometry: phosphatidylcholine prevents the detection of further species. Anal. Biochem. 2001, 289, 202-216.

Qureshi, M.N. Stecher, G.; Huck, C.; Bonn, G.K. Online coupling of thin layer chromatography with matrix-assisted laser desorption/ionization time-of-flight mass spectrometry: synthesis and application of a new material for the identification of carbohydrates by thin layer chromatography/matrix free material enhanced laser desorption/ionization mass spectrometry. Rapid Commun. Mass Spectrom. 2010, 24, 2759-2764.

Rohlfing, A.; Müthing, J.; Pohlentz, G.; Distler, U.; Peter-Katalinić, J.; Berkenkamp, S.; Dreisewerd, K. IR-MALDI-MS analysis of HPTLC-separated phospholipid mixtures directly from the TLC plate. Anal. Chem. 2007, 79, 5793-5808.

Salo, P.K. Salomies, H.; Harju, K.; Ketola, R.A.; Kotiaho, T.; Yli-Kauhaluoma, J.; Kostiainen, R. Analysis of small molecules by ultra thin-layer chromatography-atmospheric pressure matrix-assisted laser desorption/ionization mass spectrometry. J. Am. Soc. Mass Spectrom. 2005, 16, 906-915.

Salo, P.K.; Vilmunen, S.; Salomies, H.; Ketola, R.A.; Kostiainen, R. Two-dimensional ultra-thin-layer chromatography and atmospheric pressure matrix-assisted laser desorption/ionization mass spectrometry in bioanalysis. Anal. Chem. 2007, 79, 2101-2108.

Scalarone, D.; Duursma, M.C.; Boon, J.J.; Chiantore, O. MALDI-TOF mass spectrometry on cellulosic surfaces of fresh and photo-aged di- and triterpenoid varnish resins. J. Mass Spectrom. 2005, 40, 1527-1535.

Schäfer, K.C.; Dénes, J.; Albrecht, K.; Szaniszló, T.; Balog, J.; Skoumal, R.; Katona, M.; Tóth, M.; Balogh, L; Takáts, Z. In vivo, in situ tissue analysis using rapid evaporative ionization mass spectrometry. Angew. Chem.-Int. Edit. 2009, 48, 8240-8242.

Schiller, J.; Huster, D.; Fuchs, B.; Naji, L.; Kaufmann, J.; Arnold, K. Evaluation of cartilage composition and degradation by high-resolution magic-angle spinning nuclear magnetic resonance. Methods Mol. Med. 2004, 101, 267-285.

Schiller, J.; Arnhold, J.; Benard, S.; Müller, M.; Reichl, S.; Arnold, K. Lipid analysis by matrix-assisted laser desorption and ionization mass spectrometry: A methodological approach. Anal. Biochem. 1999, 267, 46-56.

Schiller, J.; Müller, K.; Süß, R.; Arnhold, J.; Gey, C.; Herrmann, A.; Leßig, J.; Arnold, K.; Müller, P. Analysis of the lipid composition of bull spermatozoa by MALDI-TOF mass spectrometry - a cautionary note. Chem. Phys. Lipids 2003a, 126, 85-94.

Schiller, J.; Süß, R.; Fuchs, B.; Müller, M.; Petković, M.; Zschörnig, O.; Waschipky, H. The suitability of different DHB isomers as matrices for the MALDI-TOF MS analysis of phospholipids: which isomer for what purpose? Eur. Biophys. J. 2007, 36, 517-527.

Schiller, J.; Süß, R.; Fuchs, B.; Müller, M.; Zschörnig, O.; Arnold, K. MALDI-TOF MS in lipidomics, Front. Biosci. 2007a, 12, 2568-2579.

Schiller, J.; Süß, R.; Fuchs, B.; Müller, M.; Zschörnig, O.; Arnold, K. Combined application of TLC and matrix-assisted laser desorption and ionisation time-of-flight mass spectrometry (MALDI-TOF MS) to phospholipid analysis of brain. Chromatographia 2003, 57(Suppl. 1), 297-302.

Schiller, J.; Süß, R.; Arnhold, J.; Fuchs, B.; Leßig, J.; Müler, M.; Petković, M.; Spalteholz, H.; Zschörnig, O.; Arnold, K. Matrix-assisted laser desorption and ionization time-of-flight (MALDI-TOF) mass spectrometry in lipid and phospholipid research. Prog. Lipid Res. 2004a, 43, 449-488.

Shariatgorji, M.; Spacil, Z.; Maddalo, G.; Cardenas, L.B.; Ilag, L.L. Matrix-free thin-layer chromatography/laser desorption ionization mass spectrometry for facile separation and identification of medicinal alkaloids. Rapid Commun. Mass Spectrom. 2009, 23, 3655-3660.

Sherma, J.; Morlock, G. Chronology of thin-layer chromatography focusing on instrumental progress. JPC-J. Planar Chromatogr.-Mod. TLC 2008, 21, 471-477.

Sherma, J. Planar chromatography. Anal. Chem. 2008, 80, 4253-4267.

Sommerer, D.; Süß, R.; Hammerschmidt, S.; Wirtz, H.; Arnold, K.; Schiller, J. Analysis of the phospholipid composition of bronchoalveolar lavage (BAL) fluid from man and minipig by MALDI-TOF mass spectrometry in combination with TLC. J. Pharm. Biomed. Anal. 2004, 35, 199-206.

Souady, J.; Soltwisch, J.; Dreisewerd, K.; Haier, J.; Peter-Katalinić, J.; Müthing, J. Structural profiling of individual glycosphingolipids in a single thin-layer chromatogram by multiple sequential immunodetection matched with direct IR-MALDI-o-TOF mass spectrometry. Anal. Chem. 2009, 81, 9481-9492.

Teuber, K.; Riemer, T.; Schiller, J. Thin-layer chromatography combined with MALDI-TOF-MS and 31P-NMR to study possible selective bindings of phospholipids to silica gel. Anal. Bioanal. Chem. 2010, 398, 2833-2842.

Tissot, B.; Gasiunas, N.; Powell, A.K.; Ahmed, Y.; Zhi, Z.L.; Haslam, S.M.; Morris, H.R.; Turnbull, J.E.; Gallagher J.T.; Dell, A. Towards GAG glycomics: analysis of highly sulfated heparins by MALDI-TOF mass spectrometry. Glycobiology 2007, 17, 972-982.

Touchstone, J.C. Thin-layer chromatographic procedures for lipid separation. J. Chromatogr. B Biomed. Appl. 1995, 671, 169-195.

Weyermann, C.; Kirsch, D.; Costa-Vera, C.; Spengler, B. Photofading of ballpoint dyes studied on paper by LDI and MALDI MS. J. Am. Soc. Mass Spectrom. 2006, 17, 297-306.

White, T.; Bursten, S.; Frederighi, D.; Lewis, R.A.; Nudelman, E. High-resolution separation and quantification of neutral lipid and phospholipid species in mammalian cells and sera by multi-one-dimensional thin-layer chromatography. Anal. Biochem. 1998, 258, 109-117.

Wilson, I.D.; Spraul, M.; Humpfer, E. Thin-layer chromatography combined with high-resolution solid-state NMR for compound identification without substance elution: Preliminary results. J. Planar Chromatogr. 1997, 10, 217-219.

Wilson, I.D. The state of the art in thin-layer chromatography-mass spectrometry: a critical appraisal. J. Chromatogr. A 1999, 856, 429-442.

Woodbury, C.P. Jr.; Fitzloff, J.F.; Vincent, S.S. Sample multiplexing for greater throughput in HPLC and related methods. Anal. Chem. 1995, 67, 885-890.

Yates, J.R.; Ruse, C.I.; Nakorchevsky, A. Proteomics by mass spectrometry: approaches, advances, and applications. Annu. Rev. Biomed. Eng. 2009, 11, 49-79.

Zhang, J.Z.: Wang, Y.; Chen, H.; Shao, H.B. TLC-SERS study on evodiamine in evodia rutaecarpa. Spectrosc. Spectral. Anal. 2007, 27, 944-947.

Zhang, Z.; Xie, J.; Zhang, F.; Linhardt, R.J. Thin-layer chromatography for the analysis of glycosaminoglycan oligosaccharides. Anal. Biochem. 2007a, 371, 118-120.

Zschörnig, O.; Richter, V.; Rassoul, F.; Süß, R.; Arnold, K.; Schiller, J. Analysis of human blood plasma by MALDI-TOF MS - Evaluation of critical parameters. Anal. Lett. 2006, 39, 1101-1113.

Contributors

María Asensio-Ramos

María Asensio-Ramos is a current Ph.D. student at the department of Analytical Chemistry, Nutrition and Food Science of the University of La Laguna in Tenerife (Canary Islands, Spain). In 2007 she finished a B.S. degree in Chemistry at the same university obtaining the prize of the best B.S. of the Faculty of Chemistry. In 2009, she obtained an M.S. (Advanced Studies Diploma) in Analytical Chemistry.

William Craig Byrdwell

Dr. Byrdwell received his undergraduate and graduate degrees from the University of Louisville, in Louisville, KY. His dissertation identified a previously 'unknown phospholipid' in the human eye lens, as well as quantification of human lens fluorophors in normal and cataractous lenses.

Dr. Byrdwell held a postdoctoral position at the USDA's National Center for Agricultural Utilization Research in Peoria, IL, where he worked on the analysis of triacylglycerols using HPLC with atmospheric pressure chemical ionization mass spectrometry (APCI-MS).

Drawn by the sun and sand in south Florida, Dr. Byrdwell left USDA to teach. As an Assistant Professor at Florida Atlantic University, he continued to pursue lipid analysis, and published the first of several reports that demonstrated the use of dual mass spectrometers, in parallel, each using a different ionization technique.

Dr. Byrdwell returned to the fertile research environment of the Agricultural Research Service; and re-joined ARS in 2005. He focused his expertise in LC-MS for Lipid Analysis on the lipid soluble vitamin D. Currently Dr. Byrdwell is working on analysis of vitamin D and triacylglycerols using a triple parallel mass spectrometry approach. Dr. Byrdwell has published more than 40 peer-reviewed articles and 8 book chapters. Pictures of Dr. Byrdwell's lab can be seen at www.Byrdwell.com, and a list of publications is at www.Byrdwell.com/publications.html.

Francesco Cacciola

Dr. Francesco Cacciola took his Ph.D. in Food and Safety Chemistry at the University of Messina in 2009 defending a thesis entitled "Employment of High Resolution HPLC Techniques for the Analysis of Complex Matrices." He is the author of 20 peer-reviewed articles, 4 book chapters, and 38 contributions in national and international symposiums. In 2008, he won the "Young Researchers award" for the high scientific productivity carried out in 2007. His research interests include conventional and innovative LC techniques (fast LC, LCxLC) and the development of coupled techniques such as LCxLC-MS, LCxLC-MS/MS for the study of natural complex matrices.

Paola Donato

Dr. Paola A. E. Donato is a contract researcher at the Campus Biomedico in Rome. In February 2004 she passed her Ph.D. examination in Pharmaceutical Sciences with a dissertation entitled "Effects of Alpha- and Beta-cyclodextrin Complexation on the Physico-chemical Properties and Antioxidant Activity of Some 3-hydroxyflavones." The main objectives of her research activity consist in the development of comprehensive (LCxLC) bidimensional chromatographic techniques for the analysis of highly complex mixtures.

Paola Dugo

Paola Dugo is full professor of Food Chemistry at the University of Messina and at the Campus Biomedico in Rome. Her research interests are vast, including the development of comprehensive chromatography, high throughput separation methods, and the use of various MS detectors (i.e., quadrupole, ion trap, high-resolution time-of-flight). Paola Dugo is actively engaged in the study of food components with a possible, beneficial biological activity.

Chiara Fanali

Chiara Fanali is Assistant Professor of Food Chemistry at Centro Integrato di Ricerca, University Campus Biomedico di Roma in Italy. In 2004 she obtained the Degree *cum summa laude* in Chemistry, Faculty of Natural Mathematics and Physical Sciences, University La Sapienza of Rome. She received her Ph.D. in Biochemistry at Catholic University of Rome. Postdoctoral research on proteomic was carried out at Catholic University of Rome. The research experience includes: proteomic studies, food analysis, two-dimensional separations based on liquid chromatography-mass spectrometry. She has published about 30 papers in international ISI Journals, two book chapters, and co-authored a patent.

Salvatore Fanali

Salvatore Fanali is Director of Research at Institute of Chemical Methodologies, Italian National Research Council in Monterotondo (Rome), Italy, and head of the Capillary Electromigration and Chromatographic Methods Unit at the same institute. His main research interests are in separation science including the development of modern miniaturized techniques (electrodriven and liquid chromatography) as well as their hyphenation with mass spectrometry. The development and study of new stationary phases for capillary electrochromatography and nano-liquid chromatography. Separation method development and applications in food, pharmaceutical, chiral environment, and biomedical analysis are currently developed. He is Editor of Journal of Chromatography A and member of the advisory editorial board of seven international scientific Journals. S. Fanali is author of about 250 publications including some book chapters. In 2009, he received the Liberti Medal in Separation Science from the Italian Chemical Society.

Beate Fuchs

Dr. Fuchs was born in 1971 in Dresden, Germany. After studies of chemistry and an intermediate trainee post at Bayer de Mexico funded by the Carl Duisberg foundation, she did her Ph.D. at the Institute for Polymer Research in Dresden investigating the modifications of fluoropolymers by ^{19}F solid-state NMR spectroscopy. Afterwards she postgraduated in toxicology. Currently, she is working at the Institute of Medical Physics and Biophysics at the Leipzig University. Her research focus is the HOCl-induced lysophospholipid generation from diacyl- and alkenyl-acyl phospholipids that is supported by the German Research Council (DFG). Further research interests are mass spectrometry and ^{31}P NMR of lipids and phospholipids, lipid oxidation and peroxidation, analysis of lipid second messenger molecules, cells, tissues, and body fluids under different clinical aspects and the online-coupling between MALDI-TOF MS and thin-layer chromatography (TLC). Dr. Fuchs is the author of about 50 journal papers, reviews, and contributions to books.

Davy Guillarme

Davy Guillarme received his Ph.D. in analytical chemistry from the University of Lyon (France) in 2004. He is now lecturer of the School of Pharmaceutical Sciences, University of Geneva, University of Lausanne (Switzerland). His interests include the development of new approaches to perform fast and ultra-fast separations in liquid chromatography and the possibility to hyphenate these techniques with alternative detection modes.

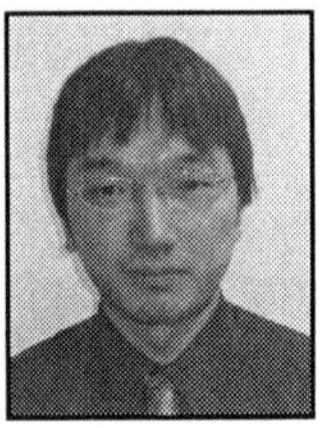

Takeshi Hara

Takeshi Hara is a Ph.D. student in Justus Liebig University of Giessen, Germany. He previously studied liquid chromatography at Kyoto Institute of Technology. His major research interest is preparation of monolithic silica columns, their characterization, and the applications to high efficiency separations.

Javier Hernández-Borges

Javier Hernández-Borges is currently a researcher at the Department of Analytical Chemistry, Nutrition and Food Science of the University of La Laguna in Tenerife (Canary Islands, Spain) where he also serves as professor. In 2005 he obtained his Ph.D. in Chemistry at that University, performing several pre-doctoral stays at the Department of Analytical Chemistry of the University of Masaryk, Brno (Czech Republic) and at the Institute of Industrial Fermentations of the CSIC in Madrid (Spain). He also performed a post-doc stay at the Institute of Chemical Methodologies of the CNR in Rome. He has published around 55 papers in international journals and books, most of them dealing with method development, in particular, with the analysis of pesticide residues in environmental and food samples by chromatographic and electromigration techniques.

Michal Holčapek

Michal Holčapek was born in 1971 in Třebíč, Czech Republic. He has obtained his Ph.D. in 1999 at the University of Pardubice, Czech Republic, where he is currently working as a full professor and the head of mass spectrometry group at the Department of Analytical Chemistry. He is also national representative of the Czech Republic in International Mass Spectrometry Foundation. His research interests are mass spectrometry and its coupling with liquid-phase separation techniques with specialization on the structural analysis of (bio)organic and organometallic compounds, mainly lipidomics, metabolomics, and drug metabolites. Professor Holčapek is author or co-author of about 90 papers in international peer-reviewed journals, 10 chapters in monographs and textbooks, and more than 200 presentations at international conferences.

Tohru Ikegami

Tohru Ikegami is Associate Professor at the Kyoto Institute of Technology. His major research interest is development of HPLC columns with novel functionalities, particularly monolithic silica columns for HILIC, chiral, and ion-exchange separations, and their applications to high-efficiency separations.

Yasushi Ishihama

Yasushi Ishihama is a Professor at the Graduate School of Pharmaceutical Sciences, Kyoto University. His major research interest is technology development for mass spectrometry-based proteomics and its application to cellular bioanalysis.

Pavel Jandera

Dr. Jandera is currently with the Department of Analytical Chemistry, Faculty of Chemical Technology at the University of Pardubice (Czechoslovaki). He received his Ing. (M.S.) degree at the University of Chemical Technology, Pardubice, Czechoslovakia in 1967, where he continued his studies and obtained CSc. (Ph.D.) degree in Analytical Chemistry in 1977. Then in 1989, he obtained DrSc. degree in Analytical Chemistry at the University of Chemical Technology, Prague, Czechoslovakia. Dr. Jandera's major research fields are chromatography—theory, instrumentation and applications, especially in High Performance Liquid Chromatography (HPLC). Prof. Jandera is author of 4 books, 15 contributed chapters in monographs, more than 180 original research papers, 220 contributions at national and international conferences and symposia, 7 patents, and 13 textbooks for graduate and undergraduate studies.

Miroslav Lísa

Miroslav Lísa was born in 1980 in Karlovy Vary, Czech Republic. He has defended his Ph.D. thesis entitled "Characterization of Triacylglycerol Composition in Natural Samples Using HPLC/MS" in 2008 at the Department of Analytical Chemistry, University of Pardubice, Czech Republic. Now he is working as a research assistant at the same institution. His main research interest is HPLC and HPLC/MS analyses of lipids in plant, animal, and human tissues and blood samples. He is the author of about 20 papers in international peer-reviewed journals and more than 60 presentations at conferences.

Kosuke Miyamoto

Kosuke Miyamoto studied the preparation and characterization of long monolithic silica columns at Kyoto Institute of Technology. He successfully prepared capillary columns of up to 500 cm long and was able to generate 1,000,000 theoretical plates or more with connected columns. Miyamoto is currently working for Yamato-Esulon, Co. Ltd.

Luigi Mondello

Luigi Mondello is a full professor of Analytical Chemistry at the University of Messina and at the Campus Biomedico in Rome. His main research interests are focused on comprehensive chromatographic techniques, fast separation methods, and mass spectrometric detection. In 2010, he was co-chairman of the 34th International Symposium on Capillary Chromatography, held in Riva del Garda (Italy). He is Editor of the Journal of Separation Science (Wiley) and Editor-in Chief of the Journal of Essential Oil Research (Allured).

Giorgia Purcaro

Giorgia Purcaro took her Ph.D. in Food Science at the University of Udine in 2008. Her research interests include food contaminants and development of conventional and comprehensive chromatography techniques. She has been co-author of 20 peer reviewed articles and 5 book chapters published by international editors. She has been co-author for 2 contributions on Chromedia (Chromatography Knowledge Base). She won the Lerici award in 2009 for the best Ph.D. thesis in the food science field. She won the Leslie Ettre award in 2010 at the 34th International Symposium on Capillary Chromatography, Riva del Garda, 2010.

Anna Rocco

Anna Rocco is a junior researcher at the Institute of Chemical Methodologies-IMC, C.N.R., in Rome. She graduated in Pharmaceutical Chemistry and Technologies (University La Sapienza, Rome) in 2003. Since 2004, she has worked at IMC where she gained experience in miniaturized separation techniques, i.e. nano-liquid chromatography, capillary zone electrophoresis, and capillary electrochromatography, and in their application for analytical aims in pharmaceutical, agro-chemical, and alimentary fields. She has published about 18 papers.

Jürgen Schiller

Dr. Schiller was born in 1965 in Bavaria, Germany. After studies of chemistry, he did his Ph.D. in Biophysics investigating the degradation of polysaccharides by reactive oxygen species. Currently, he is the leader of a working group at Leipzig University. His research focus is on (phospho)lipids, their role in different pathologies, and the investigation of lipid oxidation products generated under inflammatory conditions by using NMR, mass spectrometry, and chromatographic techniques. He is particularly interested in combining the resolving power of thin-layer chromatography with the sensitivity of MS detection. Dr. Schiller is the author of about 150 journal papers, reviews, and contributions to books.

Nobuo Tanaka

Nobuo Tanaka is Professor Emeritus at Kyoto Institute of Technology and is currently serving as an Editor of the Journal of Chromatography A and working as an Advisor for GL Sciences, Inc. His major research interest is development of high-performance HPLC columns and their applications.

Thorsten Teutenberg

Thorsten Teutenberg obtained his doctorate from the University of Bochum in 2004. His thesis was on the development of a specially designed heating system for high-temperature liquid chromatography. He subsequently joined the Institut für Energie- und Umwelttechnik e. V. in Duisburg where he became responsible for a small working group within the Department of Research Analysis focused on method development strategies for high temperature liquid chromatography. Currently, he heads a number of joint research projects, with partners from university and industry, aimed at implementing high temperature liquid chromatography as a routine method. His special interest is the understanding of the temperature dependence of physicochemical solvent properties of binary liquid mixtures used for reversed phase HPLC. Future research interests include multidimensional liquid chromatography, capillary, and nano-HPLC as well as systematic optimization strategies for an efficient method development.

Peter Quinto Tranchida

Dr. Tranchida occupies a position as Assistant Professor in Food Chemistry in the Faculty of Pharmacy (University of Messina). His main research activities are focused on the analysis of volatiles contained in complex samples, by using advanced gas chromatographic-mass spectrometric methodologies and innovative column stationary phases. Furthermore, he is actively engaged in the development of classical multidimensional and comprehensive two-dimensional GC, as well as LC-GC systems. Dr. Tranchida has co-authored approximately 70 papers and 120 oral/poster presentations in national and international meetings.

Jean-Luc Veuthey

Jean-Luc Veuthey obtained his Ph.D. in analytical chemistry from the University of Geneva (Switzerland) in 1987. He is now full professor of the School of Pharmaceutical Sciences, University of Geneva, University of Lausanne (Switzerland). His interests include the development of LC and CE hyphenated to several detection modes for the analysis of drugs and metabolites. Sample preparation and validation of the procedures are also particularly studied in his laboratory.

Index

CPSIA information can be obtained at www.ICGtesting.com
Printed in the USA
LVOW010740031111

253286LV00001B/1/P

9 781893 997660